AIR POLLUTION METEOROLOGY AND DISPERSION

S. PAL ARYA
Department of Marine, Earth, and Atmospheric Sciences
North Carolina State University

New York   Oxford
Oxford University Press
1999

Oxford University Press

Oxford New York
Athens Auckland Bangkok Bogotá Buenos Aires Calcutta
Cape Town Chennai Dar es Salaam Delhi Florence Hong Kong Istanbul
Karachi Kuala Lumpur Madrid Melbourne Mexico City Mumbai
Nairobi Paris São Paulo Singapore Taipei Tokyo Toronto Warsaw

and associated companies in
Berlin Ibadan

Copyright © 1999 by Oxford University Press, Inc.

Published by Oxford University Press, Inc.
198 Madison Avenue, New York, New York 10016
http://www.oup-usa.org

Oxford is a registered trademark of Oxford University Press.

All rights reserved. No part of this publication may be reproduced,
stored in a retrieval system, or transmitted, in any form or by any means,
electronic, mechanical, photocopying, recording, or otherwise,
without the prior permission of Oxford University Press.

Library of Congress Cataloging-in-Publication Data
Arya, S. Pal.
  Air pollution meteorology and dispersion / S. Pal Arya.
    p.   cm.
  Includes bibliographical references (p.   ) and index.
  ISBN 0-19-507398-3 (cloth)
  1. Air--Pollution--Meteorological aspects.   I. Title.
QC882.A856 1999   97-35738
551.5--DC21   CIP
                                              Rev.

9 8 7 6 5 4 3 2 1

Printed in the United States of America
on acid-free paper

To Nirmal, Niki, Sumi, Vishal

Contents

PREFACE viii
ACKNOWLEDGMENTS x

1 INTRODUCTION TO AIR POLLUTION 1
   1.1 The Air Pollution Problem 1
   1.2 Sources of Air Pollution 3
   1.3 Air Pollutants 4
   1.4 Effects of Air Pollution 14
   1.5 Regulatory Control of Air Pollution 21
   Problems and Exercises 24

2 ATMOSPHERIC STRUCTURE AND DYNAMICS 26
   2.1 Introduction 26
   2.2 Composition and Thermal Structure of the Atmosphere 26
   2.3 State Variables and Thermodynamics 29
   2.4 Atmospheric Stability 32
   2.5 Conservation Laws and Atmospheric Dynamics 33
   2.6 Large-scale Inviscid Flows 35
   2.7 Small-scale Viscous Flows 37
   2.8 Applications 40
   Problems and Exercises 40

3 ATMOSPHERIC SYSTEMS AND POLLUTANT TRANSPORT 42
   3.1 Introduction 42
   3.2 Macroscale Systems 42
   3.3 Synoptic Weather Systems 51
   3.4 Mesoscale Systems 59
   3.5 Microscale Systems 69
   Problems and Exercises 75

4 MICROMETEOROLOGY AND PLANETARY BOUNDARY LAYER 77
   4.1 Introduction and Definitions 77
   4.2 Earth-Atmosphere Exchange Processes 78
   4.3 Vertical Distributions of Thermodynamic Variables 80
   4.4 Vertical Distribution of Winds in the PBL 83
   4.5 Turbulence 84
   4.6 Gradient-transport Theories 89
   4.7 Similarity Theories 91
   4.8 Boundary-layer Parameterization for Dispersion Applications 96
   Problems and Exercises 103

5 STATISTICAL DESCRIPTION OF ATMOSPHERIC TURBULENCE 105
   5.1 Reynolds Averaging 105
   5.2 Probability Functions 105
   5.3 Autocorrelation Functions 108
   5.4 Spectrum Functions 110
   5.5 Taylor's Hypothesis 112
   5.6 Statistical Theory of Turbulence 113
   5.7 Observed Spectra and Scales 117

5.8 Effects of Smoothing and Finite Sampling 120
5.9 Lagrangian Description of Turbulence 122
5.10 Parameterization of Turbulence for Diffusion Applications 124
Problems and Exercises 125

6 GRADIENT TRANSPORT THEORIES 127

6.1 Eulerian Approach to Describing Diffusion 127
6.2 Mass Conservation and Diffusion Equations 128
6.3 Molecular Diffusion 131
6.4 Turbulent Diffusion 137
6.5 Constant K (Fickian Diffusion)-Theory 139
6.6 Variable K-Theory 143
6.7 Limitations of Gradient Transport Theories 148
6.8 Experimental Verification of K-Theories 150
6.9 Applications of K-Theories to Atmospheric Dispersion 152
Problems and Exercises 152

7 STATISTICAL THEORIES OF DIFFUSION 155

7.1 Lagrangian Approach to Describing Diffusion 155
7.2 Statistical Theory of Absolute Diffusion 155
7.3 Plume Diffusion from Continuous Sources 161
7.4 Statistical Theory of Relative Diffusion 163
7.5 Puff Diffusion from Instantaneous Releases 167
7.6 Fluctuating Plume Models 169
7.7 Experimental Verification of Statistical Theories 170
7.8 Applications to Atmospheric Dispersion and Limitations 175
Problems and Exercises 176

8 SIMILARITY THEORIES OF DISPERSION 178

8.1 Dispersion in Stratified Shear Flows 178
8.2 Lagrangian Similarity Theory for the Neutral Surface Layer 180
8.3 Lagrangian Similarity Theory for the Stratified Surface Layer 183
8.4 The Mixed-layer Similarity Theory 188
8.5 Experimental Verification of Similarity Theories 192
8.6 Applications to Dispersion in the PBL 194
8.7 Limitations of Similarity Theories 195
Problems and Exercises 195

9 GAUSSIAN DIFFUSION MODELS 197

9.1 Basis and Justification for Gaussian Models 197
9.2 Gaussian Plume and Puff Diffusion Models 198
9.3 Diffusion Experiments 200
9.4 Empirical Dispersion Parameterization Schemes 202
9.5 Further Improvements in Dispersion Parameterization 207
9.6 The Maximum Ground-Level Concentration 210
9.7 Model Evaluations and Uncertainties 213
9.8 Limitations of Gaussian Diffusion Models 216
9.9 Practical Applications of Gaussian Diffusion Models 216
Problems and Exercises 218

10 PLUME RISE, SETTLING, AND DEPOSITION 220

10.1 Momentum and Buoyancy Effects of Release 220
10.2 Plume-rise Theory and Observations 222
10.3 Gravitational Settling of Particles 226
10.4 Dry Deposition 228
10.5 Dispersion-Deposition Models 234
10.6 Applications 237
Problems and Exercises 237

11 NUMERICAL DISPERSION MODELS 239

11.1 Introduction 239
11.2 Short-range Gradient Transport Models 239
11.3 Turbulence Kinetic Energy Models 243
11.4 Higher Order Closure Models 245
11.5 Large-eddy Simulations 249
11.6 Lagrangian Stochastic Models 260
Problems and Exercises 267

12 URBAN AND REGIONAL AIR QUALITY MODELS 269

12.1 Introduction 269
12.2 Components of an Air Quality Model 269
12.3 Urban Diffusion and Air Quality Models 276
12.4 Regional Air Quality Models 281
12.5 Applications of Air Quality Models 285
Problems and Exercises 286

REFERENCES 287
SYMBOLS 300
INDEX 305

Preface

This last quarter of the twentieth century can be distinguished by the vastly increased awareness of our natural environment among people all over the world and their heightened desire to restore and preserve the high quality of their environment as an integral part of the quality of life. This is clearly reflected in the tremendous growth in media coverage as well as in the scientific and popular literature on environmental problems of our earth, atmosphere, and oceans and their likely consequences to the biosphere. In particular, the most serious prblems of our atmospheric environment, such as local and urban air pollution, regional haze, photochemical smog, acidic precipitation, stratospheric ozone depletion, and global climate change, have received tremendous public attention during the decades of the eighties and nineties. In response to the increased interest in this field, many universities and colleges now offer interdisciplinary programs in environmental sciences and/or environmental engineering. Consequently, there is an increased need for suitable textbooks for courses offered in such programs. This was indeed the primary motivation for writing this book.

Here at North Carolina State University, interdisciplinary programs in both environmental science and engineering have been developed recently. But courses on air pollution meteorology and atmospheric dispersion have been offered by the author for more than twenty years. This book is based on the material I have developed for a two-semester sequence of courses on air pollution meteorology and atmospheric dispersion. The first course serves a larger number of students with different backgrounds—seniors in meteorology and graduate students in marine, atmospheric, and other environmental sciences and engineering. The second course is a more advanced graduate-level course on atmospheric dispersion. For the first course, I mostly use the material presented in chapters 1, 2, 3, 4, 6, 9, and 10. In the more advanced graduate-level course, I cover the material presented in chapters 5, 6, 7, 8, 11, and 12. Other instructors offering only a one-semester course may use selected portions of the book depending on their students' backgrounds and interests. Covering the whole book in one course could be a formidable task, which I would not advise anybody to undertake.

The organization of the book in a sequence of twelve chapters follows what I consider to be a natural development or review of the basic concepts, theories, and models of pollutant dispersal in the atmosphere and the related atmospheric systems affecting transport, transformation, and removal of air pollutants. Chapter 1 serves as a general introduction to the air pollution problem with a brief description of its main components, such as emission sources and air pollutants, their effects, and regulatory controls. Chapter 2 gives an overview of the atmospheric structure and dynamics, including some of the fundamental concepts and equations of atmospheric thermodynamics and motion based on the conservation of mass, momentum, and energy. Distinctions between different types and scales of atmospheric motions are also brought out here. A more detailed, but largely qualitative, description of the various atmospheric circulation and weather systems and their relevance to pollutant transport and dispersal is given in Chapter 3. Both chapters 2 and 3 are primarily intended for non-meteorologists who might need some background knowledge of atmospheric structure, dynamics, and circulation systems. Similarly, Chapter 4 serves as an introduction to micrometeorology and atmospheric boundary layer—the part of the atmosphere in which most pollutants are released, transported, dispersed, and removed. Here the basic concepts and theories of turbulent exchange, mixing, and scaling are also introduced. More sophisticated statistical description and theory of turbulence are presented in Chapter 5. These early chapters (1–5) are intended to provide the necessary background for the various dispersion theories and models discussed in subsequent chapters.

The simplest and also the oldest gradient transport approach and the analytical K-theory models based on the same are described in Chapter 6. More sophisticated statistical and similarity theories of dispersion are covered in chapters 7 and 8. The complementary nature of these different diffusion theories is emphasized. Their use in conjunction with experimental diffusion data has resulted in simpler Gaussian diffusion models with empirical dispersion parameterization schemes, which are covered in Chapter 9. Regulatory applications of such models are also emphasized. Chapter 10 gives a brief overview of the effects of effluent momentum and buoyancy, gravitational settling of particles, and dry deposition on the average plume trajectory and ground-level concentrations. Here, simpler parameterizations of plume rise and dry deposition velocity used in Gaussian dispersion models are emphasized. More sophisticated short-range numerical dispersion models using a variety of turbulence closure schemes, large-eddy simulation, and stochastic or random-walk particle trajectory models are reviewed in Chapter 11. Finally, the basic components of urban and regional air quality models and their types are described in Chapter 12. The primary emphasis in both chapters 11 and 12 is on the physics and simple parameterizations of transport, diffusion, and removal processes, rather than on the numerical and computational aspects of air quality models.

I hope that instruction faculty will find the book useful as a text or primary reference material for their courses in air pollution meteorology, atmospheric dispersion, and air quality or air pollution modeling. Problems and exercises at the end of each chapter may be used for homework assignments, tests, and examinations. I also hope that students will find this easy to read and to grasp the meteorological fundamentals as well as theories and models of atmospheric dispersion. Professional air quality meteorologists and environmental scientists and engineers may also find this a useful reference for reviewing the theoretical and empirical bases as well as limitations of the particular air quality models they might be using for regulatory applications or in their research.

Raleigh, NC
June 20, 1997

S. Pal Arya

Acknowledgments

The wealth of knowledge presented in this book has been drawn from many sources, which is attested by the extensive list of references given at the end. I wish to acknowledge especially the books authored or edited by Ahrens (1994), Csanady (1973), Godish (1991), Griffin (1994), Hanna et al. (1982), Miller et al. (1983), Monin and Yaglom (1971, 1975), Panofsky and Dutton (1984), Pasquill (1974), Pasquill and Smith (1983), Seinfeld (1986), Slade (1968), Stern (1976–86), Stull (1988), and Turner (1994), from which I borrowed more heavily.

I am grateful to the following publishers or authors for giving me their kind permission to reproduce figures of which they hold copyright:

Academic Press, Inc. (Figs. 1.2, 2.1)

American Meteorological Society (Figs. 3.1, 3.14, 4.3, 4.5, 4.6, 4.12, 5.4, 5.8, 5.9, 5.12–.14, 7.7, 7.10, 8.8, 8.11–.13, 9.8, 10.2, 10.3, 11.3)

Viney P. Aneja (Fig. 1.3)

CRC Press, Inc. (Figs. 9.6, 9.7)

Elsevier Science Ltd. (Figs. 6.10, 6.12, 7.8, 7.9, 8.5–.7, 8.9, 8.10)

Kluwer Academic Publishers (Figs. 2.3, 4.4, 4.10, 4.11, 4.13, 4.15, 5.3, 6.3, 6.4, 8.1, 8.4, 11.2)

The MIT Press (Fig. 5.1)

Charles E. Merrill Publishing Co. (Figs. 3.2, 3.7–.10)

National Center for Atmospheric Research (Figs. 10.7, 10.8)

John G. Navarra (Fig. 3.5)

Oxford University Press (Figs. 5.10, 5.11)

Matthias Roth (Fig. 4.7)

W.B. Saunders Co. (Fig. 3.5)

U.S. Department of Energy (Figs. 3.12–.16, 6.11, 7.6, 7.11, 7.12, 9.1–.5, 10.9, 11.1)

U.S. Environmental Protection Agency (Figs. 12.2, 12.3)

Wadsworth Publishing Co. (Figs. 3.4, 3.6, 3.11)

John Wiley and Sons, Inc. (Figs. 1.1, 1.4, 5.2, 5.7, 5.15, 7.3, 7.4, 12.1)

The authors and dates of original articles, books, and monographs from which the figures are reproduced are mentioned in the figure captions, while further details of title, journal, and publisher are given in the list of references.

Finally, I would like to acknowledge the critical comments and corrections made by some of my students who reviewed parts of the manuscript I used in my coursepack. I also wish to acknowledge the help of Brenda Batts and Mel DeFeo for word processing of the manuscript and in going over the painstaking tasks of correcting, spellchecking, and proofreading several times. They did this patiently over the period of the several years it took me to write this book. I also want to thank LuAnn Salzillo for drafting most of the figures.

The Department of Marine, Earth and Atmospheric Sciences at North Carolina State University provided me with the release time including a sabbatical in Fall 1996 to write this book. Atmospheric Sciences Modeling Division, U.S. Environmental Protection Agency, kindly provided me with a visiting appointment and an office in Research Triangle Park where I wrote a substantial part of the manuscript. I would like to thank Mr. Frank Schiermeir, Director of ASMD, for giving me that opportunity. Partial support of the U.S. EPA's Cooperative Agreement CR 822057 is also gratefully acknowledged.

1

Introduction to Air Pollution

1.1 THE AIR POLLUTION PROBLEM

It is not easy to give a simple, comprehensive and yet concise definition of air pollution. The word *pollution* comes from the Latin *pollutus*, which means made foul, unclean, or dirty. Thus, air pollution is usually defined as an atmospheric condition in which substances are present at concentrations higher than their normal ambient (clean atmospheric) levels to produce significant effects on humans, animals, vegetation, or materials (Seinfeld, 1986). The substances present may be any natural or man-made chemical elements or compounds in gaseous, liquid, or solid state that are capable of being airborne. Although the above definition includes any airborne material, whether harmful or benign, we are primarily concerned with substances that may cause significant undesirable effects, such as objectionable or foul odor, irritation of senses, sickness and death of people, damage to vegetation leading to stunting of growth and decay, damage to materials and property, obscuration of visibility, and adverse weather and climate changes. These adverse or undesirable effects may result from sudden or short-term exposure to very high concentrations or from exposure to even small concentrations over long periods.

Air pollution is ubiquitous. Smoke, haze, dust, mist, foul-smelling and corrosive gases, and toxic compounds are present nearly everywhere, even in the most remote, pristine wilderness. Human activities have caused air pollution ever since our ancestors began building fires. But it became a serious problem only during the last 200 years when growing population and industrialization produced vast quantities of contaminants. An estimated 150 million metric tons of air pollutants (excluding carbon dioxide and wind-blown soil) are released into the atmosphere each year by human activities in the United States. The total worldwide emissions of these pollutants are around 2 billion metric tons per year.

Air pollution is not only an emissions' problem, it is also a weather-related condition or phenomenon and, as such, should be considered one of the weather hazards (Pielke, 1979). In fact, air pollution turns out to be by far the worst weather hazard if one compares the average estimated number of lives lost per year in the United States due to air pollution, 15,000, with the about 750 deaths per year due to all other major weather hazards combined. This figure for air pollution-related deaths is probably a gross underestimate since it does not include the mortality rate due to long-term exposure to secondhand smoke from tobacco smokers. Also, total deaths from air pollution cannot be estimated accurately, because poor air quality can act insidiously over a long period of time. It increases the mortality rate in not only chronically polluted urban and industrial areas, but also in much larger, seemingly unpolluted regions with fewer local sources of air pollution. Long-range and regional transport of air pollutants makes the air pollution problem of widespread concern and not merely confined to urban metropolitan areas.

1.1.1 Types of Air Pollution Problems

As a general rule, air quality tends to be worst where most air pollutants are emitted, that is, in highly industrial and traffic-congested urban areas. So, local and urban air pollution including the indoor air pollution constitute a major problem. Certain pollutants, such as tropospheric ozone, photochemical oxidants, and sulfur and nitrogen compounds, are easily transported by winds and spread over large regions. Consequently, the regional air pollution problems of ozone and acid precipitation have attracted increasing attention because these affect people, animals, vegetation, and materials over much larger areas. There are also air pollution problems of global concern which

Table 1.1 Types and Scales of Air Pollution Problems

Type of Problem		Horizontal Scale	Vertical Scale	Temporal Scale	Type of Organization
I	Indoor	10^{-2}–10^{-1} km	Up to 10^{-1} km	10^{-1}–10^{0} hr	Family/business
	Local	10^{-1}–10 km	Up to 3 km	10^{-1}–10 hr	Municipality/county
	Urban	10–10^{2} km	Up to 3 km	10^{0}–10^{2} hr	Municipality/county
II	Regional	10^{2}–10^{3} km	Up to 15 km	10–10^{3} hr	State/country
	Continental	10^{3}–10^{4} km	Up to 30 km	10^{2}–10^{4} hr	Country/world
III	Hemispheric	10^{4}–2×10^{4} km	Up to 50 km	10^{3}–10^{5} hr	World
	Global	4×10^{4} km	Up to 50 km	10^{3}–10^{6} hr	World

Source: Modified after Stern et al. (1984).

are primarily caused by increasing concentrations of the so-called greenhouse gases, such as carbon dioxide, methane, tropospheric ozone, nitrous oxide, and particulate matter. The problem of stratospheric ozone depletion due to reactions with man-made chemicals also appears to be global, but with largest effects over polar and high-latitude regions. Thus, we have significant air pollution problems on local/urban, regional, and global scales. These have to be addressed accordingly on local/urban, regional and global bases as well. The corresponding time scales may vary from minutes and hours to decades. Different types of air pollution problems, temporal and spatial scales associated with them, and appropriate organizations for addressing these problems are given in Table 1.1.

Type I problems include indoor air pollution in houses and commercial buildings; local pollution around isolated factories, power plants, waste disposal sites, and so on; and urban air pollution resulting from a variety of urban sources. These are often referred to as local and urban air pollution, examples of which are urban smog and haze. Type II problems of regional and continental air pollution are simply referred to as regional or interregional pollution. Examples are transport of sulfur and nitrogen oxides from major industrial areas to other regions where they are washed out of the air as acid precipitation, regional transport of ozone, and transport of particulates from forest fires and slash burning resulting in widespread haze.

Type III problems include hemispheric and global air pollution and their possible consequences of climate changes over periods of decades and centuries. The best known examples are the predicted global warming due to increasing concentrations of CO_2 and other greenhouse gases, the stratospheric ozone depletion leading to the formation of the so-called ozone hole over Antarctica, and stratospheric intrusion by particulates from volcanic eruptions and anthropogenic emissions from high-flying aircraft, which have a cooling effect on the global climate.

1.1.2 Components of an Air Pollution Problem

Each of the above mentioned air pollution problems has three main components: (1) emission sources that produce air pollutants; (2) the atmosphere in which transport, diffusion, chemical transformations, and removal processes occur; and (3) receptors near the ground that respond to trace amounts of air pollutants reaching them.

Estimating or measuring the emissions of pollutant species of interest from a variety of sources requires a knowledge of the chemistry of combustion and engineering aspects of the equipment design and air pollution control technology. Understanding the characteristics of the various air pollutants requires a knowledge of atmospheric chemistry, aerosol physics, and atmospheric radiation. Understanding atmospheric transport and diffusion of pollutants requires a knowledge of meteorology, environmental fluid mechanics, and turbulent exchange and mixing processes. A comprehensive understanding of chemical transformations and removal processes (e.g., dry deposition, cloud and precipitation scavenging) requires the knowledge of atmospheric chemistry, cloud and precipitation physics, and cloud microphysical processes. Finally, an understanding of the various receptors (e.g., instruments, human beings, animals, vegetation, and materials) and their response to air pollutants requires background in air quality monitoring, physiology, medicine, plant pathology, and materials science. Thus, air pollution is an interdisciplinary problem whose study and solution require interdisciplinary efforts by scientists, engineers, environmental protection agencies, legislators, and the public at large.

Figure 1.1 represents a schematic of the various components of the air pollution problem on a local/urban scale. Controls are represented at three points. Of these, control at the emission source (source control) is the most efficient, feasible, and practical. Legislative action based on the receptor response has become commonplace, particularly in developed countries. Automatic control of emissions based on detector response is rarely used. As a result of the various control measures adopted, air quality has improved appreciably over the past 25 years in most cities in western Europe, North America, and Japan. We have made significant progress in reducing air pollution in most of the United States. But many large cities have not yet met the air quality standards mandated by the Clean Air Act. In particular, the urban and regional

3 SOURCES OF AIR POLLUTION

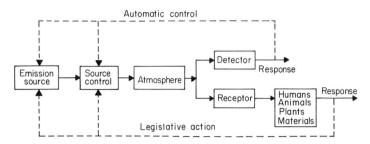

Figure 1.1 Schematic of the various components of the air pollution problem on a local/urban scale. From Seinfeld, 1986.

ozone air pollution has shown no sign of improvement but seems to be getting worse. More strict control measures on automobile and industrial emissions may be necessary for controlling this persistent air pollution problem.

In this book we will be concerned primarily with the meteorological aspects of air pollution, including atmospheric transport, diffusion, and removal processes. Other aspects of the problem are discussed only briefly and superficially in this chapter. More detailed and comprehensive reviews of these are given elsewhere (Stern, 1976–86; Stern et al., 1984; Seinfeld, 1986; Godish, 1991; Griffin, 1994).

1.2 SOURCES OF AIR POLLUTION

The various sources of air pollution can be divided into three broad categories: (1) urban and industrial sources; (2) agricultural and other rural sources; (3) natural emissions. Each of these categories can further be subdivided into several subcategories, listed in the following. This categorization is by no means standard or unique and is given here only for illustrative purposes.

1.2.1 Urban and Industrial Sources

a. *Power generation*: Conventional fossil-fuel power plants are the major source of air pollution. Vast quantities of particulate matter including fly ash (SiO_2), iron and aluminum oxides from coal, and heavy metals from oil are emitted from power-plant stacks. Major gaseous pollutants emitted are carbon monoxide (CO), carbon dioxide (CO_2), sulfur dioxide (SO_2), nitrogen oxides (NO_x), and certain hydrocarbons and volatile organic compounds (VOCs). Nuclear power plants are much cleaner in their normal operation, during which cooling tower plumes might create the occasional nuisance of fog formation and visibility reduction. But accidental releases of radioactive substances are of great concern to the public, even though the probability of a major accidental release is extremely small.

b. *Industrial facilities*: These include mining, refining, manufacturing, smelting, pulp and paper, chemical, metallurgical, pharmaceutical, and other industries that emit a wide variety of gaseous and particulate matter in the atmosphere. A great deal of industrial air pollution comes from manufacturing products from raw materials, such as iron from ore, gasoline from crude oil, stone from quarries, and lumber from trees. Considerable pollution is also emitted by industries that convert the above products into other consumer products, such as automobiles, houses, furniture, and appliances. There are also large amounts of fugitive industrial emissions that do not go through smokestacks.

c. *Transportation*: This category includes automobiles, buses, trucks, airplanes, boats, etc., which are all mobile sources. It is not practical to follow all the transportation sources as they move around, but their emissions are estimated on per unit area basis, considering traffic density, speed, emissions per vehicle, and other variables. Major pollutants produced by transportation sources are CO, CO_2, NO_x, SO_2, hydrocarbons, and VOCs.

d. *Process emissions*: These include furnaces and other processes used for heating homes, office, and commercial buildings; fireplaces, stoves, backyard barbecue grills, and open burning of refuse and leaves. Their emissions are estimated on per unit area basis. The major pollutants emitted to the atmosphere are CO, CO_2, NO_x, SO_2, hydrocarbons, VOCs, and particulate matter.

e. *Waste disposal*: Urban household, commercial, and industrial waste products are disposed of in landfills, incineration facilities, sewage treatment plants, backyard compost pits, and refuse dumps. All these constitute major sources for gaseous (e.g., CO, CO_2, CH_4, H_2S, and NH_3) and particulate air pollution.

f. *Construction activities*: These include land clearing, demolition, digging, grinding, paving, painting, and other construction-related activities. Pollutants emitted to the atmosphere are dust and other particulate matter, hydrocarbons, VOCs, CO, CO_2, and NO_x.

1.2.2 Agricultural and Other Rural Sources

The various sources of air pollution in rural areas, including agricultural operations, can be classified as follows:

    a. *Dust blowing*: Agricultural operations of ploughing, tilling, and harvesting lead to considerable amounts of

dust blowing in the wind. Emissions from tractors, harvesters, and other machinery are not considered very significant.
b. *Slash burning*: Land clearing by burning of forests, straw, wild grass, and agricultural waste products constitutes a major source of smoke and haze in the countryside.
c. *Soil emissions*: Croplands are usually treated with heavy applications of fertilizers containing nitrates and phosphates. Treated soils emit nitrogen oxides that are produced by microbial activity in the topmost soil layer.
d. *Pesticides*: Applications of pesticides to croplands through spraying by aircraft may result in their transport to residential areas.
e. *Decaying wastes*: Agricultural and animal waste products decay and release ammonia, methane, and noxious vapors to the atmosphere. Large cattle and hog farms constitute major sources of air pollution in the countryside.

1.2.3 Natural Sources

In addition to the anthropogenic sources of atmospheric pollution listed above, there are many natural sources, which may be classified as follows:

a. *Wind erosion*: Wind erosion of bare soils and desert lands by strong winds or by dust devils can result in major dust storms. Dust particles consist mainly of SiO_2, but may also include traces of heavy metals.
b. *Forest fires*: Many forest fires are ignited by lightning and emit large amounts of smoke, CO, CO_2, NO_x, and hydrocarbons.
c. *Volcanic eruptions*: Major volcanoes spew huge amounts of particulate matter, CO_2, SO_2, and other gases into the atmosphere. Some of the material is transported to high enough altitudes and stays there for months or several years to impact on the global climate.
d. *Biogenic emissions*: These include emissions from forests and marshlands. Major pollutants emitted to the atmosphere are hydrocarbons, such as terpenes and isoprene; methane, ammonia, pollen, and spores.
e. *Sea spray and evaporation*: Sea spray by wave breaking in strong winds is the major source of salt particles in the atmosphere. In general, the process of evaporation emits not only water vapor but also many trace gases from the seawater to the atmosphere.
f. *Soil microbial processes*: These include aerobic and anaerobic respiration of natural soils and vegetation resulting in the emissions of nitric oxide (NO), methane (CH_4), hydrogen sulfide (H_2S), and ammonia (NH_3) to the atmosphere.
g. *Natural decay of organic matter*: This includes decay of vegetation and other organic matter, which emits mainly CH_4, H_2S; and NH_3.
h. *Lightning*: Lightning produces large amounts of NO, which may then participate in photochemical reactions creating ozone.

1.3 AIR POLLUTANTS

Such a wide variety of pollutants is emitted to the atmosphere that a neat and compact generally acceptable classification of the same does not exist. Instead, several broad classification systems have been proposed. One based on origin divides air pollutants into two broad categories:

1. Primary pollutants emitted directly from sources and not undergoing any chemical or physical transformation.
2. Secondary pollutants formed in the atmosphere as a result of chemical reactions among primary pollutants and normally present atmospheric constituents. Several different types of chemical reactions lead to the formation of secondary pollutants such as salt particles, acid droplets, nitrates, sulfates, NO, NO_2, O_3, and a variety of oxygenated hydrocarbons (Stern et al., 1984). These are schematically depicted in Figure 1.2.

Note that many reactive species of pollutants can be present in the atmosphere in both primary and secondary forms, whereas nonreactive species can exist as primary pollutants only.

Another broad classification is based on the state of the pollutant matter. Gaseous air pollutants include all the gases that are found in the atmosphere above their normal ambient levels, while particulates include both solid and liquid particles that become airborne.

The ambient or clean atmospheric concentrations of the various gases and their approximate residence times in the atmosphere are given in Table 1.2.

The dry atmosphere is primarily composed of nitrogen (N_2), oxygen (O_2), and several inert noble gases—argon (Ar), neon (Ne), krypton (Kr), and xenon (Xe). Their relative concentrations, expressed in parts per million (ppm) by volume, have remained essentially fixed over time. The concentrations of water vapor (H_2O) are observed to be highly variable, both in time and space. It is one of the most important ingredients of weather and climate in the lower atmosphere, but is not normally considered an air pollutant. Carbon dioxide (CO_2) also has relatively high concentrations, most of which are due to natural emissions as part of the natural carbon cycle. But, the increasing ambient levels of CO_2 at the current rate of 0.5 percent per year are clearly due to anthropogenic emissions from fossil fuel burning, slash burning, and deforestation. Even though increasing levels of CO_2 might be beneficial to agriculture and forestry, as some studies have suggested, the potential consequences of climate warming due to increasing levels of CO_2, methane (CH_4), and other greenhouse gases are considered serious enough to put these gases under the category of air pollutants. Most of the other species considered air pollutants have natural as well as anthropogenic sources. Their concentrations in polluted atmospheres, such as over large urban and industrial complexes, are found to be several orders of magnitude larger than the "clean" atmospheric values given in Table 1.2.

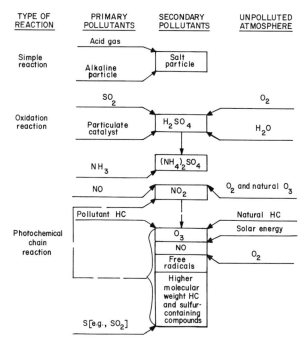

Figure 1.2 A schematic of reactions leading to the formation of secondary pollutants. From Stern et al., 1984.

1.3.1 Concentration Units

Concentrations of gaseous species in the atmosphere are most commonly expressed in terms of the volume of the species per unit volume of air (e.g., ppm by volume). Another commonly used measure for expressing concentrations of gases as well as particulates is the mass of species per unit volume of air (e.g., $\mu g\ m^{-3}$). The former is referred to as the volumetric concentration, while the latter is called the mass concentration. The conversion between the volumetric concentration, c_{vi} ppm, and the mass concentration, c_{mi} $\mu g\ m^{-3}$, where the subscript i denotes the particular species whose concentrations are referred to, can be made by expressing both c_{vi} and c_{mi} in terms of the molar concentration, c_i, defined as moles per unit volume of air. Thus, from the definitions of the various concentration measures, it follows that

$$c_{vi} = \frac{c_i}{c_a} \times 10^6\ [\text{ppm}] \quad (1.1)$$

$$c_{mi} = c_i\, m_i \times 10^6\ [\mu g\ m^{-3}] \quad (1.2)$$

where m_i denotes the mean molecular mass in grams per mole, and c_a is the molar concentration of air in

Table 1.2 Ambient ("Clean" Atmospheric) Concentrations of Normal Atmospheric and Pollutant Gases

Normal Atmospheric Gas	Average Concentration (ppm)	Pollutant Gas	Average Concentration (ppm)	Approximate Residence Time
N_2	780,840	CO_2	355	15 years
O_2	209,460	CH_4	1.7	7 years
Ar	9,340	N_2O	0.3	10 years
Ne	18.2	CO	0.05–0.2	65 days
Kr	1.1	SO_2	10^{-5}–10^{-4}	40 days
Xe	0.09	NH_3	10^{-4}–10^{-2}	20 days
He	5.2	NO_x	10^{-6}–10^{-2}	1 day
H_2	0.58	HNO_3	10^{-5}–10^{-3}	1 day
H_2O	Variable	O_3	10^{-2}–10^{-1}	

Source: Seinfeld (1986).

Table 1.3 A Classification of Gaseous Air Pollutants Based on Their Chemical Properties

Class	Primary Pollutants	Secondary Pollutants
Sulfur compounds	Sulfur dioxide (SO_2), sulfur trioxide (SO_3), hydrogen sulfide (H_2S), carbonyl sulfide (COS), carbon disulfide (CS_2), dimethyl sulfide ($(CH_3)_2S$)	Sulfur trioxide (SO_3), sulfuric acid (H_2SO_4), sulphate ion ($SO_4^=$), ammonium sulphate ($(NH_4)_2SO_4$), etc.
Nitrogen compounds	Nitric oxide (NO), nitrous oxide (N_2O), nitrogen dioxide (NO_2), ammonia (NH_3)	Nitric oxide (NO), nitrogen dioxide (NO_2), nitric acid (HNO_3), nitrate ion (NO_3^-), ammonium ion (NH_4^+)
Inorganic carbon compounds	Carbon monoxide (CO), carbon dioxide (CO_2)	Carbon dioxide (CO_2), oxygenated inorganic carbon compounds
Organic carbon compounds	Methane (CH_4), terpenes, isoprene, C1–C5 compounds classified as paraffins, olefins, and aromatics	Organic nitrates, peroxides (e.g., PAN), oxygenated hydrocarbons classified as aldehydes, ketones, and acids
Halogen compounds	Hydrogen fluoride (HF), hydrogen chloride (HCl), carbon tetrachloride (CCl_4), chlorofluorocarbons (CFCs), methyl chloride (CH_3Cl, CH_3CCl_3)	
Radioactive substances	Strontium ($^{90}S_r$), cesium ($^{137}C_s$), iodine (^{131}I), tritium (^3H) radon and radon daughters	

moles per cubic meter. The latter can be expressed in terms of atmospheric pressure and temperature, using the ideal gas law for air

$$pV = nR^*T \quad (1.3)$$

Note that, by definition,

$$c_a = \frac{n}{V} = \frac{p}{R^*T} \quad (1.4)$$

where $R^* = 0.08314$ mb m^3 K^{-1} mol^{-1} is the absolute gas constant, p is pressure in millibars, and T is the absolute temperature.

From Eqs. (1.1) through (1.4) one obtains the following relationship between volumetric and mass concentrations:

$$c_{vi} = \frac{R^*T}{p\,m_i} c_{mi} \quad (1.5)$$

where c_{vi} is the concentration of species i in ppm and c_{mi} is that in µg m^{-3}. Concentrations are also sometimes expressed in units of parts per billion (ppb) and parts per trillion (ppt) by volume, depending on the trace amounts of the species present. Concentrations of organic carbon compounds are sometimes reported as parts per million of carbon (ppmC). The concentration is obtained by multiplying the volumetric concentration of the carbon-containing species in ppm by the number of carbon atoms in the molecule. The units ppm, ppb, and so on, are also used to express mole fraction or mass fraction of species in air. The former is the ratio of moles of the species to those of air in a given volume, while the latter is the ratio of their masses. It is easy to see that

$$\text{Mole fraction} = \frac{c_i}{c_a} \quad (1.6)$$

$$\text{Mass fraction} = \frac{c_i\,m_i}{c_a\,m_a} \quad (1.7)$$

1.3.2 Further Classification of Gaseous Pollutants

A commonly used classification of gaseous pollutants found in the atmosphere is given in Table 1.3; it is based on their chemical properties.

Sulfur Compounds The major sources of sulfur-containing compounds present in the atmosphere are combustion of sulfur-containing fossil fuels and organic matter, biological decay, waste disposal, pulp and paper manufacturing, smelting processes, sea spray, and evaporation from oceans. Natural sulfur comes from H_2S released by $SO_4^=$-reducing bacteria; other biogenic sulfur species are carbonyl sulfide, carbon disulfide, and dimethyl sulfide. Carbonyl sulfide (COS) is the most abundant gaseous sulfur species in the troposphere.

Sulfur dioxide (SO_2) has been recognized as a major air pollutant emitted from anthropogenic sources (e.g., burning of high-sulfur fuels and smelting of metal sulfides). Sulfur trioxide (SO_3) is also emitted with SO_2 at about 1 to 5 percent of the SO_2 concentration. Some industries, such as sulfuric acid manufacturing, electroplating, and phosphate fertilizer manufacturing, may emit higher relative amounts. But most of the SO_3 in the atmosphere is probably formed as a secondary pollutant resulting from the relatively slow SO_2 oxidation. Sulfur trioxide rapidly reacts with water in air to form sulfuric acid (H_2SO_4) which, in turn, may react with ammonia to become sulphate ion ($SO_4^=$). Sulphate aerosol or acid mist is frequently associated with haze and poor visibility in urban atmospheres.

Sulfur dioxide is a colorless corrosive gas with a pungent, irritating odor. It has long been recognized as a major air pollutant. The oxidation of SO_2 occurs through many different mechanisms, including contact with surfaces, gas-to-particle conversion, and catalysts (Lyons and Scott, 1990). The reactions lead to the formation of sulfuric acid and sulfates as secondary pollutants which make cloud and rainwater acidic. Since 1970, sulfur dioxide concentrations in urban areas have dropped significantly because of the shift to natural gas and lower sulfur fuel oils for domestic heating and for smaller industrial power and heating plants (Urone, 1986).

There are also several reduced sulfur compounds that are of interest in air pollution. Hydrogen sulfide (H_2S) is a highly toxic, foul-smelling gas that is mostly emitted by natural sources, including anaerobic biological decay processes in swamps and bogs, volcanic eruptions, hot-water springs, and geysers. Anthropogenic sources of H_2S are sewage treatment plants, waste disposal sites, paper and pulp industry, rayon industry, coke ovens, oil refineries, and metal smelters. Other sulfur compounds of interest, primarily because of their strong odors, are methyl mercaptan (CH_3SH), dimethyl sulfide, dimethyl disulfide, carbonyl sulfide, and their higher molecular homologs. There are natural biogenic sources for most of these compounds.

Nitrogen Compounds Nitrogen forms a very stable diatomic gas, N_2, that makes up the bulk of the atmosphere and tempers the oxidative power of atmospheric oxygen. It also occurs in a large number of nitrogen compounds, which are produced by high-temperature combustion and natural processes such as bacterial fixation, biological growth and decay, lightning, volcanic activity, and forest fires. In the oxidized state nitrogen forms seven oxides and a large number of nitro, nitrite, and nitrate derivatives. In the reduced state it forms such compounds as ammonia, amides, amines, amino acids, and nitrites (Urone, 1976).

Of the seven oxides of nitrogen, only three, nitrous oxide (N_2O), nitric oxide (NO), and nitrogen dioxide (NO_2), are found in any significant concentrations. The latter two are often combined together and are referred to as nitrogen oxides and denoted by NO_x. Nitrous oxide is a colorless, nontoxic gas present in the atmosphere in relatively large concentrations. It is commonly called laughing gas because it can cause those who inhale it to laugh violently. Because of its low reactivity in the lower atmosphere, N_2O is not considered to be a pollutant of significant concern. Also, there are no significant anthropogenic sources of N_2O, aside from the fertilized agricultural soils.

Nitrogen oxides, NO_x, are important air pollutants that take part in the well-known photochemical reactions resulting in the formation of ozone (O_3) and photochemical smog. Nitric oxide is a colorless and odorless gas. In air it is oxidized rapidly by atmospheric ozone and more slowly by oxygen to form NO_2. As an anthropogenic pollutant, NO is produced largely by fuel combustion in both stationary and mobile sources such as automobiles. In high-temperature combustion, nitrogen reacts with oxygen to form NO. In hot exhaust gases, some (0.5–10%) of the NO is oxidized to NO_2. In a polluted atmosphere, NO is oxidized to NO_2 primarily through photochemical secondary reactions. Nitrogen dioxide is a reddish-brown gas with a pungent, irritating odor. It readily absorbs sunlight, more strongly in the yellow to blue end of the visible solar radiation and the near ultraviolet, to form nitric oxide and atomic oxygen. The latter is very reactive and forms ozone with oxygen and initiates a number of secondary photochemical chain reactions. Emissions of NO_x in the United States have been increasing at the rate of 10 to 15 percent per year.

Ammonia (NH_3) is considered to be a relatively unimportant pollutant. It is primarily emitted by natural bacterial processes in organic waste materials. Major industrial sources include ammonia manufacturers, petroleum refineries, nitrate fertilizer plants, large cattle feedlots, and coke ovens. Ammonia plays an important role in the reactions and fate of many other gaseous pollutants. For example, NH_3 forms ammonium (NH_4^+) salts with sulfuric, nitric, and hydrochloric acids. A large fraction of atmospheric particulates are found to contain ammonium compounds. Other secondary nitrogen compounds of interest in air pollution are nitrous acid (HNO_2), nitric acid (HNO_3), and nitrates, which are responsible for causing acidic clouds and precipitation.

Inorganic Carbon Compounds Primary inorganic compounds present in the atmosphere are carbon monoxide (CO) and carbon dioxide (CO_2), which are produced by incomplete and complete combustion of carbon-containing fossil fuels (coal, oil, charcoal, and gas), incineration of biomass or solid waste, and anaerobic decomposition of organic material and in respiratory processes of plants and animals. Mostly anthropogenic sources are responsible for the observed increasing levels of CO and CO_2 in the atmosphere.

Carbon monoxide is a colorless, odorless, nonirritating, but highly toxic gas that is considered a dangerous asphyxiant. It combines with the hemoglobin of the blood and reduces the blood's ability to carry oxygen to cell tissues. In cities, almost all of CO is anthropogenic, most of it due to motor vehicle emissions and fossil fuel burning for heating and power generation. Natural sources, such as forest and grassland fires, volcanoes, and natural gas vents, also produce huge quantities of CO, far exceeding the amount produced by all the anthropogenic pollution sources.

In the presence of hydroxyl radical, CO may be converted into CO_2 in the atmosphere. Anthropogenic emissions of CO in the United States have been decreasing primarily due to the mandatory use of catalytic afterburners in new automobiles.

Carbon dioxide is not generally considered a dangerous air pollutant. It is nontoxic and is an essential ingredient of plant and animal life cycles. Through photosynthesis, CO_2 is converted to plant tissues, with oxygen produced as a by-product. A large portion of CO_2 in the atmosphere is due to natural emissions, but anthropogenic sources have contributed to an increasing burden of CO_2 in response to increasing worldwide consumption of fossil fuels for transportation and power generation since the advent of the industrial revolution. Thus, the well-known observed increasing trend of CO_2 concentrations in the remote atmosphere is essentially due to human activities. The primary consequence of the increase is its important role in possible climate change (warming). Carbon dioxide is the most abundant of the "greenhouse" gases in the atmosphere; it has been observed to be increasing at the rate of 0.5 percent per year.

Organic Carbon Compounds The organic carbon-containing compounds found in natural and polluted atmospheres comprise a broad spectrum of hydrocarbons and oxygenated hydrocarbons. Some of these are emitted to the atmosphere as primary pollutants, while others form as products of photochemical chain reactions. Because carbon can form bonds with other elements such as hydrogen, nitrogen, oxygen, and sulfur and, at the same time, combine with itself to form a series of chain, cyclic, and combined systems, a large number of compounds are produced. Many gaseous carbon compounds are emitted to the atmosphere by natural sources, such as volcanoes, forest fires, natural gas seepage, and biological processes. In urban and industrial areas, many hydrocarbons, including volatile organic compounds (VOCs), are emitted by anthropogenic sources, such as transportation, fossil fuel-burning power plants, chemical plants, petroleum refineries, certain construction activities, solid waste disposal, and slash burning.

Methane (CH_4) is probably the most abundant organic compound. It is produced by natural wetlands and rice paddies, by rotting of dead plants in swamps, natural gas leakage, and by bacteria in the guts of termites and ruminant animals. It is relatively nonreactive and not considered important in photochemical reactions. The primary concern for methane as an air pollutant is in its increasingly important role in the possible climate warming.

Nonmethane hydrocarbons (NMHC), especially VOCs, are very important ingredients of photochemical chain reactions with nitrogen oxides (NO_x) in the presence of warm sunlight, which result in photochemical smog and ozone. Depending on their bonding, these compounds are classified into four major groups: alkanes or paraffines, alkenes or olefins, acetylenes, and aromatics (Urone, 1976; Seinfeld, 1986). There are many compounds in each group. Some of the important species of anthropogenic VOCs are benzene, toluene, formaldehyde, vinyl chloride, phenols, chloroform, and trichloroethylene. The more toxic of these are called hazardous organic compounds (HOCs).

In addition to the anthropogenic VOCs, many hydrocarbons are produced naturally by trees. The most abundant are isoprene, which is primarily emitted by deciduous trees (e.g., oak, sycamore, poplar, aspen, spruce, etc.), and α-pinene, β-pinene, and limonene, which are produced by coniferous trees (e.g., pines, firs, cypress, etc.). Biogenic emissions of hydrocarbons are likely to exceed anthropogenic emissions in heavily vegetated and forested regions. In particular, isoprene plays a major role in the production of ozone and photochemical smog because of isoprene's large emission rate and strong reactivity.

By themselves, the hydrocarbons in air have relatively low concentrations and low toxicity. They are of concern mainly because they play a major role in atmospheric chemistry at local/urban, regional, and global scales because of their reactivity with ozone, NO_2, and hydroxyl radicals. For example, in the presence of nitrogen oxides and sunlight, they react to form photochemical oxidants, including ozone, peroxyacetyl nitrate (PAN), and a number of oxygenated hydrocarbons. Nonmethane hydrocarbons also react rapidly with ozone and hydroxyl radicals, forming among other products carbon monoxide and thereby impacting on the oxidizing capacity of the atmosphere.

Oxygenated hydrocarbons are classified as alcohols, phenols, ethers, aldehydes, ketones, peroxides, and organic acids. Some minor amounts of oxygenated hydrocarbons are also emitted as solvent vapors from chemical, paint, and plastic industries. Far greater primary emissions are usually associated with automobiles and other transportation sources, power plants, and petroleum refineries. Many oxygenated carbon compounds are formed as secondary products of photochemical chain reactions. There are literally hundreds of chain reactions involving many reaction products (see, e.g., Seinfeld, 1986, Ch. 4).

The tropospheric lifetimes of volatile hydrocarbons range from several months to hours, with the most oxidized in the planetary boundary layer. For a single VOC in the presence of NO_x, CO, H_2O, and other trace substances, many chemical reactions may occur. These reactions also produce secondary pollutants, such as ozone.

Ozone and Oxidants Ozone (O_3) is a highly reactive bluish gas that is naturally formed at high altitudes in the stratosphere by photochemical reactions involving molecular and atomic oxygen in the presence of high-intensity ultraviolet radiation. Its concentration in the upper atmosphere depends on both the altitude and the latitude. Most of the stratospheric

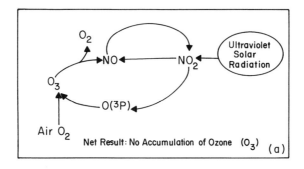

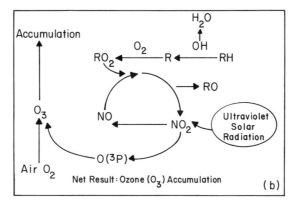

Figure 1.3 A schematic of ozone formation in the polluted troposphere: (a) The photostationary state in the absence of VOCs. (b) Ozone accumulation in the presence of VOCs.

ozone resides between 10 and 40 km with its maximum concentration around 20 km above the earth's surface. Ozone there plays a beneficial role by absorbing ultraviolet radiation from the sun and thus protecting the life on earth from the destructive effects of such radiation. In this sense, the stratospheric ozone layer has been called "our ozone shield" (University Corporation for Atmospheric Research, 1992). The strength and the effectiveness of this ozone shield have been threatened recently by anthropogenic emissions of pollutants that destroy ozone through complicated reactions. This issue will be discussed in the next section on the effects of air pollution.

Unlike the "good" stratospheric ozone, there is also the "bad" ozone in the troposphere near the ground, which is damaging to plants and materials and harmful to human health. Ozone and other oxidants such as peroxyacetyl nitrate and hydrogen peroxide (H_2O_2) are formed in polluted atmospheres as a result of a rather wide variety of photochemical reactions involving nitrogen oxides and volatile organic compounds in the presence of sunlight.

The chemistry of tropospheric ozone formation is rather complex. It is schematically shown in Figure 1.3. The production of ozone from the photodissociation of NO_2 is illustrated by the following chemical reactions:

$$NO_2 + h\nu \rightarrow NO + O(^3P) \quad (1.8)$$

$$O(^3P) + O_2 \rightarrow O_3 \quad (1.9)$$

$$O_3 + NO \rightarrow NO_2 + O_2 \quad (1.10)$$

where $h\nu$ represents the ultraviolet radiation. Nitrogen dioxide (NO_2) is photodissociated into nitric oxide (NO) and an excited state of oxygen O (^3P). The excited oxygen reacts with a diatomic oxygen molecule, producing ozone, O_3. However, this ozone reacts with NO, forming NO_2 and O_2 and closing the cycle. This simple set of repetitive reactions resulting in formation but no net accumulation of ozone is called the photostationary state; it is schematically shown in Figure 1.3(a).

In the presence of volatile organic compounds, however, the above photostationary equilibrium is disturbed, because NO is more readily converted into NO_2 by chemical reactions involving reactive hydrocarbons without consuming O_3. Reactions of VOCs with OH radicals, which normally exist in the ambient atmosphere, yield RO_2 radicals, which then compete with ozone for the oxidation of NO to NO_2. There are hundreds of photochemical chain reactions involving the wide variety of reactive hydrocarbons that exist in a polluted atmosphere. The net result is the accumulation of ozone, as shown schematically in Figure 1.3(b).

Ozone is used as an indicator pollutant for photochemical oxidation products. The total mixture, frequently referred to as smog, causes eye irritation, lachrymation, and respiratory difficulties for people walking or working outdoors.

Ozone has an acrid, biting odor that is a distinctive characteristic of photochemical smog. High concentrations of ozone and other photochemical oxidants are observed over most large cities and metropolitan areas during summer months. Harmful levels of ozone are also found to exist over large rural regions to which ozone gets transported from large urban and industrial areas. Thus, the tropospheric ozone is not merely an urban air pollution problem but also a regional problem, particularly for North America and Europe. It is by far the most persistent problem that has defied simple solutions based on tailpipe hydrocarbon emissions control.

Halogen Compounds Gaseous halogen pollutants include the elements and compounds of fluorine, chlorine, bromine, and iodine. Of these, chlorine (Cl_2), hydrogen fluoride (HF), hydrogen chloride (HCl), chlorofluorocarbons (CFCs), the Freons, and the halogenated pesticides and herbicides are encountered most frequently. There has been considerable interest in halogen compounds because of several important atmospheric phenomena, including the depletion of stratospheric ozone due to CFCs and possible contribution of HCl to acid precipitation, that have caught media and public attention. These will be discussed in a later section.

Fluorides are present in air worldwide, but their concentrations are generally low. Hydrogen fluoride (HF) is the most common inorganic fluoride gas found in air. Higher concentrations are found in areas near pollutant sources. The principal sources include the aluminum, steel, glass, brick, tile, and phosphate fertilizer industries.

The gaseous forms of chlorine and its compounds most frequently found in polluted atmospheres are elemental chlorine (Cl_2), hydrogen chloride (HCl), and the vapors of chlorinated hydrocarbon solvents, pesticides, and herbicides. Elemental chlorine is widely used in chemical and plastic industries, water and sewage treatment plants, household bleaches, and swimming pools. It is a yellowish colored gas with a strong pungent odor. It is very reactive and highly irritating, toxic gas. Of primary concern are accidental releases due to rupture of containers. Hydrogen chloride is a common air pollutant emitted by a number of industrial activities. It is not highly toxic, but may contribute to the acidity of precipitation. It is generally present in relatively low concentrations when compared with SO_2, NO_x, and other acidic gases.

Chlorinated hydrocarbons constitute a very broad range of organic compounds that have been used extensively. These include many solvents, cleaning agents, herbicides, pesticides, and fungicides. A sizable fraction of what have been classified as hazardous air pollutants (HAPs) are chlorinated hydrocarbons. Some of the more important compounds are carbon tetrachloride, methyl chloroform, methyl chloride, perchloroethylene, and trichloroethylene.

Radioactive Substances The principal anthropogenic radioactive nuclides released in the atmosphere are frequently the same regardless of whether the emission source is a nuclear reactor or a plant reprocessing spent reactor fuel. Their significance as atmospheric contaminants depends on the type of energy they emit, their radioactive half-lives, and the manner in which they are metabolized when absorbed into the body (Eisenbud, 1973, 1976).

Three of the most important fission products are strontium-90 (^{90}Sr), cesium-137 (^{137}Cs), and iodine-131 (^{131}I). Strontium-90 is considered to be the most dangerous from the standpoint of its long-term effects on public health because of its relatively long (28-yr) half-life and ready absorption by animals and humans through the food chain. Cesium-137 also has a long half-life, 30 years, and is a relatively soluble substance. But unlike ^{90}Sr, it does not become fixed in the skeleton and is more readily eliminated from the body.

Of the several radioactive isotopes of iodine produced in the fission process, ^{131}I is of principal concern because of its relatively longer half-life (8 days). When absorbed into the body, radioiodines tend to concentrate in the thyroid and can thus deliver a relatively high dose to a very small volume of tissue. Because of their volatility, the radioiodines can more readily escape to the atmosphere than most other radionuclides. Most extensive exposure to these is likely to result from their deposition on pasture lands with subsequent ingestion by grazing animals and passage to humans through the food chain (e.g., via cow's milk).

Many other fission and activation products, such as carbon-14 (^{14}C) and tritium (^{3}H), are formed in nuclear explosions and also in the operation of nuclear reactors. Some of the more important radionuclides produced by nuclear reactions and their half-lives are listed in Table 1.4. It is safe to say that there is such a great public concern over the release of radioactive material in the atmosphere, even in small amounts, that there is an effective moratorium on the planning and construction of new reactors in the United States.

There has also been concern, in recent years, about the natural sources of radioactive gases in air, particularly in houses. The soils and rocks contain naturally radioactive minerals in variable amounts, depending on local geology. Of particular concern in indoor air pollution are radon (^{222}Rn) and thoron (^{220}Rn), which are radioactive progeny of two nuclides of radium, ^{226}Ra and ^{228}Ra. Radon, with a half-life of 3.8 days, has a reasonably high probability of escaping into the atmosphere or indoor air before it decays. The atmospheric concentrations of radioactive noble gases and their daughter products depend on many geologic and meteorological factors. Their concentration in indoor air can build up, particularly if basements are not properly ventilated. Recent surveys of radon and its daugh-

Table 1.4 Some of the More Important Radionuclides Produced by Nuclear Reactions

Fission Products		Activation Products	
Radionuclide	Half-life	Radionuclide	Half-life
^{85}Kr	10.4 years	^{3}H	12 years
^{89}Sr	50 days	^{14}C	5,730 years
^{90}Sr	28 years	^{54}Mn	314 days
^{95}Zr	65 days	^{55}Fe	2.7 years
^{131}I	8.1 days	^{59}Fe	45.6 days
^{133}I	21 hours	^{60}Co	5.3 years
^{135}I	6.7 hours	^{65}Zr	245 days
^{137}Cs	30 years	^{238}Pu	86.4 years
^{140}Ba	12.8 days	^{239}Pu	24,000 years
^{144}Ce	285 days		

Source: Eisenbud (1976).

ter products in residential basements in the United States indicate unsafe levels existing in significant proportion (up to 30%) of houses in certain areas.

Another source of natural radioactivity in the atmosphere is the combustion of fossil fuels. Coal may contain 1 to 2 ppm of uranium-238 and thorium-232 and the coal ash is found to contain measurable amounts of ^{226}Ra and ^{228}Ra, some of which escape into the atmosphere. Oil burning plants also discharge ^{226}Ra, ^{228}Ra, and other radionuclides into the atmosphere (Eisenbud, 1976). Some of the radionuclides attach themselves to the inert dust particles in the atmosphere; the bulk of the radioactivity is contained on fine particles having diameters less than 0.2 μm. The natural radioactivity of atmospheric dust, due primarily to the adsorbed radon daughters, can be readily demonstrated (Eisenbud, 1976).

1.3.3 Classification of Atmospheric Particulates

Atmospheric particulates or aerosols include all liquid and solid particles, except pure water, that exist in the atmosphere under normal conditions. Most of these are a result of direct emissions as particles from the various natural and anthropogenic sources, while others form from the condensation of certain gases and vapors that are emitted into the atmosphere or are a result of chemical transformations. Thus, like gaseous pollutants, atmospheric particulates can also be broadly classified as primary and secondary aerosols. A full description of them requires specification of their concentration, size distribution, chemical composition, phase (liquid or solid), morphology, and biological activity. Several terms are commonly used in characterizing the particulate-laden cloud masses (Seinfeld, 1986):

Dust: Coarse solid particles produced by mechanical disintegration of material such as crushing, grinding, and blasting.
Smoke: Fine solid particles resulting from incomplete combustion of carbon and other combustible material.
Fog and Mist: Suspension of liquid droplets formed by condensation of vapor. A fog consists of high concentration of relatively small droplets, while a mist has relatively large droplets in small number concentration.
Haze: Combination of water droplets, dust, and smoke particles, and photochemical pollutants impeding visibility.
Smog: Combination of smoke and fog, but more commonly the combination of photochemically produced pollutant gases and particles.

Concentrations of particulate matter are usually expressed in terms of mass per unit volume (e.g., μg m^{-3}), although concentrations of certain ionic aerosol constituents are also sometimes expressed in terms of mole fraction or mass fraction (e.g., ppb mole fraction).

Sizes of atmospheric particles are expressed in several different ways. The most common measure is the actual diameter in micrometers (μm) for spherical particles. Nonspherical particles are frequently characterized in terms of the diameter of equivalent spherical particles that would have the same volume or the same mass as the actual particles. On the basis of size, atmospheric particles are usually divided between two broad categories, fine particles and coarse particles. But there is no commonly agreed on particle size that separates these two categories. Proposed sizes in the literature range from 0.1 μm to 100 μm. In view of the National Ambient Air Quality Standards (NAAQS) for particulate matter less than 10 μm in size (PM$_{10}$), 10 μm might be considered a reasonable choice for the boundary between coarse and fine particles. On the basis of the distinctive modes of idealized distributions of particle surface area and mass, 1 μm would be a better choice (Seinfeld, 1986). Considerations of health effects of fine particles, including viruses and other viable particles, would suggest even a lower limit of, say, 0.1 μm. In practice, fine and coarse fractions are considered to be those collected by the fine and coarse fractions of a dichotomous particulate sampler, the fine stage having an upper cutoff point of about 2.5 μm (Urone, 1986). Although the total suspended particulate matter (TSP) is relevant for visibility, soiling, and corrosion effects of particles, the PM$_{10}$ is considered more important for health effects. Also, particles larger than 10 μm fall out more readily through gravitational settling.

Atmospheric aerosols consist of solid or liquid particles ranging in size from a few tens of Angstroms to several millimeters. Based on their emission sources and mechanisms of formation, aerosols can be classified as primary and secondary aerosols. Primary aerosols are emitted in particulate form directly from sources and contain particles of all sizes mentioned above. Secondary aerosols are particles produced in the atmosphere from gas-phase chemical reactions that generate condensable species. These are mainly submicron-sized fine particles (Seinfeld, 1986).

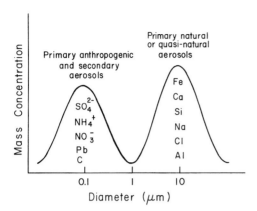

Figure 1.4 A schematic of idealized aerosol mass distribution showing a typical segmentation of chemical species into fine and coarse particle fractions. From Seinfeld, 1986.

The size, number, and chemical composition of atmospheric aerosols are changed by several mechanisms, such as nucleation, condensation, coagulation, and aggregation, until ultimately they are removed by natural processes. The primary mechanisms for particle removal are dry deposition by gravitational settling and inertial impaction on surfaces, and wet deposition following cloud and precipitation scavenging. Since fine and coarse particles have different sources and formation mechanisms, different transformation and removal mechanisms, and also different chemical and physical properties, it is important to categorize and treat them separately. Therefore, the distinction between fine and coarse particles based on their origin, transformation, and removal mechanisms is a fundamental one in any discussion of aerosol physics and chemistry (Seinfeld, 1986). The existence of a bimodal distribution of aerosol mass with distinct fine and coarse modes around 0.1 and 10 μm, respectively, has been clearly demonstrated in atmospheric aerosol measurements. The minimum around 1 μm separates the two fractions. The idealized particle mass distribution is schematically shown in Figure 1.4.

The two modes also show distinctive chemical compositions of particulate matter. Coarse particles consist mainly of crustal material, such as iron, calcium, silica, sodium, and aluminum. Fine particles, on the other hand, mostly contain sulphate, ammonium and nitrate ions, lead, carbon-containing material including soot, condensed organic matter, and certain toxic compounds. These distinctive chemical properties of fine and coarse particle fractions are schematically shown in Figure 1.4 together with their origin. For more detailed descriptions of the chemical composition of atmospheric particulates, the reader may refer to books by Stern (1976–86), Seinfeld (1986), and other authors dealing with aerosol physics and chemistry.

Major natural sources of atmospheric particulates are soil and rock debris, sea spray, wild fires, volcanic eruptions, and reactions between natural gaseous emissions. Anthropogenic sources of the particulate matter can be divided into four broad categories: (1) fuel combustion and industrial processes, (2) industrial process fugitive particulate emissions, (3) nonindustrial fugitive emissions, and (4) transportation sources (Seinfeld, 1986). According to the estimates by U.S. Environmental Protection Agency (1982), nonindustrial fugitive emissions (roadway dust from paved and unpaved roads, wind erosion from croplands, agricultural activities, etc.) in the United States, on an annual basis, far exceed the particulate emissions from industrial and transportation sources. But, the impact of nonindustrial fugitive dust emissions is limited to rural areas, because the emissions are mostly large particles that settle to the ground a short distance from the source. In urban areas, local emissions from industrial and transportation sources are more important. Most of the particulates from transportation sources come from vehicle exhausts. These are generally smaller than 1 μm in diameter and are composed primarily of lead halides, sulfates, and carbonaceous matter. Particulate matter from other fuel combustion sources also fall into the category of fine particles, but may contain a large variety of chemical compounds, depending upon the type of fuel used and the type of combustion process involved. Particulates from other industrial processes and fugitive emissions contain a much wider range of sizes and chemical composition. On the basis of chemical composition and biological viability of atmospheric particulates, these can be classified into three categories: (1) trace metal particles, (2) nonviable carbonaceous particles, and (3) viable particles.

Trace Metal Particles Atmospheric particulate matter is analyzed routinely for more than forty trace elements, most of them metals. For example, the results from a trace element analysis of 130 samples taken in summer from the Washington, DC, area indicate abundance (concentrations more than 0.1 μg m^{-3}) of such metals as iron, aluminum, magnesium, calcium, potassium, sodium, lead, bromium, and titanium and measurable amounts of many other metals in urban particulates (Seinfeld, 1986).

Trace element analyses of particulate samples are sometimes used to characterize emissions from different types of sources, such as sea salt, soil, limestone, coal, and oil. Certain elements are selected as fingerprint elements for characterizing these sources, for example, Na for sea salt, Mn for soil, Ca for limestone (Seinfeld, 1986).

Carbonaceous Particles Carbonaceous particles in the atmosphere consist of two major components: (1) graphite black carbon and (2) organic material. Black carbon particles are produced as a result of combustion processes and are therefore primary aerosols. Organic particulate matter is emitted directly from certain sources and is also produced as a secondary aerosol from atmospheric reactions involving gaseous organic precursors. The total primary carbonaceous material (the sum of black carbon and primary organics) is often called soot, which is produced by a variety of mobile and stationary sources.

Carbonaceous particulate matter is a mixture of many classes of compounds (see e.g., Seinfeld, 1986). Primary organic compounds can be classified as alkanes, alkenes, aromatics, and polycyclic aromatics. Oxidized hydrocarbons, such as acids, aldehydes, ketones, quinones, phenols, and esters, may be emitted directly by combustion sources or produced as a result of oxidation reactions in the atmosphere. Some of these have been identified as carcinogens or toxics, which might pose a long-term health hazard. Organic compounds of relatively low molecular weight and high vapor pressure are distributed between solid (particulate) and vapor phases in the atmosphere. Emissions from high-temperature combustion processes contain larger proportions of organic compounds in the vapor phase. Some of the compounds emitted as vapor condense on the surfaces of particles as they cool.

Carbonaceous particles in air are found everywhere, with the mean carbon concentrations ranging from about 1 $\mu g\ m^{-3}$ in remote areas to 23 $\mu g\ m^{-3}$ in large cities like Los Angeles. Thus, anthropogenic sources account for most of the carbonaceous particulate matter in urban atmospheres. About 80 percent of the elemental carbon mass is associated with particles of diameter less than 2.5 μm. In these the carbon derived from fossil fuels can be distinguished from those derived from living organic matter such as wood by measuring the ratio of the radioactive isotope ^{14}C to the total carbon mass (Seinfeld, 1986).

Viable Particles There is considerable interest in the dispersal of biologically viable organisms as particulates in the atmosphere, because of their importance to human health as well as to plant life. Jacobson and Morris (1976) have reviewed the types of viable particles present in air, their sizes and other characteristics, sources, and effects. These are divided into three broad categories—pollen, microorganisms, and insects. A comprehensive review of the meteorology of airborne organisms is given by Pedgley (1982).

Pollen grains are released into the atmosphere from weeds, grasses, and trees. They range in size between 10 and 100 μm and may not travel long distances before they deposit. Most pollens are quite hygroscopic and their deposition rates depend on humidity. Pollen releases are highly seasonal; most tree pollen is shed during spring and early summer, grass pollen during midsummer, and weed pollen during late summer and fall. Pollination is an important biological process in the propagation of plant life. But, pollens also cause allergic responses in many sensitized humans, and soiling of clothes and materials. In particular, pollinosis (hay fever) severely affects 5 to 10 percent of the population of the United States. It is mostly caused by pollen from species of *Ambrosia* (ragweed). The major source of pollinosis in Europe is grass pollen.

There are a wide variety of airborne microorganisms, such as algae, protozoa, fungi, yeasts, molds, rusts, spores, bacteria, and viruses, that are associated with humans, animals, and plant life. Algae are chlorophyll-bearing plants of about 0.5 μm diameter that originate primarily from cultivated and uncultivated soils. Protozoa are primitive animals with no tissues, ranging in size from 2 μm to several centimeters. Airborne algae and protozoa are of concern because they tend to concentrate radioisotopes and pesticides and could be carriers of pathogenic bacteria and viruses. Algae also commonly occur in house dust (Jacobson and Morris, 1976).

Fungi are practically ubiquitous in soil and airborne dust. Many species of fungi produce spores whose dispersive properties depend on humidity. Very high concentrations (approaching $10^5\ m^{-3}$) of spores have been observed in certain regions in the United States in summer months. The most common type of spores is one-celled hyaline type. Concentrations of these and other spores and yeast show strong diurnal variations with distinct daytime peaks. Spore release from vegetation depends on temperature, relative humidity, precipitation, mean winds, and turbulence within and just above the canopy. Most of the larger spores are usually deposited on the ground within 100 m of their source. Smaller ones ($<10\ \mu m$) can be carried by winds and dispersed over large distances (>100 km) from the source region. Atmospheric dispersion plays an important role in the spread of plant diseases from infected areas.

Of much greater importance to human health are the airborne bacteria and viruses. Bacteria are found everywhere and, being small (0.3–15 μm), they can be transported long distances by winds. Major sources of airborne bacteria are soil, water, sewage treatment, and waste disposal facilities. Trickling filter sewage treatment plants and spray irrigation of sewage effluent produce biological aerosols that can be carried by winds several kilometers downwind of such plants. Air samples collected over large cities have shown significant concentrations of bacteria (e.g., 100–1000 bacteria m^{-3} over New York City). Viruses have not been detected in significant concentrations in outdoor air. But direct dissemination of bacteria and viruses from people in indoor air is of particular concern. These are released from activities involving the respiratory tract, such as sneezing and coughing. Viruses can be aerosolized during sneezing and coughing and can be carried onto small droplets or dust particles and transmitted to other persons breathing the air carrying the virus-laden particles (Jacobson and Morris, 1976). There is now an increased recognition that indoor air is a major source of airborne allergies and diseases (Urone, 1986).

1.3.4 Air Toxics and Hazardous Air Pollutants

A wide variety of substances have been demonstrated as or have the potential of being toxic to humans. The original United States Clean Air Act of 1970 required that the U.S. Environmental Protection Agency develop standards to control the emissions of hazardous air pollutants. Initially only a few substances like arsenic, asbestos, benzene, beryllium, mercury, and vinyl chloride were promulgated as HAPs. That list was later expanded to many more potentially toxic air pollutants (Urone, 1986). The most recent comprehensive amendments to the Clean Air Act in 1990 list 189 substances as HAPs. These are subject to national ambient standards for hazardous air pollutants based on maximum achievable control technology. Many toxic air pollutants are considered to be carcinogenic, mutagenic, or teratogenic. In addition to the long recognized hazardous chemicals mentioned above, HAPs include all kinds of fungicides, herbicides, and insecticides, dioxin, toxic hydrocarbons, polynuclear aromatics, and other chemicals. A majority of cancer cases have been attributed to environmental pollutants including many HAPs.

1.4 EFFECTS OF AIR POLLUTION

Our concern about air pollution is essentially a reflection of the accumulating evidence that air pollutants adversely affect the health and the welfare of human beings. Extensive effects research has established that air pollutants affect the health of humans and animals, damage vegetation and materials, reduce visibility and solar radiation, and affect weather and climate. Although some of the effects are direct, specific, and measurable, such as damages to vegetation and materials and visibility reduction, many other effects are indirect and more difficult to measure, such as health effects on human beings and animals. Comprehensive reviews of the effects of air pollution have been given elsewhere (see e.g., Stern, 1976, 1986; Stern et al., 1984; Godish, 1991). Here we give only a brief summary of these effects.

1.4.1 Effects on Human Health

Accidental releases of poisonous and toxic gases from rupture of storage tanks and other industrial accidents are the incidences of extreme air pollution that can kill many people and animals in the vicinity of the accident. The worst industrial air pollution disaster of this kind occurred in 1986 in Bhopal, India, where heavy poisonous gas released from a fertilizer plant during nighttime strong inversion conditions drifted in the surrounding residential area, killing more than 3000 people and seriously injuring many more.

Aside from accidental releases, extreme air pollution episodes, during which concentrations of air pollutants in particular urban-industrial areas can reach excessively high levels for periods of several hours to several days, can cause extreme discomfort, diseases, and even deaths among the most vulnerable part of the population (small children, elderly, and sick). Some of the well-known and well-documented examples of extreme air pollution episodes in the past occurred in Meuse Valley, Belgium, in 1930; Donaro, Pennsylvania, in 1948; Poza Rica, Mexico, in 1950; London, England, in 1952; and New York in 1966. In each of these episodes, persistent thermal inversions and weak and stagnant winds were responsible for poor atmospheric dispersion of then largely unregulated industrial and urban air pollution. Confinement of air pollution to river valleys made it worse. Nearly 4000 excess deaths were attributed to the worst air pollution disaster in London, while 168 excess deaths resulted from the 1966 New York episode. Thousands of people in the affected areas became ill. From these recorded air pollution episodes, it is clear that acute exposure to relatively high air pollution levels over even a short period of time can result in death and sickness. Such acute exposure conditions are now rare due to air pollution control programs implemented in most of the developed countries, but inhabitants of megasize industrial cities in developing countries, such as Mexico, Greece, India, and China, still bear the brunt of occasionally acute and often chronic exposure conditions.

Chronic exposure is defined as the long-term (e.g., lifetime) exposure to commonly encountered air pollution levels in an urban-industrial complex. With the general trend of increased urbanization and industrialization, the world's urban population is being subjected to chronic air pollution exposure and is growing much faster than the general population. Therefore, it is of great interest to understand the possible adverse health effects of both the acute and chronic exposures to various air pollutants individually, as well as in various combinations. Long-term exposures to low ambient concentrations may not cause acute illness or deaths, but these exposures have been implicated in many other adverse health effects, such as respiratory and cardiovascular diseases; irritation of the eyes, nose, and throat; aggravation of an existing disease such as asthma; and impairment of performance of work and athletic activities.

Health effects of the various air pollutants have been studied following three different approaches: epidemiological, clinical, and toxicological. Epidemiological studies are systematic statistical analyses of data from the community at large or a diseased group, which attempt to identify associations between health effects and air pollution. They are useful in identifying both the acute and chronic effects of air pollution on human health. The association between daily mortality and acute exposure to high levels of air pollutants has been well documented for the air pollution episodes mentioned earlier. The association between adverse health effects and long-term exposure to low levels of pollutants is less certain and is confounded by many coexisting factors (Godish, 1991, Ch. 5). Clinical studies also focus on human subjects and provide data on the effects of air pollutants under controlled laboratory conditions. The exposure levels as well as their physiological effects may be quantified reasonably well, but such studies are necessarily limited to only short-term exposures to low-level pollutant concentrations, so that the effects are small and reversible.

Toxicological studies are more often conducted on animals or simpler cellular systems, but sometimes also on humans. Such studies are conducted to determine the toxicity of a substance when administered in varying doses (the dose is a function of the concentration and the duration of exposure). Different responses may occur when the same dose of a toxic substance is administered in different concentrations and different exposure durations. The response to increasing concentrations may not be linear, so that extrapolation to long-term exposure to low-level concentrations may not be easy to make. Even more uncertain is the extrapolation from animal response to human health effects. Ethical and legal constraints limit the use of human volunteers in toxicological studies. But the health effects of long-term exposure of industrial workers to higher levels of air pollution in the work place may be determined from clinical studies of diseased workers. Such occupational exposure studies have their own limitations in that the overall level and mixture of pollutants may be quite different from those of the ambient urban environment and the population of industrial workers may not be representative of the general population.

Health-effect studies have identified a number of adverse effects associated with commonly encountered air pollutants. For example, eye irritation is associated with high concentrations of photochemical oxidants, such as peroxyacetyl nitrates (PANs), acrolein, formaldehyde and other oxygenated hydrocarbons, aldehydes, and particulates. Respiratory diseases, such as chronic bronchitis, pulmonary emphysema, lung cancer, bronchial asthma, and respiratory infections, are other common health hazards of community air pollution. Epidemiological studies have demonstrated that exposure to fine particulates is associated with increased incidence of respiratory illness and chronic bronchitis, decreased pulmonary function, and increased mortality rates (U.S. Environmental Protection Agency, 1982). Even short-term exposures to particulate matter and a variety of gaseous air pollutants have been associated with an increased rate of asthma attacks. Other ambient air pollutants, such as sulfur oxides (SO_x), are also identified to play a role in chronic bronchitis. There is little epidemiological evidence linking ambient air pollutants to pulmonary emphysema. However, some animal studies suggest that chronic exposures to NO_2 may initiate this lung disease. Lung cancer is a more serious affliction that has been linked to tobacco smoke and occupational exposures to asbestos, arsenic, radioactive gases (e.g., radon), and dusts. A causal role of ambient air pollution in the development of some lung cancers is also suggested by several epidemiological studies. How-ever, the evidence is mixed and there are many confounding factors that make it difficult to establish a direct association between ambient air pollutants and lung cancer (Godish, 1991, Ch. 5). Benzo(a)pyrene and other polycyclic aromatic hydrocarbons in urban air are identified as potent carcinogens.

For more detailed descriptions of the health effects of criteria pollutants (CO, SO_2, ozone, hydrocarbons, lead, and particulate matter) and hazardous air pollutants, including air toxics, the reader should refer to Goldsmith and Friberg (1977), Seinfeld (1986), and Godish (1991).

1.4.2 Effects on Vegetation and Animals

A large number of food, forage, and ornamental crops as well as trees have been identified to be damaged by air pollutants. The effects are in the form of leaf damage, stunting of growth, decreased size and yield of fruits, and wilting and destruction of flowers. Some plant species, such as blue grass, pinto bean, and spinach, are so sensitive to specific air pollutants that they can be useful in monitoring air quality. Air pollutants identified as responsible for major damage are sulfur dioxide, ethylene, acid mists, fluorides, ozone, particulate matter, and a number of organic oxidants (e.g., PAN). Of somewhat lesser concern are nitrogen dioxide, chlorine, hydrogen chloride, ammonia, and mercury.

Crops and Ornamental Plants Injury or damage to a plant species may depend not only on the pollutant concentration and the exposure duration, but also on other factors, such as age, nutrient balance, soil conditions, temperature, humidity, and sunlight. For a given plant species and air pollutant, no significant effects may be observed at low levels of exposure. With increasing levels of exposure, however, a series of potential injuries may occur, including biochemical alterations, physiological response, wilting, and eventual death. Systematic studies of exposure and effects of selected air pollutants under controlled conditions have been conducted in phytotrons. These have contributed much to our present understanding of the biochemical, physiological, and visible effects of the so-called phytotoxic pollutants. These are summarized in Table 1.5. The effects range from slight reduction in yield to extensive visible injury, depending on the level and duration of exposure. The pollutants of major concern are SO_2, O_3, PAN, NO_2, HF, ethylene, chlorine, ammonia, hydrogen chloride, hydrogen sulfide, and sulfuric acid.

The economic costs of air pollution damage to crops and ornamental vegetation are difficult to estimate. Recent estimates indicate that crop losses due to air pollution in the United States run into several billions of dollars. On regional and larger scales, ozone and acid deposition are considered to account for more than 90 percent of these losses. On the local scale, however, individual point sources, such as power plants and metal smelters, can be associated with specific crop losses in their vicinity.

Forests On regional and continental scales, extensive damages to certain forests in Europe and the United States have been attributed to air pollution, es-

Table 1.5 Effects of Various Air Pollutants on Plants

Pollutant	Symptoms and Types of Injury	Maturity and Parts of Leaves Affected
Sulfur dioxide	Bleached spots, bleached areas between veins, chlorosis	Middle-aged leaves most sensitive, mesophyll cells
Ozone	Flecking, stippling, bleached spotting, pigmentation; conifer needle tips become brown and necrotic	Oldest leaves most sensitive, palisade or spongy parenchyma
Peroxyacetyl nitrate (PAN)	Glazing, silvering, bronzing on lower surface of leaves	Youngest leaves most sensitive, spongy cells
Nitrogen dioxide	Irregular, white or brown collapsed lesions on intercostal tissue and near leaf margin	Middle-aged leaves most sensitive, mesophyll cells
Hydrogen fluoride	Tips and margin burns, leaf abscission; narrow brown-red band separates necrotic from green tissue	Youngest leaves most sensitive, epidermis and mesophyll cells
Ethylene	Sepal withering, leaf abnormalities, flower dropping, abscission	Older leaves most affected
Ammonia	"Cooked" green appearance becoming brown or green or drying	Mature leaves most sensitive
Hydrogen sulfide	Basal and marginal scorching	Youngest leaves most affected
Sulfuric acid	Necrotic spots on upper surface	All leaves affected
Chlorine	Bleaching between veins, tip and margin burn, leaf abscission	Mature leaves most sensitive, epidermis and mesophyll cells
Hydrogen chloride	Acid-type necrotic lesion, tip burn on fir needles, leaf margin necrosis on broad leaves	Oldest leaves most sensitive, epidermis and mesophyll cells

Source: Modified after Stern et al. (1984).

pecially acid deposition. However, firm evidence for the cause and effect relationship is lacking, because there are many confounding factors, besides air pollution, that can explain much of the observed damage. In some cases, even beneficial effects of certain air pollutants acting as nutrients to forest ecosystems have been pointed out. W. H. Smith (1981) has classified the relationships of air pollutants with forests into three categories, depending on the pollutant dose levels. Under low-dose conditions, forest ecosystems act as sinks for atmospheric pollutants and in some instances as sources. The interactions of air pollutants with forests result in little or no effects on the nitrogen and carbon cycles. In some cases, these interactions may be beneficial to the forest ecosystem.

At the intermediate-dose level, the forest–pollutants interactions can result in measurable effects on forest ecosystems. These effects consist of a reduction in forest growth, change in tree species, and change in their susceptibility to pests. Field studies as well as controlled laboratory experiments show sulfur dioxide to be an inhibitor of forest growth. Photochemical oxidants, primarily ozone and PAN, have been found to alter the forest species composition and their susceptibility to pests (Stern et al., 1984).

In the third category of interactions at high-dose levels, the effects are large enough to be observed by the casual observer. High-dose conditions are almost always associated with large point-source emissions, such as from fossil-fuel power plants and smelters. The pollutants most often involved are sulfur dioxide and hydrogen fluoride, and high-dose exposures to these can cause severe injury or destruction of exposed forest.

Domestic Animals Earlier studies of the effects of severe air pollution episodes indicate that in addition to the deaths and illnesses suffered by human population, domesticated animals such as cattle, dogs, and cats were also affected. Autopsies of animals revealed evidence of pulmonary edema. Breathing toxic air pollutants is not, however, the major mode of pollutant intake by animals; ingestion of pollution-contaminated feeds is the primary mode. In the case of cattle, we should be primarily concerned about the deposition of air pollutants on vegetation or forage that serves as their feed and subsequent effects of ingested pollutants on the animals and the food products obtained from them (e.g., milk and meat) for human consumption. There is increasing concern about the possible effects of exotic chemicals, including pesticides, herbicides, and fungicides, that are ingested by meat and dairy cattle with their feed and forage.

The air pollutants of greatest concern to the food chain involving domesticated animals are fluoride, lead, heavy metals, and other particulate matter. Fluoride causes acute or chronic fluorosis in forage-consuming livestock animals such as cattle, sheep, horses, and pigs. Fluorides may be emitted either in gaseous or particulate form by phosphate fertilizer plants, aluminum smelters, and other manufacturing plants. Lead poisoning of animals resulting from the ingestion of lead dust-contaminated forage may also be acute or chronic, depending on the exposure level. It is usually associated with point sources that process minerals and emit mineral dusts. Other heavy metal elements in mineral dusts, such as arsenic, selenium, and molybdenum, may also poison livestock animals.

1.4.3 Effects on Materials and Structures

Air pollutants affect materials by soiling or through chemical reactions. Damage to structural metals, building stones, surface coatings, fabrics, rubber, leather, paper, and other materials occurs extensively. The total annual loss from these and the related cleaning and protective activities in the United States has been estimated at several billion dollars. The material damage is mainly attributed to acid mists, oxidants of various kinds, hydrogen sulfide, and particulate matter. Materials can be affected by both physical and chemical processes. Physical damage may result from soiling due to dust deposition and also from the abrasive effect of wind-blown particulate matter. Chemical damage to materials is a more serious and pervasive problem, because certain pollutants readily react with materials after coming in direct contact with them. Effects of air pollutants on different types of materials have been described in detail by Yocom and Upham (1977), Stern et al. (1984), and Godish (1991); here we give only a brief summary of these effects.

Metals The principal effect of air pollutants on metals is corrosion of their surfaces with some alteration to their physical, chemical, and electrical properties. The rate of corrosion depends on moisture, temperature, and pollutant levels in air. Air pollution-induced corrosion of ferrous metals, which account for about 90 percent of all metal usage, is of particular concern. The various studies have shown that the rates of corrosion of metals are much greater in highly polluted industrial environments than in clean rural environs. The main agent of corrosion appears to be sulfuric acid, which is produced from the oxidation of SO_2 or SO_x in the presence of water vapor or liquid water. Nonferrous metals such as zinc, copper, nickel, and brass are also corroded by SO_x, H_2S, and other acidic pollutants. Pollution-induced corrosion of metals used in electric and electronic equipment can cause serious problems.

Building Materials The most pervasive and visible effect of air pollution on buildings is the soiling and staining of their surfaces by sticky particulate matter. The associated cleaning and repainting costs in the United States probably run into billions of dollars. In addition to being soiled, building materials such as marble, limestone, and dolomite are chemically eroded by various acidic air pollutants. Air pollution-related chemical erosion of priceless, irreplaceable historical monuments, sculptures, and other structures is a serious problem all over the world, particularly in highly polluted cities. The principal air pollutants responsible for chemical deterioration are SO_2 and CO_2, which in the presence of moisture form sulfuric and carbonic acids. These acids react with surface materials to form water-soluble substances that are eventually leached out by rain. Famous monuments such as Cleopatra's Needle in New York; Acropolis near Athens, Greece; Colosseum in Rome; Taj Mahal in Agra, India; and many other historic buildings and marble statues are in various stages of chemical deterioration due to urban and industrial air pollution.

Many building surfaces are protected by paints whose appearance and protection effectiveness depend on environmental factors such as moisture, temperature, sunlight, and air pollutants (e.g., H_2S, SO_x, NH_3, O_3, and particulates). Effects of air pollutants on paints include soiling, discoloration, loss of gloss, and reduced adhesion and strength. Once the paint coating is significantly damaged, the underlying surface may be subjected to attack by air pollutants and natural weathering processes.

Fabrics, Leather, and Paper The major effects of air pollution on fabrics are soiling and weakening of textile fibers. Atmospheric particulates are responsible for soiling fabrics, while acidic pollutants weaken them. Air pollutants may also react with fabric dyes and cause discoloration or fading. Fading has been associated with SO_2, NO_2, O_3, and particulate matter.

Sulfur dioxide also affects the appearance and composition of leather and paper. It is absorbed by leather and the cellulose fiber in paper and is converted to sulfuric acid in the presence of moisture. The presence of acid makes these materials somewhat brittle, causing significant deterioration. This is of great concern to libraries where leather-bound books must be kept for long-term archival storage and usage. Preventive measures now include storage in SO_2-free air.

Rubber It has been known for some time that atmospheric ozone cracks rubber products under tension. The extent of cracking depends on the ozone concentration, the type of rubber, and the amount of rubber tension. Rubber cracking, particular of automobile tires, was one of the first effects of photochemical smog observed in Los Angeles. The number of cracks as well as their depth can be related to ambient ozone concentrations. Unsaturated natural and synthetic rubbers are especially vulnerable to ozone, which attacks the carbon–carbon double bonds of these materials. Rubbers made of saturated compounds are more resistant to ozone cracking.

1.4.4 Atmospheric Effects

Air pollutants frequently cause widespread haze and fog, reducing visibility and solar radiation near the ground; alter the near-surface energy balance; and possibly change local weather and climate. In addition, the important problems of stratospheric ozone depletion, acid deposition, and climate change are attributed to air pollution. These are discussed only briefly here; for more detailed descriptions the reader should refer to, among others, Robinson (1977), Stern et al. (1984), and Godish (1991).

Visibility Reduction The most noticeable effect of air pollution on properties of the atmosphere is the reduction in visibility caused by the absorption and scattering of light by airborne particles. The most obvious to the casual observer is the visibility reduction associated with industrial plumes and urban smog. More widespread regional hazes are also often caused by air pollution dispersed over large regions under fair-weather, stagnant, high-pressure, and warm conditions. The atmospheric visibility is influenced by many factors, which can be broadly classified as the optical characteristics of the illumination source (e.g., solar angle, spectrum, and intensity of radiation as altered by cloud cover and atmosphere), the viewed objects, the intervening atmosphere, and the observer (U.S. Environmental Protection Agency, 1979). Quantitatively, visibility is expressed in terms of the maximum horizontal distance at which a standard object is visible to an observer.

Our ability to see objects clearly in a clean atmosphere is limited by Rayleigh scattering by air molecules, scattering by natural aerosols, and the curvature of the earth's surface. At sea level, the horizontal visual range through a particle-free atmosphere is about 330 km. Natural aerosols and the associated fogs, rain, snow, wind-blown dust, and natural hazes can reduce the visibility to a few tens of meters. However, our primary concern is the reduction in visibility due to anthropogenic particulate material. This is primarily due to scattering of light by particles and to a lesser extent to its absorption by particles and gaseous pollutants such as NO_2. Particulate absorption of light is significant only when large dust and dark soot particles are present. Nitrogen dioxide absorbs short-wavelength blue light, giving a yellow to reddish-brown appearance.

Aerosols composed of fine (0.1–1.0 μm) solid or liquid particles are responsible for most of the light scattering. Visibility is inversely proportional to the scattering or extinction coefficient. Scattering coefficient, and hence visibility, depends on the size, chemical composition, and hygroscopicity of particles. It also depends on a number of meteorological variables, which determine the number concentration and size distribution of aerosol particles. In particular, wind speed, temperature, and relative humidity strongly influence the visibility of a polluted atmosphere.

A number of local and regional visibility problems have been identified. The problem of photochemical smog can be attributed to local industrial and urban pollution in many cities during warm months. The problem of regional haze during the summer months in the eastern and midwestern United States and in other parts of the world appears to have grown progressively worse. This is caused by long-range transport of pollutants from a variety of large and small sources undergoing chemical and photochemical transformations in the presence of abundant sunlight, warm temperatures, high relative humidities, and stagnant high-pressure systems. Most optically active components of such hazes are sulfate aerosols. The visibility reduction in pristine mountain areas of scenic beauty has been of greatest concern to public as well as regulatory agencies. In particular, open vistas in national parks have to be protected against the long-range transports of pollution.

Radiative Effects The gaseous as well as particulate air pollutants in the atmosphere can significantly alter the radiation balance of the near surface layer in highly polluted urban atmospheres. At the ground a significant reduction in solar radiation may occur due to its absorption and scattering by particles. Particles are most effective in reducing solar radiation when solar altitude or angle is low. Decreases of 10 to 20 percent in solar radiation due to air pollution have been observed over large cities. At night, a highly polluted layer near the ground under inversion conditions will absorb as well as emit more longwave radiation. The net effect may be a marked cooling of the polluted layer.

Fog Formation and Precipitation Atmospheric particles can serve as nuclei for the condensation of water vapor, which is an important factor in the formation of fog and clouds, as well as in precipitation. However, only a small fraction of the total atmospheric aerosols serve as cloud condensation nuclei (CCN). The frequency of fog formation has been observed to be higher in cities than in the countryside, in spite of the fact that air temperatures tend to be higher and relative humidities tend to be lower in cities compared with the countryside. The explanation for this lies in the mechanism of fog formation and the effect of air pollutants on the same. With high concentrations of SO_2, the formation of sulfuric acid on the surface of particles in a humid environment leads to the formation of small droplets of fog that would not otherwise have formed.

Increased precipitation has also been observed downwind of some cities, which has been attributed to air pollution. The mechanism of precipitation and the role of increased number of CCN due to pollutant particulate matter are very complex and it is not clear that enhanced precipitation is the significant and unambiguous effect of urban air pollution.

Acidic Deposition The primary pollutants such as SO_2 and NO_2 undergo chemical transformations during their long-range transport and produce secondary acidic species in vapor or aerosol forms. These are eventually entrained in clouds, making the resulting rain or snow more acidic than that in cleaner environments. Acidic deposition includes both the wet (related to precipitation) and dry deposition of acidic air pollutants to soil and water surfaces. It is more general than the commonly used term *acid rain*. The acidic deposition problem was first recognized in Sweden in the mid-1960s and much later in North America. Studies conducted in Europe, North America, and some other locations indicate that acidic deposition is a widespread problem with significant ecological effects on lakes and certain forests.

The acidity of precipitation is commonly measured in terms of its pH, which refers to the negative log of the hydrogen ion [H^+] concentration. Due to the dissolution of CO_2 in rainwater, a pH of 5.7 is used as a reference for assessing the pollution-related acidity of precipitation. The northeastern region of the United States is characterized by the most acidic precipitation, with an average value of pH $\simeq$ 4.3. But much smaller values of pH <3 have been observed in localized episodes of acidic precipitation. Increased precipitation acidity is found to be due to sulfuric (65%) and nitric (30%) acids. The major sources are large power plants that burn high-sulfur coal and emit huge amounts of SO_x and NO_x, and ever-increasing mobile sources of transportation that emit NO_x.

The ecological effects of acidic deposition in the United States were assessed during the National Acid Precipitation and Assessment Program (NAPAP). Regional acidic deposition was not found to be significantly correlated with the acidity of surface water in lakes and rivers. Soil chemistry, geology, and land use were found to be more important factors. The best documented effects have been the depletion of fishery resources in freshwater oligotrophic lakes and streams in North America and Europe. Acidic deposition also has the potential to cause significant damage to terrestrial vegetation, particularly high-elevation forests. A causal role of acidic deposition in widespread damage to certain mountain forests has been suggested, but it is difficult to prove in the light of other confounding factors.

Stratospheric Ozone Depletion In the late 1960s and early 1970s, the spectre of stratospheric ozone depletion caused by anthropogenically produced air pollutants was raised by a number of scientists. The initial concern was expressed about the proposed building and extensive use of high-altitude supersonic transport (SST) planes for commercial aviation. In assessing the possible impact of the projected commercial fleet of SSTs, some scientists predicted that the reactive nitrogen compounds from SST exhaust might cause significant destruction of stratospheric ozone. It was largely because of this concern that the United States decided against allowing SSTs to be built and used by U.S. companies. The effect of a few transcontinental SST flights that are operating now is not considered to be significant.

The other, more serious, threat to the stratospheric ozone layer was implied by the extensive use of halogen compounds such as chlorofluorocarbons (CFCs), which were commonly used as propellants in aerosol spray cans, as refrigerants in refrigerators and air conditioners, as cleaning agents for printed circuits, and as blowing agents for making foam and styrofoam products. Because CFCs were extremely stable in the lower atmosphere, they could drift up into the stratosphere where they would break apart after being bombarded by the sun's high-energy radiation. Molina and Rowland (1974) hypothesized that the resulting chlorine and its compounds (e.g., ClO, HCl) from the breakup of CFCs would spell severe trouble for the ozone layer. According to their predictions, each chlorine atom could destroy 100,000 ozone molecules, implying that decades of CFC use could cause substantial decline in stratospheric ozone levels. These dire predictions of ozone destruction were verified first by observations of stratospheric ozone over Antarctica by a British team, which began noticing declining springtime ozone concentrations in the late 1970s and accelerating sharply thereafter. Subsequent springtime observational expeditions to Antarctica as well as satellite data confirmed substantial reductions in the total column ozone over Antarctica over increasingly large areas of the continent. The so-called ozone hole in which springtime ozone concentrations over Antarctica were observed to be less than 50 percent of the normal level has been shown to be widening since its inception or discovery in the early 1980s. The ozone hole forms only in Antarctica because of this region's unique weather conditions. However, observations over the Arctic and other high-latitude regions have revealed the presence of chlorine and bromine compounds and some evidence of ozone destruction also in the northern hemisphere, mainly in winter. The problem of stratospheric ozone depletion attracted even more attention when an international panel of scientists announced that ozone levels had dropped by measurable amounts not only in winter and spring but also in summer. In response to this threat, eighty-one countries signed a treaty in 1989 in Helsinki, Finland, agreeing to phase out production of CFCs by the end of the century.

The possible effects of stratospheric ozone depletion are decrease in the stratospheric temperatures and increase in the amount of ultraviolet radiation reaching the ground. Stratospheric temperatures have been observed to decrease globally by about 1.7°C between altitudes of 25 and 55 km since 1979. Their possible influence on stratospheric dynamics and global atmospheric circulations are not clear. More serious are the potential effects of increased ultraviolet radiation of wavelengths less than 320 nm, particularly more energetic UV-B (290–320 nm) radiation, on biological systems. The major effects of UV-B radiation on humans are sunburn and skin cancer. It has been estimated that a 1 percent decrease in the total column ozone would increase the downward flux of UV-B radiation by 2 percent. Thus, a depletion of the stratospheric ozone layer by a few percent over the heavily

Table 1.6 Major Radiatively Active Greenhouse Gases

	Carbon Dioxide CO_2	Methane CH_4	Nitrous Oxide N_2O	Chlorofluorocarbons CFCs	Tropospheric O_3
Principal anthropogenic sources	Fossil fuels; deforestation	Rice culture, cattle; fossil fuels; biomass burning	Fertilizer; land use conversion	Refrigerants; aerosols; industrial processes	Hydrocarbons (with NO_x); biomass burning
Principal natural sources	Balanced in nature	Wetlands	Soils; tropical forests	None	Biogenic hydrocarbons
Atmospheric lifetime	50–200 yr	10 yr	150 yr	60–100 yr	Weeks to months
Present atmospheric concentration (ppb) at surface	353,000	1720	310	CFC-11: 0.28 CFC-12: 0.48	20–40 ↑
Preindustrial concentration (ppb) at surface	280,000	790	288	0	10
Present annual rate of increase	0.5%	0.9%	0.3%	4%	0.5–2.0%
Relative contribution to the anthropogenic greenhouse effect	60%	15%	5%	12%	8%

Source: University Corporation for Atmospheric Research (1990).

populated mid-latitude regions of Europe, Asia, and North America would be expected to cause a significant increase in skin cancer. Such dire predictions have not been confirmed by observations so far. Of more immediate concern are the potential effects of larger ozone depletions in the high latitude regions on marine life, especially phytoplankton and animals feeding on the same. Increased UV-B radiation on a large scale has the potential to affect adversely an enormous number of animal and plant species.

Climate The possible effects of air pollutants on local climate, particularly of cities, have already been mentioned. Of greater concern are the potential effects on global climate due to the observed and projected increases in radiatively active gases such as carbon dioxide, methane, nitrous oxide, CFCs, and tropospheric ozone, as well as particulates in the atmosphere. Table 1.6 lists the major radiatively active gases, their principal sources, atmospheric lifetimes, present concentrations, and annual rates of increase, and their estimated relative contribution to the anthropogenic greenhouse effect. Although CO_2 contributes to nearly 60 percent of the greenhouse effect at present, the increasingly important role of the other greenhouse gases in future climate change scenarios must be considered, because they are accumulating at faster rates than CO_2.

The potential effects of global climate warming (tropospheric) in response to the rapidly increasing ambient concentrations of the so-called greenhouse gases has been a subject of intense scientific and public debate. General circulation models (GCMs) of the atmosphere have been used to predict the impacts of the projected doubling of CO_2 on climate by the middle of the next century. All such models predict that a doubling of CO_2 from 300 to 600 ppm will result in global tropospheric warming and stratospheric cooling. Different model formulations and parameterizations of physical processes, however, lead to somewhat different quantitative predictions. Global average surface temperature is predicted to increase by about 1.5°C in response to the more realistic gradual increase of CO_2 on the way to its doubling; the predicted response to an instantaneous doubling is much larger. Inclusion of other greenhouse gases would also increase the predicted global climate warming to 3°C. The observed global surface temperature in the past century increased by about 0.6°C, with the highest temperatures in recent years. But, the observed record does not show a steady increase and is marked by slight cooling from 1940 to 1965. It is not surprising, then, that the greater concern in the 1960s was about global cooling.

Some scientists have suggested that increase in atmospheric aerosols, particularly those of volcanic origin and injected in the stratosphere, can significantly cool the global climate. Aerosols can scatter incoming solar radiation and absorb and radiate longwave radiation back to the surface. Backscattering of incoming solar radiation can effectively increase the earth's albedo and result in global cooling. Absorption of longwave radiation and subsequent downward emission by aerosols would have the opposite effect of warming. The former mechanism appears to dominate, so that the net effect of increased aerosol loading in the upper atmosphere would be global cooling. This hypothesis has been supported by the observed cooling of the global atmosphere for several years immediately following major volcanic eruptions.

Although anthropogenic aerosols account for less than half of the atmospheric aerosols, the increasing trend in the former, particularly stratospheric sulfuric acid aerosols, can cause slight warming of the stratosphere and cooling of the troposphere. This effect is opposite that of the greenhouse gases. The net effect of the increasing burden of the anthropogenically produced air pollution on global climate cannot be predicted accurately with the current climate models. It is even more speculative to consider the possible consequences of such a climate change. But, all kinds of dire predictions about the melting of glaciers and Antarctica ice cap, rising sea levels, flooding of low-level coastal areas, destruction of estuaries, changes in regional temperature and rainfall patterns and consequential impacts on agriculture, forests, and biological systems have been made. Uncertainties in climate modeling may have been partly responsible for the generation of a wide range of scenarios of climate change effects.

1.5 REGULATORY CONTROL OF AIR POLLUTION

1.5.1 Common Law Approaches

Health and environment effects of air pollution have been used to justify its regulatory control through legislation. Air pollution control laws and regulations are at the heart of modern air quality management. Historically, a general or common law approach to air quality concerns has been taken. Outright prohibitions against certain activities generating air contaminants which harmed the health and safety of citizens have been adopted for centuries. A second approach has been to allow for civil suits against the offending source (e.g., a power plant or a factory) based on public nuisance created by certain pollutants emitted by the source. The nuisance may be in the form of foul smells, eye irritation, and/or other sensory effects. A nuisance is an unreasonable interference with the use and enjoyment of one's property, which may result from an intentional or negligent act. Particulate-laden smoke from an external source entering one's property could also be considered to violate trespass laws. However, these common law approaches have been found to be inadequate for controlling the air pollution problems of a developing society.

1.5.2 Air Pollution Control Laws

More effective approaches to abate air pollution rely on laws and regulations with the specific objective of reducing emissions and, hence, improving air quality. These may include taxation, land-use controls, source-specific emission standards, and air quality standards based on human health and welfare. Taxes on certain fuels and materials that produce harmful air pollutants may help to reduce their use, while tax credits may help the development and use of relatively cleaner alternative fuels and materials.

The land-use control or zoning regulations, administered at the municipal or county level, attempt to separate the industrial and commercial sources from the receptors (people) in residential areas. They also control urban and industrial development, and hence, emissions over the urban or county area.

Air pollution control or more general environmental protection laws at national and state levels are the backbone of modern air quality management in most industrial and some developing countries. These include emission standards for various types of sources; regulations for fuels, fuel additives, and certain hazardous materials; and ambient air quality standards. Each country has different sets of environmental laws and regulations. As an example, here we give a brief historical development and description of the air pollution control laws in the United States.

1.5.3 Historical Developments in the United States

Air pollution control in the United States began with the passage of smoke control ordinances by the cities of Chicago and Cincinnati in 1881. Similar ordinances were passed later in other major cities. The primary focus of these was the abatement of dense smoke from the combustion of coal. In the 1930s and 1940s, the cities of Pittsburgh and St. Louis enacted tougher smoke control ordinances when the earlier efforts failed to improve air quality. Until the 1950s, there was little or no involvement of federal and state governments in the legislative effort to control air pollution.

The advent of the nuclear energy industry and increasing public concern over the safety from accidental releases of radioactive material to the atmosphere provided the initial thrust toward air pollution control through federal legislation. The U.S. Atomic Energy Commission and its successor Nuclear Regulatory Commission enacted regulatory control measures for the design and operation of nuclear power plants. More general air pollution-related federal legislation was enacted in 1955 authorizing the Public Health Service to conduct air pollution research and training programs and to provide technical assistance to state and local governments. This also affirmed the responsibility of state and local governments in controlling air pollution. Amendments to this legislation in 1960 and 1962 authorized special studies on the health effects of motor vehicle emissions.

1963 Clean Air Act As air quality in the urban and industrial regions of the United States continued to deteriorate, the first comprehensive air pollution control legislation in the form of the Clean Air Act was passed in 1963. This act considerably broadened the role of the federal government in air pollution abatement by providing grants for air pollution control programs as well as for research and training, setting of federal enforcement authority, and development of air quality criteria for the protection of public health and welfare. Its amendments in 1965 led to the establishment of the emission standards for motor vehicles and also the creation of the erstwhile National Air Pollution Control Administration (NAPCA).

Air pollution control efforts were further strengthened by enactment of the Air Quality Act in 1967, which instituted a regional approach to air pollution control. The act required the federal government to designate air quality control regions and to issue air quality criteria and control strategies for specific (criteria) pollutants. But the progress in implementation of the act remained slow while air quality continued to worsen.

1970 and 1977 CAA Amendments Because of the heightened public concern about air quality, 1970 amendments to the Clean Air Act led to replacement of the NAPCA by the U.S. Environmental Protection Agency (EPA), which was designated as an independent federal agency to lead the nation's effort in protecting our environment, including the quality of air we breathe. The EPA was given significant federal enforcement authority and responsibility for setting uniform National Ambient Air Quality Standards (NAAQS), stringent new automobile emission standards for new and modified sources and for hazardous air pollutants. Immediate designation of air quality control regions and development of state implementation plans (SIPs) to achieve NAAQS were also mandated.

Since the major air quality goals set in 1970 amendments, including the meeting of NAAQS by 1975, could not be met without major economic disruptions, 1977 amendments to the Clean Air Act provided a mid-course correction to the original goals, by relaxing the timetable for their achievement and allowing some growth in "dirty air" regions. These also introduced the new concept of the prevention of significant deterioration (PSD) for the designated Class I and Class II areas. Only very small incremental increases of air contaminants are permitted in the Class I areas of originally very good air quality, scenic beauty, and wilderness, such as national parks.

One of the major goals of the PSD legislation is the visibility protection for Class I areas. The 1977 amendments also gave authority to EPA to regulate stratospheric ozone-destroying chemicals.

1990 CAA Amendments The current, most comprehensive air pollution control legislation in the United States was passed in 1990 as the Clean Air Act Amendments (CAAA). These introduced sweeping changes in air quality management and redirected the entire regulatory effort with respect to criteria pollutants, hazardous air pollutants, and regional and global air pollution problems. There are eleven major "titles" to the CAAA dealing with such issues as attainment of the NAAQS, mobile sources, hazardous air pollutants, acidic deposition, stratospheric ozone, permits, enforcement, and research. Only a brief review of some of these issues is given here; for more details the reader should refer to Air and Waste Management Association (1991), Godish (1991), and Griffin (1994).

1.5.4 Title I—Attainment of NAAQS

The National Ambient Air Quality Standards for the so-called criteria pollutants are given in Table 1.7. The primary standards are designed to protect public health; secondary standards protect public welfare. These are based on the air quality criteria, which have been developed from all the relevant scientific information on the health and welfare effects of individual pollutants. In response to new research findings, NAAQS have undergone some revisions since their original promulgation in 1970; these are included in Table 1.7. The latest revision to the ozone standard was made in 1997. The proposal to change the PM_{10} standard to the $PM_{2.5}$ standard is currently under intensive debate and study.

The major provisions of Title I are focused on the criteria pollutants, such as ozone, carbon monoxide, and fine particulate material (PM_{10}), for which NAAQS have not been attained in certain areas of the country. Nonattainment classification is based on the severity of the air pollution problem. Compliance deadlines, reduction of VOCs, and offsets depend on the nonattainment classification. Within two years of state's failure to develop an adequate SIP, U.S. EPA must impose a federal implementation plan (FIP) for mandatory attainment of NAAQS.

The role of NO_x and VOCs in ozone nonattainment areas is required to be studied. Major NO_x sources may have the same control requirements as major sources of VOCs unless significant benefits cannot be achieved from additional NO_x controls. The mandatory use of oxygenated fuels in wintertime and enhanced vehicle inspection programs are required in many CO-nonattainment areas. Control measures are also prescribed for the PM_{10} nonattainment areas.

1.5.5 Title II—Mobile Sources

Tighter tailpipe emission standards for hydrocarbons, CO, and NO_x are required for cars and trucks. Fuel quality is regulated so that gasoline volatility and diesel sulfur content are reduced; lead in motor vehicle fuel is banned. Oxygenated fuels are required in most CO-

23 REGULATORY CONTROL OF AIR POLLUTION

Table 1.7 National Ambient Air Quality Standards*

Pollutant	Averaging Time	Primary Standard	Secondary Standard	Measurement Method
Carbon monoxide	8 hr	10 mg/m^3 (9 ppm)	Same	Nondispersive infrared spectroscopy
	1 hr	40 mg/m^3 (35 ppm)	Same	
Nitrogen dioxide	1 yr	100 μg/m^3 (0.05 ppm)	Same	Colorimetry
Sulfur dioxide	1 yr	80 μg/m^3 (0.03 ppm)		Pararosaniline method or equivalent
	24 hr	365 μg/m^3 (0.14 ppm)		
	3 hr		1300 μg/m^3 (0.5 ppm)	
Ozone	8 hr	157 μg/m^3 (0.08 ppm)	Same	Chemiluminescent method or equivalent
Hydrocarbons (corrected for methane)	3 hr (6–9 AM)	160 μg/m^3 (0.24 ppm)	Same	Flame ionization detector using gas chromatography
Particulate matter (PM$_{10}$)†	1 yr	50 μg/m^3	Same	Size-selective samplers
	24 hr	150 μg/m^3	50 μg/m^3	
Lead	3 mo	1.5 μg/m^3	Same	Atomic absorption

*Standards, other than those based on the annual average, are not to be exceeded more than once a year.
†U.S. Environmental Protection Agency has proposed revisions of this standard in terms of PM$_{2.5}$ (particulate matter less than or equal to 2.5 μm in diameter) concentrations of 15 and 65 μg/m^3 for the 1 yr and 24 hr periods, respectively.

nonattainment areas, whereas reformulated gasoline will be required in severe ozone areas. A clean fuel vehicle pilot program has been established in California. Ultimately, clean fuel vehicles will be required for centrally fueled fleets in many areas. On the basis of a study of mobile source-related air toxics, emissions of benzene and formaldehyde are regulated.

1.5.6 Title III—Hazardous Air Pollutants

A total of 190 hazardous air pollutants (HAPs), including air toxics, are listed. These include many metal compounds (e.g., arsenic, mercury, lead, etc.), organics (e.g., benzene, dioxins, pesticides, PCBs, etc.), acids, radionuclides, and fine mineral fibers (e.g., asbestos). A list of HAP source categories and a schedule of their regulation must be published by EPA. Maximum achievable control technology (MACT) emission standards must be developed for existing and new listed sources. An incentive is given for voluntary emission reductions. Eight years after MACT standards are established, standards to protect against the residual health and environmental risks must be promulgated, if necessary.

Standards to prevent accidental releases of toxic chemicals are required. EPA must publish a list of 100 toxic chemicals and their threshold quantities for evaluating the impact of potential accidental releases.

A study of area source emissions and a national urban air toxics strategy to reduce cancer risks from these sources by 75 percent are required by EPA. Regulation of source categories accounting for 90 percent of the thirty most hazardous area source pollutants is mandated. Regulations are also required for all types of municipal waste combustors.

1.5.7 Title IV—Acid Rain

Focus is on reducing acidic deposition through reductions in annual SO$_2$ emissions of 10 million tons and NO$_x$ emissions of 2 million tons from their 1980 levels. Annual SO$_2$ emissions from utilities are to be capped at 8.9 million tons by January 1, 2000. Reductions will take place in two phases. Phase I includes large units of 100 MW or greater, while Phase II beginning in 2000 will cover thousands of smaller utilities that burn coal or oil and also some industrial facilities. Affected sources are allocated allowances based on required emission reductions and past energy use. An allowance is worth one ton of SO$_2$ emissions per year and is fully marketable (it may be banked or sold). Sources must hold allowances equal to their level of emissions or face a stiff penalty. Alaska and Hawaii are excluded from these provisions. Incentive allowances are allocated to energy conservation and renewable energy projects, while certain clean coal technology demonstration projects may be exempted from regulatory requirements.

1.5.8 Title V—Permits

Five-year renewable operating permits are required for all the major sources covered by the Clean Air Act. States must develop operating permit programs which must be approved by EPA. Permits must address all emission points at a source and list the pertinent CAA requirements, a compliance schedule, and any monitoring and reporting requirements. These are subject to reviews by the state, the EPA, and the public. Judicial review is the final step if objections are made by citizens during public hearings to issuing a permit to a new or existing source.

1.5.9 Title VI—Stratospheric Ozone

Focus is on stratospheric ozone-depleting chemicals, which are divided into two classes. Complete phase-out is required of the Class I substances (CFCs, halons, carbon tetrachloride, and methyl chloroform) by the year 2000 and of the Class II chemicals containing hydrochlorofluorocarbons (HCFCs) by 2030, in accordance with the Montreal Protocol. The actual phase-out of CFC manufacturing was moved up to 1995 based on federal government approval of revisions to that protocol.

Safe alternatives or substitutes to Class I and Class II chemicals must be published by EPA. Recycling and safe disposal of these chemicals is also required. Also, warning labels are required on all products containing the ozone-depleting chemicals. The ozone-depleting potential of methane and control measures to reduce methane emissions must be studied and reported.

1.5.10 Title VII—Enforcement

Focus is on the enhanced enforceability of air pollution control regulations under the Clean Air Act and its latest amendments. A broad array of new enforcement activities ranging from traffic tickets to criminal prosecution are provided to better match the penalty to the severity of the violation. Criminal violations include felonies, knowing endangerment, and negligent endangerment in connection with air toxics. Provisions of stiff penalties and punishments for violations are provided. The ability to prove and adequately punish violators is strengthened because the burden of proof is on the defendant once the government shows that a violation has occurred.

1.5.11 Title VIII—Miscellaneous Provisions

Under these miscellaneous provisions are included a program to control air pollution from sources on the outer continental shelf, establishment of a program to monitor and improve air quality in regions along the Mexican border, and monitoring and research on the impairment of visibility in Class I areas. Visibility-related research includes assessment of sources affecting visibility, adaptation of regional air quality models, and basic studies of atmospheric chemistry and physics pertaining to visibility.

1.5.12 Title IX—Clean Air Research

The focus of clean air research is on developing improved techniques for measurement of individual air pollutants and their complex mixtures; addressing urban and regional ozone pollution problems; studying short- and long-term health effects of air pollutants, with an emphasis on HAPs; improving our understanding of the effects of air pollutants on ecosystems; developing better predictive models for the dispersion of accidental releases of dense gases; developing control technologies and strategies for air pollution from stationary and area sources; studying potential environmental effects associated with alternative fuels; and continuing the acid precipitation research program.

PROBLEMS AND EXERCISES

1. a. In what respects is air pollution a weather hazard?
   b. Discuss the effects of weather on urban and regional ozone and photochemical smog problems.
2. Draw a schematic of the regional acid deposition problem showing major sources, acidic pollutants, and receptors.
3. Discuss the relative contributions (in a qualitative sense) of the following types of sources to the regional air pollution problems of tropospheric ozone and acid deposition: (a) fossil fuel power plants; (b) automobiles and other transport vehicles; (c) agricultural operations; (d) Biogenic emissions.
4. Differentiate between primary and secondary air pollutants and give two examples each of air pollutants that can be categorized as: (a) only primary; (b) only secondary; (c) both primary and secondary.
5. The following National Ambient Air Quality Standards were specified originally in mass concentration units:

Pollutant	Averaging Time	Primary Standard
CO	8 hr	10 mg/m^3
NO$_2$	Annual	100 μg/m^3
SO$_2$	Annual	80 μg/m^3
O$_3$	1 hr	235 μg/m^3

Express the same in volumetric units of parts per million, using a reference air temperature of 25°C and a reference air pressure of 760 mm of mercury.
6. Express the 1 ppm by volume concentration of the following gases in μg m^{-3} at the standard air temperature and pressure: (a) CO$_2$; (b) CH$_4$; (c) N$_2$O; (d) NH$_3$.
7. Concentrations of aerosol constituents are sometimes expressed as parts per billion (ppb) (mole fraction) for

comparison with gaseous pollutant concentrations in ppb. Assuming a reference temperature of 20°C and a reference pressure of 1000 mb, express the 1 ppb (mole fraction) of the following aerosols in $\mu g\ m^{-3}$: (a) nitrate ion NO_3^-; (b) ammonium ion NH_4^+.

8. A pollutant cloud containing SO_2 at a concentration of $C\ \mu g\ m^{-3}$ engulfs a building for t_1 minutes after which it lifts off and is replaced by clean air. If the air in the building is recycling every T minutes and any incoming pollutant is well mixed throughout the building of volume V_B, use the principle of mass conservation to derive an expression of SO_2 concentration in the building, C_b, and schematically plot C_b versus t during the following two periods: (a) SO_2 accumulation period $t \leq t_1$; (b) SO_2 decay period $t > t_1$. (Hint: Solve the resulting ordinary differential equations with appropriate initial conditions.)

9. a. Which sulfur and nitrogen compounds are most abundant in a polluted urban atmosphere?

b. Which of the inorganic carbon compounds is of greatest concern to the world community and why?

10. a. Discuss the importance of volatile organic compounds and NO_x in the formation of urban smog.

b. Under what situations may the hydrocarbon control strategy not be effective in reducing ozone and why?

11. a. How do CFCs released near the ground reach the stratosphere and accumulate there?

b. Discuss the likely role of CFCs in the stratospheric ozone depletion and the possible mechanisms involved in the large depletion in polar regions.

12. a. How would you distinguish between dust and smoke particles and how do they differ from the photochemical smog?

b. What are the principal mechanisms by which fine and coarse mode particles are produced? Point out the other differences (besides size) between the fine and coarse mode particles.

13. a. What are the potential acute or chronic health effects of the criteria air pollutants SO_2, CO, and O_3?

b. What is the significance and basis for classifying the hazardous or toxic air pollutants?

14. Indicate the principal air pollutants that might be responsible for the following effects on vegetation and materials: (a) flecking, stippling, and bleaching of leaves; (b) corrosion of metals; (c) cracking of rubber; (d) disintegration of paper and leather.

15. a. Discuss the significance of local and regional air pollution on visibility reduction.

b. What are the possible effects of air pollution on local weather and climate of cities?

16. a. What are the potential effects of increasing CO_2, CH_4, and CFCs on global climate?

b. What is the significance of increasing levels of stratospheric aerosols to possible climate change in the near future?

17. a. On what basis are the National Ambient Air Quality Standards specified? Why are these restricted to only seven criteria pollutants?

b. Which standard is most difficult to be attained in certain cities and urban complexes and why?

18. a. Which of the 1990 Clean Air Act Amendments is most sweeping in terms of bringing in more pollutants under control using the state of the art technology?

b. What will be the focus of air pollution research under the 1990 CAAA?

2

Atmospheric Structure and Dynamics

2.1 INTRODUCTION

An important component of the air pollution problem mentioned in Chapter 1 is the atmosphere in which transport and diffusion of pollutants, as well as their chemical transformations and removal processes, occur. Meteorology is the study of the structure, thermodynamics, and dynamics of the atmosphere, particularly the lower parts of the atmosphere in which most of the active weather phenomena and processes take place. Since transport and fate of pollutants in the atmosphere are intimately related to meteorology, some of the meteorological fundamentals will be briefly reviewed in this and the next chapter. The primary objective is to introduce these fundamentals to those readers who are not meteorologists. More traditional weather-oriented meteorologists may also find this introduction to air pollution meteorology useful in that the relevance of the various meteorological parameters to air pollution will be emphasized wherever it is considered appropriate. We will briefly describe the composition and thermal structure of the atmosphere, defining its characteristic strata or layers and reviewing the basic thermodynamics and state variables; introduce the concepts of local and nonlocal static stability; and discuss the fundamental equations of the conservation of energy, mass, and momentum. Atmospheric motions and dynamics are governed by the same conservation equations, as are the pollutant transport and diffusion processes.

2.2 COMPOSITION AND THERMAL STRUCTURE OF THE ATMOSPHERE

As shown in Table 1.2, dry air is composed of about 78.1 percent by volume nitrogen, 20.9 percent oxygen, 0.93 percent argon, and trace (<0.1% total) amounts of other noble gases, carbon dioxide and air pollutants. In addition, the atmosphere contains water in gaseous, liquid, and solid forms.

Water vapor is the most variable but thermodynamically an important constituent, whose mixing ratio varies from near zero in cold dry upper atmospheric air to about 4 percent in hot and humid marine tropical air. Due to the presence of water vapor, moist air is lighter than dry air. Small water particles (less than 10^{-3} m in diameter) exist as fog and clouds, while larger solid or liquid particles fall out as various forms of precipitation. Water vapor as well as fog and clouds all play significant roles in radiation budget, energy transfer, and weather-related processes in the atmosphere.

2.2.1 Radiative Effects of Atmospheric Constituents

The major constituents of dry air are radiatively quite inactive. However, oxygen does have some absorption in red and orange parts of the solar spectrum. Especially at high altitudes, oxygen absorbs short ultraviolet radiation strongly, thus preventing it from reaching the earth's surface. Of the other gases, the most important for radiative heating and cooling of the atmosphere are water vapor, carbon dioxide, ozone, nitrous oxide, and methane. Their individual absorption spectra as well as the combined absorption spectrum are shown in Figure 2.1.

Water vapor is mostly concentrated at low levels (<10 km) and has several strong absorption bands in the infrared. The bands at wavelengths below 4 μm are important because they prevent more than 10 percent of the solar radiation from reaching the ground. The larger absorption bands above 4 μm prevent a very large fraction of the terrestrial infrared radiation from going out into space. Together with CO_2 and

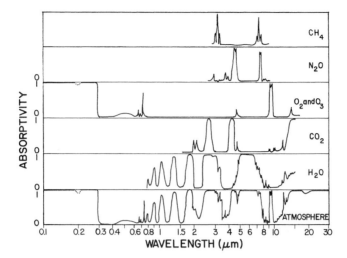

Figure 2.1 Absorption spectra for important atmospheric gases with the total absorption spectrum represented by the lower curve. From Fleagle and Businger, 1980.

other "greenhouse" gases, water vapor keeps the earth and the lower part of the atmosphere considerably warmer than they would be without this vapor (primarily H_2O and CO_2) blanket.

Radiative effects of clouds and fog are also very important, because they absorb, scatter, and reflect solar radiation and reradiate the energy as infrared radiation. Clouds also absorb a large portion of the infrared radiation emitted by the earth's surface, as well as that coming from the atmosphere above. The net effect of clouds on the global energy budget and their possible role in climate change are highly uncertain.

Carbon dioxide has several narrow absorption bands in the near infrared and a wide one at higher wavelengths between 13 and 17 μm. It is considered to be the most important greenhouse gas, whose global concentration variations in the past have been linked to climate warming and cooling episodes. The most recent observed increase of about 0.5°C in the average surface temperature during the past century has also been attributed mostly to the increasing CO_2 levels in the atmosphere (Intergovernmental Panel on Climate Change, 1995).

With more rapidly increasing levels of methane, tropospheric ozone, and nitrous oxide in the last few decades, it has been estimated that these other radiatively active greenhouse gases, combined together, may account for as much climate warming in the near future as CO_2 (University Corporation for Atmospheric Research, 1990). Somewhat more uncertain is the role of chlorofluorocarbons (CFCs) which are also radiatively active greenhouse gases and, at the same time, responsible for some depletion of stratospheric ozone. Ozone has an absorption band in the infrared, but it absorbs more strongly the ultraviolet radiation coming from the sun, thus protecting life on earth from the dangerous effects of such radiation. Ozone absorbs radiation strongly between 0.20 and 0.32 μm wavelengths. Below 0.29 μm, the absorption is almost total, so that this part of the shortwave radiation does not reach the ground but warms up the atmosphere above 10 km, with maximum effect near 50 km.

There is very little ozone near the ground, with concentration reaching up to 0.5 ppm only over highly polluted urban/industrial areas in summer. But in the upper atmosphere, higher ozone concentrations, on the order of 1 ppm, occur all over the globe.

Ozone is formed in the upper atmosphere principally by the photodissociation of molecular oxygen. This occurs in the presence of ultraviolet radiation of wavelengths less than 0.24 μm. Absorption of this radiation excites the electrons of the oxygen molecule into orbits of high energy in which two atoms cannot stay bound together and the molecule dissociates. Each of the resulting atomic oxygen quickly combines with molecular oxygen to form ozone. The reaction needs the help of other molecules in air. Ozone so formed is also destroyed both by radiation and by collision with free atomic oxygen and is converted back to oxygen. In this photostationary state of equilibrium, a unique ozone concentration profile with a maximum concentration between 20 and 30 km is established. The vertical distribution of ozone in the upper atmosphere depends on latitude and season, both of which determine the intensity of incoming solar radiation. Overall, nearly 97 percent of ozone in the atmosphere resides above 10 km.

2.2.2 Atmospheric Strata

The layered structure of the atmosphere can be classified in several different ways. But the vertical distribution of temperature, schematically shown in Figure 2.2, provides the most distinct and commonly

28 ATMOSPHERIC STRUCTURE AND DYNAMICS

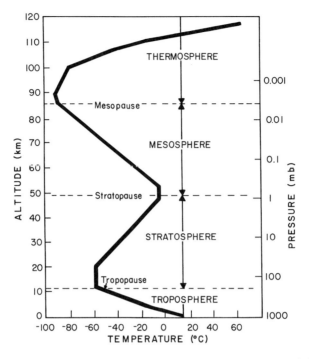

Figure 2.2 The average temperature profile through the atmosphere and the characteristic atmospheric layers.

used division of the atmosphere in terms of the following strata or layers:

1. *Troposphere*: The layer closest to the ground extending to an altitude of 9 to 16 km, depending on the latitude, in which the temperature decreases with height. The average lapse rate (the rate of decrease of temperature with height) in the troposphere is about 6.5°C km^{-1}. This is due to the radiative heating of the earth's surface (both land and oceans) by the sun and subsequent transfer of this energy to the troposphere by convection and longwave radiation. The vertical convection as well as horizontal and vertical advections of momentum, heat energy, and moisture keep the troposphere relatively well mixed. Most of the weather is formed in the troposphere. Also, almost all the air pollutants emitted from near-surface sources are transported, dispersed, transformed, and removed in this part of the lower atmosphere. The only exceptions are the relatively inert CFCs and other halogen compounds that find their way to the stratosphere and those pollutants directly injected into the stratosphere (e.g., by volcanoes, high-altitude aircraft, and rockets). The upper boundary of the troposphere is called the tropopause. Its altitude varies with latitude and longitude, on the average, sloping downward from the equator toward the poles. Tropopause height also shows significant seasonal variations. The temperature of the tropopause is typically 220 K, but varies considerably with latitude, longitude, and the change of seasons. The relatively low temperatures in the upper troposphere and at the tropopause preserve water on earth.

2. *Stratosphere*: The layer above the tropopause extending up to 50 km in altitude in which temperature increases with height, in the average. The lower part of the stratosphere in middle and high latitudes is often characterized by an isothermal (constant temperature) profile. The temperature in the upper region increases with height as a consequence of the ozone formation at higher altitudes and the absorption of shortwave radiation by this ozone. Temperature typically reaches its maximum value of about 270 K at an altitude of about 50 km. This marks the upper boundary of the stratosphere, which is called the stratopause. Because of the strong temperature inversion, there is very little mixing in the stratosphere. Any pollutants injected in this part of the upper atmosphere can be transported by winds around the globe without much mixing in the vertical. The lack of clouds and precipitation in this region also precludes any efficient removal of such pollutants. For this reason, stratospheric aerosols are considered to be an important ingredient that may significantly influence the radiation and energy budgets and, hence, climate.

3. *Mesosphere*: The region of the atmosphere extending from 50 km to 85 km over which temperature again decreases with increasing altitude. The height of the minimum temperature of about 170 K defines the upper boundary of the mesosphere, called the mesopause. The coldest temperature in the atmosphere occurs at the mesopause.

4. *Thermosphere*: The uppermost layer of the atmosphere extending above the mesopause in which temperature increases again rapidly with increasing height. This is due

to the conversion of solar energy into sensible heat as a result of photodissociation of O_2 and photoionization of N_2 and atomic oxygen. The temperature, as defined by the kinetic energy of molecules, may reach as high as 1500 K at altitudes on the order of 1000 km (Williamson, 1973). However, the molecular densities in the thermosphere are so low that an object placed there would not become hot. The lower part of the thermosphere extending up to 400 km is also called the ionosphere, because here atoms of oxygen and molecules of nitrogen become photoionized. The ionosphere is an important region for transmission and reflection of radio waves. Spectacular displays of aurora borealis (northern lights) and aurora australis (southern lights) also occur in this part of the atmosphere.

In this text on air pollution meteorology and dispersion, we are primarily concerned with the troposphere. Further stratification of the troposphere into the atmospheric surface layer, the planetary boundary layer, and the free atmosphere will be discussed in Chapter 4.

2.3 STATE VARIABLES AND THERMODYNAMICS

The basic state variables involved in atmospheric thermodynamics are pressure (p), temperature (T), density (ρ), and specific humidity (q) or water vapor mixing ratio. Their commonly used definitions and units in meteorology are briefly described below. We also give an overview of the governing equations of state and other thermodynamic relations.

2.3.1 Atmospheric Pressure

Pressure at a point in a fluid is commonly defined as the force per unit area that is uniformly directed in all directions. Atmospheric pressure can be regarded as the weight of the atmosphere per unit horizontal area. The conventional meteorological unit is millibar (mb), which is equivalent to 1000 dynes cm^{-2}, 100 pascals (Pa), or 100 newtons (N) m^{-2}. The average sea-level pressure of the standard atmosphere is 1013.25 mb. In public weather forecasts, pressure is also given in terms of the height (expressed in inches or millimeters) of the equivalent mercury column of the same weight per unit area as that of the atmosphere (the approximate conversions are: 1 in. Hg = 33.86 mb; 1 mm Hg = 1.333 mb). Pressure is usually measured with a mercury barometer or an aneroid barometer.

2.3.2 Air Temperature

Temperature is a thermodynamic property that is often confused with heat. One must distinguish between the internal energy or heat content of a body (e.g., air parcel) and the sensible heat that can be transferred. The former is related to the body's absolute temperature, whereas the latter depends on the temperature difference or gradient. For air temperature, one can use the more rigorous definition based on the kinetic theory of gases, according to which temperature is directly proportional to the mean kinetic energy of the molecules. Then, the absolute air temperature represents the degree of hotness of the air parcel relative to that at absolute zero at which molecular agitation will cease altogether. For this reason, the absolute temperature, expressed in Kelvin (K), is the most fundamental unit, which should be used in all thermodynamic relations involving temperature. More commonly used units of degree Celsius (°C) or degrees Fahrenheit (°F) in daily practice may easily be converted to absolute units in K (e.g., °C = 5/9(°F − 32), K = °C + 273.2). Air temperatures are measured using a variety of sensors such as mercury thermometer, thermistor, thermocouple, and resistance thermometer.

2.3.3 Air Density

Density is defined as the mass of a substance per unit volume. Like densities of other gases, the air density in the atmosphere varies with pressure and temperature. In addition, moist air is a mixture of dry air and water vapor and its density varies with water vapor mixing ratio or specific humidity. The recommended SI units of density are kilograms per meter, with a typical dry air density of 1.29 kg m^{-3} at the standard sea-level pressure (1013.25 mb) and temperature (273.2 K).

In meteorology, often the inverse of density, called the specific volume α ($\alpha = 1/\rho$), is also used. But we will generally avoid its use in this text.

2.3.4 Atmospheric Humidity

Many moisture variables are used meteorologically to express the water content of the air. To avoid largely superfluous and often confusing terminology, we will primarily use the specific humidity (q), which is defined as the ratio of the mass of water vapor to that of air. It is commonly expressed in parts per thousand, or grams per kilogram, although dimensionless fractions should be used in all thermodynamic relations involving q. When an air parcel is saturated with water vapor, its specific humidity at saturation will be denoted by q_s. The saturation specific humidity depends mainly on temperature and, to a much smaller extent, on atmospheric pressure; q_s increases with increasing temperature. The ratio q/q_s defines the relative humidity (usually expressed in percent). When $q \geq q_s$, water vapor is assumed to condense on cloud condensation nuclei in the form of liquid water or ice, depending on the temperature.

2.3.5 Thermodynamic Relations

All the state variables discussed so far are not entirely independent. They are related to each other through the equation of state, the hydrostatic equation, the first law of thermodynamics, and other thermodynamic relations. Here we give only a brief overview of these without going through their derivations.

The Equation of State   A fundamental form of the equation of state for an ideal gas has been given earlier as Eq. (1.3) in Chapter 1. In meteorological literature, the more common form using the air density ρ, in place of the mole density n/V, is frequently used, that is,

$$p = \frac{R^*}{m}\rho T = R\rho T \tag{2.1}$$

in which R^* is the absolute gas constant ($R^* = 8.314$ J K^{-1} mole^{-1}), R is the specific gas constant, and m is the mean molecular mass of air (for dry air, $m = 28.966$ g mol^{-1} and $R = 287.04$ J K^{-1} kg^{-1}).

Hydrostatic Equation   The pressure at any height z is due to the weight of air above. Consequently, the rate of change of pressure with height obeys the hydrostatic equation

$$\frac{\partial p}{\partial z} = -\rho g \tag{2.2}$$

Equation (2.2) is strictly valid only for the atmosphere at rest. From the equations of motion for the real atmosphere in motion, to be given later, we will see that the hydrostatic balance represented by Eq. (2.2) is only an approximation which turns out to be quite good for large-scale atmospheric flows. For some of the mesoscale and microscale flows, however, the effect of vertical accelerations that make the pressure nonhydrostatic has to be considered. In the thermodynamic relations developed in this section, however, we assume the atmosphere to be in hydrostatic equilibrium.

First Law of Thermodynamics   The first law of thermodynamics is essentially a statement of conservation of energy. When applied to a parcel of air, the law states that an increase in the internal energy, dU, of the parcel can occur only through the external addition of heat, dH, to the parcel and by performing work, dW, on the parcel. This can be expressed mathematically, for a unit mass parcel,

$$dU = dH + dW \tag{2.3}$$

Experimental evidence indicates that the internal energy per unit mass for an ideal gas is proportional to its absolute temperature, so that $dU = c_v dT$ for a constant volume process, and $dU = c_p dT$ for a constant pressure process. Here, c_v and c_p denote specific heat capacities at constant volume and pressure, respectively. Their difference is given by the specific gas constant, that is, $c_p - c_v = R$ (for dry air, $c_p \simeq 1005$ J K^{-1} kg^{-1}).

Expressing the work done on the air parcel in terms of pressure and volume, one can derive several different expressions of the first law of thermodynamics (see e.g., Hess, 1959; Fleagle and Businger, 1980). Here we give only the one involving changes in temperature and pressure:

$$dH = c_p dT - \frac{dp}{\rho} \tag{2.4}$$

which states that an addition of heat to the parcel would change its temperature and/or pressure.

Adiabatic Process   A process involving no heat exchange between the parcel and its environment is called an adiabatic process. Air parcels moving up and down in the atmosphere in response to vertical motions may not exchange any significant sensible heat and are often assumed to follow the adiabatic process. If an air parcel is lifted upward, its pressure will decrease in response to the decreasing pressure of its surroundings, so that its temperature will also decrease due to the resulting expansion of the parcel. Assuming that no heat is exchanged ($dH = 0$), that is, the parcel follows an adiabatic process, the rate of change of parcel temperature can be obtained from Eqs. (2.2) and (2.4) as

$$\left(\frac{dT}{dz}\right)_{ad} = -\frac{g}{c_p} \tag{2.5}$$

This rate of decrease of temperature with height is commonly referred to as the adiabatic lapse rate, Γ, which is approximately equal to 0.0098 K m^{-1} or 9.8 K km^{-1}. This will also be the temperature lapse rate in an adiabatic atmosphere.

The relationship between the relative changes in temperature and pressure in a parcel moving adiabatically or in an adiabatic environment are also given by Eqs. (2.2) and (2.4):

$$\frac{dT}{T} = \left(\frac{R}{c_p}\right)\frac{dp}{p} \tag{2.6}$$

Integration of Eq. (2.6) yields the Poisson equation

$$T = T_0\left(\frac{p}{p_o}\right)^k \tag{2.7}$$

in which T_0 is the temperature corresponding to the reference p_o and $k = R/c_p \simeq 0.286$.

Potential Temperature   Potential temperature is defined as the temperature an air parcel would have if it were brought down to a reference sea-level pressure of 1000 mb adiabatically from its initial position. Equation (2.7) may be used to relate the potential temperature θ to the actual temperature T as

$$\theta = T\left(\frac{1000}{p}\right)^k \tag{2.8}$$

where p is expressed in millibars. Potential temperature is a characteristic property of an air parcel that is invariant (conserved) during adiabatic movements of the parcel. Then, a parcel may be identified by its potential temperature. Also, in an adiabatic layer of the atmosphere, potential temperature must remain constant with height. Such a layer is commonly referred to as a mixed layer.

Equation (2.8) may also be used to relate the potential temperature gradient to the actual temperature gradient; to a good approximation,

$$\frac{\partial \theta}{\partial z} = \frac{\theta}{T}\left(\frac{\partial T}{\partial z} + \Gamma\right) \simeq \frac{\partial T}{\partial z} + \Gamma \quad (2.9)$$

This approximation is particularly useful in the lower troposphere or the atmospheric boundary layer where θ and T may not differ by more than 10 percent. An integral version of the above relationship,

$$\Delta \theta \simeq \Delta T + \Gamma \, \Delta z \quad (2.10)$$

in which Δs represent differences between two vertical levels, is frequently used. In an adiabatic atmosphere, $\partial \theta / \partial z = 0$, so that $\partial \theta / \partial z$ is a measure of the departure of the actual temperature profile or gradient from adiabatic conditions. In general, the atmosphere deviates from the simple adiabatic state, because many convective, radiative, advective, and chemical processes affect the thermal structure of the atmosphere.

On the basis of actual temperature gradient or lapse rate, relative to the adiabatic lapse rate Γ, an atmospheric layer can variously be characterized as follows:

1. Superadiabatic, when $-\partial T/\partial z > \Gamma$
2. Adiabatic, when $-\partial T/\partial z = \Gamma$
3. Subadiabatic, when $-\partial T/\partial z < \Gamma$
4. Isothermal, when $\partial T/\partial z = 0$
5. Inversion, when $\partial T/\partial z > 0$

Later, these characterizations will be related to atmospheric stability.

The above thermodynamic relations are applicable to both dry and moist atmospheres, so long as water vapor does not condense and one uses the appropriate values of thermodynamic properties for the moist air. In practice, it is found to be more convenient to use constant dry-air properties such as R and c_p, but to modify some of the relations or variables for the moist air.

2.3.6 Modifications for Moist Air

The moisture in the form of water vapor in air can be considered to exert a vapor pressure (e) that is a small fraction of the total pressure p. From the equations of state for the moist air and the water vapor alone, it can easily be shown that

$$q \simeq \frac{m_w}{m_d}\frac{e}{p} = 0.622\frac{e}{p} \quad (2.11)$$

where m_w and m_d are the mean molecular masses of water vapor and dry air, respectively. For a given temperature, the actual water vapor pressure is always less than the saturation vapor pressure; the latter is related to temperature through the Clasius–Clapeyron equation (Hess, 1959; Ch. 4)

$$\frac{de_s}{e_s} = \frac{m_w L_e}{R^*}\frac{dT}{T^2} \quad (2.12)$$

where L_e is the latent heat of vaporization or condensation. Integration of Eq. (2.12), with the empirical condition that at the freezing point ($T = 273.2$ K), $e_s = 6.11$ mb, yields

$$\ln\frac{e_s}{6.11} = \frac{m_w L_e}{R^*}\left(\frac{1}{273.2} - \frac{1}{T}\right) \quad (2.13)$$

in which e_s is in millibars. A plot of e_s versus T is called the saturation water vapor pressure curve, which is given in many meteorological textbooks and monographs (see e.g., Hess, 1959).

Using Dalton's law of partial pressures, the equation of state for moist air can be written in the form (Fleagle and Businger, 1980)

$$p = R\rho T_v \quad (2.14)$$

in which T_v is the virtual temperature defined in terms of the actual temperature and specific humidity as

$$T_v = T(1 + 0.61\, q) \quad (2.15)$$

This is the temperature that dry air would have if its pressure and density were equal to those of moist air. The virtual temperature is always more than the actual temperature and the difference between the two, $T_v - T \simeq 0.61qT$, may become as large as 7 K over warm tropical oceans, but typically is less than 2 K over mid-latitude land surfaces.

The specific humidity being always very small (<0.04), the specific heat capacity of water vapor is not significantly different from that of dry air. Therefore, changes in state variables of moving parcels in an unsaturated atmosphere are not appreciably different from those in a dry atmosphere, so long as the processes remain adiabatic. Consequently, the adiabatic lapse rate given by Eq. (2.5) is also applicable to moist unsaturated air parcels moving in a moist environment. However, the virtual potential temperature defined as

$$\theta_v = T_v\left(\frac{1000}{p}\right)^k \quad (2.16)$$

is more appropriate than the potential temperature, θ, as the property to be conserved during adiabatic movements of moist air parcels. An approximate relationship between the gradients of virtual potential temperature and virtual temperature is

$$\frac{\partial \theta_v}{\partial z} \simeq \frac{\partial T_v}{\partial z} + \Gamma \quad (2.17)$$

which is similar to Eq. (2.9).

Our earlier definition of the mixed layer should also be modified, so that $\partial \theta_v/\partial z = 0$ in the mixed layer. The potential temperature may also remain nearly constant in a mixed layer, while specific humidity often shows a larger variation and is not well mixed over the layer depth. Also, in the presence of moisture, the virtual temperature gradient or lapse rate should be compared with the adiabatic lapse rate for characterizing the various atmospheric layers as superadiabatic, adiabatic, or subadiabatic.

A moist air parcel rising in a moist atmosphere may eventually become saturated as it expands and cools. A part of the water vapor may condense and the latent heat of condensation is released. During the saturation condition, the parcel temperature changes according to the moist adiabatic or saturation lapse rate given by (Hess, 1959)

$$\Gamma_s = -\left(\frac{\partial T}{\partial z}\right)_{\text{sat.}} = \frac{g}{(c_p + L_e dq_s/dT)} \quad (2.18)$$

where dq_s/dT is the slope of the q_s vs. T curve. Note that the moist (saturation) adiabatic lapse rate is always less than the dry adiabatic lapse rate. Unlike Γ, Γ_s is quite sensitive to temperature, because dq_s/dT is an increasing function of temperature. The average lapse rate in the troposphere falls between Γ_s and Γ.

2.4 ATMOSPHERIC STABILITY

The variations of temperature and specific humidity with height lead to the variations of density in the vertical, that is, density stratification of the atmosphere. An upward or downward moving parcel in such a stratified environment would be acted upon by a buoyancy force or acceleration, depending on the density difference between the parcel and the local environment. The buoyant acceleration can be expressed, using the Archimedes principle and the equation state for the moist air, as

$$a_b = g\left(\frac{\rho - \rho_p}{\rho_p}\right) = g\left(\frac{T_{vp} - T_v}{T_v}\right) \quad (2.19)$$

where the subscript p refers to the parcel. Equation (2.19) is quite general as it is valid for any parcel irrespective of where it came from.

Considering a small displacement, Δz, of the parcel from its initial equilibrium position, where density of the parcel is equal to that of the environment, the buoyant acceleration on the parcel can also be expressed as

$$a_b \simeq -\frac{g}{T_v}\left(\frac{\partial T_v}{\partial z} + \Gamma\right)\Delta z = -\frac{g}{T_v}\frac{\partial \theta_v}{\partial z}\Delta z \quad (2.20)$$

Qualitatively, a stratified environment is called statically stable if the buoyancy force on any vertically displaced parcel acts to bring the parcel back to its equilibrium position, unstable if the buoyancy force acts to move the parcel farther away from its equilibrium position, and neutral if there is no buoyancy force at all acting on the parcel. In terms of the buoyant acceleration on an upward moving parcel, the criteria are $a_b < 0$ for a stable environment, $a_b > 0$ for an unstable environment, and $a_b = 0$ for a neutral environment. A quantitative measure of static stability or instability, often used in meteorology, is the static stability parameter

$$s = \frac{g}{T_v}\frac{\partial \theta_v}{\partial z} \quad (2.21)$$

which is based on the approximate relation (2.20). An advantage of using s or $\partial \theta_v/\partial z$ as a stability parameter is that it is independent of the origin and thermodynamic properties of different parcels and depends only on the vertical stratification of the environment. However, it is essentially a local parameter that does not account for the wide range of accelerations acting on different air parcels, originating from different height levels, before they arrive at the particular level where s is defined. Considering only the parcels originating in the immediate neighborhood (implied by the small displacement assumption) makes the definition of static stability somewhat narrow and restrictive. Yet, this has been the traditional view of static stability in meteorology, according to which a statically stable environment is characterized by $s > 0$ or $\partial \theta_v/\partial z > 0$, while an unstable atmosphere must have $s < 0$, or $\partial \theta_v/\partial z < 0$.

This traditional view of stability has several limitations and is not very satisfactory, especially when static stability is used as a measure of turbulent mixing and diffusion, for example, in air pollution and boundary layer meteorology. For example, it is well known that in the bulk of the convective mixed layer, $\partial \theta_v/\partial z \simeq 0$, or is slightly positive. Its characterization as a neutral or slightly stable layer, according to the traditional definition of stability, would be very misleading, because the convective mixed layer has all the attributes of an unstable layer, such as upward heat flux, strong mixing, and a large thickness. It would be correctly recognized as an unstable layer if one considers the positive buoyant accelerations on warm air parcels originating in the much warmer superadiabatic layer near the surface. Observations in the convective boundary layer confirm that there are large thermals or updrafts of warm air originating near the surface, and there are compensating downdrafts of relatively colder air sinking through the mixed layer. In both the updraft and downdraft, buoyancy acts on parcels arriving from different levels to move them farther away from their original positions, suggesting an overall unstable characterization for the whole boundary layer. Thus, the original criterion of static stability in terms of buoyant acceleration on parcels coming from all possible initial positions appears to be basically sound. It is only when one considers the alternative criterion, in terms of the local lapse rate, $\partial \theta_v/\partial z$, or s, that ambiguities arise in certain situations.

2.4.1 Nonlocal Stability

To remove any ambiguities in the characterization of static stability, Stull (1988, 1991) has proposed a nonlocal interpretation of static stability. For the nonlocal characterization of static stability, it is suggested that

33 CONSERVATION LAWS AND ATMOSPHERIC DYNAMICS

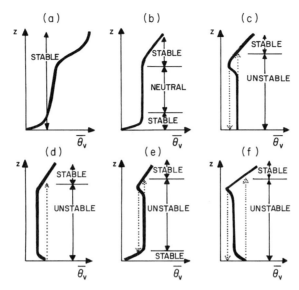

Figure 2.3 Nonlocal stability characterizations for the various hypothetical virtual potential temperature profiles. From Stull, 1988.

environmental sounding should be taken over a deep layer between the surface and the level where vertical parcel movements are likely to become insignificant (e.g., the tropopause). To determine the stability of the various layers, air parcels should be displaced up and down from all possible starting points within the whole domain. In practice, one needs to consider parcels starting from only the relative minima and maxima of virtual potential temperature in the sounding. Air parcel movement (up or down) from the initial position should be based on parcel buoyancy and not on the local lapse rate. The parcel buoyancy at any level is determined by the difference in virtual temperatures of the parcel and the environment at that level. Buoyant motion is associated with warm parcels that rise and cold parcels that sink. Displaced air parcels that would continue to move farther away from their starting level should be tracked all the way to the level where they would become neutrally buoyant ($T_{vp} = T_v$), neglecting possible overshoot due to parcel momentum. After parcel movements from all the salient levels (minima and maxima in the θ_v profile) have been tracked, static stability of different portions of the sounding domain should be determined in the following order (Stull, 1991):

1. *Unstable*: Those regions in which parcels can enter and transit under their own buoyancy. Note that individual parcels need not traverse the whole unstable region. For example, if a subregion is traversed by one air parcel, and a partially overlapping region is traversed by a different buoyant air parcel, then the whole region formed as the union of those two subregions is identified as "unstable."
2. *Stable*: Those regions of subadiabatic lapse rate that are not unstable.
3. *Neutral*: Those regions of adiabatic lapse rate that are not unstable.
4. *Unknown*: Those top or bottom portions of the sounding that are apparently stable or neutral, but that do not end at a material surface, such as the ground surface, or the tropopause. Above or below the known sounding region might be a cold or warm layer, respectively, that could provide a source for buoyant parcels.

The above method and procedure for determining the nonlocal stability is illustrated in Figure 2.3 for the various hypothetical soundings (θ_v profiles). Note that, in some of these cases, traditional characterizations of stability based on the local lapse rate or $\partial \theta_v / \partial z$ would be quite different from the broader, nonlocal characterizations used here. The latter are based on the buoyant movements (indicated by vertical dashed lines and arrows) of air parcels (indicated by dots at their starting positions).

The nonlocal characterization of static stability recommended here is consistent with the empirical evidence that unstable and convective atmospheric boundary layers have strong and efficient mixing, while the stable boundary layer and other inversion layers have relatively weak mixing associated with them.

2.5 CONSERVATION LAWS AND ATMOSPHERIC DYNAMICS

Dynamics of any fluid flow, such as that of air in the atmosphere, are governed by the fundamental laws of conservation of mass, energy, and momentum. Following a brief statement of these laws, we will give the governing differential equations based on the same, without going through their derivation. For detailed derivations of the conservation equations and descriptions of atmospheric dynamics based on the same, the reader should refer to other books on fluid mechanics, dynamical meteorology, or geophysical fluid dynamics (Haltiner and Martin, 1957; Hess, 1959; Pedlosky, 1979; Kundu, 1990; Holton, 1992; Dutton, 1995).

2.5.1 Conservation of Mass

The law of mass conservation states that the total mass in a system must be conserved. The law can also be applied to the mass of each component or species, so long as the chemical reactions between different species resulting in any gain or loss of the particular species under consideration are accounted for.

Continuity Equation When applied to the bulk air mass, in an elementary volume, at any instant of time, the law of mass conservation yields the continuity equation

$$\frac{\partial \rho}{\partial t} = -\left[\frac{\partial(\rho u)}{\partial x} + \frac{\partial(\rho v)}{\partial y} + \frac{\partial(\rho w)}{\partial z}\right] \quad (2.22)$$

where u, v, and w are the velocity components in x, y, and z directions, respectively, using a Cartesian frame of reference. The above equation is simply a mathematical statement of the conservation of mass of air, so that the local rate of change in density is equal to the divergence of mass in all the directions.

The simplifying assumption of incompressibility of flow leads to the simpler continuity equation

$$\frac{\partial u}{\partial x} + \frac{\partial v}{\partial y} + \frac{\partial w}{\partial z} = 0 \quad (2.23)$$

which requires the divergence of velocity to vanish in an incompressible flow. Although air is strictly not an incompressible fluid, Eq. (2.23) is found to be generally valid for the small-scale (microscale and mesoscale) atmospheric motions in the lower atmosphere that we are likely to deal with in air pollution meteorology and dispersion modeling.

Diffusion Equation From the application of the principle of conservation of mass to a particular species, one can derive (see Chapter 6, Section 6.2) an equation for the concentration of species i

$$\frac{Dc_i}{Dt} = D_i \nabla^2 c_i + R_i + S_i \quad (2.24)$$

where D_i is the molecular diffusivity of the species in air, R_i and S_i represent the rates of change of concentration with time due to chemical reactions and presence of sources and sinks, respectively.

The total derivative D/Dt and ∇^2 operators in Eq. (2.24) are defined as

$$\frac{Da}{Dt} = u\frac{\partial a}{\partial x} + v\frac{\partial a}{\partial y} + w\frac{\partial a}{\partial z} + \frac{\partial a}{\partial t} \quad (2.25)$$

$$\nabla^2 a = \frac{\partial^2 a}{\partial x^2} + \frac{\partial^2 a}{\partial y^2} + \frac{\partial^2 a}{\partial z^2} \quad (2.26)$$

where a can be any variable.

When Eq. (2.24) is applied to water vapor, with its mixing ratio replaced by specific humidity, we have

$$\frac{Dq}{Dt} = D_w \nabla^2 q + S_w \quad (2.27)$$

where S_w represents the rate of change of specific humidity in an elementary volume due to evaporation or condensation within the volume.

2.5.2 Conservation of Energy

The principle of conservation of energy states that the overall energy of a system is conserved, although it can be converted into different forms, such as heat or kinetic energy. Here we are primarily concerned with heat energy. In the atmosphere, as in any other transparent fluid medium, heat is transferred by three different mechanisms or modes—conduction, convection, and radiation.

The transfer of heat from molecule to molecule within a substance is called conduction. In a gas, such as air, conduction occurs through molecular agitation and collisions. Heat is transferred from the region of high temperature to that of low temperature. The rate of heat transfer or heat flux per unit area per unit time is proportional to the temperature gradient, the proportionality coefficient defining the thermal conductivity of the substance. In the atmosphere, conduction is not important except in the immediate vicinity (within a few millimeters) of the earth's surface including water, soil, vegetation, and so on.

The transfer of heat by movement of fluid (such as air) is called convection. In the atmosphere, heat is transferred by horizontal motions (this is often called the advection of heat), as well as by vertical motions. The vertical heat transfer by convection near the earth's surface is always accompanied by conduction in the immediate vicinity of the liquid or solid elements of the surface. But, for simplicity, we will refer to the entire process as convection.

The convective heat transfer between the earth and the atmosphere is further divided into two types, sensible heat and latent heat. Sensible heat is the heat energy we can sense and measure, while latent heat is the hidden warmth associated with a change of state of matter (e.g., liquid water into water vapor, or vice versa). The heat energy associated with evaporation, condensation, freezing, and melting processes is called latent heat. It is an important source of atmospheric energy and its horizontal and vertical distributions. For example, water vapor evaporated from warm tropical oceans can be carried into middle and higher latitude regions where it condenses and releases its latent heat energy.

The third mode of energy transfer is radiation, which does not require any substance (solid, liquid, or gas) as a carrier. Radiative energy travels in the form of waves similar to those involved in other forms of electromagnetic wave propagation, but is confined to a certain wavelength range. One can also think of radiation as streams of photons that transfer their energy when absorbed by a substance. All substances emit radiation whose wavelengths and intensity (energy) depend on their temperature and emissivity. The maximum radiation at any temperature is emitted by a blackbody with an emissivity of 1. The higher the temperature, the greater is the emitted radiant energy (radiation) and the shorter are the wavelengths of the emitted radiation. The reader can find the fundamentals of radiative transfer, including radiation laws, elsewhere (Liou, 1980; Goody and Yung, 1990).

In meteorology we often distinguish between shortwave radiation (0.15–4 μm) and longwave radiation (3–100 μm). The former term is applied to direct solar radiation, as well as to its diffused forms due to scattering from air molecules and atmospheric particulates and reflection from the earth's surface and clouds. The sun has an effective radiative surface temperature of about 6000 K and hence emits high intensity radiation at short wavelengths. In comparison, the earth and the atmosphere have much lower temperatures and emit longwave radiations of relatively low intensity. All these forms of radiation are found to be equally important, however, when one considers the energy budget of the earth-atmosphere system, either on a local scale or on the global scale. Some aspects of this energy budget will be discussed in later chapters.

The application of the law of energy conservation to the heat energy within an elementary volume yields the following differential equation for temperature:

$$\frac{DT}{Dt} = \kappa \nabla^2 T + S_H \quad (2.28)$$

where κ is the thermal diffusivity and S_H is the source term representing the rate of temperature change due to any condensation or evaporation within the elementary volume. Here, we have neglected the smaller contributions due to the dissipation of kinetic energy and due to compressibility of air, using the so-called Boussinesq's approximations. The term representing radiative warming or cooling is also not included. A similar equation can also be used for potential temperature, that is,

$$\frac{D\theta}{Dt} = \kappa \nabla^2 \theta + S_H \quad (2.29)$$

2.5.3 Conservation of Momentum

The law of the conservation of momentum states that the overall momentum must be conserved for any solid or fluid in motion. This is often identified as Newton's second law. Its application to the fluid flow in an elementary volume yields the well known Navier–Stokes equations of motion:

$$\frac{Du}{Dt} = -\frac{1}{\rho}\frac{\partial p}{\partial x} + fv + \nu \nabla^2 u \quad (2.30)$$

$$\frac{Dv}{Dt} = -\frac{1}{\rho}\frac{\partial p}{\partial y} - fu + \nu \nabla^2 v \quad (2.31)$$

$$\frac{Dw}{Dt} = -\frac{1}{\rho}\frac{\partial p}{\partial z} - g + \nu \nabla^2 w \quad (2.32)$$

where ν is the kinematic viscosity and f is the Coriolis parameter that depends on the latitude ϕ

$$f = 2 \Omega \sin \phi \quad (2.33)$$

The significance of the various terms in the above equations of motion can be explained. The left-hand side represents the inertia terms, which include local and advective accelerations. On the right-hand side, the first term in each equation is the pressure gradient term, the second term is the Coriolis term, which is neglected in the equation of vertical motion and is replaced by gravitational acceleration, and the last terms are viscous terms. The various terms can also be considered to represent inertia, pressure gradient, Coriolis, gravity, and viscous forces per unit mass of a fluid element. It is a rather complicated force balance involving all the relevant forces. The above set of equations of motion also implies the usual Boussinesq's assumptions. Note that, according to Eq. (2.32), the use of hydrostatic equation can be justified only under certain restricted conditions when vertical inertia forces or accelerations (Dw/Dt) are small enough to be neglected. This generally is the case for most large-scale flows in the atmosphere. But, in small-scale flows over topography including wave-like motions and those involving deep convection, Dw/Dt must be considered. For modeling atmospheric flows, the hydrostatic assumption is often made in order to simplify the set of equations and their numerical solution. Further simplifications can be made depending on the nature and type of the flow to be modeled.

2.6 LARGE-SCALE INVISCID FLOWS

For theoretical treatment, fluid flows are commonly divided into two broad categories, inviscid and viscous flows. In an inviscid flow, the effects of fluid viscosity and molecular diffusivities are completely ignored, that is, the fluid is assumed to have no viscosity. Inviscid flows are smooth, streamlined, and orderly, with no friction between adjacent fluid layers. Consequently, there is no mixing and no transfer of momentum, heat, or mass across these layers or streamlines. Such properties can only be transferred along streamlines through advection. The assumption of inviscid flow allows the viscous terms in the equations of motion and the molecular diffusion terms in other equations to be neglected, leaving a much simpler set of equations. For example, the resulting Euler's equations of motion for an inviscid flow are:

$$\frac{Du}{Dt} = -\frac{1}{\rho}\frac{\partial p}{\partial x} + fv \qquad (2.34)$$

$$\frac{Dv}{Dt} = -\frac{1}{\rho}\frac{\partial p}{\partial y} - fu \qquad (2.35)$$

$$\frac{Dw}{Dt} = -\frac{1}{\rho}\frac{\partial p}{\partial z} - g \qquad (2.36)$$

The inviscid flow theory obviously has some limitations. It is found to result in some serious dilemmas (e.g., no drag on a body in an inviscid flow) and inconsistencies when applied anywhere close to solid surfaces and density interfaces where effects of viscosity are too strong to be neglected. Outside the viscous boundary layers, interfacial layers, and mixing layers, however, the fluid viscosity can safely be ignored and the inviscid flow theory provides a good approximation to real atmospheric flows. Extensive applications of the theory are given in books on dynamic meteorology and geophysical fluid dynamics (see e.g., Pedlosky, 1979; Holton, 1992; Dutton, 1995). Here we introduce only some of the well-known wind systems based on Euler's equations for horizontal motion.

2.6.1 Geostrophic Winds

The simplest type of atmospheric flow is the unaccelerated, inviscid (frictionless) flow resulting from a simple balance between the pressure gradient and Coriolis forces. This is called the geostrophic flow. Geostrophic winds can be defined from Eqs. (2.34) and (2.35), neglecting the inertia terms, as

$$u_g = -\frac{1}{\rho f}\frac{\partial p}{\partial y}; \quad v_g = \frac{1}{\rho f}\frac{\partial p}{\partial x} \qquad (2.37)$$

These imply that the geostrophic wind vector (G) must be parallel to the isobars with low pressure to the left in the northern hemisphere ($f > 0$) and to the right in the southern hemisphere ($f < 0$). Thus, geostrophic winds blow clockwise about high-pressure centers (highs) and counterclockwise about lows, in the northern hemisphere. The reverse is true in the southern hemisphere. Also, the magnitude of the geostrophic wind is directly proportional to the horizontal pressure gradient.

In much of the atmosphere, geostrophic winds provide a good approximation (say, within ± 10%) to the actual winds. The major exceptions to this are: (a) the lowest layer of the troposphere, called the planetary boundary layer (PBL), in which viscous or friction effects become important; (b) near the equator where the Coriolis force becomes small ($|f| \to 0$); and (c) the regions of strong curvature where accelerations are large.

2.6.2 Thermal Winds

The horizontal pressure gradients or geostrophic winds may vary with height in response to horizontal temperature gradients. The vertical gradients of geostrophic winds, that is, geostrophic wind shears, are given by the thermal wind equations:

$$\frac{\partial u_g}{\partial z} \simeq -\frac{g}{fT}\frac{\partial T}{\partial y}; \quad \frac{\partial v_g}{\partial z} \simeq \frac{g}{fT}\frac{\partial T}{\partial x} \qquad (2.38)$$

which can be derived from Eq. (2.37) in conjunction with Eqs (2.1) and (2.2) (see e.g., Hess, 1959, Ch. 12). The above thermal wind equations are somewhat approximate in that the vertical temperature gradient terms have been ignored. They imply that the geostrophic shear vector $\partial G/\partial z$ must be parallel to the isotherms, with colder air to the left in the northern hemisphere and to the right in the southern hemisphere. Also, the magnitude of geostrophic wind shear is directly proportional to the horizontal temperature gradient. Note that a horizontal gradient of only 1°C per 100 km in middle latitudes will cause a geostrophic wind shear of about 0.0033 s^{-1}, which amounts to a change in geostrophic wind of 3.3 ms^{-1} per 1000 m change of height. Thus, the normal (climatological) decrease of temperature in going toward the poles causes an appreciable shear in the west winds. Consequently, there are very strong west winds at the top of the troposphere over most of the globe outside of the tropics.

The changes in geostrophic wind components over an increase in height of Δz can be calculated from the finite-difference form of thermal wind equations:

$$\Delta u_g = -\frac{g\Delta z}{fT}\frac{\partial T}{\partial y}; \quad \Delta v_g = \frac{g\Delta z}{fT}\frac{\partial T}{\partial x} \qquad (2.39)$$

which can be used to estimate large-scale geostrophic winds at the top of the PBL from the estimated horizontal pressure and temperature gradients from a surface chart depicting isobars and isotherms.

The geostrophic and thermal winds in meteorological literature are often expressed in terms of the height (h) of constant-pressure surface, because most weather maps represent constant-pressure (e.g., 850 mb) rather than constant-level surfaces. Along any constant-pressure surface, $dp = 0$, and we can express

$$u_g = -\frac{g}{f}\frac{\partial h}{\partial y}; \quad v_g = \frac{g}{f}\frac{\partial h}{\partial x} \qquad (2.40)$$

$$\Delta u_g = -\frac{g}{f}\frac{\partial(\Delta h)}{\partial y}; \quad \Delta v_g = \frac{g}{f}\frac{\partial(\Delta h)}{\partial x} \qquad (2.41)$$

where Δh is the thickness of the layer between two pressure surfaces p_1 and p_2, and Δu_g and Δv_g are the corresponding thermal wind components. The hypsometric equation can be used to express Δh in terms of p_1, p_2, and the average temperature of the layer. Equation (2.40) implies that the geostrophic wind is proportional to the slope of the constant-pressure surface.

In a barotropic atmosphere, geostrophic winds are constant with height, implying essentially parallel constant-pressure surfaces and no significant horizontal temperature gradients. More often, though, the atmosphere is baroclinic; thermal winds or geostrophic wind shears are a measure of its baroclinicity.

In strongly curved flows about low- and high-pressure systems, the centripetal force alters the normal geostrophic balance between pressure gradient and Coriolis forces. Consequently, the actual wind can be significantly more or less than the geostrophic wind in strongly cyclonic or anticyclonic flows. The so-called "gradient wind" provides a much better approximation to the actual winds in such curved flow systems (Hess, 1959; Holton, 1992).

2.7 SMALL-SCALE VISCOUS FLOWS

The other broad category of fluid flows, viscous flows, are commonly encountered in engineering and biological systems and, to some extent, also in our natural environment. All viscous flows can broadly be classified as laminar or turbulent flows, although an intermediate category of viscous flows undergoing transition from laminar to turbulent or vice versa has also been recognized.

A laminar flow is generally characterized as a smooth, streamlined, orderly, and very slow motion in which adjacent layers (*laminae*) of fluid slide past each other with very little mixing and transfer of properties (primarily, by molecular agitation). Some of these characteristics are quite similar to those of the inviscid flows discussed earlier. Still, large differences between inviscid and laminar viscous flows arise due to the effects of viscosity and molecular transfers of momentum, heat, and mass in the latter. Viscosity causes viscous stresses everywhere in the flow and makes the fluid stick to solid surfaces (also liquid surfaces for gaseous flows). It is also responsible for the viscous dissipation of the kinetic energy of motion, which is constantly converted into heat.

The addition of viscous and other molecular diffusion terms to the governing set of partial differential equations changes their order, boundary conditions, and solutions drastically. For laminar flows, the Navier–Stokes equations of motion and related scalar diffusion equations can easily be solved analytically for a number of simpler flows and numerically on computers for any complex flow situations. Consequently, an extensive literature in fluid mechanics exists that deals with all kinds of laminar flow problems encountered in engineering, physical sciences, and biological sciences. The possible applications of this to atmospheric environment are limited to very thin (of the order of millimeters) laminar or viscous sublayers that form at the boundaries of solid and liquid surfaces (e.g., smooth ice, snow and water surfaces, plant leaves, raindrops, snowflakes, and suspended particles). We will not consider such specialized thin-film flows in this text (for these the reader may refer to Lowry, 1970; Gates, 1980; Monteith and Unsworth, 1990), except in the context of gravitational settling of particles in the atmosphere, to be discussed in Chapter 10.

2.7.1 Turbulent Flows

In sharp contrast to inviscid and laminar viscous flows, turbulent flows are highly irregular, almost random, and unpredictable. These are also highly rotational, dissipative, and diffusive (mixing) motions, unlike those of two- or three-dimensional waves. Turbulence generally refers to the apparently chaotic nature of many fluid flows that is manifested in the form of irregular, almost random fluctuations in velocity, temperature, and scalar concentrations around their mean values. These irregular fluctuations in a turbulent flow can be observed as functions of time at fixed points in space, or as functions of spatial coordinates at any fixed time. In the atmosphere, turbulence is ubiquitous near the surface in the so-called surface layer. It is also present intermittently, if not continuously, throughout the planetary boundary layer (PBL). In the free atmosphere above the PBL, however, turbulence only occurs in vigorous convective systems and also in certain mixing layers with strong wind shears. Turbulence in the cloud-free regions of the upper troposphere and lower stratosphere, where most aircraft fly, is commonly referred to as clear air turbulence.

Near the earth's surface, atmospheric turbulence is manifested in the flutter of flags, leaves of trees, and blades of grass; the swaying of tree branches; the irregular movements of chimney smoke and dust particles; the generation of ripples and waves on water surfaces; and a variety of other visible phenomena. In common meteorological terminology, it is referred to as gustiness of wind. In the upper part of the PBL, turbulence is manifested by irregular movements of kites and tethered balloons, spreading of smoke from tall stacks, and fluctuations in temperature and refractive index affecting transmission of sound, light, and radio waves (e.g., in remote sensing of turbulence by acoustic sodar, lidar, or radar).

The PBL turbulence is also characterized by a very wide range of length and time scales of motion. Length scales vary from the order of a millimeter (10^{-3} m) to several times the PBL thickness (10^2–10^4 m). The corresponding time scales range from the order of a millisecond (10^{-3} s) to several hours (10^4 s). Thus, there is more than a millionfold range between small (micro) and large (macro) scales of turbulence in the PBL. A much larger range of scales can exist in deep convective systems and extensive regions of clear air turbulence. Large-scale turbulent motions contain most of the kinetic energy and are primarily responsible for turbulent exchanges of momentum, heat, and mass, as well as for the dispersion of pollutants in the atmosphere. Micro or small-scale motions are primarily responsible for the dissipation of kinetic energy, which is passed over to them in a cascade-like process involving the whole range of intermediate scales. The transfer of energy from large scales to smaller and still smaller scales, down to microscales, is a highly nonlinear process that is not well understood.

2.7.2 Modeling of Turbulence

Here we discuss, rather qualitatively, several different techniques of modeling turbulence that have been proposed and used in certain applications involving turbulence and diffusion. Most mathematical models of turbulence are based on the Navier–Stokes equations of motion, either in the original instantaneous form or after an appropriate averaging. No general solution to this highly nonlinear system of equations is known and there is little hope of finding one through purely analytical methods. However, approximate numerical solutions through the use of computers have been obtained for a variety of turbulent flows. These, in conjunction with the various finite-difference and finite-element computational techniques, constitute the fast emerging field of computational fluid dynamics.

Direct Numerical Simulation (DNS) In this age of supercomputers, it is natural to attempt direct solution of the instantaneous Navier–Stokes equations on a three-dimensional grid over the flow domain of interest. However, the amount of computational effort required to make a fundamental attack on most practical problems involving atmospheric turbulence and diffusion is simply prohibitive and outside the range of the biggest and fastest supercomputers currently available. The difficulty comes from the fact that an enormous number of grid points (10^{15}–10^{21}) is required to resolve the smallest scales of motion that are involved in energy dissipation and, at the same time, make the computational domain large enough to permit an accurate specification of boundary conditions and resolution of the largest scales of interest that are important in turbulent exchange and mixing processes. So far, the DNS technique has been applied only to low Reynolds-number flows with the ratio between the largest and smallest scales limited to a few hundreds. In the not too distant future, some of the environmental flows might become amenable to this approach, as computer speed and power continue to grow exponentially.

Large-eddy Simulation (LES) A computationally more feasible but less fundamental approach, which has been extensively used in engineering as well as environmental applications, is large-eddy simulation (LES). It attempts to faithfully simulate only the larger eddies or scales of motion, lying between the smallest grid size and the largest dimension of the computational domain. The smaller, subgrid scales are not resolved, but their important contributions to energy dissipation and minor contributions to turbulent transfer processes are usually parameterized using simpler subgrid scale models.

The origins of the LES technique lie in the early global weather prediction and general circulation models in which the permissible grid spacing could hardly resolve large-scale atmospheric structures and the effects of small-scale processes had to be parameterized. Three-dimensional large-eddy simulation of turbulent flows, including the PBL, started with the pioneering work of Deardorff (1970a, 1972). For a long time, successful applications of LES were limited to unstable and convective boundary layers with vigorous turbulence (Deardorff, 1974; Sommeria, 1976; Nieuwstadt et al., 1992). But more recently, shear-dominated neutral and stable PBLs have also been simulated (Andren et al., 1994; Moeng and Sullivan, 1994; Andren, 1995). A major problem is an accurate representation of the near-surface shear layer. Simulation of weak, sporadic, and intermittent turbulence associated with weak winds and strong stable stratification also remains a challenging problem for LES. With increasing computer speed and power and recent improvements in subgrid scale modeling and parameterization, LES is expected not only to remain the best tool for research, but it may also become the most widely used computational approach for modeling environmental turbulence, diffusion, and air quality.

With increasing grid resolution (reduced size), LES should, in principle, become DNS when the grid size becomes small enough to resolve the smallest microscale of turbulence. However, in practice, the difference between LES and DNS is more than that of scale separation or grid resolution. Direct numerical simulation is based on the "primitive" Navier–Stokes equations of motion, which have to be solved simultaneously at every grid point, at every small instant of time. Large-eddy simulation utilizes the same equations, but averaged over the finite grid volume. This amounts to applying a filter to remove all the subgrid scale variability. The volume averaging also generates new unknowns, which are functions of the filtered-out subgrid scale variables. These must be parameterized in terms of the resolved-scale variables in order to close the set of equations. Numerical integrations of both the systems of equations (instantaneous and volume averaged) yield highly irregular variables as functions of time and space over the simulation domain. Average statistics, such as means, variances, and higher moments, are then calculated from these "simulated turbulence" data.

It is obvious that only DNS can give a faithful simulation of actual turbulence including small-scale fluctuations. Although LES resolves only large-eddy structures, it can also give accurate estimates of those turbulence statistics that are not significantly influenced by small-scale motions. If the main objectives of LES are to estimate these statistics and simulate turbulent transfer processes (e.g., dispersion of pollutants) governed by large-eddy motions, there may not be any merit in reducing the grid size below some optimum size, depending on the atmospheric stability. For this reason, LES is likely to become increasingly attractive as computational resources increase and costs decrease. Direct numerical simulation will, most probably, remain a research tool in the foreseeable future.

Turbulence Closure Models In most practical applications involving engineering and environmental fluid flows, only the mean fields (velocity, temperature, concentration, etc.) are of interest. Sometimes, turbulence statistics may also be desired, but more detailed turbulence structure, including instantaneous fluctuations of variables in time and space, are rarely needed. For these applications, LES might be considered too expensive and unnecessary. Computationally more efficient and cheaper modeling approaches are based on the Reynolds-averaged equations of motion and scalar diffusion. This averaging procedure was originally suggested by Osborne Reynolds in 1895. He also derived the ensemble-averaged equations, which are based on certain averaging rules or conditions.

If f and g are any two fluctuating variables, or functions of such variables, with their mean values denoted by $\bar{f}$ and $\bar{g}$, respectively, and c is a constant, the Reynolds averaging conditions are:

$$\overline{(f+g)} = \bar{f} + \bar{g} \tag{2.42}$$

$$\overline{cf} = c\bar{f}; \quad \overline{\bar{g}f} = \bar{g}\bar{f} \tag{2.43}$$

$$\overline{\frac{\partial f}{\partial s}} = \frac{\partial \bar{f}}{\partial s}; \quad \overline{\int f \, ds} = \int \bar{f} \, ds \tag{2.44}$$

where $s = x, y, z$, or t, and overbar denotes ensemble averaging. In general, only ensemble averaging, rather than time or space averaging, satisfies the above averaging conditions. Especially, the conditions (2.44) are not satisfied by the time and space averages that are often used in practice. These conditions imply that the operation of differentiation or integration is commutable with that of averaging. The conditions (2.43) imply that the average of a fluctuating variable multiplied by a constant or an average quantity is equal to the average of that variable multiplied by the constant or average quantity. Thus, averages can be treated as constants.

Expressing each variable as a sum of its mean and fluctuation (e.g., $u = \bar{u} + u'$, $\theta = \bar{\theta} + \theta'$, etc.) in the instantaneous equations of continuity, momentum, heat, and other scalars, averaging them term by term while applying the above averaging conditions, we can derive the following Reynolds-averaged equations (see e.g., Arya, 1988, Ch. 9):

$$\frac{\partial \bar{u}}{\partial x} + \frac{\partial \bar{v}}{\partial y} + \frac{\partial \bar{w}}{\partial z} = 0 \tag{2.45}$$

$$\frac{D\bar{u}}{Dt} = -\frac{1}{\rho_0}\frac{\partial \bar{p}}{\partial x} + f\bar{v} + \nu\nabla^2\bar{u} - \left(\frac{\partial \overline{u'^2}}{\partial x} + \frac{\partial \overline{u'v'}}{\partial y} + \frac{\partial \overline{u'w'}}{\partial z}\right) \tag{2.46}$$

$$\frac{D\bar{v}}{Dt} = -\frac{1}{\rho_0}\frac{\partial \bar{p}}{\partial y} - f\bar{u} + \nu\nabla^2\bar{v} - \left(\frac{\partial \overline{v'u'}}{\partial x} + \frac{\partial \overline{v'^2}}{\partial y} + \frac{\partial \overline{v'w'}}{\partial z}\right) \tag{2.47}$$

$$\frac{D\bar{w}}{Dt} = -\frac{1}{\rho_0}\frac{\partial \bar{p}}{\partial z} - g + \nu\nabla^2\bar{w} - \left(\frac{\partial \overline{w'u'}}{\partial x} + \frac{\partial \overline{w'v'}}{\partial y} + \frac{\partial \overline{w'^2}}{\partial z}\right) \tag{2.48}$$

$$\frac{D\bar{\theta}}{Dt} = \kappa\nabla^2\bar{\theta} + \bar{S}_H - \left(\frac{\partial \overline{\theta'u'}}{\partial x} + \frac{\partial \overline{\theta'v'}}{\partial y} + \frac{\partial \overline{\theta'w'}}{\partial z}\right) \tag{2.49}$$

$$\frac{D\bar{q}}{Dt} = D_w\nabla^2\bar{q} + \bar{S}_w - \left(\frac{\partial \overline{q'u'}}{\partial x} + \frac{\partial \overline{q'v'}}{\partial y} + \frac{\partial \overline{q'w'}}{\partial z}\right) \tag{2.50}$$

$$\frac{\partial \bar{c}}{Dt} = D\nabla^2\bar{c} + \bar{R} + \bar{S} - \left(\frac{\partial \overline{c'u'}}{\partial x} + \frac{\partial \overline{c'v'}}{\partial y} + \frac{\partial \overline{c'w'}}{\partial z}\right) \tag{2.51}$$

When these equations are compared with the corresponding instantaneous equations, one finds that most of the terms, except for the turbulent transport terms in parentheses, are similar and have the same interpretation or physical significance in the two sets of equations. An important consequence of averaging the instantaneous equations is the appearance of new, unknown, turbulent transports or covariances, which are averages of the products of two fluctuating variables around their means. A more detailed physical interpretation of these covariances, representing turbulent transports or fluxes, will be given later in Chapter 4. Here, suffice to say that these are the most important quantities in a stratified turbulent shear flow, which cannot be ignored. Their presence in the Reynolds-averaged equations constitutes the fundamental closure problem of turbulence (there are many more unknowns than the number of equations), which has been a major stumbling block in developing a rigorous and general theory of turbulence. The closure problem arises only from averaging of the nonlinear terms that are included in D/Dt terms, in the governing equations; the original instantaneous equations form a closed set. This is the price we pay for developing equations for average quantities, which can be solved only after introducing certain closure assumptions or hypotheses constituting a turbulence closure model. Many semiempirical closure theories or models of turbulence have been proposed, but none of them has proved to be entirely satisfactory. Some of the better known theories and models of turbulent diffusion will be discussed in later chapters.

Several other fundamental differences between the instantaneous and Reynolds-averaged equations can be pointed out. The former deal with instantaneous (mean plus fluctuation) variables varying rapidly and irregularly in time and space, whereas the latter deal with mean variables that are comparatively well behaved and vary rather slowly and smoothly. The instantaneous equations are closed but almost impossible to solve for turbulent flows in our atmospheric environment. The Reynolds-averaged equations are fundamentally unclosed, but relatively easy to solve after making appropriate closure assumptions. These can also be much simplified by neglecting the molecular diffusion terms outside of very thin molecular sublayers and also other terms on the basis of considerations of stationarity, symmetry, and homogeneity of flow.

2.8 APPLICATIONS

In this section we discuss some of the applications of the basic concepts of atmospheric structure, thermodynamics, and dynamics introduced in this chapter. We will focus primarily on applications to dispersion of pollutants in the atmosphere.

The peculiar thermal structure of the atmosphere tells us that much of the action (transport and diffusion) is limited to the troposphere, because strong stability of the stratosphere does not permit diffusion and mixing of pollutants from near-surface sources to higher levels. While the whole depth of the troposphere eventually becomes involved in long-range transport and dispersion, most of the action on time scales of the order of a day or less is further confined to the planetary boundary layer (PBL). The height and structure of the PBL, which undergo strong diurnal variations, are thus most important for short-range air pollution and dispersion problems. The air pollution meteorologist must become familiar with the basic state variables, their variations in the troposphere including the PBL, and the various thermodynamical relations between them. In particular, the concepts and quantitative measures of local and nonlocal stability are extremely important, because they are intimately related to diffusion and mixing processes. The importance of identifying mixed layers and inversions to the dispersion of pollutants cannot be overemphasized.

Transport and dispersion of pollutants are intimately connected with mean winds and turbulence in the lower atmosphere. Therefore, one must be familiar with the dynamics of small-scale motions in the PBL as well as large-scale motions in the free troposphere. These are based on the conservation equations for mass, energy, and momentum. The governing equations can be much simplified for essentially inviscid motions outside the PBL and other mixing (turbulent) layers. Large-scale geostrophic and thermal winds are involved in the long-range transport of pollutants in the free troposphere. However, the dynamics of small-scale turbulent flows in the PBL are far more complex. A variety of approximate numerical modeling approaches, as well as semiempirical turbulence closure models, are available. Dispersion theories and models to be described in later chapters are essentially based on the fundamental concepts and dynamical equations for turbulent flows introduced here.

PROBLEMS AND EXERCISES

1. There are substantial differences in the spectral characteristics and total amount of solar radiation that is received at the top of the atmosphere and that reaches the ground. Discuss these differences and the factors that are responsible for the same.

2. Discuss the implications and consequences of the absorption of longwave (infrared) radiation in passing through the atmosphere. What are the two most radiatively important gases, which are responsible for atmospheric absorption?

3. Discuss the physical and/or chemical basis for the observed average thermal structure of the following atmospheric strata: (a) troposphere; (b) stratosphere.

4. a. What is the importance of tropopause in the long-range dispersion of pollutants from anthropogenic sources?

   b. What is the significance of the PBL to short-range dispersion from near-surface sources?

5. a. Show that Eq. (2.1) is consistent with the equation of state given in Chapter 1.

   b. Calculate and compare the densities of dry air at temperatures of -40, 0, and $40°C$ for the standard sea-level pressure of 1013 mb.

6. a. Derive the expression (2.5) for the dry adiabatic lapse rate, Γ.

   b. Explain why the moist or saturated adiabatic lapse rate must be smaller than Γ.

7. a. From the definition of potential temperature, θ, derive an expression for $\partial\theta/\partial z$.

   b. How would you convert a vertical temperature sounding or profile to the corresponding θ profile when the surface pressure is given (say, 1000 mb)?

8. a. Assuming that moist air follows the ideal gas law and using Dalton's law of partial pressures, derive Eq. (2.14) and explain why the virtual temperature is a useful state variable.

   b. Calculate the density of moist air at a sea-level pressure of 1000 mb when its temperature is $40°C$ and specific humidity is 25 g kg^{-1}. What would be the density of dry air at the same temperature and pressure?

9. a. Show that, to a good approximation,

$$\frac{\partial T_v}{\partial z} \simeq \frac{\partial T}{\partial z} + 0.61T\frac{\partial q}{\partial z}$$

   b. Hence, discuss the significance of specific humidity gradient in determining local static stability near land and ocean surfaces.

10. a. Considering different densities of air parcel and environment, show that the acceleration on the parcel due to buoyancy is given by Eq. (2.19).

b. Assuming a small vertical displacement, Δz, of the parcel from its equilibrium position where parcel and environmental densities are the same, derive Eq. (2.20) for the buoyant acceleration.

11. a. What are the limitations of the local static stability parameter s?

b. Explain why the mixed layer of the convective boundary layer in which $\partial \theta_v/\partial z \simeq 0$, should not be identified as a neutral layer. What should be its proper overall stability characterization and why?

12. Calculate the virtual temperature of the saturated air near the sea surface (pressure = 1000 mb) when its temperature is (a) 10°C; (b) 40°C.

13. The following temperature and specific humidity measurements were made from a radiosonde during the Wangara PBL experiment:

Height (m)	Temperature (°C)	Specific Humidity (g kg^{-1})
10	14.4	3.3
50	13.5	3.1
100	12.9	3.1
200	11.8	3.3
300	10.5	3.2
400	9.4	3.2
500	8.5	3.1
600	7.5	3.1
800	5.4	3.0
1000	3.4	2.9
1200	1.4	2.3
1400	2.2	1.0
1600	1.4	0.7
1800	0.8	0.7
2000	0.1	0.7

a. Calculate and plot the virtual temperature as a function of height.

b. Assuming a surface pressure of 1000 mb, calculate and plot the virtual potential temperature as a function of height.

c. Characterize the various layers on the basis of local static stability.

d. What should be the overall (nonlocal) stability characterization of this sounding?

14. a. Calculate the geostrophic wind at the surface when isobars run east-west with the low pressure toward north, pressure gradient is 3 mb per 100 km, and air density is about 1.25 kg m^{-3}. The latitude of the observation site is 30°N.

b. If the observed temperature gradient is 1.5°C per 100 km toward the west, calculate the thermal wind between the surface and the 2000-m height level. What would be the geostrophic wind speed and direction at 2000 m?

15. Using appropriate schematics with isobars and isotherms as parallel straight lines, show that in the northern hemisphere the geostrophic wind is expected to turn cyclonically with height in the case of cold air advection and anticyclonically in the case of warm air advection.

16. Using the Reynolds averaging conditions and the usual decomposition of a variable into its mean and fluctuating parts, show that

$$\overline{uw} = \bar{u}\bar{w} + \overline{u'w'}$$
$$\overline{u\theta} = \bar{u}\bar{\theta} + \overline{u'\theta'}$$

17. a. Writing the total derivative term explicitly and using the continuity equation, show that

$$\frac{D\theta}{Dt} = \frac{\partial \theta}{\partial t} + \frac{\partial(u\theta)}{\partial x} + \frac{\partial(v\theta)}{\partial y} + \frac{\partial(w\theta)}{\partial z}$$

b. Taking a term by term average, show that averaging of the above inertia terms yields new unknowns in the form of covariances.

18. Expressing horizontal pressure gradients in terms of u_g and v_g, simplify the Reynolds-averaged equations of motion for a stationary (time-independent) and horizontally homogeneous planetary boundary layer.

3

Atmospheric Systems and Pollutant Transport

3.1 INTRODUCTION

In the previous chapter we discussed some of the fundamentals of atmospheric thermodynamics and dynamics and reviewed the governing equations for viscous and inviscid fluid flows that are applicable to small- and large-scale atmospheric motions, respectively. Traditionally, atmospheric motions and related phenomena have been classified according to their horizontal dimensions into three broad categories, macroscale, mesoscale, and microscale. Macroscale motions, processes, and phenomena have horizontal length scales of the order of 1000 km or larger (but limited by the circumference of the earth) and time scales of the order of a day or longer. In the vertical, they may extend through the whole troposphere (here we will not consider stratospheric circulations because they are not so relevant to pollutant transport). On the other extreme, the microscale motions and processes have horizontal length scales of 5 km or less and time scales of the order of an hour or less. The vertical scale is limited by the depth of the PBL, mixing layer, convective cloud, or any other system with which they might be associated. The smallest microscale is determined by the size of smallest turbulent eddies that can survive the viscous dissipation mechanism, and is of the order of a millimeter; the corresponding time scale is of the order of a millisecond. Between the two extremes of macroscales and microscales lie the wide range of intermediate synoptic scales and mesoscales. Some atmospheric scientists consider synoptic scales as the lower end of macroscales, but meteorologists interested in day-to-day weather and its evolution with time ranging from a few hours to a week, often deal with synoptic scale systems separately. These are, of course, influenced by the larger macroscale or global scale systems. Synoptic systems, in turn, interact with and influence the smaller mesoscale systems.

Atmospheric transport and diffusion are governed by atmospheric motions or systems covering the whole range of scales. For example, pollutants from car exhaust or a smoke stack are first dispersed by microscale motions in the PBL. Subsequently, their transport and diffusion are influenced by mesoscale circulations associated with urban heat islands, land and sea breezes, mountain and valley winds, thunderstorms, and other mesoscale systems. Horizontal transport winds are rarely straight, uniform, and nondivergent over distances exceeding tens of kilometers and travel times exceeding several hours. In mesoscale systems, dispersion of pollutants is determined by two- or three-dimensional mean flow, as well as by spatially inhomogeneous turbulence. For large travel times (of the order of a day or longer), synoptic scale systems, such as cyclones and anticyclones, determine the mean pathways or trajectories for pollutant transport, while the embedded mesoscale and microscale motions within the large-scale systems determine the horizontal and vertical spreads (dispersion) about the mean trajectories. Most of the pollutants are removed by dry deposition and wet removal processes (e.g., precipitation scavenging) operating at meso and larger scale weather systems. Some may be carried around the globe and are influenced by the global or general circulation patterns. A brief review of macroscale, synoptic, mesoscale, and microscale systems that might be relevant or important to the dispersion of pollutants in the atmosphere is given here. Small-scale turbulence and diffusion processes in the planetary boundary layer that are especially important in short-range dispersion and maximum ground level concentrations will be covered in greater detail in later chapters.

3.2 MACROSCALE SYSTEMS

Before we discuss the various macroscale systems in the atmosphere, including global climate and general circulation, we consider first the global energy budget

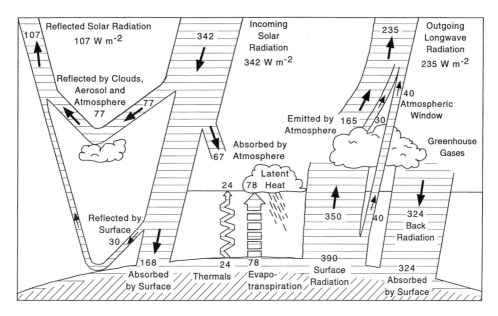

Figure 3.1 Schematic representation of the annual global mean energy budget in the earth–atmosphere–sun system. From Kiehl and Trenberth, 1997.

of the earth–atmosphere system, which provides the physical basis or mechanism for large-scale atmospheric structure and motion.

3.2.1 Global Energy Budget

The sun is the primary source of energy for the earth–atmospheric system. It radiates energy approximately as if it were a blackbody with a surface temperature of about 5800 K. Almost all (99%) of the solar radiation is in the form of shortwave radiation. The average flux of solar radiation received at the top of the atmosphere perpendicular to the sun's rays remains fairly constant around 1368 Wm^{-2}; it is called the solar constant. Since the circular area of the solar beam intercepted by the earth is πR^2, where R is the mean radius of the earth, and the energy falling within this circle is spread over the area of the earth's sphere, $4\pi R^2$, in 24 hours, the globally averaged solar radiation reaching the top of the atmosphere is 342 W m^{-2}. The actual radiation received from the sun over a particular location depends on the latitude, the time of day, and the day of the year, and can easily be calculated from astronomical tables or expressions for the solar altitude angle. For the global energy budget, the average amount of 342 Wm^{-2} is more appropriate. For considering how this solar radiation gets redistributed among various components of the earth–atmosphere system, the above figure is assumed to be 100 percent, or 100 units.

Figure 3.1 gives a schematic representation of the annual global mean energy budget of the earth–atmosphere–sun system (Kiehl and Trenberth, 1997). It includes the redistribution of the incoming solar radiation as it passes through the atmosphere and reaches the earth's surface. On the average, about 67 Wm^{-2} (20%) is absorbed by atmospheric gases and clouds (the absorption by clouds is much smaller than that by gases); 107 Wm^{-2} is reflected or scattered back to space, implying a global albedo of about 31 percent; and 168 Wm^{-2} (49%) is absorbed at the earth's surface.

The relative distribution of longwave radiation exchange among the earth–atmosphere system is also an important component of the overall energy budget. Figure 3.1 indicates that, on the average the earth radiates out about 390 Wm^{-2}, of which only 40 Wm^{-2} actually escape out to space and the remaining 350 Wm^{-2} are absorbed by atmospheric gases and clouds. The atmosphere radiates about 195 Wm^{-2} out to space and about 324 Wm^{-2} downward to the earth's surface. Thus, the total outgoing longwave and shortwave radiation from the earth–atmosphere system exactly matches the incoming solar radiation received by the system. This implies a state of equilibrium for the system on time scales of tens of years, which are used in the determination of climate.

For the global energy budget of the earth's surface alone, one must consider the sensible and latent heat fluxes in addition to those of shortwave and longwave radiations. The net radiation (R_N) of about 102 Wm^{-2} gained by the surface must be balanced by the sum of sensible and latent heat fluxes from the surface to the atmosphere that is, $R_N = H + H_L$. The individual components H and H_L have not been determined precisely, but the latent heat flux (H_L) is estimated to be more than 75 percent of the total convective heat flux ($H + H_L$) to the atmosphere. The above energy budget implies that there is no loss to or gain of energy from the subsurface medium in the long-term average.

Considering the global energy budget of the atmosphere alone, the atmosphere gains a total of about 519 Wm^{-2} of energy flux from terrestrial radiation (350 Wm^{-2}), solar radiation (67 Wm^{-2}), and convection (102 Wm^{-2}). It also loses the same amount through longwave radiation to the surface (324 Wm^{-2}) and that to space (195 Wm^{-2}).

3.2.2 General Circulation and Climate

The above-mentioned average global radiation and energy budgets of the earth–atmosphere system and its individual components do not give any clue to direct or indirect mechanisms for large-scale average atmospheric motions, called the general circulation. These become more clear when one considers the latitudinal dependence of incoming and outgoing radiations and their difference. On the average, equatorial or tropical regions receive more energy from the sun (incoming radiation) than they lose to space in the form of outgoing radiation. Polar regions, on the other hand, lose more energy than they receive from the sun. Thus, the difference between the incoming and outgoing radiations, that is, the radiation imbalance, is a strong function of latitude, with positive values in low-latitude regions and negative values in high-latitude regions. This imbalance would cause very strong temperature gradients from poles toward the equator if it were not for transfer of heat poleward from the equatorial regions by atmospheric winds and, to a smaller degree, by ocean currents. Even with this planetary scale heat transfer, we have quite large surface temperature differences (of the order of 50°C) between the equator and the poles. As a consequence of climatological horizontal temperature gradients, strong westerlies are set up as thermal winds in the upper troposphere (see e.g., the thermal wind equations in Chapter 1).

A more direct thermally induced circulation might be expected from the differential heating of the earth's surface by the sun. Such circulations are commonplace phenomena on the mesoscale. For example, in the absence of strong large-scale steering flow, the temperature contrast between cold sea surface and warm land sets up a sea breeze. This direct thermal circulation is often explained by the fact that heated air over warm land rises and is replaced by that from the sea where colder air sinks. The return flow from land to sea at higher levels completes the circulation. The same phenomenon on a larger scale is responsible for monsoon circulations over many areas. On the global or hemispherical scale, however, the temperature contrast is not so sharp, except at the frontal boundaries between different air masses, and the earth's rotation around its axis and the resultant Coriolis force greatly alters the expected meridional circulation between the equator and the poles. Meridional (north-south) motions are deflected by the Coriolis force to become more zonal (west-east) and, instead of a single cell hemispheric circulation, we observe a more complex general circulation pattern over the globe.

Three-cell Circulation Model An idealized, hypothetical three-cell model of the atmosphere's general circulation is depicted in Figure 3.2. It maintains the expected rising motion over the equatorial belt and sinking motion over the poles, giving rise to surface low and high pressures over equatorial and polar regions, respectively. In addition, subtropical highs and subpolar lows are associated with the near-surface sinking and rising motions of the middle cells.

Other features of the general circulation model, especially the near-surface winds in various latitude belts, are also schematically shown in Figure 3.2. The horizontal flow near the earth's surface is shown in the center of the diagram, while the meridian circulation, both at the surface and aloft, is depicted around the periphery. The meridional (north-south) component has an order-of-magnitude slower speed than the zonal (west-east) component. The equatorial belt is characterized by warm air, weak horizontal pressure gradients, and light easterly winds, called doldrums, over the tropical oceans. These are associated with the intertropical convergence zone (ITCZ) at the boundary of the southeast and northeast trade winds. Due to convergence and intense convection near the surface, warm air rises, often condensing into towering cumulus clouds and thunderstorms that release tremendous amounts of latent heat in the atmosphere. The release of latent heat makes the air more warm and buoyant. This rising air reaches the tropopause, which provides a barrier to further upward motion and forces the air to move toward higher latitudes. The Coriolis force, however, deflects this flow toward the right in the northern hemisphere and to the left in the southern hemisphere, causing westerly winds aloft in both the hemispheres. These winds often attain speeds of more than 100 knots and are commonly known as subtropical jets or jet streams.

As warm air moves poleward from the tropics, it loses heat by radiation and cools. The cold air sinks, and this sinking, being aided further by convergence aloft, causes the surface pressure to increase in the regions of subtropical highs. The subsiding air also warms and dries out, producing generally clear skies and warm surface temperatures. It is in these regions of subtropical highs, also called the horse latitudes, that we find major deserts of the world. A part of the sinking motion in the Hadley cell circulation around 30°N and 30°S latitudes, after reaching the surface, moves back toward the equator. During this movement, however, the Coriolis force deflects the air, causing it to blow from the northeast in the northern hemisphere and from the southeast in the southern hemisphere. This explains the presence of steady trade winds over the tropical ocean areas on both sides of the ITCZ.

Some of the sinking air around the horse latitudes moves poleward and deflects toward the east, resulting

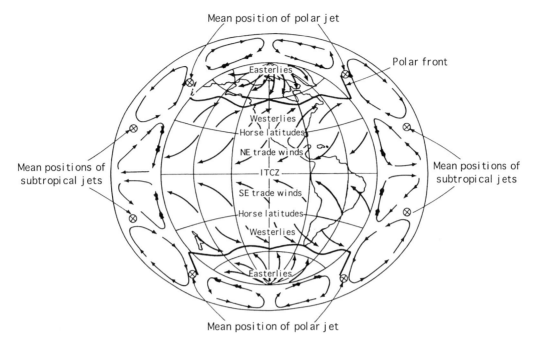

Figure 3.2 Idealized three-cell general circulation of the atmosphere and the associated surface winds and pressures over a uniformly covered rotating earth. ITCZ, intertropical convergence zone. From Miller et al., 1983.

in approximately westerly flow near the surface in middle latitudes. These winds are called the prevailing westerlies, and exist over 30° to 60° latitude belts in both the hemispheres. As the mild air from mid-latitude regions travels poleward, it encounters cold polar air mass coming down toward lower (subpolar) latitudes. These two air masses of contrasting temperatures form the so-called polar front at their boundary. The polar front is a low-pressure zone, called the subpolar low, where surface air converges and rises, causing the development of storms along the front. A part of this rising air, upon reaching the tropopause, returns toward middle latitudes, thus completing the Ferrel cell circulation. The other part returns toward poles as part of the weaker polar cell circulation.

Behind the polar front, the general flow near the surface is northeasterly because the cold air from the poles is deflected by the Coriolis force toward the east. The lower part of the polar cell circulation is also called polar easterlies. In winter, the polar front sometimes moves down to middle and subtropical latitudes, producing intense cold air outbreaks over these regions. Along the front, the portion of the rising air moving toward the poles at higher altitudes is deflected by the Coriolis force to become more westerly aloft. This air eventually reaches the poles where it sinks to the surface and flows back toward the polar front, thus completing the weaker polar cell circulation.

We can summarize the above description of the three-cell general circulation model by noting that there are two major zones of low pressure and rising air in the tropics and in the region of polar front. One would expect major storms and precipitation systems to be associated with these zones. Complementing these are the major zones of high pressure and descending air—in horse latitudes and near the poles. These would be characterized by clear skies and relatively little precipitation. Near the surface, easterly tradewinds extend from the equator to the subtropical high, the westerlies from the subtropical high to the polar front, and the polar easterlies from the polar front to the poles, A qualitative confirmation of the three-cell model is provided by the observed mean meridional circulation. The best agreement is obtained over the oceans, particularly in the southern hemisphere where continental areas are relatively small. But there are also many limitations of this conceptual model. It is only qualitative and assumes a uniform water-covered earth's surface. The model does not account for the actual distribution of major surface features (continents and oceans), each with its distinct topography, albedo, and other surface characteristics. The continents develop thermally induced wind and pressure systems that are not represented in the model. Other departures of the observed circulation from the symmetrical, idealized patterns in the model have been pointed out (see e.g., Byers, 1974).

Numerical General Circulation Models Quantitative numerical models, called the general circulation models (GCMs), have extensively been used to simulate the physical processes of the atmosphere and its interactions with the earth's surface. These are dynamical models in which primitive equations of motion and thermodynamics for the atmosphere are numerically solved at millions of grid points distributed around the globe and in the vertical from the surface to the top of the stratosphere. Horizontal and vertical grid spacings are typically 200 km and 1 km, respectively, with a smaller spacing near the surface for a better resolution of the surface features and processes. Subgrid scale features and processes, such as radiation, clouds, and turbulent exchange processes in the surface layer and the PBL, are parameterized. For simulating the present climate and general circulation, the GCMs are typically run over a period of several decades. Comparisons of the model results with observations indicate that the GCMs simulate the behavior of the real atmosphere quite well and describe the major circulation features as well as their seasonal and latitudinal changes. Their applications to simulating future climate changes in response to increases in greenhouse gases and particulate matter in the atmosphere are somewhat controversial, but have provided the primary impetus for the initial development and subsequent refinements of the GCMs. The more sophisticated general circulation and climate models dynamically couple the atmospheric and oceanic circulations and also use very sophisticated parameterizations of land surface processes.

Observed Features of the General Circulation and Climate The observed features of the general circulation of the atmosphere have been analyzed from the extensive, archived global meteorological data obtained from weather satellites, surface observations, and upper air soundings. By averaging with respect to time for a month, a season, a year, or longer, one can recognize the various transitory, semipermanent, and permanent features of the general circulation.

When averages are taken over a year or longer, the most prominent features of the observed general circulation are the easterly trade winds in low latitudes, particularly near the surface, and the prevailing westerlies in middle latitudes and aloft. Besides the average winds, a host of other meteorological variables such as surface pressure, incoming solar radiation, net radiation, sensible and latent heat fluxes, temperature, humidity, cloudiness, and precipitation are used to represent the global climate. Traditionally, a typical averaging period of 30 years has been used for obtaining climatological averages. Comprehensive reviews of the global climate are given elsewhere (Koppen and Geiger, 1955; Flohn, 1969; Palmen and Newton, 1969; Gates, 1972; Baumgartner et al., 1982; Landsberg and Essenwanger, 1985). Here we discuss only a few salient features.

Mean zonal winds over the earth in summer and winter are shown in Figure 3.3 in which isotachs (lines of constant wind speed) represent wind speeds in meters per second. The prevailing motion in most of the troposphere is from west to east. The westerlies become more pronounced and steadier with increasing height, attaining maximum wind speeds between 10 and 12 km above the surface. There are two major exceptions to the westerly flow, the tropical trade winds and a shallow layer of polar easterlies. North-south and vertical velocities, averaged around latitude circles, are much smaller than the zonal winds. Longitudinal variations in the low-level winds are caused by the presence of continents and also by the presence of major topographical features on the same. For example, the low-level flow is often blocked and steered by major mountain ranges. Differential heating between land and ocean surfaces also gives rise to thermally induced circulations, such as monsoons.

Global distributions of the net radiation and other components (sensible and latent heat fluxes) of the surface energy budget show not only strong north-south gradients, but also considerable east-west variability due to variations in surface albedo, temperature, and moisture over different land and ocean surfaces (see e.g., Baumgartner et al., 1982). Everywhere ocean surfaces have a higher net radiation than neighboring continents in the same latitude. The maxima in R_N are located in the areas of tropical oceans centered over the equator. The maxima in the latent heat flux are also located over the tropical oceans, but centered around $\pm 20°$ latitudes. Sensible heat flux is primarily supplied by tropical continents; the contribution from ocean surfaces is much smaller but still significant, considering the much larger area covered by oceans.

Global distribution of the average sea-level temperature reflects the expected poles-to-equator gradients (Baumgartner et al., 1982). But these are also modified by continental land masses, particularly in the northern hemisphere. The north-south gradients are weakest over the tropical oceans and strongest over Antarctica. Global distributions of water vapor pressure, relative humidity, cloudiness, and precipitation show complex patterns, reflecting the presence of major land masses (continents) and major mountain ranges on the same, oceans, large seas, and major ocean currents. These patterns together with geographical features provide the primary basis for physical climate classifications (Koppen and Geiger, 1955; Baumgartner, et al., 1982; Landsberg and Essenwanger, 1985).

When the averages of meteorological variables are taken by months or seasons, many other semipermanent features of the atmosphere's general circulation can be identified. For example, from the average sea-level pressure distributions and surface wind-flow patterns for the months of January and July, shown in Figure 3.4, one can identify regions of semipermanent highs and lows (Ahrens, 1994). In the northern hemisphere in winter (January) there are four semipermanent high pressure systems. Two of these—the Pacific high and the Bermuda high—are located over the oceans; the other two are located on the continents (North America and Asia). In the southern hemisphere also, four high-pressure systems show up over the subtropical oceans. These subtropical highs are semiper-

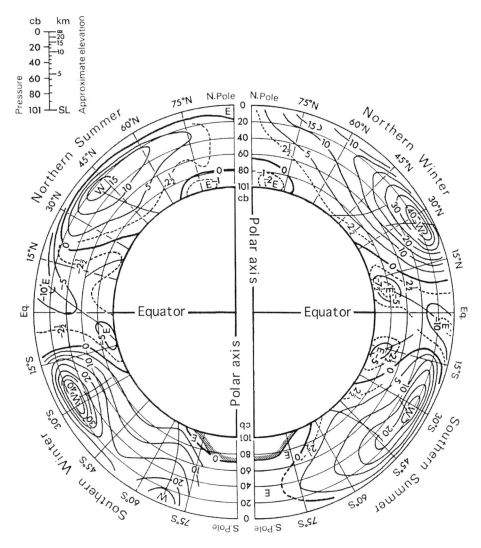

Figure 3.3 Mean zonal winds over the summer and winter. Regions of easterly winds are hatched and isotachs are given in meters per second. From Flohn, 1969.

manent features, since they show up in the July circulation also with only minor shifts in their positions. These subtropical anticyclones develop in response to the upper-level convergence. High-pressure systems over continents appear to be more transitory; these are essentially formed as a result of intense cooling of the interior land surfaces.

Among the semipermanent low-pressure systems are the Icelandic and Aleutian lows, which cover the regions where numerous storms tend to converge, especially in winter. In the southern hemisphere, the subpolar low forms a continuous trough that completely encircles the globe. Thermal lows forming on warm continents are only seasonal (summer) features. Most of these form along the intertropical convergence zone, whose position changes seasonally. Distinct monsoon circulations are also associated with some of the thermal lows (e.g., the one over India). Low-pressure systems are the regions of cyclonic activity where air rises and condenses to form clouds and precipitation. Over the tropical oceans, some of these rapidly intensify to form tropical storms and hurricanes or typhoons.

Upper-level Flow and Jet Streams Long-term averages of upper-level observations of heights, temperatures, and so on, on constant pressure (e.g., 850, 500, and 300 mb) surfaces yield the basic features of the general circulation aloft. The average height fields for January and July, for example, indicate essentially westerly flow, except for a few closed circulations around subtropical highs and polar/subpolar lows (see e.g., Palmen and Newton, 1969; Neiburger et al., 1982,

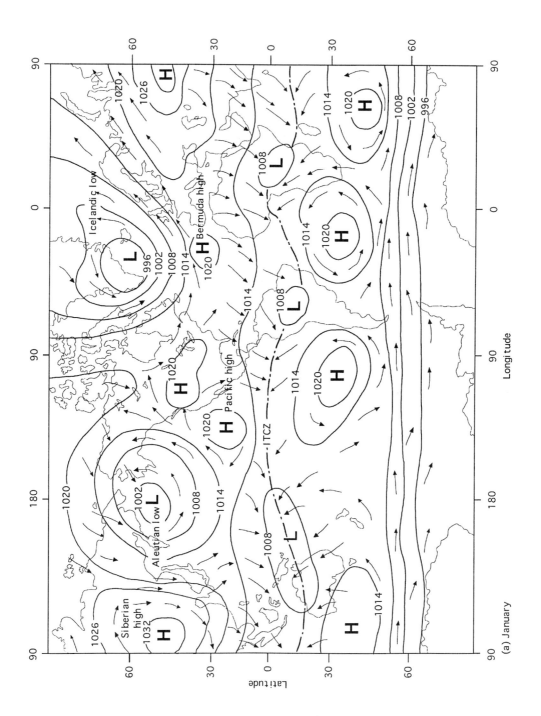

(a) January

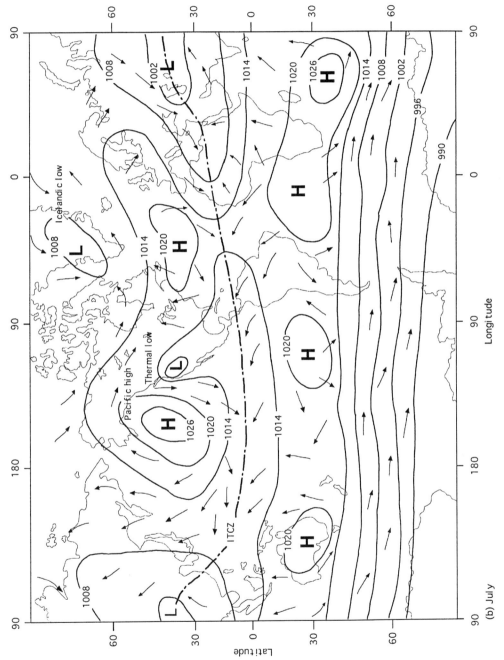

Figure 3.4 Average sea-level pressure distribution and surface wind flow patterns for the months of (a) January, and (b) July. From Ahrens, 1994.

Ch. 9; Ahrens, 1994, Ch. 11). In the southern hemisphere there are only slight deviations from the zonal flow, but in the northern hemisphere the wave pattern has larger amplitudes, with troughs over North America, eastern Europe, and off the east coast of Asia.

The deviations from purely zonal flow are obviously due to topographic features of the earth's surface. Significant orographic influences are found to be associated with major mountain ranges such as Himalayas, Rocky Mountains, Andes, and Alps. The influence of thermally induced circulations due to land and sea temperature contrasts is also considered equally important. The large observed seasonal and local variations in the total (sensible plus latent) heat fluxes from oceans and continents indicate the importance of thermal influences on the deviations from the zonal flow pattern and on the locations of troughs and ridges in the observed longwave pattern of constant pressure heights.

Orographic influences are partly mechanical and partly thermal in nature. Some aspects of the mechanical influence of topography as an obstacle in the flow have been demonstrated in the laboratory experiments involving obstacles in a rotating hemispherical shell. With westerly flow, a train of waves is generated whose wavelength depends on the latitude of the obstacle (Palmen and Newton, 1969).

Thermal effects on the formation of high-pressure systems over or near large mountain ranges during the summer season have also been noted, particularly in the northern hemisphere. In the southern hemisphere, a more symmetrical zonal mean circulation is obtained because of the limited west-east extent of mountain ranges and the smallness of land areas. In both the hemispheres, the westerly winds extend to much lower latitudes in winter than in summer. The average isotachs show concentrated areas of strong westerly winds; these are the average positions of what are known as jet streams.

Discovered by high-flying bomber pilots near the end of the Second World War, a jet stream is defined as a strong narrow current, concentrated along the quasi-horizontal axis in the upper troposphere or in the stratosphere, characterized by strong wind shears and featuring one or more wind speed maxima (Reiter, 1967). Jet streams in the atmosphere are huge rivers of fast-flowing air from west to east thousands of kilometers long, several hundred kilometers wide, and a few kilometers deep. Wind speeds in the central core of a jet stream often exceed 100 knots and occasionally exceed 200 knots. Like rivers, they meander in the horizontal with ridges and troughs separated by large distances. Surface weather changes dramatically in response to their large-scale meanders or oscillations.

Jet streams normally form at the upper boundaries of the meridional circulation cells, as shown in Figure 3.2. The jet stream situated about 13 km above the subtropical high is called the subtropical jet stream, while that situated nearly 10 km above the polar front is known as the polar jet. Both are found near the tropopause, where mixing between tropospheric and stratospheric air takes place. Due to their large meanders, the two jet streams come close to each other and sometimes merge at the locations of the semipermanent troughs of middle-latitude westerlies. Where the two jet streams merge, the warm air from low latitudes is brought into juxtaposition with the cold air mass from higher latitudes. Consequently, the meridional temperature gradient is much enhanced. Since the westerly winds in the upper troposphere are roughly proportional to the horizontal temperature gradient or the baroclinicity of the atmosphere, the regions of the mean polar troughs are also the locations of the strongest west wind, which can reach speeds of more than 200 knots and occasionally 300 knots. This region of strong wind is called a jet maximum or a jet streak, and is an important factor in the development and intensification of near-surface storm systems.

A jet stream moves around the globe from west to east with large north and south swings (meanders) from its mean position. Pollutants such as volcanic dust and aerosols injected high into the jet stream can be transported rapidly around the world in a few days to a week. Jet streams are also involved in the significant global transfer of heat. On the eastern side of the trough, warmer air is carried poleward, while on the western side, the more northerly flow brings cold air toward low latitudes.

There are two primary mechanisms for the formation of jet streams. The first involves the existence of a polar front (for the polar jet) or a subtropical front (for the subtropical jet) at the boundary of the two air masses. The associated steep gradient in temperature leads to strong westerly winds in the convergence zone aloft near the tropopause. The second mechanism involves the conservation of angular momentum, which is the product of the mass, the velocity, and the radius of curvature of a moving parcel. For a poleward-moving parcel in Hadley cell circulation, for example, the decrease in the radial distance from the axis of rotation must be compensated by an increase in speed. The air parcel must, therefore, move faster toward the east (west wind) as the parcel moves away from the equator. Hence, the conservation of angular momentum of poleward-moving air leads to the generation of strong westerly wind and the formation of the subtropical jet stream. Both the above mechanisms usually operate simultaneously.

In addition to the westerly polar and subtropical jet streams, there is an easterly tropical jet stream that forms in the summer near the tropopause above Africa, India, and Southeast Asia. Its winds are easterly, because the jet forms on the equatorward side of the upper-level subtropical high. The formation of this easterly tropical jet appears to be partly related to the heating of air over elevated land masses, such as the Tibetan Plateau, compared with that over the ocean to the south. This produces a north-to-south temperature gradient and, hence, strong easterly winds aloft near the 15°N latitude.

3.2.3 Macroscale Motions and Dispersion

The long-term averaged macroscale systems, such as the general circulation, semipermanent highs and lows, and jet streams, that we have discussed in this section are expected to affect the long-range transport of pollutants released into the atmosphere by natural and anthropogenic sources. By the time macroscale transports become effective, however, pollutants from individual sources have undergone successive diffusion and dispersion by microscale, mesoscale, and synoptic scale systems. In this process they are mixed throughout the depth of the troposphere, as they also spread horizontally over the dimensions of synoptic scale systems and, at the same time, undergo the usual transformation and removal processes. Consequently, individual puffs and plumes lose their identities as they merge into a regional haze or a polluted air mass. Most of the pollutant species are eventually washed out and deposited. Only some of the more stable and inert species remaining in the troposphere are transported by macroscale systems and thus dispersed over the whole northern or southern hemisphere, if not the whole globe.

Machta (1958) suggested that synoptic and global circulations would tend to distribute an effluent uniformly in the northern hemisphere after 14 days. His theoretical calculations also indicated that pollutants emitted in the northern hemisphere tend to stay in the northern hemisphere with little transport across the equator. A verification of this hypothesis is provided by observations during an ocean cruise, of average Freon concentrations in the marine atmospheric surface layer as a function of latitude (Lovelock et al., 1973). Both the data and theoretical calculations indicated a similar concentration distribution with the maximum value around 50°N, concentration decreasing toward the equator and attaining smaller values in the southern hemisphere (see e.g., Pendergast, 1984). The observed distribution in the northern hemisphere, which is the main source region for Freon (or CFCs), is still far from the uniform concentration suggested by Machta (1958).

Worldwide surveys of average carbon monoxide concentrations near the surface also indicate a large drop near the equator in going from the northern to the southern hemisphere, which is consistent with the hypothesis of little or no interhemispheric transport, particularly in the lower troposphere. The trade winds on both sides carry air toward the equator. This converging air from the two hemispheres is more likely to mix at higher levels above the ITCZ. Thus, higher up in the upper troposphere, interhemispheric differences may not be as large as near the surface. Still, the general circulation in the upper troposphere and stratosphere being more nearly zonal, interhemispheric transport and mixing are not expected to be strong. Over long enough periods, however, relatively inert pollutants emitted largely in the northern hemisphere are likely to find their way into the farthest and most remote regions of the southern hemisphere. This has been amply verified by aircraft measurements made in the stratosphere over Antarctica since 1985, as part of the broader study of the stratospheric ozone depletion in that area. While the ozone depletion mechanism is primarily chemical, atmospheric dynamics and the resulting macroscale motions and circulation systems provide the necessary transport mechanism for carrying the CFCs and other halogen species from the primary source areas in the northern hemisphere.

A comprehensive understanding of macroscale systems is necessary for studying the tropospheric and stratospheric chemistry involving many radiatively active species, such as carbon dioxide, methane, ozone, water vapor, nitrous oxide, chlorofluorocarbons, and aerosols, which are considered important in determining the global climate. General circulation models are used in conjunction with atmospheric chemistry models for studying possible climate changes in response to increasing concentrations of the chemical species due to human activities.

3.3 SYNOPTIC WEATHER SYSTEMS

Within the general circulation and the associated macroscale systems, when viewed on time scales of less than a few weeks, are buried the large areas of more or less homogeneous air masses, formed over different source regions, acquiring their special characteristics. Air masses are separated by narrow areas of discontinuity or transition, called fronts. Along these fronts often form the middle-latitude cyclones and other low-pressure systems. There are also other tropical and extratropical cyclones and anticyclones, which are sometimes called large eddies in the atmosphere. All these synoptic scale systems greatly influence and determine the day-to-day weather and, hence, are commonly called synoptic weather systems. We will only give a brief description of these systems and their relevance to long-range transport and fate of air pollutants. For more comprehensive reviews of synoptic weather systems, the reader should refer to other meteorology books and monographs (e.g., Palmen and Newton, 1969; Byers, 1974; Houghton, 1977).

3.3.1 Air Masses

An air mass is a large body of air whose properties (particularly, temperature and moisture content) remain more or less homogeneous over horizontal distances of the order of thousands of kilometers, at least in the zonal (east-west) direction. For an air mass to form, air must remain long enough over its source region, which is typically a very large, flat, and homogeneous area with uniform surface properties (e.g., temperature and moisture content). This means that the air must stagnate or move very slowly over the source region. The longer the air remains over its source region, the more likely it will acquire the properties of the surface below through the processes of radiation, vertical convection, and turbulent mixing. Thus, the ideal source regions for air masses to form are the extensive flat and uniform areas dominated by stationary high-pressure systems. These include ice- and snow-covered high-latitude plains in winter, and extensive desert areas and subtropical oceans in summer. The middle latitudes, where surface temperature and moisture vary considerably, are not good source regions. Instead, the mid-latitude belt is a transitional zone where different air masses move in, clash, and produce constantly changing weather systems.

The principal characteristics of source regions that determine the air-mass properties are the surface temperature and its character as land or sea. As an indirect measure of their surface temperature, source regions and their characteristic air masses are classified by latitude belts, that is, Arctic (or Antarctic), polar (actually, subpolar), tropical (actually, subtropical) and equatorial, and the surface character is designated as continental or maritime. On the basis of these characterizations, air masses are generally classified and designated as follows:

Arctic or Antarctic	A or cA
Continental Polar	cP
Maritime Polar	mP
Continental Tropical	cT
Maritime Tropical	mT
Equatorial	E or mE

Note that the first letter (in lowercase) in the abbreviated name specifies the continental (c) or maritime (m) character and the second letter (in capital) specifies the latitude belt of the source region. Arctic (or antarctic) air is typically continental in its properties, and equatorial air is typically maritime. Therefore, c and m are often dropped from the designations of these two air masses. The remaining four air masses are given the usual two-letter abbreviated designations. Sometimes, a third letter, k or w, is added to indicate whether the air mass is colder or warmer than the surface over which it has moved. For example, cPk designates an air mass that originally formed over the continental polar (actually subpolar) source region, but subsequently moved over a warmer surface and, hence, is colder (designated by k) than the underlying surface.

Properties of Air Masses Air masses generally acquire the properties of their source regions. The temperature and moisture distributions in the vertical can be used to characterize different air masses. Potential temperature and mixing ratio are more conservative air mass properties. Here, we give a brief description of air mass properties in a qualitative sense only; for more quantitative descriptions the reader should refer to general books on meteorology.

Arctic (or antarctic) air masses form over polar regions in winter when the polar high (anticyclone) is present over these regions. These are characterized by extremely low temperatures and very low specific humidities (typically, less than 0.1 g kg^{-1}). They are very stable near the ground, with strong inversions extending up to several kilometers above the surface. There is very little mixing or exchange of heat between the earth and the atmosphere in the presence of such strong surface inversions. During the austral winter, an antarctic air mass is the coldest air mass, because of the extremely low temperatures over the South Pole.

Continental polar air masses form in the northern hemisphere when anticyclones stagnate over Siberia, Alaska, and northern Canada. In winter, they are similar in properties to arctic air masses, but not quite as cold and dry. In summer, they are moderately cold, with much less stability and more variable humidity.

Maritime polar air masses form over oceans at high (55–65°) latitudes (e.g., North Pacific, Gulf of Alaska, and North Atlantic). They are cool, moist, and unstable. The mP air is milder (warmer) than the cP air in winter; the former is colder than the latter in summer. Continental polar air masses, after moving over the relatively warmer oceans at the same latitudes, eventually acquire the characteristics of the mP air, particularly in the lower layer modified by the ocean surface.

Continental tropical air masses form over subtropical land areas of North Africa, southwestern United States, northern Mexico, and desert regions of Asia, which are dominated by subtropical highs. They are hot, dry, and unstable near the surface, but stable aloft. Clouds are rarely present in the cT air mass because of the very low humidity.

Maritime tropical air masses form over the warm oceans on the equatorial side of the subtropical anticyclones. The air is warm, moist, and unstable near the ocean surface. But above the shallow marine boundary layer, there is an inversion in which the air is hot and dry because of subsidence.

The maritime equatorial air mass forms near the intertropical convergence zone, particularly in the doldrum regions of stagnant air. It is warm, moist, and usually unstable to high levels, except over the eastern parts of the oceans where ocean surface is relatively cold due to upwelling of deep water.

Air Mass Modification When an air mass moves out of its source region, in response to the general circulation or short-term changes in the same, it is modified from below by the characteristics of the new surface. The modification of air mass takes place quite rapidly in the surface layer, continues through the

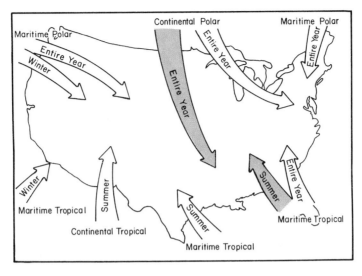

Figure 3.5 Schematic of the polar and tropical air masses moving into the continental United States during particular seasons and the entire year. From Navarra, 1979.

outer part of the planetary boundary layer (PBL), and eventually extends to most of the free troposphere as the air spends a long enough time over the new surface. The ratio of the depth of the modified layer to the fetch (distance traveled on the new surface) is typically 0.1 for the neutral and unstable PBLs, 0.01 for the stable PBL, and probably much less than 0.01 for the free atmosphere. The primary mechanisms for air-mass modification are turbulent mixing, which is essentially confined to the PBL; the vertical convective activity including deep convection; and instability of the interface between the modified and original air masses.

For the air-mass modification to occur over the entire depth of the troposphere, the air would have to travel thousands of kilometers away from its source region and spend several days over the new surface. For example, a continental polar air mass moving over the warmer ocean would eventually become a maritime polar air mass. Many other examples of air-mass modification are given in books on general meteorology (see e.g., Navarra, 1979; Neiburger et al., 1982).

Air Masses of the Continental United States The continental United States is visited by primarily four air masses—continental polar (cP), maritime polar (mP), continental tropical (cT), and maritime tropical (mT). These air masses influence and largely determine the synoptic scale weather in the United States. Typical locations and directions of their intrusion during particular seasons and the entire year are schematically shown in Figure 3.5.

The continental polar air mass in winter moves from Canada into the interior plains of the United States, continuing southward in response to the southward swing of the polar jet stream. As it moves south, the air mass is designated as cPk because it is colder than the underlying surface. Except for the shallow unstable PBL developed during the shorter daytime period, this air stays very cold, dry, and stable, even after modification. Consequently, generally fair weather prevails over much of the land under the influence of the quasi-stationary or stagnant high. Considerable air-mass modification occurs, however, when cP air mass passes over the warm bodies of water such as the Great Lakes and Atlantic Ocean. Rapid moistening of the modified air results in heavy snow showers, called the lake-effect snows, on the downwind sides of the lakes. Mountain ranges, such as the Rockies and Appalachians, act as barriers to the movement of cP air. Consequently, the western and southeastern regions of the United States are often spared from most of the cold air outbreaks that occur in the middle portion.

The western and northwestern coastal regions of the United States are frequently visited by maritime polar (mP) air masses originating from the North Pacific. This cool and moist air is strongly heated and moistened from below as it travels to lower latitudes over the ocean before moving over land. When atmospheric pressures are relatively low and falling, the air modification produces cumulonimbus clouds and moderate-to-heavy showers. On the other hand, if the air pressure is relatively high, as in summer, warming and moistening generally produce stratocumulus clouds and light scattered showers. In winter, as the modified mP air moves inland, coastal mountains force it to rise. Consequently, much of its water vapor condenses to form clouds, causing rain showers at lower elevations and heavy snowfalls at higher levels in the coastal mountains and other mountains farther to the east. During summer, the mP air moving toward lower latitudes usually develops stratocumulus clouds. As it crosses a zone of cold water (California current) that hugs the shoreline, the mP air is cooled and low stratus clouds and fog develop in the air. Fog and low stratus are commonly encountered along the west coast in summer.

Somewhat less frequent invasion of the mP air occurs over the east coast where the cold air from the North Atlantic moves inland in response to lower pressures over land compared with those over the ocean in summer. The Atlantic mP air masses are usually much colder than their Pacific counterparts. They also have relatively low specific humidity and thin stratocumulus clouds associated with them. The warming of the air, as it moves inland, dissipates the clouds. Thus, along the northeast coast in summer, the mP air from the Atlantic often brings in clear and cool weather. The same air mass in winter is much colder and is brought in by cold highs (anticyclones) slowly drifting to the east (north of New England states). A stationary front forms along the boundary between this cold mP air mass and the warm mT air mass to the south coming from the Gulf of Mexico. North of this front, cold and damp northeasterly winds provide generally cloudy, rainy, or snowy weather. Under appropriate upper atmospheric conditions, storms originating in the south travel along the stationary front and intensify off the coast of the Carolinas into Atlantic northeasters. These winter storms cause heavy snowfalls over most of the Atlantic coast states north of Florida and gale-force winds over the coastal waters.

The southern parts of the United States are frequently under the influence of tropical air masses, particularly in summer. The only source regions for the hot and dry continental tropical (cT) air masses in summer are the arid or desert regions of northern Mexico and the adjacent southwestern United States. Here the cT air mass develops in response to intense solar heating and dry convection to heights of 3 km or more. The subsidence over the large region of stationary high further makes the near surface and the PBL air warmer and the air aloft more stable. The prevailing conditions in this air mass are hot, dry, and clear. The air mass is usually confined to its source region. Occasionally it moves out to the Great Plains and stagnates there, causing severe drought conditions.

The maritime tropical (mT) air masses are more frequently encountered in the southeastern part of the United States throughout the year and occasionally in southern California in winter. These air masses are warm and moist and produce much rain, particularly in the coastal regions. In winter, the Pacific mT air usually stays far south of California. Occasionally, weak, upper-level southerly flow brings this warm and humid air into the southwestern United States. Middle and high-level clouds develop over most of the affected region. But rising motion over the coastal and other mountain ranges in southern California usually produces heavy rains and mud slides.

The mT air originating from the Gulf of Mexico and Caribbean Sea influences much of the weather in the summer east of the Rockies. In winter, the mT air is usually confined to the Gulf and extreme southern states. Occasionally, however, a slow-moving storm system over the central plains brings the mT air northward. As the warm, moist air passes over the very cold ground surface, fog and low clouds form in the early morning, dissipate by midday, and reform in the evening. Thus, the mT air mass usually brings in mild winter weather in the region, but only for a few days. When the storm system over the central plains stalls, the mT air can bring unseasonably warm and mild weather to much of the eastern half of the United States.

In summer, the mT air masses are generally responsible for the hot and humid conditions over the eastern half of the United States. As this marine air moves over the hot continent, it warms, rises, and condenses into cumuliform clouds, which produce afternoon showers and thunderstorms. Under the influence of the Bermuda High, the mT air dominates the summer weather in the southeast. A weak but persistent flow around an upper-level high can spread this air all the way to the Great Plains and to the southern and central Rockies, where it causes afternoon showers.

3.3.2 Fronts

The concept of a front as the sharp boundary between two different air masses was introduced by Norwegian meteorologists in 1918 (during World War I). A front is the transition zone between two air masses of different densities. Since density differences between the air masses are caused by temperature and specific humidity differences, fronts are usually characterized by sharp gradients in both temperature and specific humidity across the transition zone, which is typically 10 to 100 km across. The upward extension of the boundary between the two air masses is called a frontal surface or frontal zone. Due to large differences in air densities across the frontal surface, this surface cannot be vertical, but slopes upward with the heavier cold air mass forming a wedge below the lighter warm air. The denser air forces itself underneath the lighter air in a wedge-like fashion. Average slopes of frontal surfaces are found to be generally quite small (say, between 1/300 and 1/50).

The process of front formation is referred to as frontogenesis. In the middle latitudes, fronts often extend for hundreds or sometimes thousands of kilometers horizontally along the surface. Waves and associated wave cyclones often develop and move along the fronts, resulting in dramatic changes in weather. As the air masses across a front become mixed and indistinguishable, the front practically disappears; this process is called frontolysis.

Types of Fronts In general, fronts can form between any two air masses that are brought together by their respective air motions to clash at their boundary. In reality, the air circulations in low latitudes are such

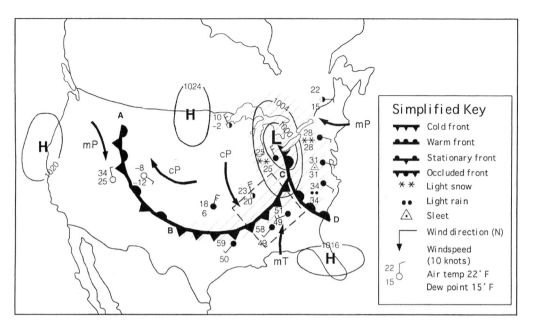

Figure 3.6 A simplified weather map showing major surface pressure systems, air masses, and fronts. From Ahrens, 1994.

that sharp fronts do not ordinarily develop between equatorial and tropical air masses. Temperature contrasts between these two air masses are not large enough to cause significant gradients in temperature across their boundary.

From a geographical point of view, the principal front is the polar front, which is the boundary between polar (actually, subpolar) and tropical air masses. In the middle and upper troposphere, the polar front is generally found to be continuous extending all around the northern (or southern) hemisphere. The polar front is associated with and follows the meanderings of the polar jet stream (Palmen and Newton, 1969, Ch. 7; Byers, 1974, Ch. 10). On the ground surface, however, the polar front is not continuous around the globe, but is interrupted by regions where the transition between the two air masses is gradual and not sharp enough to be distinguished as a front. Over North America, the polar front is usually located near the southern border of the United States in winter and to the north of the Great Lakes in summer. The front frequently occurs between mP and cP air masses and also between these polar and tropical air masses. The front between arctic (or antarctic) and polar air masses is called the arctic front. It too is not continuous around the earth. Over North America, the arctic front runs through central Alaska and northern Canada. Both the polar and arctic fronts move with the air mass associated with them. Thus, their average positions undergo large seasonal variations.

Fronts are further subdivided according to the direction and the speed of their movement. A cold front represents a situation in which cold air is replacing warm air at the surface. On a surface weather map, it is usually represented by a thick blue line with a set of triangles on the warmer side of the line, pointing toward the direction of frontal motion. A warm front is one in which warm air is replacing cold air at the surface. It is represented on a map by a thick red line, with a set of half circles on the colder side of the line. A stationary front is one with no forward motion toward either side; the air masses may, however, move parallel to the front. On a weather map, it is represented by a heavy line with alternating sets of triangles and half-circles pointing toward warm and cold air masses, respectively. A fourth type of front, called an occluded front, results when a cold front overtakes a warm front. It is also represented by a thick line with alternating sets of triangles and half-circles, but both pointing toward the same direction of frontal motion. These four different types of fronts are shown on a considerably simplified weather map in Figure 3.6, without the usual color coding used on original manuscript maps. Note that the line sector AB represents a stationary front, BC a cold front, CD a warm front, and CL an occluded front. All four types are present in the particular synoptic situation shown in Figure 3.6; sometimes only one or two fronts occur at the same time.

General Characteristics of Fronts Here we give only a brief, qualitative description of the general characteristics commonly associated with almost all fronts. More detailed accounts of temperature, moisture, wind fields, and weather phenomena associated with each type of front can be found elsewhere (Palmen and Newton, 1969; Byers, 1974; Navarra, 1979; Ahrens, 1994).

Fronts usually form at the outer boundaries of high-pressure systems (anticyclones). But they extend all the way to the center of a low-pressure system or cyclone. Thus, the center of a cyclone is found at one end of the front. These features can be clearly seen in the synoptic weather map of Figure 3.6.

At the surface, fronts occur mostly along low-pressure troughs. Thus, isobars at a front are generally U- or V-shaped. Moving toward the front from either the cold or the warm air side the pressure drops. Conse-quently, the surface pressure at a station always falls as a front approaches and rises after the front passes the station.

Fronts form only between air masses of different temperature and moisture conditions, so that there is a sharp density difference between them at the same level. The frontal surface or zone always slopes upward, with the denser cold air wedged underneath the lighter warm air. The slope of the frontal zone is steeper for a cold front, with average values between 1/150 and 1/50, than that for a warm front (average slope between 1/300 and 1/100).

The wind shift associated with the passage of a front in the northern hemisphere is always such that the wind should veer (turn clockwise) with time. This is always the case in a moving trough (low-pressure system) that passes over an observation station. Since winds move around the lows in a counterclockwise rotation, one can easily show that as a front passes a station, the wind must turn in a clockwise direction (veer) with time. However, a veering wind does not always mean a frontal passage.

There are also significant wind speed shears associated with a front. The wind through a front is such that the component parallel to the front, taken positive with cold air to the left, increases to the right. This type of shear is cyclonic, representing positive vorticity (Byers, 1974). In most cases, the wind components parallel to the front on the two sides of the front are in opposite directions (positive in warm air and negative in cold air). The vorticity is always positive (cyclonic) in the frontal zone.

The wedge of cold air under the warm moist air results in some lifting of the latter. Consequently, clouds and precipitation develop before and during the passage of a front. Decreasing intensity of showers and then clearing occur after the frontal passage. Long periods of steady precipitation are often associated with stalled or stationary fronts when warm and moist air is continuously supplied by a stationary high.

With the above mentioned general characteristics of fronts, the following criteria are often used to locate fronts on a surface weather map:

1. Low-pressure trough extending away from a surface low or cyclone
2. Sharp changes in temperature and specific humidity over a relatively short distance, with maximum gradients centered on the trough
3. Sudden shifts in wind direction and wind speeds, with cyclonic shear in the northern hemisphere
4. Cloud and precipitation patterns

3.3.3 Cyclones and Anticyclones

Embedded in the average general circulation of the atmosphere are the synoptic scale eddies, called cyclones and anticyclones, which are associated with constantly moving low- and high-pressure systems. These large-scale rotating systems or vortices tend to be masked by the long-term averaging used to analyze the general circulation. But a synoptic analysis of pressure, temperature, moisture, and flow fields for any particular day and their evolution from day to day reveals these synoptic scale eddies. Compared with the size of the macroscale circumpolar vortex (about 10,000 km in diameter), the cyclonic and anticyclonic eddies range from several hundred to a thousand kilometers in horizontal dimensions. Depending on the latitude belt in which they are formed, these eddies or whirls are further subdivided into middle-latitude systems, tropical storms, and polar lows.

Middle-latitude Systems The largest and most numerous of the synoptic systems are the middle-latitude cyclones and anticyclones that move in a general easterly direction around each hemisphere. These dominate the atmospheric flow and day-to-day weather over much of the earth between 30° and 70° latitudes. They are also the most significant transporters of heat and moisture from low to high latitudes, as well as efficient exchangers of atmospheric momentum. The middle-latitude storms that produce much of the winter precipitation are the cyclonic systems. Around their low-pressure centers (lows), the air circulates in a counterclockwise direction in the northern hemisphere and in a clockwise direction in the southern hemisphere. The circulating air masses are generally heterogeneous in temperature and moisture, with sharp transition zones or fronts separating cold, dry air from warm, moist air. These cyclonic storms go through a complex life cycle, lasting from several days to several weeks.

The polar front theory, originally proposed by Norwegian meteorologists in 1918 and later refined by others, is used to explain the development of a wave cyclone along a front. Schematic depiction of the genesis and early development of the cyclone are shown in Figure 3.7. It relates the formation of the cyclone (cyclogenesis) to the undulations of the frontal boundary (frontal wave). Figure 3.7(a) shows a segment of the polar front as a stationary front. It represents a low-pressure trough between high-pressure systems located on both sides of the front. Cold dense air to the north and warm lighter air to the south flow parallel to the front but in opposite directions. This type of flow sets up a strong cyclonic wind shear, giving both

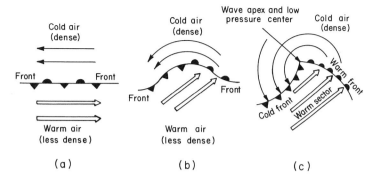

Figure 3.7 Genesis and early development of a wave cyclone in the northern hemisphere. From Miller et al., 1983.

the air streams an angular velocity or a spin. Under such conditions, part of the frontal boundary will begin to turn about some point along the front, forming a frontal wave, as shown in Figure 3.7(b). Different types of initially small wave disturbances such as gravity shear and inertial waves can be visualized. Only for certain wavelengths (say, between 600 and 3000 km), the wave disturbances become dynamically unstable and they grow further beyond the stage shown in Figure 3.7(b).

As the wave grows, it develops a kink at its apex around which a low-pressure system forms. This is still an early development stage of the wave cyclone. Both the warm and cold air currents move in a cyclonic pattern around the low. Figure 3.7(c) shows this developed wave with a cold front pushing southward and a warm front moving northward. The region of warm air between the two fronts is called the warm sector. A narrow band of precipitation forms alongside the two fronts, as cold air pushes the warm air upward along the cold fronts and overrunning occurs ahead of the warm front. The early stages of the wave cyclone, shown in Figure 3.7(a) through (c), take between 12 and 24 hours. The cyclone is generally steered toward the east or northeast by the winds aloft. The central pressure drops with time and winds blow more vigorously.

Subsequent development of the wave may take several more days; these later stages are shown in Figure 3.8. As the wave becomes steeper, the faster moving cold front comes closer to the warm front. Eventually, the cold front overtakes the warm front, resulting in an occluded front near the cyclone center. This is the most intense and mature stage of the cyclone development, shown in Figure 3.8(b). Since cold air lies now on both sides of the occluded front, the storm begins to dissipate. The warm sector is still present but is far removed from the center of the storm, cutting off an important source of energy to the storm. Gradually, the fronts begin to dissolve as the kinetic energy of the system is dissipated by friction and winds subside. The final dissipating stage is shown in Figure 3.8(c). The entire life cycle of a wave cyclone can last from a few days to more than a week.

Sometimes, a new wave called a secondary low forms at the triple point where cold, warm, and occluded fronts meet, moves eastward, and intensifies

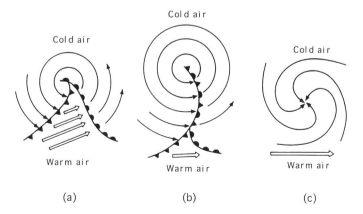

Figure 3.8 Later stages in the development and dissipation of a wave cyclone in the northern hemisphere. From Miller et al., 1983.

into a new cyclonic storm. A succession of such storms can make a family of wave cyclones. Wave cyclones usually occur in families of three to five (Miller and Thompson, 1970).

The above described sequence of events in the life cycle of wave cyclones is, of course, an idealization. In fact, few cyclonic storms would faithfully follow this simplified polar front model. Nevertheless, the model provides a good basis for studying and understanding the middle-latitude cyclones. On a typical winter day, about ten of these wave cyclones are present in the winter hemisphere, making them the most ubiquitous weather producers.

Initial development and further strengthening of a cyclone is called cyclogenesis. Certain regions show a natural propensity for cyclogenesis. In the United States, these regions include the eastern slopes of the Rockies, the Great Basin, the Gulf of Mexico, and the Atlantic Ocean off the coast of the Carolinas. The so-called northeasters develop and rapidly strengthen over the warm waters of the Gulf Stream, which hugs the coast of the Carolinas up to Cape Hatteras where it takes an eastward turn. These fast-developing and northeasterly moving winter storms often dump a lot of snow along the east coast of the United States and Canada. When a cyclone develops rapidly with its central pressure falling in excess of 24 mb in 24 hours, the term *explosive cyclogenesis* or *bomb* is often used to characterize it. There have been several cases of bombs, severely affecting life and property in the Atlantic states, in the last 20 years. For further details of the vertical structure of these storms and necessary ingredients for explosive cyclogenesis, the reader should refer to Ahrens (1994, Ch. 13).

In contrast to wave cyclones, anticyclones are high-pressure systems in which circulation (wind flow is clockwise in the northern hemisphere and counterclockwise in the southern hemisphere) is relatively weaker. In these systems, air masses are generally homogeneous in temperature and moisture. The diverging flow near the surface causes slow subsidence of air, which suppresses the formation of clouds and precipitation except for the convection on a local or mesoscale. Subsidence also helps in the formation of low-level inversion and reduction of the PBL thickness or mixing height. Thus, middle-latitude anticyclones are usually fair-weather systems. These are harbingers of good weather, but generally bad air quality. Air pollutants are generally confined to a shallow mixing layer where they are cooked by sun and subjected to photochemical reactions leading to smog. The most severe air pollution problems arise during stagnant high-pressure systems.

Tropical Cyclones Tropical cyclones, as their name implies, are cyclonic eddies that are formed over the warm tropical oceans between 5° and 20° latitudes and generally travel from east to west. These are the major rain- and wind-producing storms, which spend most of their lifetimes over open oceans. Occasionally, however, they have devastating attacks on tropical islands that fall on their way and on certain continental coastal areas where they make landfall, weaken, and eventually dissipate. Depending on the surface pressure and maximum sustained winds, these cyclonic eddies are variously classified as tropical disturbances, tropical depressions, tropical storms, and hurricanes (in the Atlantic), or typhoons (in the Pacific), or simply cyclones (in the Indian Ocean). Here we will use the term *hurricane* for the mature, well-developed tropical cyclone with maximum sustained winds in excess of 64 knots and discuss some of the salient features of this storm. For further details about the structure of hurricanes and their formation and dissipation processes, the reader should refer to books on tropical meteorology (Palmen and Newton, 1969; see also Byers, 1974; Ahrens, 1994).

A hurricane is an extremely low-pressure system with a typical central pressure of 950 mb (the lowest pressure ever recorded was 870 mb), horizontal diameter of 500 km, and the vertical extent of 15 km (up to the tropopause). There is a well-defined pattern of swirling winds and spiraling rain bands as a function of radial distance from the center. In the center, there is an area of relative calm and broken clouds or clear skies and no precipitation. This is called the eye of the hurricane, which can easily be spotted from satellite pictures. Away from the eye, clouds align themselves in spiraling bands resulting in rain bands at the surface. Surface winds, blowing counterclockwise, increase in speed with increasing distance from the eye, attaining maximum speeds in the so-called eye wall of towering clouds. The precipitation is also heaviest within the eye wall, which typically extends from 10 to 50 km from the center. Several cloud and rain bands follow with areas of light precipitation between them. Clouds from the eye wall spread horizontally hundreds of kilometers in all radial directions, as the rising warm air is deflected by the tropopause. The track of a hurricane can be followed by satellite photos, but it is not easy to predict more than 12 hours in advance. Abrupt changes in track may occur as a hurricane approaches the continental land mass.

When a hurricane or tropical storm moves on to the land, much of its energy source in the form of the latent heat of evaporation and condensation is cut off. In addition, the increased frictional drag of the land surface helps dissipate the storm, particularly its wind speeds. There is still very heavy precipitation associated with the weakened storm. For obvious reasons, tropical cyclones are less frequent and not as important in air pollution meteorology as their middle-latitude counterparts.

Polar Lows Somewhat similar to tropical cyclones, but less intense, are the low-pressure systems that form over the subpolar oceans on the poleward sides of the polar fronts. These cyclonic storms are called polar lows and have a typical diameter of 500 km or less. Some polar lows have a comma-shaped cloud band, while others have a spiral of convective cloud bands that swirl around a clear area (eye) at the center. The eye of a polar low resembles that of a tropical cyclone. There are also other similarities between polar lows and hurricanes; both have a warmer central core, strong winds, and heavy precipitation (in the form of snow in the polar low).

In the North Atlantic, polar lows typically form during the winter from November through March. They usually form along the arctic front where the cold and dry arctic air mass (cA) meets the relatively warmer and moist marine polar air (mP). The storm gains most of its energy from the potential energy associated with the rising of warm air and the sinking of cold air along the front. Additional energy is provided by the transfer of sensible and latent heats from the warmer ocean surface to the air above. The storm's development is enhanced if a jet streak moves over the surface low or an upper trough lies to the west of the storm. At its mature stage, a polar low may attain a central pressure of 980 mb or lower. It dissipates rapidly after moving over land. Because of their location over high-latitude ocean areas, far away from the sources of air pollution, and their low frequency of occurrence, polar lows are not considered significant in air pollution meteorology and dispersion.

3.3.4 Synoptic Systems and Dispersion

Synoptic systems, such as fronts, cyclones, and anticyclones, affect the dispersion of pollutants released in the lower part of the atmosphere on time scales of the order of a day or longer. By this time, the pollutants have been thoroughly mixed over the maximum mixing depth of the daytime PBL. They have undergone considerable diffusion by PBL turbulence and additional dispersion by any mesoscale systems that might be present. Synoptic systems further tend to spread them over the much greater horizontal extents of their circulations. However, vertical transport and mixing in the free atmosphere above the PBL are confined to rather limited areas of upward motion and turbulence associated with convective systems. Pollutants are vented out of the mixed layer by deep penetrative convection, as well as by upward motion associated with fronts and cyclones. At the same time, they are also scavenged by clouds and rain and are wet deposited on the ground. Thus, fronts and cyclones with their strong winds, clouds, and precipitation bands serve more to remove the pollutants from the atmosphere and dilute their concentrations. Only a few very stable and inert gases such as freon and other CFCs can escape this scavenging process.

Synoptic systems of greatest concern to the air pollution meteorologist are the stationary or quasi-stationary fair-weather high-pressure systems. As mentioned earlier, the subsiding motion associated with these tends to confine the pollutants to a shallow mixing layer near the surface. Stagnation of such systems for a period of several days and longer also results in their slow recirculation around the high and hence increasing concentrations with time. Almost all the air pollution episodes in history have been associated with stagnant conditions under high-pressure systems that stalled over an urban/industrial area.

3.4 MESOSCALE SYSTEMS

Weather and atmospheric circulation systems of horizontal scales between 5 and 1000 km are called the mesoscale systems. Most of these are associated with fixed geographical features and do not travel in space. They are the result of influences on the air flow by different geographical and topographical configurations, such as land and sea, city and countryside, and mountain and valley. The resulting mesoscale circulations or systems include monsoon circulations, sea and land breezes, urban heat-island-induced circulations, mountain and valley winds, mountain wake vortices, and lee waves. Other mesoscale systems are formed more or less randomly from convection over heated land and water surfaces, or they spawn from larger storms and weather systems. Consequently, they move with the large-scale flow and the mother storm to which they are attached. The moving mesoscale systems include dust devils and water spouts, individual convective clouds and cloud streets, thunderstorms, squall lines, tornadoes, and gravity waves. Here we give a brief description of only the former type of mesoscale systems that are tied to geographical and topographical features of the earth's surface, because they are more important in air pollution meteorology and dispersion.

In Chapter 2 we explained that atmospheric motion is largely a result of the balance between inertia, pressure gradient, Coriolis (rotational force), and friction force. The friction force is significant only in small-scale motions in the PBL and is generally ignored in the dynamics of large-scale flows and frequently ignored also in most of the mesoscale systems. But, in mesoscale flows, inertia forces become more important, and a simple geostrophic balance, like that of large-scale flows, does not exist. However, horizontal pressure gradients still remain the primary driving mechanism for these flows.

Following Atkinson (1981), we divide the topographically induced mesoscale circulation systems into two broad types: (1) thermally induced circulations and (2) mechanically induced circulations. Specific examples of each of these are briefly discussed in the following.

3.4.1 Thermally Induced Circulations

Since pressure gradients in the atmosphere are intimately related to density and/or temperature gradients, through the equation state and the hydrostatic equation, one can expect a strong influence of the horizontal temperature gradients or differences associated

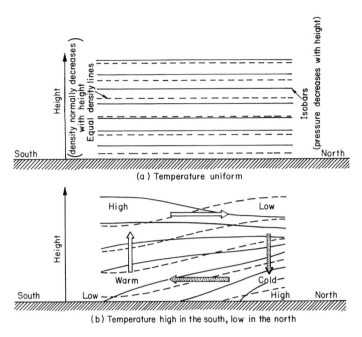

Figure 3.9 Schematics of isobars and equal density lines and thermal circulation: (a) Uniform temperature. (b) Nonuniform temperature with low temperature in the north and high temperature in the south. From Miller et al., 1983.

with surface inhomogeneities on the resulting mesoscale circulations. Dynamically, this influence can be explained by the equations of motion introduced in Chapter 2. However, here we will give only a qualitative and physical explanation of this influence by considering a basic, single-cell thermal circulation induced by a temperature difference or gradient in the horizontal. This direct thermal circulation can help explain most of the thermally driven mesoscale systems to be introduced later.

Basic Thermal Circulation Consider the idealized, schematic vertical cross sections of the atmosphere above a flat and homogeneous surface, as shown in Figure 3.9. In the top section (a), temperature is assumed to be uniform in the horizontal, so that density would also be uniform. In the absence of any large-scale pressure gradient, isobars and isotherms will be horizontal and parallel to the surface and there will be no wind. Suppose the surface to the north becomes colder than that to the south due to the differential heating or cooling at the two ends. In turn, the air over the warm surface heats up and expands more rapidly than that over the cold surface. This situation, shown in Figure 3.9(b), will lead to density gradients such that the air density at any level will increase from south to north. According to the equation of state, cold air is denser than warm air at the same reference pressure. Due to the approximate hydrostatic conditions, the vertical gradient of pressure must be greater in the cold air than that in the warm air. Consequently, in the cold dense air, isobars tend to bunch closer together, while in the warm less dense air, they spread farther apart. This is schematically shown in Figure 3.9(b). The resulting pressure gradients are from north to south near the surface and in the reverse direction aloft. This means that, at a constant level high above the ground surface, pressure is higher over the warm surface than over the cold surface. Since horizontal wind flow is generally from high to low pressure, a slight movement of air from south to north is produced aloft.

Horizontal convergence in the region of the upper-level high produces downward vertical motion, and divergence from the upper-level low produces upward vertical motion. These vertical motions are consistent with the notion that less dense air over the warm surface rises and dense air over the cold surface sinks. These motions result in further strengthening of the surface low over the warm land and the surface high over the cold ground. The resulting pressure gradient forces the near-surface air to flow from cold to warm end of the surface, thus completing the thermal circulation. If this air is moist enough, the upward vertical motion over the warm region may produce clouds and possible precipitation, while the subsiding motion will suppress cloud formation over the cold region.

The circulations driven by horizontal temperature contrasts or gradients, in which warm air rises and cold air sinks, are called thermal circulations. The regions of surface low and high pressures created as a result of warming and cooling of the atmosphere are called thermal lows and thermal highs, respectively. In general, these are shallow systems with circulations extending no more than a few kilometers above the surface. Since they are maintained by horizontal temperature gradients, thermal circulations weaken as temperature gradients become weak. In the following we give some specific examples of thermal circulations in the lower atmosphere.

Sea and Land Breezes A sharp thermal contrast often occurs at the coastline due to large differences in thermal properties of land and sea surfaces and the diurnal heating and cooling cycle. Because of the large heat capacity of the coastal sea, the sea surface temperature does not change by more than 2°C between day and night. The land, on the other hand, heats up rapidly during the day and cools at night, especially under clear-sky and fair-weather conditions with weaker large-scale atmospheric flow. A diurnal range of 10° to 20°C in surface air temperatures over land is typically observed in coastal areas. Consequently, large temperature differences occur between land and sea across the coastline. These temperature contrasts induce mesoscale thermal circulations, known as land and sea breezes, similar to that of Figure 3.9(b).

During a clear, sunny day, the land heats up much more quickly than the adjacent seawater, and this intense heating leads to the development of a shallow thermal low over land. As warm air rises over land and cool air sinks over coastal waters, a shallow thermal high forms over the latter. Consequently, a sea breeze circulation develops that blows colder air from the sea toward the land near the surface and warmer air from the land to the sea at higher levels. This sea breeze is similar to the basic thermal circulation shown in Figure 3.9(b) if half of the surface to the right (north) may be considered as the colder sea surface.

In a sea breeze circulation, strongest winds usually occur at the coastline where temperature gradient is strongest and they diminish in going farther inland. Sea breezes are strongest in the afternoon when the land surface temperature is maximum. At this time in summer, sea breezes may extend several tens of kilometers inland in coastal regions. The depth of sea breeze circulation, including the upper seaward flow, is typically several hundred meters in middle latitudes, but it may reach several kilometers in the tropics. The surface winds are usually gusty and constantly shifting in direction.

The leading or forward edge of the sea breeze is called the sea breeze front. The front moves inland with time after the initiation of the sea breeze before noon and it reaches its maximum distance around the time of the maximum temperature over the land. As the sea breeze front passes a certain location inland, a sudden drop in temperature and increase in relative humidity are observed. A change in wind speed and sudden shift in wind direction also occur, especially on the east coasts of continents. Sometimes fog or low stratus clouds develop in the sea breeze and light precipitation may result over the region of the strongest upward motion near the sea breeze front. If the rising air is sufficiently moist and unstable, a line of cumulus clouds will form along the sea breeze front. In summer, thunderstorms may also develop along this front.

Sea breezes in Florida produce abundant rainfall in summer. Across the peninsula, the sea breezes blow in from the Atlantic Ocean and the Gulf of Mexico in opposite directions. The convergence of these very moist air streams, coupled with the daytime heating and convection, produces extensive cumulus clouds and thunderstorms over land. Over coastal waters, sinking motion associated with thermal circulation produces clear skies.

Sea breeze decreases in intensity and ceases entirely by late evening as the contrast between land and sea temperatures vanishes. Later at night, the land cools more readily and quickly than the coastal sea, producing a temperature contrast between the warmer sea and the colder land. Consequently, thermal circulation is reversed, with the colder air from the land blowing over the sea near the surface and the warmer marine air blowing over land aloft. This land breeze circulation is similar to the thermal circulation of Figure 3.9(b) if the left half of the surface were taken as the warm sea surface.

The land breeze circulation is not as strong, either in intensity or height, as the daytime sea breeze circulation, because the temperature contrasts between land and sea are generally much smaller at night. The circulation also develops rather slowly, starting just before midnight and reaching its maximum intensity near sunrise. It has a stabilizing influence on the lower atmosphere over land and a destabilizing effect over the sea. Since no significant changes in weather occur due to land breezes, except for some increased cloudiness and possible precipitation offshore, land breezes are hardly noticed by people living in coastal areas. However, the reversal of wind direction from the daytime sea breeze has important implications for the transport and diffusion of pollutants from the coastal sources. In the absence of strong synoptic flow, pollutants are transported offshore by the land breeze, where they are likely to be scavenged by clouds and water surface.

Similar thermal circulations are observed in the coastal areas of large lakes. There, the daytime breeze from the lake is called the lake breeze, while the land breeze is often much weaker. Inland penetration of lake breezes is usually not more than a few kilometers and their depths are also smaller as compared with sea breezes. But even these smaller thermal circulations have important influences on local environments around large lakes and rivers.

The sea and land breeze circulations are observed to occur more frequently and with greater intensity in the tropics than in the middle and high latitudes. Many climatological studies of tropical sea/land breezes have been conducted (see e.g., Atkinson, 1981, Ch. 5).

Sea breezes in the tropics often penetrate several hundreds of kilometers inland. With the larger horizontal and vertical extents of these circulations, the earth's rotational or Coriolis effects also become significant.

Other factors affecting sea and land breeze circulations that we have not mentioned so far are the large-scale geostrophic or gradient wind, atmospheric stability, and topography of land near the coast. The effects of these factors on sea/land breezes have been investigated (see e.g., Atkinson, 1981, Ch. 5). In particular, strong onshore gradient winds normal to the coastline generally suppress or overwhelm the purely thermal circulation. With weaker onshore gradient winds, however, sea breezes are observed to develop earlier in the morning and extend farther inland. Offshore winds, on the other hand, tend to suppress both the intensity and inward penetration of sea breeze but sharpen the sea breeze front. With an offshore wind, sea breeze forms later in the day, is much shallower, and retreats much earlier than a breeze forming in the calm atmosphere. Gradient winds parallel to the coastline may not affect the development of sea breeze, if one considers only the wind component normal to the coastline.

The effect of atmospheric stability on sea breeze development has been observed qualitatively. The most favorable condition for sea breeze formation is that of greatest instability. A more unstable atmosphere increases the intensity of sea breeze circulation and extends it farther both horizontally and vertically. A stable atmosphere over land, on the contrary, dampens the mechanism for the vertical circulation and, hence, suppresses the sea breeze.

Topography may influence the sea breeze circulation in many ways. Hills with equatorward-facing slopes heat up more than the flat land and may accentuate the sea breeze. Once the sea breeze begins to develop, hills and valleys may greatly affect its speed and direction. Gaps on coastal ranges produce channeling of flow. A low-level, strong marine inversion may force the breeze around hills rather than over them. Landward hill slopes may also affect the intensity of breeze farther inland. Vegetation cover of the coastal land, including hills, determines to what extent the land is heated by the sun. Dry, barren coastlines heat up much more quickly than moist, vegetated surfaces and thereby cause stronger sea breezes.

Monsoon Circulation The monsoon system is another example of direct thermal circulation, but on a much larger, subcontinental scale. The circulation changes direction seasonally, rather than diurnally. The seasonal reversal of monsoon circulation is especially well developed in eastern and southern Asia where winds blow from one direction in summer and from the opposite direction in winter. The mechanism of circulation is essentially the same as in the case of sea land breezes; it is also explained in our description of the basic thermal circulation. The seasonal monsoon circulation is schematically shown in Figure 3.10.

During the winter, the air over the continent becomes much colder than the air over the ocean. A shallow thermal high develops over the continent and a thermal low over the ocean. Pressure gradients force the winter monsoon winds to blow from land to sea, while the subsiding air over the continent causes generally fair weather and dry season over eastern and southern Asia. Here, the winter monsoon means clear skies and light winds blowing from land to sea throughout the winter season, except for some short periods of disturbed weather.

In summer, the monsoon circulation is reversed as air over the continent becomes much warmer than that over the ocean. A shallow thermal low develops over the interior of the continent. The heated air over the low-pressure region rises and the cool moist marine air rushes in to replace it. The convergence of the marine air with the westerly flow of dry continental air causes the former to rise; further lifting is provided by coastal hills and inland mountains. This lifting, accompanied by condensation, results in heavy rains and thunderstorms. A constant supply of the moist air associated with steady summer monsoons produces heavy and steady rains, which are interrupted only by brief periods of monsoon break. Thus the summer monsoon in southern and eastern Asia generally means wet season with winds blowing from the sea to land.

The summer monsoon is much more intense than the winter monsoon and is extremely important for providing much needed water for reservoirs and rivers and soil moisture for agricultural production in this heavily populated region. The circulation is further enhanced by the latent heat released during condensation. Rainfall is enhanced by weak westward or northwestward moving low-pressure systems called monsoon depressions. These depressions usually form and track along the monsoon trough; their formation is aided by an upper-level jet stream.

Monsoon circulations also exist in other regions of the world, such as Africa, Australia, and South America, where large contrasts in temperature develop between continents and oceans. In the United States, monsoon-like circulations exist along the Pacific coast and in the southwestern region, especially in Arizona and New Mexico. However, most North American monsoons are usually obscured and frequently interrupted by migratory cyclones and fronts.

Urban Heat Island Circulation On a much smaller scale, direct thermal circulations are also associated with urban heat islands whenever there is a significant horizontal (usually, radial) temperature gradient between the city center and the surrounding countryside and the large-scale flow is weak. The difference between the urban and rural air temperatures (ΔT_{u-r}) is called the urban heat island intensity. The maximum difference generally occurs at nighttime; the urban-rural temperature differences during the day are much smaller. In the absence of large-scale gradient flow, the urban heat-island-induced thermal circulation de-

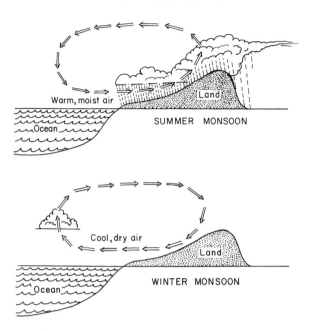

Figure 3.10 Schematics of summer and winter monsoon circulations. From Miller and Thompson, 1970.

pends on the size (diameter) of the city, the heat island intensity, ΔT_{u-r}, the ambient atmospheric stability, and the time after sunset (see e.g., Lu and Arya, 1995).

The intensity of an urban heat island depends on many factors, such as the size and energy consumption of the city, geographical location, time of day, month or season, and synoptic conditions. The maximum intensity for a given city occurs on clear and calm nights, a few hours after sunset, during the winter season. Some attempts have been made to correlate the maximum nocturnal heat-island intensity with the city population and the near-surface wind speed (see e.g., Oke, 1987). But such correlations are not universal and are found to depend on other factors, including geographical location, latitude, and climate of the city.

The circulation is essentially caused by the rising of warm air over the city and the sinking of cold air over the surrounding countryside. The near-surface air from rural surroundings converges toward the thermal low that forms over the city, while at upper levels the air diverges away from the city. During the daytime, this circulation may extend up to the base of the lowest inversion, which may then acquire a dome shape with the maximum mixing height at the city center. Since mixing is generally restricted to the mixed layer below the inversion, it is also referred to as the mixed-layer dome or dust dome, in which dust, smoke, and haze from urban emissions accumulate during stagnant conditions. Some observational evidence for the occurrence of the urban mixed-layer dome has been presented by Spangler and Dirks (1974). Observations show significant increases (up to 40%) in the mixed-layer height in going from the rural surroundings to the city center even when $\Delta T_{u-r} < 1°C$.

A radially symmetrical circulation would be expected to occur only over a circular city in the absence of gradient wind. More frequently, even light gradient winds over irregularly shaped urban heat islands can cause more complex thermal circulation patterns whose centers may not coincide with city centers. Presence of topography may also greatly influence these thermal circulations. Many observational, experimental, and numerical modeling studies of urban heat islands and urban boundary layers have been reported in the literature (see e.g., Oke, 1987; Arya, 1988; Lu and Arya, 1995).

Mountain and Valley Winds Due to the diurnal heating and cooling of mountain slopes, thermal circulations often develop along these slopes. The lower portions of these circulations, called mountain and valley winds or breezes, are schematically shown in Figure 3.11.

During the day, solar radiation warms the mountain slopes or valley walls, which in turn warm the air in contact with them. Due to convective mixing, air gets heated up to several hundred meters above the sloping surface. This heated air, being less dense than the air at the same elevation above the valley floor, rises as an upslope wind and is called a valley wind. It is only the lower (near-surface) limb of the mountain-valley thermal circulation in which return flow toward the valley occurs aloft and cold dense air sinks over the valley floor. A more rigorous explanation of

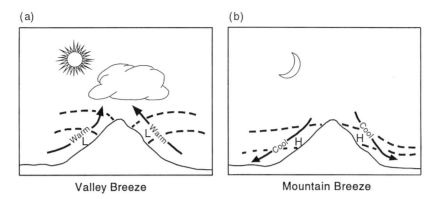

Figure 3.11 Schematics of (a) valley winds and (b) mountain winds associated circulations. Dashed lines represent constant pressure surfaces. Modified after Ahrens, 1994.

the mechanism of valley wind, similar to that used in the formation of the basic thermal circulation of Figure 3.9(b), is given by Atkinson (1981).

During the night, the mountain slopes cool more quickly by outgoing radiation than the valley floor. The air in contact with the slopes and up to a depth of several tens of meters cools through conduction and turbulent mixing. This cooler dense air flows down the slope and is called a slope or a mountain wind. Again, it is only the lower limb of thermal circulation in which less dense warmer air over the valley floor rises and the return flow aloft is from the valley to the mountain. Thus, the nighttime mountain wind circulation is the reverse of the daytime valley wind circulation, as shown in Figure 3.11.

In mountainous areas, the upslope valley winds begin early in the morning, shortly after the beginning of sunshine on the slope, reach a maximum speed by midday, and die down before sunset. The downslope mountain winds begin shortly after sunset, strengthen throughout the night, and reach their peak strength just before sunrise. The mountain and valley winds are most pronounced on clear summer nights and days when the large-scale gradient winds are weak. The near-surface winds also depend on the magnitude and direction of slope, configuration and depth of the valley, and vegetation cover on the slopes. Mountain and valley winds are best developed in deep valleys with mountain slopes facing the midday sun. Thus, in the northern hemisphere, the valley winds are particularly well developed on south-facing slopes. When these have sufficient moisture, the upslope flow results in the formation of cumulus clouds above the mountain tops. In summer, showers and even thunderstorms commonly occur in mountainous areas, especially in early afternoon when valley winds reach their maximum strength.

Early observations of mountain (slope) and valley winds and their climatology have been reviewed by Atkinson (1981). More recent and systematic field studies of these winds and their influence on dispersion in complex terrain have been conducted in the United States under the project ASCOT (Atmospheric Studies in Complex Terrain) (Orgill and Schreck, 1985). High frequencies of slope winds are observed in high latitudes, particularly in winter months. Downslope (mountain) winds are generally stronger and more steady than the upslope (valley) winds. But the former are confined to a shallower (∼50 m) layer near the surface, while valley winds extend up to several hundred meters. There is some observational evidence that the depth of the valley-wind system is nearly equal to the height of the ridge line. The presence of cross-valley gradient winds is found to have a strong influence on the intensity and evolution of mountain and valley winds. Gradient winds of speeds above 10 m s^{-1} usually prevent the development of any local thermal circulations.

Katabatic Winds The term *katabatic* (fall) winds is generally used for downslope winds that are much stronger than mountain breezes of typically a few meters per second. The ideal setting for these strong, cold, downslope winds is an elevated plateau where the air is chilled over the winter snow by radiative cooling, becomes very dense, and comes rushing down the surrounding slopes through gaps and saddles in the mountains. A surface thermal high forms over the plateau, making the horizontal pressure gradient near the edge of the plateau strong enough to cause the cold air to flow across the isobars. The resulting downslope drainage winds are forced by gravity, like water flowing down the slope, and are called katabatic winds. Most of the time, these have moderate speeds of 3 to 10 m/s^{-1}. Occasionally, however, katabatic winds become very strong (>25 m/s^{-1}) and gusty when forced by a migratory cyclone or anticyclone and confined through a narrow canyon or gap through the mountains.

Katabatic winds are observed in many mountainous regions of the world, particularly in Alaska, Greenland, Antarctica, Alps, Himalayas, and Rockies. In some locations where they become particularly strong and violent, they have been given local names, such as *bora* in the Austrian Alps and *mistral* in the Rhone Valley of France. Wind speeds of more than 100 knots have been observed in some of these katabatic flows.

3.4.2 Mechanically Forced Circulations

Atmospheric circulations resulting from air flow over low hills, valleys, and mountains exist over the whole range of atmospheric scales of motion. Here we are primarily concerned with the mechanically forced mesoscale circulations, including upslope and downslope winds, mountain and valley wakes, and lee waves. Obviously the horizontal and vertical extents of a circulation are intimately related to the height, slope, and shape of the forcing obstacle (topography). But the ambient atmospheric conditions, such as winds and atmospheric stability, also play an important role. Only a limited number of field studies of atmospheric flows over isolated hills and two-dimensional ridges have been made, because of the large expense and logistical difficulties encountered in such studies (Karacostas and Marwitz, 1980; Taylor and Teunissen, 1987). More extensive and systematic experimental investigations have been conducted in fluid modeling facilities, such as wind tunnels, water channels, and towing tanks, using scale models of topography in simulated environments (Meroney, 1990). A recent, comprehensive review of the literature is given by Bains (1995). Here we give only a brief and qualitative description of some of the features that might be considered important in air pollution meteorology and dispersion.

Upslope and Downslope Winds Flow patterns over and around hills depend strongly on the stability of the approach flow (ambient atmosphere). When the approach flow is neutral or unstable, it tends to go over the hill rather than around it. An important feature of such a flow is that it speeds up (accelerates) in going over the upwind slope to the hill top. For quantifying this effect, a speed-up factor is often defined as the ratio of the wind speed at some height above the hill to the wind speed at the same height above the flat surface in the approach flow. While the speed-up factor decreases with increasing height, its maximum value near the surface is found to range from just over 1 to about 3, depending on the hill shape, slope, and aspect ratio. The largest speed-up factors are observed over three-dimensional hills of moderate slope. In approaching an isolated hill or a two-dimensional ridge, the flow near the surface first decelerates slightly before reaching the upwind base of the hill and then accelerates in going over the hill slope. The latter phenomenon is commonly observed by mountain climbers who experience the strongest winds upon reaching the mountain top. Of even greater significance are the observed adiabatic cooling of air with increasing height and condensation of moisture into clouds above the lifting condensation level. Cloud condensation and possible precipitation are obviously enhanced by moist ambient air and increasing convection over the heated slope in the afternoon.

Equally interesting phenomena occur on the lee side of hills and mountains. On small and gentle hills, slight adiabatic warming compensates for the slight cooling experienced by the upslope flow on the windward side. However, the downslope winds are decelerated much more than their speed-up in going to the top of the hill. Thus, winds near the downwind base are much weaker than those at the hill top and also somewhat weaker than the winds at the upwind base at the same heights above the ground.

Much larger changes in temperature and relative humidity occur when moist, neutral, and unstable approach flows impact on large hills and mountains. These flows have turbulent boundary layers with typical heights of 1 km or more. When condensation and precipitation occur on the upwind slope, the lee-side flow becomes considerably drier and warmer as it reaches the downwind base. But winds there are still expected to be somewhat weaker than those on the upwind side. There is considerable mixing in the relatively deeper PBL on the lee side. The presence of a strong inversion capping the PBL at a relatively low height above the hilltop can, however, accelerate the flow over the hill and on the downwind side (see e.g., Hunt and Simpson, 1982). On steep lee-side slopes, the flow near the surface may not continue to be downslope; instead, it separates from the surface and reattaches further downstream on the hill slope or even beyond the foot of the hill on the flatter ground. The separated flow region is always characterized by recirculating type of flow with weaker upslope winds near the surface and downslope winds in the upper portion. It is also called the recirculating cavity, which often forms on the lee side of steep hills and ridges. Favorable locations for the flow separation to occur are the salient edges where hill slope increases abruptly, narrow and precipitous hilltops, and deep gorges and canyons.

Upslope and downslope flows are quite different when the ambient atmosphere is stably stratified. Effects of topography on the flow are considerably modified in the presence of stable stratification. The stratification effects are generally described in terms of the Froude number

$$F = \frac{U_o}{NH} \tag{3.1}$$

where U_o is the characteristic velocity of the approach flow (say, at $z = H$), H is the height of the mountain, and N is the Brunt–Vaisala frequency related to the static stability of the ambient approach flow, that is,

$$N = \left(\frac{g}{T_o}\frac{\partial \theta}{\partial z}\right)^{1/2} \tag{3.2}$$

Physically, the Froude number is the ratio of inertia to gravity forces; it has an inverse-square relationship to the bulk Richardson number, that is, $F \sim \text{Ri}_B^{-2}$. Thus, F may be considered as a measure of atmospheric stability ($F \ll 1$ for a strong stability or stratification, and $F \gg 1$ for a weak or near-neutral stratification).

An important aspect of the stably stratified flow approaching a hill or mountain is the increasing tendency of fluid to go around rather than over the topography with increasing stratification (decreasing F). This is because fluid particles may not possess sufficient kinetic energy to overcome the potential energy associated with stable stratification and to lift them over the topography. Whether a given parcel in the approach flow would go over or around the hill depends on the height of the parcel relative to the hill height, the Froude number F, the initial lateral displacement of the parcel from the stagnation streamline, and possibly other flow and topographical parameters. Parcels approaching the hill at low heights are likely to go around, while those near the hilltop may go over the hill. For this, the concept of a dividing streamline that separates the flow passing around the sides of the hill from that going over it has been introduced. For the simpler case of a uniform, inviscid, linearly stratified approach flow, the height of the dividing streamline is given by (Hunt and Snyder, 1980)

$$H_s = H(1 - F) \qquad (3.3)$$

Note that Eq. (3.3) implies that under very strong stratification ($F \ll 1$), most of the flow passes around the hill, while under weaker stratification ($F \geq 1$), all the flow goes over the hill, similar to the case of neutral flow. Laboratory experiments and limited observations of smoke plumes impacting on hillsides have confirmed the validity and usefulness of the dividing streamline concept, even though Eq. (3.3) may not always give the true height of the dividing streamline (see e.g., Bains, 1995; Snyder et al., 1985).

In the case of a strongly stratified ($F \ll 1$) flow approaching perpendicular to a very long (nearly two-dimensional) mountain ridge, there is distinct possibility of the upstream blocking of the fluid, because fluid parcels do not have sufficient kinetic energy to go over the ridge, except those near the top of the ridge. A region of nearly stagnant fluid may be found below the height of the dividing streamline, especially near the centerline and away from the edges where the flow can go around the ridge. Such a blocked flow may not acquire a steady state, because the upstream edge of the blocked flow usually propagates as a density front. Topographically blocked flows have important implications for local weather, as well as for the dispersion of pollutants from upwind sources.

In stably stratified flows over hills and mountain ridges, a variety of flow patterns may develop on the lee side, depending on the Froude number, hill slope and shape, presence of elevated inversion, and wind shear (see e.g., Hunt and Simpson, 1982; Bains, 1995). Perhaps the most spectacular of these are the strong downslope winds as gravity flows, which are sometimes followed by a hydraulic jump. These flows are analogous to the passage of water over a dam spillway, as it rushes down the spillway at very high speeds and creates a spectacular hydraulic jump at the base of the dam. Similar phenomena have been observed in experiments on stratified air flow over hills and also in the real atmosphere (see e.g., Klemp and Lilly, 1975; Hunt and Simpson, 1982; Bains, 1995).

Strong downslope winds often occur when very moist and warm winds are forced up the slopes of tall mountains. Often, these orographically forced flows produce strong, warm, and dry downslope winds, such as chinook, foehn, and Santa Anna. These are not forced by gravity, as in the case of cold air drainage, but are driven by the large-scale flow across a mountain range. When the prevailing flow of warm and moist air is forced up the mountain, the air expands and cools and often condenses to form clouds and precipitation on the windward slope of the mountain. With most of the moisture removed from the air and the latent heat of condensation added to the same, the drier and warmer air begins to descend on the leeward slope. On its way down the slope, the air is further warmed due to compression at the dry adiabatic rate of $9.8°C/km$, which is larger than the moist adiabatic cooling following saturation and condensation on the upwind side. When it arrives at the downwind base, the air is much warmer and drier than that on the same height on the upwind side.

Similar cases of cold downslope winds, called bora, have been observed in different parts of the world, but most notably along the coast of Yugoslavia (Atkinson, 1981, Ch. 3; Bains, 1995). Some of the boras start rather abruptly, while many others develop gradually over several hours as the synoptic scale wind changes to the appropriate direction (perpendicular to the mountain ridge).

Except for the relative temperature change they bring to the downwind bases of mountains, cold (bora) and warm (foehn) downslope winds are similar in many respects. They often start suddenly and abruptly and are strong, gusty, and fairly constant in direction. They all have low relative humidities. Their diurnal and monthly frequencies reveal nighttime and winter maxima. The approach flows typically associated with downslope winds are normal to the mountains and usually contain a stable layer at a critical height (Atkinson, 1981, Ch. 3). Even though the early thermodynamic explanation of the warmth and dryness of foehn may be considered satisfactory, the actual mechanism of the downslope flow is still not clear. Several different mechanisms and theories of downslope winds have been proposed in the literature (Atkinson, 1981, Ch. 3), but we will not go into these here. In general, strong downslope winds are not of much concern to an air pollution meteorologist. Like other wind storms, they provide most rapid transport and efficient dispersion of pollutants.

3.4.3 Lee Waves

A commonly observed flow phenomenon in stably stratified flows over hills and other obstacles are the lee waves, which are produced by the disturbance of the stratified flow by an obstacle in the vertical direction. When an air parcel is forced by the obstacle to rise, a restoring force due to gravity tends to bring it back to its original (equilibrium) level. In fact, due to its downward momentum or the inertia force, the parcel tends to overshoot the equilibrium level and is then further subjected to the buoyancy force to raise it back toward the original level. Thus, the tussel between gravity (buoyancy) and inertia forces produces vertical oscillations in a stratified airstream once it is disturbed by an obstacle. The nature of the fundamental vertical oscillation in a stratified fluid was first analysed by Brunt (1927), who derived an expression for the period or the frequency of the oscillation. The Brunt–Vaisala frequency, defined in Eq. (3.2), is also the natural frequency of topographically produced lee waves, whose period is $2\pi/N$ and wavelength is given by

$$\lambda = \frac{2\pi \bar{u}}{N} \tag{3.4}$$

where $\bar{u}$ is the horizontal velocity of the airstream. In general, $\bar{u}$ and N are functions of height, so that the wavelength of lee waves varies with height, but it is independent of the size and shape of the obstacle. With typical values of $T = 250$ K, $\partial\theta/\partial z = 0.015$ K m^{-1}, and $\bar{u} = 20$ m s^{-1}, $\lambda \simeq 10$ km, a frequently observed value in lee waves (Atkinson, 1981, Ch. 2).

Since N is essentially dependent on static stability ($N = s^{1/2}$), it is clear from Eq. (3.4) that strong stability and low wind speeds would generate lee waves with short wavelengths whereas weak stability and strong winds would favor long wavelengths. Observations of lee waves in the atmosphere have been reviewed by Atkinson (1981) and more recently by Bains (1995). Lee waves have also been observed and studied in stratified flow wind tunnels, channels, and towing tanks (see e.g., Hunt and Simpson, 1982).

Often spectacular clouds with a definite periodicity associated with lee waves downwind of mountain ranges can be seen by any observer. Glider pilots often take advantage of these strong wave motions. Observations of wavy flow and the associated thermal structure have been taken by surface-based instruments, constant-level balloons, glider and power aircraft, radars, lidars, and satellites. As a result of extensive observations since 1950, we now know that lee waves occur in all parts of the world where there is some surface relief or topography to disturb the stable air flow. Observations have confirmed the expected linear relationship between wavelength and mean wind speed and the inverse square-root relationship between wavelength and stability. A wide range of wavelengths from 1 km to 70 km, but with the bulk of them lying between 5 and 20 km, have been observed. The corresponding wave amplitudes range from tens of meters to several kilometers. Very large amplitude waves, with peak to trough vertical distance of 6 to 7 km, have been observed in the stratosphere in the lee of the Colorado Rockies; a hydraulic jump is often associated with such waves.

Vertical velocities associated with mountain lee waves are usually in the range of 2 to 6 m s^{-1}, but extreme values of 15 m s^{-1} have also been reported.

The occurrence of mountain lee waves primarily depends on atmospheric stability and wind speed. A necessary requirement, at least for strong waves, is strong stability at levels comparable to the mountain height. Lee waves are also favored by strong winds perpendicular to the mountain, as well as by strong wind shears. Decrease of wind speed with height (negative wind sheer) generally suppresses lee wave development above the level of the maximum winds.

Depending on the approach flow speed at the height of the mountain, as well as the wind distribution as a function of height, the air flows over and behind mountain ridges have been classified into four different categories: (a) laminar streaming; (2) standing eddy streaming; (3) wave streaming; (4) rotor streaming (see e.g., Atkinson, 1981, Ch. 2). In light winds the flow is characterized by a smooth shallow solitary wave over the ridge. With somewhat stronger winds, a semipermanent larger amplitude solitary wave with a standing eddy of recirculating flow underneath on the lee side of the mountain is formed; this is called eddy streaming. With still stronger winds increasing with height, a lee wave system develops downstream. Rotors can form near the surface under the crests of large-amplitude lee waves. But a more favorable condition for the rotor streaming flow to occur is a strong wind maximum extending through a limited vertical depth, comparable with the height of the mountain, followed by decreasing winds aloft.

Experimental studies of inviscid, uniform, stratified flows over two-dimensional hills in a channel have shown that flow separation can occur on the lee slope of the hill under certain conditions, depending on the Froude number. This separation of flow from the hill surface is actually induced by lee waves and their associated rotor-like vortices near the surface. Since flow reverses and recirculates in the region of flow separation, pollutants from any source in that region would be trapped, resulting in very high concentrations on the lee-side slope and base of the hill. At higher levels, lee waves may break and produce turbulence to cause efficient mixing.

Mountain-wake Circulations The most important mechanically forced circulations associated with topography, so far as air pollution dispersion is concerned, are the so-called wake flows and vortices. The region of flow immediately behind and downwind of an obstacle is called the wake. According to some broader definitions, the wake also includes any regions of flow separation on the sides and in the lee of the obstacle. Sometimes this complicated region of flow separations and reattachments and the associated recirculating cavities is called the near wake. The phenomena of boundary layer flow separation and reattachment on curved and sloping surfaces have been experimentally studied by fluid dynamists and aerodynamists. Wakes behind idealized bluff bodies or obstacles such as cylinders, spheres, discs, and flat plates, placed in a uniform free stream flow, have also been extensively studied in wind tunnels and water channels in the laboratory. Wakes of isolated hills and elongated ridges in the atmospheric boundary layer have been studied in meteorological wind tunnels and other environmental fluid modeling facilities (Arya et al., 1987; Synder et al., 1991). But only a few field experiments with limited scopes and objectives have been conducted (Mason and Sykes, 1979; Karacostas and Marwitz, 1980; Mason and King, 1984; Salmon et al., 1988).

Most of our limited knowledge of the near wakes of hills and ridges in the atmosphere comes from the laboratory experiments in fluid modeling facilities. When the approach flow is a deep turbulent boundary layer, as often occurs in the atmosphere during the daytime, the flow usually separates from the hill sides, but a distinct Karman vortex street with a regular frequency of vortex shedding is usually not detected. The flow separates in the lee of the hill only when the hill slope is steeper than about 1/4. The steeper the hill slope, the larger the region of flow separation. The latter also depends on the aspect ratio of the hill; the larger the aspect ratio, the greater the extent of the recirculating flow, also called cavity. The largest cavity, extending to about ten times the hill height (H), is found for a very steep two-dimensional ridge (Arya and Shipman, 1981). For an axisymmetrical hill, this region of recirculating flow reduces to only about $2H$. The vertical extent of the cavity also depends on the height, the slope, and the aspect ratio of the hill.

When the approach flow is stably stratified (e.g., a shallow stable boundary layer, or an inviscid uniform flow), the lee-side flow separation is generally suppressed by stratification. But it can be induced by lee waves and associated rotors under appropriate conditions (Hunt and Simpson, 1982). The possible influence of wind shear on this is not clear. Since the flow usually separates from the hillsides, the near wake region is almost always turbulent even when the approach flow may not be turbulent. This has important implications for the dispersion of pollutants at nighttime from sources near the downwind base of a hill.

The mean flow and turbulence in the near wakes of hills are strongly influenced by regions of flow separation and associated shear layers that generate a lot of turbulence. These near wake flows are characterized by weak and variable winds and high turbulence intensities. They are typically confined to downwind distances of a few hill heights. At large distances, the wake expands in both the lateral and vertical directions, while the hill-induced flow perturbations decay with increasing distance from the hill. Eventually, the wake relaxes back to the undisturbed approach flow and may not be distinguished as a separate entity. Hill-induced perturbations in mean flow and turbulence in the far wake can typically be detected up to $10H$ for axisymmetrical hills and much farther downstream for nearly two-dimensional ridges (see e.g., Arya and Shipman, 1981). Satellite photographs of vortex streets downwind of small, isolated mountain islands indicate that their wakes often extend several hundred kilometers, that is, more than $15H$ in the downwind direction (Atkinson, 1981, Ch. 4).

Vortex streets typically occur in strongly stratified flows with a low-level inversion, so that fluid goes around the hill more readily than over it. Experimental and theoretical studies of wakes in stratified flows have been reviewed by Turner (1973), Lin and Pao (1979), and Bains (1995). Most of these studies are considered only uniform, laminar, linearly stratified flows over and around idealized two- or three-dimensional hills. Consequently, for a given hill shape and size, wake phenomena are described primarily as functions of Froude number or Richardson number. The effects of wind shear, large surface roughness, and complex irregular topography, such as those commonly encountered in nature, are hardly understood. A few field studies of near-surface flow around isolated hills and ridges have been conducted, but measurements did not extend into the far wake region.

3.4.4 Mesoscale Systems and Dispersion

Mesoscale systems and circulations described in this section have profound effects on transport and diffusion of pollutants when they are released from sources into any of these systems. In general, thermal circulations such as sea and land breezes, urban heat island circulation, and mountain and valley winds are best developed when large-scale gradient winds are weak. Under such conditions, stack plumes will follow the local circulation once they are trapped into it, as schematically shown in Figure 3.12. Note that pollutants can accumulate with time in closed thermal circulations. Often, such mesoscale circulations represent the worst conditions resulting in highest ground level concentrations. At times, the flow patterns and weather associated with mesoscale features may also be beneficial and lead to improved air quality. For example, strong downslope winds in the lee of mountains would cause efficient dispersion of pollutants from sources on their top and downwind bases. The presence of a recirculating flow region on the lee side, on the other hand, can trap the pollutants and lead to high concentrations. These situations are also schematically shown in Figure 3.12.

To correctly represent the pollutant transport in a mesoscale system, one has to know the whole flow field associated with the system. Since diffusion is caused by spatially inhomogeneous turbulence, the latter should also be known. Such a detailed depiction of mean flow and turbulence and the associated dispersion is possible only through the use of mesoscale numerical models (Pielke, 1984). We will discuss some aspects of numerical modeling in later chapters. It should be pointed out, however, that meteorological data collected at one point (say, airport) far away from a source is hardly suitable for describing dispersion when it may be strongly influenced by a mesoscale

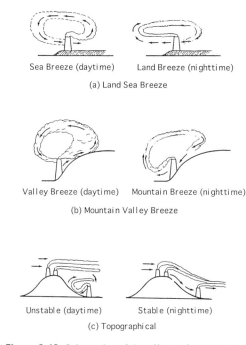

Figure 3.12 Schematics of the effects of mesoscale systems on smoke stack plumes: (a) sea and land breezes; (b) mountain and valley winds; (c) topographically forced circulations. Modified after Pendergast, 1984.

system. Thus, a knowledge of mesoscale systems is necessary for the air pollution meteorologist.

3.5 MICROSCALE SYSTEMS

Coming down to the lower end of the atmospheric scales of motion, we now consider the microscale motions and processes that are confined to the atmospheric boundary layer. Since most of the air pollutants are originally emitted from sources at the ground or within the PBL, the importance of the PBL winds and turbulence to their short-range dispersion cannot be overemphasized. In this section, we give only a brief qualitative description of the mean flow and turbulence characteristics of the PBL and their dependence on the characteristics of the underlying surface and the larger scale flow in which the PBL is embedded. In particular, the possible effects of small-scale surface inhomogeneities on the boundary layer and the associated microscale circulations will be mentioned. More detailed and quantitative descriptions of mean flow and turbulence in the PBL over a homogeneous surface will be given in chapters 4 and 5.

3.5.1 The Idealized PBL over a Uniform Surface

As explained in Chapter 2, large-scale horizontal pressure and temperature gradients near the surface set up the geostrophic or gradient winds in the lower atmosphere. Over small scales of concern in micrometeorology (say, up to 10 km), these large-scale flows can essentially be considered horizontally uniform or homogeneous. If the underlying surface is also uniform and flat, the planetary boundary layer developed on it as a result of the interaction of the atmosphere with the surface will also be horizontally homogeneous. Such an idealized PBL may evolve gradually with time in response to the similar evolution of the large-scale systems. Even when the large-scale flow is steady, the PBL evolves diurnally as a result of the diurnal heating and cooling of the surface. Diurnal changes in the PBL become insignificant or minimal only over open ocean surfaces or extensive snow- and ice-covered surfaces over the polar regions during winter. Over a short period of the order of an hour, even a diurnally evolving PBL over land is often assumed to be quasi-stationary when one considers its short-term average properties (e.g., mean wind speed, direction, turbulence intensities, and the PBL height). This idealized (horizontally homogeneous and quasi-stationary) PBL has been the subject of many theories, models, and experiments and will be described more completely in chapters 4 and 5. Here we will mention only some of the more important features of the PBL that influence pollutant transport and diffusion.

The distribution of mean winds with height in the PBL essentially determines the speeds and directions at which pollutants will be transported at different heights. Since both wind speed and direction vary with height above the surface, a vector depiction showing wind vectors at different heights is often used to show this combined change. Figure 3.13 shows average wind vectors constructed from the data obtained for a number of summer nights and days during the Great Plains Field Program (Lettau and Davidson, 1957). A smooth curve joining the end points of velocity vectors is called a wind hadograph.

It can be noted from Figure 3.13 that changes in wind speed and wind direction with height in the bulk of the PBL are much smaller during the daytime than at night. The PBL heights are not indicated in this figure, but can be estimated to be about 1200 m and 400 m for day and night, respectively. During the mid-day and afternoon periods, the heating of the surface by solar radiation and the associated convective mixing cause the PBL to grow deep and make winds nearly uniform with height. Thus, wind shears in the convective boundary layer (CBL) remain insignificant, except very close to the surface where winds increase rapidly with height. At night, the radiative cooling of the surface and the associated surface inversion suppress turbulent mixing and make the PBL very shallow. In this nocturnal stable boundary layer (SBL) both the wind speed and the wind direction vary significantly with height, with a distinct wind-speed maximum occurring near the top of the PBL. This is also referred to as the low-level nocturnal jet in the literature.

On a short-term basis, both wind speed and wind direction in the PBL have been observed to fluctuate

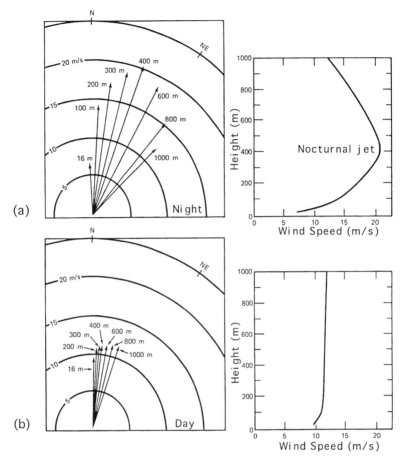

Figure 3.13 Average wind vectors and wind speed profiles constructed from the pilot balloon soundings during the Great Plains Field Program: (a) Night. (b) Day. From Pendergast, 1984.

more or less randomly with time and space. This continuous fluctuation or turbulence is a fundamental characteristic of the PBL that causes the diffusion of pollutants released into it. At any fixed point in space, the instantaneous velocity vector can be divided into a mean velocity vector of the type represented in Figure 3.13 and the superimposed fluctuating velocity vector with components in all three directions. It will be shown later that the diffusion of passive scalars, including pollutant species, is intimately related to the statistics of turbulent velocity fluctuations. In the PBL, turbulence is produced by two mechanisms, the first, involving wind shear, produces mechanical or shear-generated turbulence, and the second, involving buoyant convection, produces convective turbulence. The former is dominant in the case of strong winds blowing over very rough surfaces and in near-neutral stability conditions. The latter is dominant in the case of light or calm winds blowing over a heated surface, that is, in convective conditions. In the nocturnal stable boundary layer, shear is the only turbulence-producing mechanism, while negative buoyancy tends to suppress turbulence. The mean wind shear and shear-generated turbulence are strongest near the surface and generally decrease with increasing height above the surface. The convective turbulence is distributed more uniformly with a broad maximum in the middle of the convective boundary layer.

The top of the PBL is usually defined as the level where the PBL turbulence disappears or becomes insignificant. The interface between the PBL and the free atmosphere is fairly sharp instantaneously, but it becomes somewhat blurred in the average. Since pollutants released anywhere in the PBL are transported and diffused through the whole depth of the PBL, the PBL depth is also called the mixing depth or height. The mixing height varies diurnally in response to the diurnal heating and cooling of the surface. It increases throughout the day, becomes maximum (typically, 1 km) in the late afternoon, and then rapidly collapses to its typical nighttime value of the order of 200 m. Actually, at night mixing height rarely stays constant,

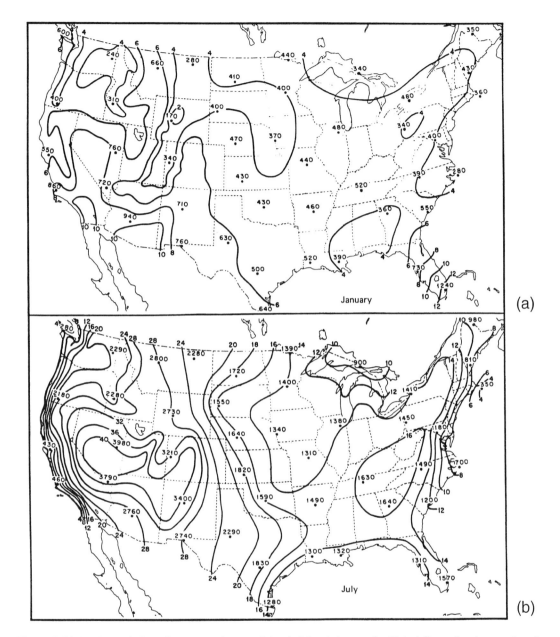

Figure 3.14 Spatial variation of mean maximum mixing heights (m) over the United States for the months of (a) January and (b) July. From Holzworth, 1964.

but usually oscillates around its mean value. The maximum mixing height during the daytime also shows large seasonal and spatial variations, illustrated in Figure 3.14 for the United States. The contours of the mean maximum mixing heights in Figure 3.14 are based on the upper-air sounding data from forty-five locations for the months of January and July. Mean afternoon (00 GMT) mixing heights are largest over the desert areas of the southwest in summer and smallest during the winter season. The largest spatial gradients occur in coastal areas, especially in summer.

3.5.2 Internal Boundary Layers

Quite often, the land surface on which the PBL is developing is not uniform in its roughness (surface cover), temperature, or moisture content. Surface inhomogeneities may be due to different surface covers, such as soil and water, rural countryside and urban development, forests and open areas, and also due to the presence of topographical and man-made features, such as low hills, valleys, fences, and buildings. Since mean flow and turbulence in the PBL are strongly influenced by the characteristics of the underlying surface, one would expect modifications of the former in response to surface inhomogeneities. The modified layer near the surface reflects the characteristics of the new surface and it grows with downwind distance from the line of discontinuity. Such a modified layer growing within the larger PBL is called an internal boundary layer (IBL). In the lower atmosphere, internal boundary layers form and develop due to changes in surface roughness, temperature, and/or moisture content. Over a patchwork of different agricultural fields, for example, one can visualize multiple internal boundary layers forming and growing in succession. At large heights, the individual effects of different surfaces may not be discernible, and the upper part of the PBL represents the integrated influence of many upwind fields or surfaces.

Experimental and theoretical studies of the growth of internal boundary layers following a sudden change in surface roughness or temperature indicate that $h_i \propto x^p$, where h_i is the IBL height and the exponent $p \simeq 0.8$ for the roughness change and $p = 0.5$ for the temperature change. The mean flow and turbulence are characterized by the local surface only within the IBL, while those above the IBL are characterized by the surface upwind of the discontinuity. Internal boundary layers also develop immediately downwind of the separated flow regions behind obstacles and topographical features, such as vertical fences, buildings, and steep hills. These internal boundary layers are often considered as parts of the far wakes of these obstacles.

3.5.3 Obstacle Wakes

Among the topographically forced mesoscale systems we considered the wakes behind large hills and ridges. Similar wakes form behind small hills and ridges as well as shelterbelts, buildings, and other structures. These are considered as microscale systems, because individual obstacle effects may not extend farther than 10 km, the upper boundary of microscales. Small hill wakes are qualitatively similar to large (mesoscale) hill wakes. The Coriolis effects of earth's rotation can safely be ignored in the former. The effects of density stratification may also not be as important as in mesoscale systems. These effects are generally ignored in building wakes, which are dominated by shear-generated mechanical turbulence.

A schematic of the modification of the wind profile and streamlines by a two-dimensional building normal to the flow is given in Figure 3.15. It also shows the characteristic flow regions including those of streamline displacement, recirculating cavity, and wake. More complex microscale flow systems including a horseshoe vortex and a pair of vortices occur around three-dimensional buildings (see e.g., Hosker, 1984; Arya, 1988).

Flows around obstacles are generally characterized by separated flow regions on the sides, at the top, and

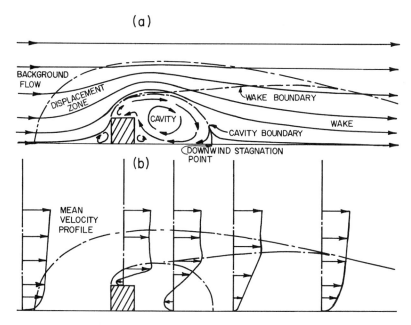

Figure 3.15 Schematic of the various flow regions around a two-dimensional wall or a building: (a) Mean streamline pattern. (b) Mean velocity profiles at various locations along the flow. From Halitsky, 1968.

behind the obstacle and more extensive far wakes in which flow slowly relaxes to its undisturbed characteristics. The extent of the large recirculating cavity in the lee depends on the height and aspect ratios (length/height and length/width) of the obstacle. For rectangular-shaped buildings, empirical relations for cavity dimensions have been obtained from wind tunnel experiments (see e.g., Hosker, 1984).

3.5.4 Microscale Systems and Dispersion

Except for the exhaust emissions from high-flying aircraft and rockets, almost all air pollutants from natural and anthropogenic sources are initially emitted within the PBL. Therefore, their short-range transport and diffusion are essentially determined by mean wind distribution and turbulence within the PBL and the associated microscale systems. Mean winds determine the average speeds and directions of pollutant transports at different heights. Turbulence determines the rates of horizontal and vertical spreads around the centers of pollutant clouds. Depending on the type of release (instantaneous or continuous), turbulent diffusion spreads the material from an individual source in the form of a puff or a plume whose dimensions increase with increasing diffusion or travel time. Eventually, pollutants are mixed throughout the PBL as they are transported to larger and larger distances and dispersed by successively larger scale motions and systems. Thus, the PBL winds and turbulence are vital to the lower atmosphere's capacity of dispersing millions of tons of pollutants that are emitted by natural and anthropogenic sources every day.

For modeling dispersion in the idealized horizontally homogeneous PBL, one needs to know the vertical distributions of mean wind and turbulence, as well as the PBL height. In simpler models, only the mean wind speed and direction at the actual or effective release height, an estimate of mixing height, and some indirect measure of turbulence such as atmospheric stability are required. In other applied dispersion models, mean winds, turbulence, and the mixing height are parameterized using semiempirical and theoretical relations for the PBL. In more sophisticated numerical dispersion models, the numerically computed wind and turbulence fields are used.

Only numerical dispersion models can properly account for the inhomogeneous mean flow and turbulence in the PBL over small-scale surface inhomogeneities including small hills and buildings. For sources located near hills and buildings, it is important to know how the flow and dispersion might be affected by these obstacles. In particular, if pollutants are released or entrained into a recirculating cavity region, very large concentrations are likely to result. Buildings and hills generally increase ground-level concentrations (g.l.c.) from sources located in their wakes. A simple quantitative measure of this obstacle effect is given by the terrain amplification factor, defined as the ratio of the maximum g.l.c. in the presence of the obstacle to that without the obstacle. Wind tunnel experiments have provided empirical estimates of terrain amplification factors for sources located in the vicinity of idealized hills, valleys, and buildings (Arya et al., 1987; Snyder et al., 1991; Thompson, 1993).

3.5.5 Micrometeorology and Characteristic Plume Shapes

Visible smoke plumes from industrial stacks have been observed and photographed under different micrometeorological conditions (see e.g., Gifford, 1968). The characteristics of horizontal spreading and plume meandering are more difficult to observe because they require a vantage point high above the plume. Aircraft and satellites provide such a vantage, while only a limited view is available to ground-based observers. However, the vertical cross sections of smoke-stack plumes can easily be observed and photographed from the ground. The vertical appearance of a plume can offer considerable information on the wind and thermal structure of the lower atmosphere in which the plume is diffusing. Some qualitative correspondence between the atmospheric PBL structure and typical observed plume shapes is schematically shown in Figure 3.16. For this purpose, observed plume shapes are frequently characterized as follows:

a. *Fanning*. As the name implies, the plume acquires the shape of an angular fan, with a large spread in the horizontal and very little spread, if any, in the vertical. This characteristic shape occurs typically at night in a very stable boundary layer with strong surface inversion and weak and variable winds. Strong stability suppresses vertical turbulence and diffusion, while variable wind direction produces a highly meandering plume in the horizontal. This plume meander causes large horizontal spread in the average, especially in the lateral direction. The vertical shear of the horizontal wind direction, which is quite large at night, causes additional horizontal spreading of the shallow plume.

Sometimes, little horizontal spreading occurs at night because of the steady winds with little wind-direction shear and weak turbulence. These conditions produce ribbon-like plumes that can be observed for long distances, because the smoke remains highly concentrated in these plumes. Strong nighttime surface inversions may be considered favorable for elevated releases, because their fanning or ribbon plumes stay aloft and have little impact on a flat ground surface. But they can cause severe impacts on hills and tall buildings that are in their way.

b. *Fumigation*. As the name implies, the plume material gets rapidly fumigated down to the ground level due to downward mixing. This situation occurs shortly after sunrise when the nocturnal inversion is dissipated from below due to surface heating and is slowly replaced by an unstable layer that grows up to the top of the plume. As soon as the plume is entrained into the mixed layer, strong convective turbulence mixes it down while the capping

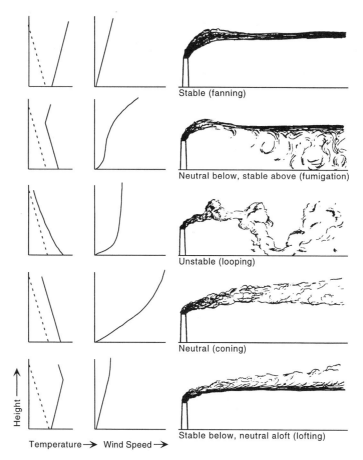

Figure 3.16 Schematic depictions of instantaneous plume patterns in the vertical and the corresponding wind speed and temperature profiles (modified after Slade, 1968).

inversion above forms an effective lid on the upward mixing. This fumigation condition is only transitory, but often leads to the highest ground-level concentrations on a short-term (say, 30 min.) basis. As the mixing height continues to grow well above the stack height, the material is spread over the deeper CBL and ground-level concentrations decrease. It is no longer considered fumigation.

Plume fumigation may also occur steadily and continuously when a narrow plume released from a coastal or lakeshore stack in flow encounters the internal boundary layer developing on the warm land. Such coastal or lakeshore fumigation conditions may persist for hours in a sea breeze or a lake breeze.

c. *Looping*. Plume looping typically occurs in very unstable and convective conditions during midday and afternoon. These conditions often lead to a deep convective boundary layer in which large thermally induced circulations in the form of updrafts and downdrafts span the whole depth of the CBL. These large convective eddies take the plume material in successively upward and downward motions, causing the looping behavior in the vertical. Similar looping behavior also occurs in the horizontal due to horizontal eddy motions. On a short-term or instantaneous basis, a downward loop may cause high concentrations at the ground at relatively short distances from the source. Looping is also responsible for some interesting phenomena associated with the average plume dispersion from surface and elevated sources in the CBL; these will be discussed later.

d. *Coning*. As the name implies, the plume looks like a cone with more or less equal spread in both the vertical and the lateral directions. This usually occurs under cloudy and windy conditions with near-neutral stability and adiabatic temperature profile. Such conditions are frequently associated with the passage of a cyclonic storm or a front, and hence may occur during the day or at night. The near-neutral stability conditions associated with morning and evening transition periods are more transitory and do not necessarily produce coning plumes.

e. *Lofting*. A lofting plume stays aloft the surface inversion, as well as the stable boundary layer. This occurs shortly after the transition from unstable to stable conditions near sunset. When a strong surface inversion develops up to the plume level it prevents the material from diffusing downward and concentrates it into a thin layer at the top of the inversion. Above the inversion, vertical diffusion and plume spreading occur in the residual mixed layer in which turbulence has not been completely dissipated. Sometimes, a strong shear layer forms in the lower part of the residual layer, which may continue to generate turbulence. Thus, the lofted plume may stay thin or become quite thick, depending on the turbulence intensity at that level. Depending upon the stack height and the rate of deepening of the inversion layer, the lofting condition may be very transitory or it may persist for several hours or even throughout the night.

f. *Trapping*. This condition is not shown in Figure 3.16, but a plume can become trapped in different ways. On a flat terrain, stack plumes released in an unstable or convective boundary layer are subjected to vertical diffusion until they occupy the whole PBL. Thus, they are trapped between the earth's surface and the capping inversion at the top of the PBL. In trapped plumes the vertical concentration still decreases with distance due to horizontal diffusion or spreading. Trapping can lead to very high ground-level concentrations when the inversion height is low because of strong subsidence, weak winds, or slow heating of the surface. The g.l.c. is inversely proportional both to the mixing height and the average wind speed in the PBL. Trapping must be taken into account in estimates of diffusion at large distances from any source.

More severe plume trapping occurs in steep valleys and gorges, where plumes can get trapped between the valley floor, the side walls, and the capping inversion. If the air flow is up or down the valley axis, plume trapping may result in nearly uniform but high concentrations that increase with decreasing height of the capping inversion. For a V-shaped valley, the g.l.c. can be shown to be inversely proportional to the square of the inversion height.

PROBLEMS AND EXERCISES

1. a. What is the commonly used classification of atmospheric systems according to length and time scales? Give typical ranges of horizontal and vertical length scales for each class of atmospheric systems.
   b. How is this classification related to the different types of air pollution problems discussed in Chapter 1?
2. On the basis of the global average energy budget of the sun-atmosphere-earth system given in Figure 3.1, tabulate the various components of (a) the atmospheric energy budget and (b) the earth's surface energy budget, in terms of incoming and outgoing energy fluxes as percentages of incoming solar radiation.
3. a. What are the major consequences of the imbalance between incoming and outgoing radiations as a function of latitude?
   b. Why are the equatorial regions of the earth not becoming warmer and the polar regions not becoming colder as a result of the above imbalance?
4. a. Discuss the limitations of the idealized three-cell general circulation model of the atmosphere.
   b. Explain why the winds in the upper troposphere generally blow from west to east in both the hemispheres.
5. a. Discuss the importance of the general circulation of the atmosphere to the long-range transport and dispersion of pollutants on global scales.
   b. Which features of the general circulation are most important in global air pollution problems and why?
6. a. Describe all the modifications that can take place as an mP air mass moves eastward from the Pacific Ocean and travels across the United States in winter.
   b. What are the major air masses that affect weather and air quality along the east coast of the United States in summer?
7. Draw plan and side views of a typical cold front, warm front, and occluded front. Include in each diagram relative temperatures on each side of the front and areas of precipitation.
8. a. What are the necessary ingredients for a wave cyclone to develop explosively into a big storm?
   b. Why do wave cyclones dissipate after they become occluded?
9. a. Discuss the importance of synoptic weather systems to the transport and dispersion of pollutants.
   b. What type of synoptic conditions would lead to the worst air quality over a region?
10. a. Explain how a sea breeze circulation begins in the morning and matures in the afternoon.
    b. How would transport and dispersion of pollutants from a coastal stack be affected by the sea breeze circulation when large-scale geostrophic winds are weak?
11. a. Discuss the possible effects of urban heat-island-induced circulation on dispersion from sources in the city center at nighttime, especially when the geostrophic flow is weak.
    b. How does the dispersion in the city differ from that in the surrounding countryside, particularly at night? Consider the differences due to different stabilities and mixing heights.
12. a. In explaining the mountain and valley winds, we considered only the primary thermal circulation normal to the valley axis. Discuss the possible influences of a valley sloping up or down along its axis.
    b. Discuss the importance of thermally induced along-valley flows to dispersion from a source at the mouth of a steep mountain valley.
13. a. Discuss the phenomenon of flow separation with particular regard to air flow over and around an isolated mountain or a long ridge.
    b. How would flow separation influence the dispersion of pollutants from sources at the base and top of the mountain?
14. a. How does thermal stratification affect the air flow over and around an isolated hill?
    b. For an elevated source located at the base of an isolated hill, describe the meteorological conditions (wind speed, wind direction, stability, and mixing height) that might lead to the highest g.l.c.
15. An industrial smoke stack of 100 m is emitting a thin plume toward an isolated bell-shaped hill of 200 m height when the ambient wind speed is 10 m s^{-1} at 200 m and above and the air temperature at the hilltop is 17°C.

Describe different regimes of air flow over and around the hill, as well as plume dispersion, as the surface inversion begins to form and intensify as a result of radiative cooling of the surface following the evening transition.

16. a. Distinguish between the near wake and the far wake of a steep hill or mountain.

   b. What are the qualitative effects of stratification on flow separation and the far wake of a steep hill? How do these affect dispersion from downwind sources?

17. Define an idealized PBL and schematically show how wind speed and wind direction vary with height in such a PBL under different stability conditions.

18. a. Show how a step change in surface roughness and/or surface temperature may lead to the formation of an internal boundary layer.

   b. Give two examples of internal boundary layers and how they might influence the plume dispersion from an elevated source located upwind of a step change in the surface property.

19. a. If a fume hood is vented at the top of a low building, discuss the flow regimes that might influence its subsequent dispersion and ground-level concentrations.

   b. Qualitatively explain how building and small hill wakes might influence dispersion from nearby sources.

20. a. Schematically show how plume fumigation may occur near a shoreline.

   b. Describe the most favorable and the least favorable meteorological conditions for high ground-level concentrations due to an elevated source.

4

Micrometeorology and Planetary Boundary Layer

4.1 INTRODUCTION AND DEFINITIONS

Short-range dispersion of pollutants released from near-surface sources is essentially determined by small-scale motions and processes occurring in the lowest layer of the atmospheric, called the planetary boundary layer (PBL) or the atmospheric boundary layer (ABL). The PBL is formed as a consequence of interactions between the atmosphere and the underlying land or water surface over short time scales of the order of an hour to a day. Sharp variations in wind speed, temperature, and scalar concentrations often occur near the surface due to these air–surface interactions. The physical and thermal properties of the underlying surface, in conjunction with the dynamics and thermodynamics of the lower atmosphere (troposphere), determine the PBL structure including its depth, wind and temperature distributions, transport, mixing and diffusion properties, and energy dissipation. For a given air mass and synoptic weather situation, surface roughness and topography determine the frictional resistance or drag that the surface exerts on the atmosphere and, hence, the exchange of atmospheric momentum to the surface. Similarly, thermal properties of the surface, such as albedo, emissivity, and moisture content, determine the thermodynamical properties of the PBL and related exchanges of heat and water vapor between the earth and the atmosphere. A schematic of the PBL as the lower part of the troposphere and the underlying land and water surfaces is given in Figure 4.1

The PBL thickness is quite variable in both time and space, particularly over land surfaces, ranging from several tens of meters to a few kilometers. It is commonly referred to as the mixing depth or mixing height in air pollution literature, because pollutants released from near-surface sources are quickly mixed up to the top of the PBL and are usually confined to the PBL due to lack of mixing in the inversion layer above. Largely in response to the strong diurnal cycle of heating and cooling of land surfaces in fair-weather conditions, boundary-layer thickness and other characteristics also display strong diurnal variations. In particular, the mixing depth typically varies from a low value of the order of 100 m (range 20–500 m) during nighttime and early morning hours to its maximum value of the order of 1 km (range 0.2–5 km) in the late afternoon. Diurnal variations of the mixing height and other meteorological variables are found to be much smaller over large lakes, seas, and oceans, because of the small diurnal changes of the water surface temperature due to large heat capacity of the mixed layer in water.

Other temporal variations in the boundary layer depth and structure often occur as a result of the evolution and the passage of mesoscale and synoptic scale systems. Generally, the boundary layer becomes thinner under the influence of large-scale subsidence (downward vertical motion) and low-level horizontal divergence associated with the passage of a high-pressure system (anticyclone). On the other hand, the PBL can grow to be very deep and merge with towering clouds in disturbed weather conditions that are associated with low-pressure systems. It is often difficult to even define and distinguish the PBL top under these conditions; the cloud base is generally used as an arbitrary cutoff for boundary layer studies. For diffusion applications, however, air pollutants may not be considered to be confined to the subcloud layer, since these can be transported to large heights in the troposphere through cloud-venting processes.

Spatial variations of the boundary layer depth and structure occur as a result of changes in land use and topography of the underlying surface and also due to spatial variations of large-scale meteorological variables. The latter are usually ignored in boundary-layer

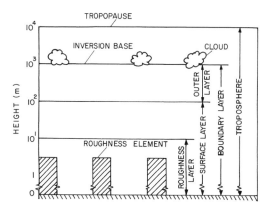

Figure 4.1 Schematic of the planetary boundary layer (PBL) as the lower part of the atmosphere.

studies, except insofar as they determine the external forcing parameters (e.g., geostrophic winds and thermal winds) for the PBL. Spatial changes in land use produce variations in surface roughness, temperature, moisture content, and radiation properties of the surface, which can lead to a myriad of the so-called internal boundary layers, each associated with a different type of surface. For the sake of simplicity, small-scale inhomogeneities of the surface are almost always ignored and the surface is characterized by some bulk, spatially averaged properties, such as roughness, temperature, and so on. The desired length scale for surface characterization may vary over a wide range (say, 100 m to 100 km), depending on the scale of interest.

Micrometeorology or boundary-layer meteorology is a branch of meteorology that deals with microscale, small-scale, or local-scale phenomena and processes occurring in the atmospheric boundary layer. It also deals with the important exchanges of heat, mass, and momentum occurring continuously between the atmosphere and the earth's surface. Vertical distributions of meteorological variables such as wind, temperature, and humidity, as well as trace gas concentrations and their role in the energy balance near the surface, also come under the scope of micrometeorology. Those aspects of micrometeorology and the PBL that are particularly relevant to the dispersion of pollutants in the lower atmosphere are reviewed in this chapter. A more detailed theoretical and experimental description of the statistical properties of irregularly (randomly) fluctuating, small-scale turbulent motion will be given in the next chapter.

Turbulence characterizes the apparently chaotic nature of fluid flows, which is manifested in the form of highly irregular, almost random, temporal and spatial fluctuations in velocity, temperature, and scalar concentrations around their mean values. The atmospheric turbulence can be felt and seen in the gustiness of winds that causes flags to flutter, blades of grass and branches of trees to sway, and smoke from chimneys and stacks to meander and disperse. The atmospheric motion in the PBL being always turbulent (intermittently, if not continuously), micrometeorologists and air pollution meteorologists must have some general understanding of turbulent exchange and mixing processes in fluid flows, particularly those of the atmosphere.

4.2 EARTH-ATMOSPHERE EXCHANGE PROCESSES

4.2.1 Exchange of Energy

The ultimate source of atmospheric energy and motions is the sun. Only a small part of solar radiation is directly intercepted by the atmosphere, primarily by clouds and aerosols. Most of it reaches the earth's surface and is partly absorbed in a thin layer of the subsurface medium, vegetation, and surface structures and is partly reflected back, depending on the surface albedo. The exchange of energy between the earth's surface and the atmosphere is primarily through turbulent motions in the lower part of the PBL, called the surface layer (see Fig. 4.1) Some of it is in the form of direct or sensible heat due to the temperature difference between the surface and the air above. The other form of energy transfer is the latent heat of evaporation or condensation from or to the surface, which is also governed by turbulent motions in the surface layer. Turbulent exchanges of heat and water vapor dominate over molecular exchanges except in the immediate vicinity (within a few millimeters) of material surfaces, such as soil, concrete, plant leaves, water, and so on. The primary mode of heat transfer through an interfacial molecular sublayer is conduction, similar to that in solids. Outside the molecular sublayer, however, the primary mode of heat transfer is convection through air motions. Sensible heat is normally transferred down the mean temperature gradient, that is, from a warm surface to colder air and vice versa. Similarly, water vapor or latent heat is transferred down the mean gradient of specific humidity or mixing ratio, that is, from a moist surface to drier air and vice versa.

Equally important and comparable to the above turbulent exchanges of energy are the radiative exchanges between the earth's surface and the atmosphere. A fraction of the shortwave radiation (wavelengths 0.15–4.0 μm) received by the surface is reflected back by the surface, depending on its shortwave reflectivity or albedo. More important, the surface continuously emits longwave (wavelengths 3–100 μm) radiation depending on its emissivity and the skin surface temperature. This is commonly referred to as the terrestrial radiation. Radiative properties of natural surfaces have been discussed in more detail and also tabulated in other books on micrometeorology (Arya, 1988; Stull, 1988) and microclimatology (Rosenberg et al., 1983; Oke, 1987). The other component of the surface radiation balance is the longwave atmospheric radiation received by the surface. The atmosphere absorbs much of the longwave terrestrial radiation and a substantial part of the longwave solar radiation. At the same time, it emits longwave radiation, part of which is received and absorbed by the surface. An important aspect of atmospheric radiation is that absorption and emission of radiation by various gases occur in a series of discrete wavelengths or narrow bands of wavelengths. All regions of the atmosphere contribute to the longwave radiation received at the surface, the bulk of it coming from the PBL.

Using the principle of conservation of energy, the various forms of energy exchanges between the earth's surface and the atmosphere can be combined in an expression of the energy budget for an ideal surface, which is assumed to be flat, bare, and opaque to radiation,

$$R_N = H + H_L + H_G \quad (4.1)$$

where each term represents an energy flux (energy per unit time, per unit area of the surface) normal to the surface; R_N is the net radiation received by the surface; H and H_L are the sensible heat flux and the latent heat flux to the atmosphere, respectively; and H_G is the heat flux to the ground or subsurface medium. Here we use the sign convention that radiative fluxes directed toward the surface and other energy fluxes directed away from the surface are positive and balance each other. An ideal surface is assumed to be a thin plane with no heat absorption capacity.

For a finite-thickness interfacial layer with finite mass and heat capacity, a more appropriate energy balance equation is

$$R_N = H + H_L + H_G + \Delta H_S \quad (4.2)$$

in which the storage term ΔH_S represents the rate of change of internal energy of the layer per unit horizontal area of the surface, which may be expressed as

$$\Delta H_s = \int_0^D \frac{\partial (\rho c T)}{\partial t} dz \quad (4.3)$$

Here the integral is over the layer depth D, the product ρc is the volumetric heat capacity, and T is the absolute temperature of the layer medium. Note that ΔH_s is positive when the layer is warming with time and vice versa. Equation (4.2) is found to be more appropriate for crop and forest canopies, urban and suburban canopies, and water surfaces, which do not meet the requirements of an ideal surface. In vegetative canopies, the energy storage term may be quite important and the latent heat exchange is due to the combined effects of evaporation from the soil surface and transpiration from leaves; their combination is called evapotranspiration. For an urban canopy, ΔH_S includes the rate of heat storage in street canyons, as well as by all the exposed surfaces of buildings and structures.

The energy balance equation (4.1) or (4.2) tells us how the net radiation received at the surface or at the top of a surface canopy might be partitioned between the various other energy exchanges with the atmosphere and the subsurface medium. Over land surfaces

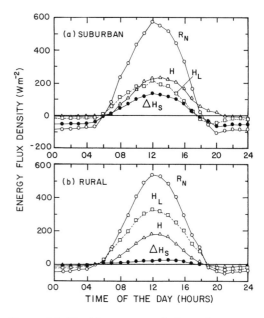

Figure 4.2 Monthly averages of the various energy fluxes at suburban and rural sites in Greater Vancouver, Canada, during summer. From Oke, 1987.

and under clear skies, all the energy fluxes show strong diurnal variations in response to the diurnal variation of net radiation, which is dominated by the incoming solar radiation in daytime ($R_N > 0$) and by the difference between atmospheric and terrestrial radiations at right ($R_N < 0$). The nighttime radiative and other energy fluxes are generally an order of magnitude smaller than and opposite to the sense (direction) of the daytime fluxes. This is illustrated in Figure 4.2 showing monthly averages of the various energy fluxes (here H_G is combined with ΔH_S) for suburban and rural sites in Greater Vancouver, Canada, during summer (see Oke, 1987, for further details of measurements and for an explanation of urban-rural differences).

The surface energy balance equation (4.1) is often used for estimating the sum of sensible and latent heat fluxes from measurements or calculations of the net radiation and the ground heat flux. Sometimes, H_G is simply assumed to be proportional to R_N with the proportionality coefficient determined empirically for a given site. Further separation of the estimated total turbulent energy flux into sensible and latent heat fluxes can be made if their ratio, called the Bowen ratio $B = H/H_L$, can be estimated independently (e.g., using the Bowen ratio method to be discussed later).

4.2.2 Exchange of Mass

The most important exchanges of mass or material between the earth's surface and the atmosphere are those of water; carbon dioxide, methane, and other greenhouse gases; particulate matter, and other pollutants.

The atmosphere receives virtually all of its water vapor through evaporation from soil and water surfaces and evapotranspiration from vegetation. This water vapor transfer is an important component of the surface water budget and the hydrologic cycle. It is also an essential ingredient of all weather involving fog, haze, clouds, and precipitation. We have already shown in the previous section that the latent heat flux associated with evapotranspiration ($H_L = L_e E$, where L_e is the latent heat of vaporization and E is the rate of evapotranspiration) is also an important component of the surface energy budget. Some of the water vapor is condensed onto cloud condensation nuclei to form clouds and returns to the earth in the form of precipitation. Vapor and liquid water often react with gaseous pollutants to form new species, such as sulfuric and nitric acids, in clouds and in precipitation. Precipitation scavenging and washout are important processes in the periodic cleansing of the atmosphere. Unfortunately, the same processes also lead to the acidification and pollution of lakes and reservoirs. On the beneficial side, some pollutants washed out in rain act as fertilizers for soil and vegetation.

Carbon dioxide is another important gas involved in the atmosphere–biosphere exchanges. Its atmospheric concentration has steadily been rising due to ever increasing use of fossil fuels in energy production and other industry, home heating and cooking, and forest clearing. This, in conjunction with similar increasing trends in the concentrations of methane and nitrogen oxides, threatens a global climate change (warming) caused by the so-called greenhouse effect. Climatically less important but more harmful toxic substances are released into the atmosphere from widespread fertilizer and pesticide applications to lawns and vegetation.

Turbulent exchange of particulate matter, including dust, salt from sea spray, pollen, and pathogens, to the atmosphere has important implications for weather, climate, biological processes, and spread of plant diseases. Particulates play an important role in the formation of haze, fog, and clouds and, hence, on the radiation balance near the surface.

4.2.3 Exchange of Momentum

Turbulent exchange of atmospheric momentum to the earth's surface is a consequence of the frictional drag exerted by the surface. The atmospheric boundary layer also forms as a consequence of this momentum exchange and may be thought of as exerting an equal but opposite drag on the surface. Air, being a viscous fluid, must come to rest right at the surface. This has important implications for the wind distribution near the surface, which always increases with height in the lower part of the boundary layer. Both the low-level wind shear and the surface drag increase with increasing surface roughness, as well as with increasing wind speed at the top of the PBL or some other reference height (say, 10 m). Another consequence of this momentum exchange and wind shear is the generation of turbulence, which in turn influences other turbulent exchange and mixing processes in the boundary layer.

Over water surfaces, such as lakes and oceans, wind drag generates surface waves and currents in water; in coastal areas it causes storm surges and beach erosion. Over land surfaces, atmospheric drag exerts wind loads on vehicles, buildings, and other structures. Surface drag and momentum exchange are also responsible for dissipating a substantial part of the large-scale atmospheric kinetic energy. This is vividly manifested in the rapid weakening and subsequent demise of hurricanes and other oceanic storms as they move inland.

4.3 VERTICAL DISTRIBUTIONS OF THERMODYNAMIC VARIABLES

Vertical distributions (profiles) of thermodynamic variables, such as air temperature, specific humidity, and density, are very important in determining the variations of wind speed and wind direction with height in the PBL, as well as its turbulent structure and mixing properties. Quite often, the boundary layer height or mixing depth is also determined from vertical sounding of temperature and humidity. Here, an overbar will be used to denote the appropriate mean or average value of the variable; different types of averages will be defined later.

4.3.1 Temperature Profiles

The variation of mean air temperature ($\overline{T}$) with height in the PBL depends on many factors, such as the ground surface temperature, air temperature at the top of the PBL, divergence of sensible heat flux and net radiation, warm or cold air advection, and local heating and cooling associated with phase changes of H_2O. Quantitatively, it is given by Eq. (2.28) for the conservation of energy in an elementary volume. In fair weather (no phase changes) and horizontally homogeneous (no advection) situations, that equation is simplified to

$$\frac{\partial \overline{T}}{\partial t} = \frac{1}{\rho c_p}\left(\frac{\partial R_N}{\partial z} - \frac{\partial H}{\partial z}\right) \quad (4.4)$$

in which the radiative flux-divergence term is likely to be significant only near the surface during nighttime calm or weak wind conditions. For all practical purposes, warming or cooling of air in the boundary layer is essentially due to the divergence of sensible heat flux, which in turn depends on the surface heat flux (H_o) and the PBL depth (h). Integrating Eq. (4.4) with respect to z and recognizing that the turbulent heat exchange must vanish at the top of the PBL, one obtains a simple expression for the surface heat flux

$$H_o = \rho c_p \int_0^h \frac{\partial \overline{T}}{\partial t} dz = \rho c_p h \left(\frac{\partial \overline{T}}{\partial t}\right)_m \quad (4.5)$$

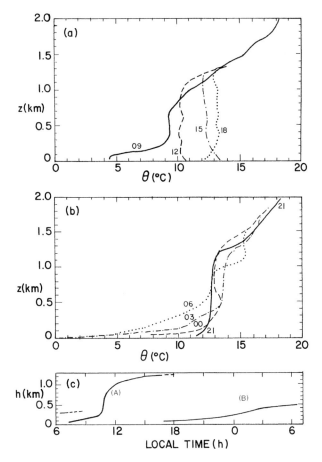

Figure 4.3 Diurnal variation of potential temperature profiles and the planetary boundary layer (PBL) height during (a) Day 33 and (b) the night of Days 33–34 of the Wangara Experiment. (c) Curve A is the PBL height during daytime and curve B is the surface inversion height at night. After Deardorff, 1978.

where $(\partial \overline{T}/\partial t)_m$ is the mean rate of warming of the PBL. The above expression can be used for determining the surface heat flux from temperature soundings, which are also used to estimate the PBL depth. This method is more useful and valid for the more vigorous daytime boundary layer over the heated surface than it is for the weaker nocturnal boundary layer over the cooled surface, because advection and radiative terms are found to become more significant in the latter.

A typical sequence of radiosonde-measured profiles of potential temperature, which is more appropriate than the actual temperature for identifying the layered structure of the PBL, during the course of a day and night in fair-weather conditions is shown in Figure 4.3. These were obtained during the 1967 Australian Wangara Experiment. Shortly after sunrise, the surface warms up and begins to transfer heat to the boundary layer, which warms up and grows with time as a result of this heat input, convective mixing, and entrainment of the nonturbulent free-atmospheric air at the top of the PBL. The daytime convective boundary layer (CBL) has a three-layered structure (see Fig. 4.3(a)) with a shallow surface layer in which potential temperature decreases with height, followed by a much deeper mixed layer in which potential temperature is nearly uniform, and a shallow transition layer of temperature inversion. The topmost transition layer gradually merges into the free atmosphere, which usually has a smaller potential temperature gradient than the transition layer. From temperature soundings alone, the top of the PBL is not as distinct as the top of the mixed layer, which is also the base of inversion; the height of the inversion base (z_i) is often used as an approximation for the PBL height (h), although the former can be substantially (5–25%) smaller than the latter. The often-neglected transition layer is characterized by weak turbulent mixing and diffusion, strong wind speed and direction shears, and entrainment of free atmospheric air into the CBL.

The nighttime profiles in Figure 4.3(b) show that, shortly after sunset, as the surface begins to cool, the boundary layer transfers heat to the surface and therefore the PBL also cools with time. The surface inversion layer initially grows with time and may reach a maximum height of several hundred meters in early morning hours. However, the surface inversion height (h_i) may not correspond to the mixing height (h) of the stable boundary layer (SBL) based on significant turbulence activity. The inversion height is more easily detected from temperature soundings (Fig. 4.3) than the actual mixing height; one needs remote sensing instruments such as sodar and lidar to monitor the latter. Remote and in-situ measurements of turbulence in the SBL indicate that the nocturnal stable boundary layer is relatively shallow. The SBL is also subdivided into a very shallow (depth 0.1 h) surface layer of large potential temperature gradient and constant heat flux, and an outer layer of much weaker, intermittent, and patchy turbulence that is often mixed with or associated with internal gravity waves. The SBL is often capped by the remnant of the previous afternoon's mixed layer of more or less uniform potential temperature. Mixing properties of this layer depend on the mean wind shear and the potential temperature gradient in the layer. This relatively thick (compared with the SBL) layer is also called the residual layer. The residual layer should be considered part of the atmospheric boundary layer on time scales longer than a day, because it is involved in the heat and momentum exchange processes between the atmosphere and the earth's surface. Consequently, meteorological variables show strong diurnal variations in the PBL, including the residual layer.

A schematic of the diurnal variation of the boundary layer height and structure in fair-weather conditions is shown in Figure 4.4. The surface layer is typically 10 percent of the boundary layer (CBL or SBL) and is characterized by large gradients of temperature and wind speed, a constant wind direction, and nearly constant fluxes of heat and momentum above the tops of dominant surface roughness elements. The daytime mixed layer is characterized by uniform and vigorous turbulent mixing, which is caused by surface heating and convection. The nocturnal stable boundary layer, on the other hand, is much shallower and is characterized by weaker turbulent mixing associated with wind shear and wave activity. One should recognize the distinction between the nocturnal mixing layer and the surface inversion layer, which have somewhat different evolution histories. The former is more appropriate in modeling dispersion of pollutants from near-surface sources. The residual layer is essentially disconnected from the surface at night, but it may still have some intermittent and patchy turbulence, probably confined to certain shear layers or regions of wave breaking.

4.3.2 Specific Humidity Profiles

The vertical distribution of water vapor mixing ratio or specific humidity in the PBL depends on the moisture content of the surface, evaporation rate, specific humidity of air mass above the PBL, advection of moisture, and local sources associated with phase changes of H_2O. Quantitatively, it is given by Eq. (2.27), the conservation of moisture in an elementary volume. In the absence of significant advection of moisture and local sources, such as fog and clouds in the PBL, specific humidity and moisture flux are simply related as

$$\frac{\partial \overline{q}}{\partial t} = -\frac{1}{\rho} \frac{\partial E}{\partial z} \quad (4.6)$$

$$E_o = \rho \int_0^h \frac{\partial \overline{q}}{\partial t} dz = \rho h \left(\frac{\partial \overline{q}}{\partial t} \right)_m \quad (4.7)$$

Equation (4.7) may be used for estimating the rate of evaporation at the surface from humidity soundings in the PBL, particularly when significant evaporation or evapotranspiration is taking place. An observed se-

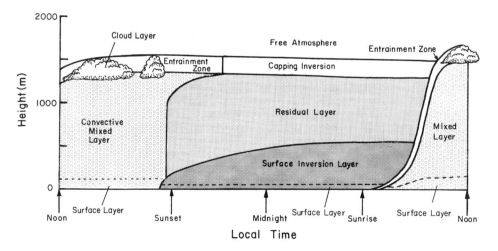

Figure 4.4 Schematic of the diurnal variation of the planetary boundary layer height and structure in fair-weather conditions. Modified after Stull, 1988.

83 VERTICAL DISTRIBUTION OF WINDS IN THE PBL

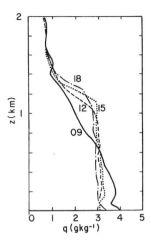

Figure 4.5 Diurnal variation of specific humidity profiles during Day 33 of the Wangara Experiment. From Andre et al., 1978.

quence of specific humidity profiles from the Wangara Experiment (Day 33) is shown in Figure 4.5. During this period of draught, the surface was rather dry, with little or no growth of vegetation. Note that, with little evaporation from the surface, specific humidity profiles are nearly uniform in the daytime convective boundary layer; $\bar{q}$ decreases with time in response to the increasing depth of the mixed layer. In other situations, where there is plenty of moisture available for evaporation, the diurnal variation of humidity profiles might be quite different from those in Figure 4.5.

4.4 VERTICAL DISTRIBUTION OF WINDS IN THE PBL

4.4.1 Factors Determining the Wind Distribution

The magnitude and direction of near-surface winds and their variations with height (wind shears) in the PBL are of utmost interest in air pollution meteorology, because they determine the mean transport speed and direction of pollutants released from near-ground sources. The so-called ventilation factor is the product of mean transport velocity and mixing height. The most important factors determining the wind distribution in a fairweather PBL over a homogeneous surface are the large-scale horizontal pressure and temperature gradients (alternatively, geostrophic and thermal winds), the surface friction and roughness characteristics, the earth's rotation, and thermal stratification (stability) caused by the diurnal heating and cooling cycle. In disturbed weather conditions, horizontal and vertical advections of momentum, as well as cloud and precipitation processes, are additional factors. Also, over a nonuniform or complex terrain, surface topographical features strongly influence the wind distrib-

ution in their vicinity, creating a variety of local circulations, such as separated flow and vortex (cavity) regions, upslope and downslope flows, and regions of flow acceleration (e.g., on a hill or ridge top) and deceleration (e.g., in the wake of a hill or a building).

The PBL flow is essentially driven by large-scale horizontal pressure gradients or geostrophic winds as defined by Eq. (2.37). Geostrophic winds are independent of height in a barotropic PBL, but vary with height in more general baroclinic PBLs, in response to horizontal temperature gradients. Geostrophic wind shears are given by the thermal wind equations (2.38). Even small horizontal temperature gradients can cause significant geostrophic wind shears (in both speed and direction) across the PBL. It can be shown that in the northern hemisphere, geostrophic wind turns cyclonically (backs) with height in the presence of cold air advection, and it turns anticyclonically (veers) with height in the presence of warm advection. Actual winds within the PBL may also be expected to back or veer with height in response to the backing or veering of geostrophic winds, but not to the same extent. The relationship between geostrophic and actual winds in the PBL depends on the surface friction, as well as on momentum exchange and mixing processes in the PBL. For a given ambient geostrophic flow, the very presence of a land or water surface at which relative wind speed must vanish because of the viscosity effect, leads to the typical boundary-layer wind profile with wind speed increasing with height, at least in the lower part of the PBL (see e.g., Fig. 4.6). Conse-quently, momentum is transferred down layer by layer and ultimately to the surface. Turbulence provides an efficient mechanism for this exchange of momentum which, in turn, causes the so-called frictional force or stress on every fluid element in the PBL. Because of the presence of the frictional force, the simple geostrophic balance between the pressure gradient and Coriolis forces that normally occurs outside the PBL cannot be expected to hold within the PBL (see e.g., Arya, 1988, Ch. 6).

In the absence of local and advective accelerations, the equation of mean horizontal motion becomes

$$2\bar{\rho}\Omega \times (\bar{V} - \bar{G}) = \frac{\partial \tau}{\partial z} \qquad (4.8)$$

in which Ω is the earth's rotational velocity vector, $\bar{V}$ and $\bar{G}$ are the actual and geostrophic wind vectors, and τ is the shear stress vector. The above equation also expresses the balance between the Coriolis force directed normal to $\bar{V}$, the pressure gradient force directed normal to $\bar{G}$, and the friction force directed normal to the ageostrophic wind $(\bar{V} - \bar{G})$, all forces being expressed per unit volume of the fluid element. This force balance requires that wind speed and direction must change with height in response to the normal decrease of the friction force with increasing height.

4.4.2 Observed Wind Profiles

The diurnal cycle of heating and cooling of the surface causes diurnally varying thermal stability, which

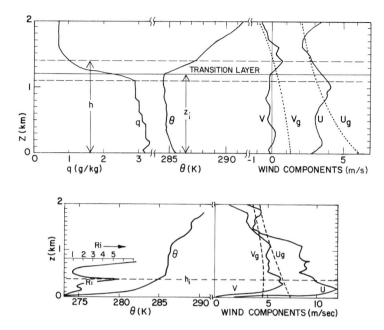

Figure 4.6 Measured wind, potential temperature, and specific humidity profiles in the planetary boundary layer. (a) convective conditions on Day 33 of the Wangara Experiment. (b) very stable conditions on the night of Day 34 of the Wangara Experiment. Richardson number profile is also shown. From Deardorff, 1978.

strongly affects turbulent mixing and momentum exchange processes and, hence, the wind distribution in the PBL. This overwhelming influence of stability can be seen in the observed wind profiles shown in Figure 4.6 for the typical daytime convective conditions and the typical nighttime stable conditions, both taken under the same synoptic situation dominated by high pressure, on Day 33 of the Australian Wangara Experiment. The corresponding potential temperature and specific humidity profiles are also given. Here $\bar{u}$ and $\bar{v}$ are the mean (based on the spatial averaging of five pibal soundings) horizontal wind components in x and y directions, respectively, with the x axis parallel to the near-surface wind. Note that, despite the large geostrophic wind shears present at the time of these observations, wind profiles in Figure 4.6(a) show the characteristic features of nearly uniform wind in the convective mixed layer and large gradients in the shallow surface layer below and the transition layer above the mixed layer. There is little, if any, change in wind direction in the CBL, which has a typical midday depth of about 1200 m.

In contrast with the CBL, the wind profiles in Figure 4.6(b) for the nighttime stable conditions show a much shallower SBL that is characterized by strong wind shears and a low-level jet of supergeostrophic winds. The occurrence of the low-level jet is a commonly observed feature of the nocturnal boundary layer. The nose of the jet often coincides with the top of the SBL to which boundary-layer turbulence extends. Note that this is often well below the top of the surface inversion layer. Thus, one needs to distinguish between the PBL mixing height (h) and the surface inversion height (h_i). Sometimes the distinction between the two is blurred by the occurrence of turbulence and mixing due to the strong negative shear in the upper part of the low-level jet. This mixing region may not strictly be a part of the SBL, but it is probably as effective in the dispersion of pollutants as the SBL. For applications in dispersion modeling, the mixing height should also include any significant turbulent mixing layer above and adjacent to the SBL. Remote sensing devices, such as acoustic sounder (sodar) and lidar, can help identify such mixing layers and, hence, provide an estimate of the mean mixing height over an appropriate sampling time. In the absence of continuous monitoring, however, periodic soundings of temperature, humidity, and winds might be used to estimate the mixing height.

4.5 TURBULENCE

4.5.1 Generation and Maintenance of Turbulence

The criteria for the likely presence or absence of turbulence in a stably stratified shear layer are given in terms of the shape of the wind profile and the gradient Richardson number

$$\mathrm{Ri} = \frac{g}{\bar{T}_v} \frac{\partial \bar{\theta}_v}{\partial z} \left| \frac{\partial \overline{V}}{\partial z} \right|^{-2} \tag{4.9}$$

which may be considered as the ratio of buoyancy to shear forces. Richardson number is also an appropriate dynamic stability parameter, because a necessary condition for dynamic instability (exponential growth of wavelike disturbances introduced in the flow) of a stably stratified flow is that $Ri < Ri_c = 0.25$. Another necessary condition for the instability of a boundary-free shear layer is the presence of an inflection point in the velocity profile. The latter condition is normally satisfied in the SBL, as well as in the upper shear layer associated with the nocturnal jet. Therefore, the primary criterion for the dynamic instability of the lower part of the atmosphere is that $Ri < 0.25$. Wind shear is the primary cause of dynamic instability of an otherwise statically stable layer ($Ri > 0$). On the other hand, any statically unstable layer ($Ri < 0$) is also dynamically unstable; here the primary instability mechanism is gravitational.

The presence of one or more of the above instability mechanisms often leads to irregular, nearly random, three-dimensional, fluctuating motions called turbulence. The subsequent maintenance of turbulence in a stably stratified layer also requires the continued presence of wind shear so that Ri remains below some critical value. The Richardson-number criterion for this is not so well established theoretically or observationally. The estimates of the critical Richardson number range from 0.2 to 0.5 (Arya, 1972; Stull, 1988). A large positive value of $Ri > Ri_c$ is indicative of weak and possibly decaying turbulence, or a nonturbulent environment. For example, the average Ri profile in Figure 4.6(b) indicates that turbulent mixing is most likely confined to a very shallow ($h < 100$ m) stable boundary layer in which $Ri < 1$.

4.5.2 Averages and Fluctuations

In a turbulent flow, such as the PBL, velocity, temperature, and other variables vary irregularly in time and space around their mean values over a wide range of time and length scales. Following the original suggestion of Reynolds (1894), it has been a common practice to consider these variables as sums of their mean (denoted by an overbar) and fluctuating parts (denoted by prime), that is

$$u = \bar{u} + u'; \; v = \bar{v} + v'; \; w = \bar{w} + w'$$
$$\theta = \bar{\theta} + \theta'; \; q = \bar{q} + q'; \; c = \bar{c} + c' \quad (4.10)$$

This is also known as Reynolds' decomposition.

There are basically three different types of averages used in theory and practice—temporal, spatial, and ensemble averages. The most commonly used in the analysis of observed data from fixed points in space is the time average of a variable obtained as an arithmetic mean of the instantaneous values of the variable observed at a number of different times over a desired sample duration or period T. For a continuous record (time series), the arithmatic mean is replaced by an appropriate integral of the variable with respect to time, divided by T. Time averages are frequently used for meteorological variables, as well as for pollutant concentrations measured from fixed instruments placed on the ground or mounted on masts, towers, or tethered balloons.

Spatial averages can be obtained similarly as arithmetic means of the values of any variable observed by an array of sensors covering a line, area, or volume at a particular instant in time. The number of observations used for spatial averaging is often limited by the number of sensors available. Horizontal line and area averages might be more meaningful, considering the vertical inhomogeneity of the lower atmosphere, particularly the surface layer. Averages over fixed volumes in space can be obtained, however, from the appropriate scanning of remote sensors such as radars, lidars, and sodars.

A combination of time and space averaging is used for observations from a moving platform, such as an instrumented van or an aircraft. Instrumented aircraft can provide more meaningful spatial averages of meteorological variables at different levels in the PBL than a fixed array or network of instrumented towers.

In meteorological and air pollution applications, the use of time or space averaging is somewhat subjective, because such averages invariably depend on the choice of sampling time T or spatial dimensions (L_1, L_2, etc.) over which the averages are taken. The choice of optimum sampling duration or spatial dimension is not so clear in the atmosphere with a more or less continuous spectrum of scales of motion (micro, meso, and macro scales). Ideally, the sampling duration or sampling length should be sufficiently long to ensure stable averages and to incorporate the effects of largest scales of interest in turbulence and diffusion. On the other hand, these should not be too long to mask what may be considered as real trends of mean variables with time (e.g., diurnal variations) or space (e.g., urban-rural differences). Sampling or averaging times frequently used in atmospheric PBL studies range from 10^3 to 10^4 s, depending on the height of observation, the PBL depth, and stability. In air pollution and dispersion applications, however, sampling time for determining mean winds and turbulence statistics must be consistent with the sampling time for concentration and dispersion parameters, because atmospheric dispersion occurs at all scales.

Finally, the type of averaging that is almost always used in theory, but rarely in practice or experiments, is the ensemble or probability average. It is the arithmetic mean of a large number (approaching to infinity) of realizations (values) of a variable obtained by repeating the experiment over and over again, under exactly the same general conditions. It is obvious that ensemble averages would be nearly impossible to obtain under varying weather conditions in the atmosphere, over which we have little or no control. In controlled laboratory environments such as meteorological wind tunnels and water channels, experiments are often repeated to reduce random experimental errors, but ensemble averages of turbulent and diffusion variables are rarely taken, due to the time and expense involved in repeating the experiment many times. The number of realizations over which ensemble averaging is done is severely limited (order of 10).

Ensemble averages, rather than time or space averages, are used in theory, because the former do not depend on subjectively chosen sampling period or lengths and they follow all the Reynolds' averaging rules or conditions used for obtaining the equations of mean variables from those of instantaneous variables. We illustrated this procedure in Chapter 2, where the rules of averaging were also given.

The fact that different types of averages are used in theory and experiments requires that we should know about the necessary conditions for the equivalence of different types of averages. These conditions have been investigated thoroughly by fluid dynamists and statisticians and are given in the form of the ergodic theorem (Monin and Yaglom, 1971): For stationary random fields (variables), time averages converge to ensemble averages as the averaging interval (sampling duration) becomes infinitely large ($T \rightarrow \infty$). Similarly, for spatially homogeneous random fields, spatial averages converge to ensemble averages as the spatial averaging intervals become infinitely large ($L_1 \rightarrow \infty$, etc.).

The concept of stationarity introduced above implies the independence of all the mean variables and turbulence statistics on time. Similarly, the concept of spatial homogeneity implies the independence of all the mean flow and turbulence statistics on distance in all the directions of homogeneity. Homogeneity may be restricted to one or two dimensions (e.g., horizontal homogeneity); homogeneity in all directions in space is also called isotropy. Isotropic turbulence, being the simplest type of turbulent flow, has been the subject of intensive theoretical and experimental investigations. It is quite obvious that, due to natural variability of atmosphere in time and space, the necessary conditions of stationarity (for time averaging) and homogeneity (for spatial averaging) cannot be strictly satisfied by atmospheric flows. Quasi-stationarity over a limited time and quasi-homogeneity in certain directions over a limited space is the best one can hope for in simpler situations of undisturbed weather and homogeneous surface. Under these conditions, one might expect an approximate equivalence of temporal or spatial averages used in experiments or observations, and ensemble averages used in theories and models. Quantitative estimates of the effects of finite sampling interval on certain statistics will be given in the next chapter.

Irrespective of the type of averaging used, the deviations of instantaneous values of any variable from its mean are called turbulent fluctuations, for example,

$$u' = u - \bar{u};\ v' = v - \bar{v};\ w' = w - \bar{w};$$
$$\theta' = \theta - \bar{\theta};\ q' = q - \bar{q};\ c' = c - \bar{c} \qquad (4.11)$$

By definition, averages of fluctuations must be zero ($\overline{u'} = 0,\ \overline{\theta'} = 0$, etc.) so that there are compensating negative fluctuations for all the positive fluctuations on the average. This is illustrated in Figure 4.7, which

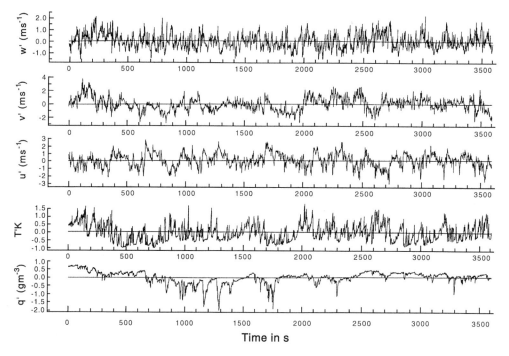

Figure 4.7 Measured time series of velocity, temperature, and absolute humidity fluctuations at a suburban site in Vancouver, Canada, during moderately unstable conditions. From Roth, 1990.

depicts the observed hour-long time series of velocity, temperature, and absolute humidity fluctuations at a suburban site in Vancouver, Canada, at a height of 27.4 m above the ground level (estimated effective height ≃ 19 m), during moderately unstable conditions. The corresponding mean values of variables at that time were

$$\bar{u} = 3.66 \text{ m s}^{-1}; \bar{v} = \bar{w} = 0;$$
$$\bar{T} = 294.6 \text{ K}; \bar{q} = 8.3 \text{ g m}^{-3}$$

Note the differences in the character of traces of horizontal and vertical velocity components, temperature, and humidity fluctuations. These tend to be more or less similar, with an approximate symmetry between positive and negative fluctuations, under strong winds and near-neutral stability. In unstable and convective conditions, on the other hand, buoyant plumes and thermals often cause strong asymmetry between positive and negative fluctuations, particularly in w', θ', and q'.

The relative magnitudes of mean and fluctuating parts of meteorological variables depend on the type of variable; momentum, heat, and moisture fluxes at the surface; the PBL depth and stability; the height of observation; and other factors. In a typical daytime, unstable surface layer, the magnitudes of vertical velocity fluctuations are much larger than the mean vertical velocity, the magnitudes of horizontal velocity fluctuation are of the same order or less than the mean wind speed, and the magnitudes of temperature fluctuation are less than 1 percent of the mean absolute temperature (see e.g., Fig. 4.7). The relative magnitudes of turbulent fluctuations generally decrease with increasing PBL stability.

4.5.3 Variances and Turbulence Intensities

The vertical profiles of mean variables, such as temperature, humidity, and wind components, can tell much about the mean structure of the boundary layer or some other turbulent flow, but little or nothing about its turbulent structure and the exchange processes occurring in the flow. Different types of statistical methods and techniques are used to analyze observed turbulence series of the type shown in Figure 4.7 and, hence, to provide different statistical measures of turbulence structure. In this section we discuss only some of the simpler measures; more sophisticated statistical descriptions of turbulence will be covered in Chapter 5.

The simplest measure of fluctuation levels are the variances or mean-square fluctuations $\overline{u'^2}$, $\overline{v'^2}$, $\overline{w'^2}$, $\overline{\theta'^2}$, and so on. The turbulent kinetic energy per unit mass combines all the velocity variances into one parameter

$$\bar{e} = \frac{1}{2}(\overline{u'^2} + \overline{v'^2} + \overline{w'^2}) \quad (4.12)$$

Other related measures are the standard deviations or root-mean-square fluctuations

$$\sigma_u = (\overline{u'^2})^{1/2}; \sigma_v = (\overline{v'^2})^{1/2}; \sigma_w = (\overline{w'^2})^{1/2} \quad (4.13)$$

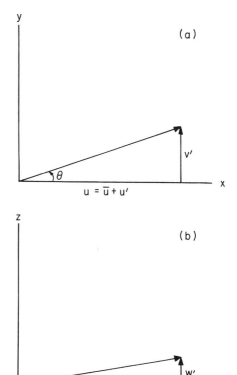

Figure 4.8 Schematic of the instantaneous velocity vector, wind direction, and velocity fluctuations in (a) the x–y plane and (b) the x–z plane.

and turbulent intensities

$$i_u = \frac{\sigma_u}{\bar{V}}; i_v = \frac{\sigma_v}{\bar{V}}; i_w = \frac{\sigma_w}{\bar{V}} \quad (4.14)$$

which are measures of relative fluctuation levels in different directions (velocity components). Note that the mean wind speed, rather than the particular component mean velocity, is used in the definition of turbulence intensities.

When the x axis is oriented with the mean horizontal wind and the z axis in the vertical, u', v', and w' are called the longitudinal, lateral, and vertical velocity fluctuations, respectively. Similarly, i_u, i_v, and i_w are designated as longitudinal, lateral, and vertical turbulence intensities. From the simple geometry of the instantaneous velocity vector depicted in Figure 4.8, one can see that the lateral and vertical velocity fluctuations are related to the fluctuations of horizontal and vertical wind directions (alternatively, azimuth and elevation angles) as

$$\tan \theta' = \frac{v'}{\bar{u} + u'}; \tan \phi' = \frac{w'}{\bar{u} + u'} \quad (4.15)$$

When turbulence intensities are much smaller than 1 (say, ≲0.3), writing $\tan \theta' \simeq \theta'$ and $\tan \phi' \simeq \phi'$ in

Eq. (4.15), squaring and taking averages of both sides, and neglecting higher-order terms, one can show that

$$\sigma_\theta \simeq \frac{\sigma_v}{\bar{u}}; \ \sigma_\phi \simeq \frac{\sigma_w}{\bar{u}} \quad (4.16)$$

Thus, standard deviations of horizontal and vertical wind-direction fluctuations, expressed in radians, are approximately equal to lateral and vertical turbulence intensities, respectively. We will show later that these turbulence parameters have special significance in determining the horizontal and vertical dispersion of pollutants.

4.5.4 Covariances and Turbulent Fluxes

Unlike variances and turbulence intensities that are statistical measures of fluctuation levels in individual variables, the covariance between two fluctuative variables is defined as the average of their product. For example, $\overline{u'w'}$, $\overline{v'w'}$, $\overline{\theta'w'}$, and so on are the covariances between different velocity components and between temperature and vertical velocity fluctuations. The covariance indicates the degree of common relationship or correlation between the two variables involved. If the variables are independent of each other, their covariance would be zero. Covariances may be positive or negative, depending on whether the two variables are positively correlated (positive excursions in one correspond to positive excursions in the other) or negatively correlated (positive excursions in one correspond to negative excursions in the other and vice versa), on the average. Such correlations need not be perfect all the time. For example, a casual inspection of the turbulence data in Figure 4.7 would show that, on the average, temperature and humidity fluctuations are positively correlated with each other and with vertical velocity fluctuations ($\overline{\theta'w'} > 0$, $\overline{q'w'} > 0$), while the longitudinal velocity fluctuations are negatively correlated with vertical velocity ($\overline{u'w'} < 0$).

A better measure of correlation between two variables is provided by the linear correlation coefficient, which is nothing but a normalized covariance, for example,

$$r_{uw} = \frac{\overline{u'w'}}{\sigma_u \sigma_w}, \ r_{qw} = \frac{\overline{q'w'}}{\sigma_q \sigma_w}, \ \text{etc.} \quad (4.17)$$

Note that, according to Schwartz's inequality, $|\overline{u'w'}| \leq \sigma_u \sigma_w$, and so on, so that all correlation coefficients must lie between -1 and 1. The values of ± 1 indicate the perfect positive or negative correlation. The more typical magnitudes of correlation coefficients in the atmospheric surface layer fall in the range of 0.3 to 0.6, which imply significant correlations between apparently randomly fluctuating variables.

Covariances involving fluctuating motion in a turbulent flow can also be interpreted as turbulent fluxes or transports of some scalar. The flux of any scalar in a given direction is defined as the amount of the scalar carried per unit time per unit area normal to that direction. For example, consider the vertical mass flux of pollutant with a variable concentration c in the flow. It is obvious that the velocity component in the direction of flux is responsible for the scalar transport. Considering the amount of material through a unit horizontal area per unit time and averaging it, the total vertical flux of the material, on the average, is given by

$$\overline{cw} = \overline{(\bar{c} + c')(\bar{w} + w')} = \bar{c}\bar{w} + \overline{c'w'} \quad (4.18)$$

Thus, the total scalar flux, in the average sense, can be represented as the sum of the mean transport ($\bar{u}\bar{w}$) by mean flow and the turbulent transport ($\overline{c'w'}$). The latter, also called the turbulent flux, is often the dominant transport term, except perhaps in the direction of mean flow. Turbulent transports of pollutants are of considerable importance in the atmospheric boundary layer. If the scalar under consideration is the potential temperature θ or enthalpy $\rho c_p \theta$, the corresponding vertical turbulent flux is $\overline{\theta'w'}$, or $\rho c_p \overline{\theta'w'}$. In this way, the covariance can be interpreted as the turbulent heat flux (in kinematic units) in the vertical direction. Similarly, $\overline{q'w'}$ represents the moisture flux ($E = \rho \overline{q'w'}$) in the vertical direction. Horizontal scalar fluxes can be defined in the same manner (e.g., $\overline{c'u'}$ and $\overline{c'v'}$ are the turbulent mass fluxes in x and y directions, respectively).

Turbulent fluxes of momentum involve covariances between velocity fluctuations in different components. For example, by symmetry, $\rho \overline{u'w'}$ may be considered as the vertical flux of u momentum and also as the horizontal (x direction) flux of the vertical momentum. Since the rate of change (flux) of momentum is equal to the force per unit area, or the stress, turbulent fluxes of momentum can also be interpreted as turbulent stresses, that is,

$$\tau_{zx} = -\rho \overline{w'u'}, \ \tau_{zy} = -\rho \overline{w'v'}, \ \text{etc.} \quad (4.19)$$

There are, in all, nine stress components acting on a cube-like fluid element, of which three (τ_{xx}, τ_{yy}, and τ_{zz}) are normal stresses (acting normal to the faces) and six (τ_{xy}, τ_{xz}, etc.) are shearing stresses (acting parallel to the faces of the element). When expressed in terms of variances and covariances, these are called the Reynolds stresses, which generally have much larger magnitudes than the corresponding mean viscous stresses in a turbulent flow. The Reynolds stresses, and fluxes are the most important quantities in a turbulent flow; these are also the primary cause for most difficulties encountered in theoretical treatments of turbulent flows. Only in a fully homogeneous (isotropic) field of turbulence, the covariances between fluctuations of different variables vanish at every point in the field, because of the symmetry conditions.

In atmospheric flows, with which we are primarily interested, the presence of wind shears and/or thermal stratification invariably leads to vertical inhomogeneities in turbulence and, hence, to turbulent transports or fluxes. In particular, the vertical fluxes of momentum, heat, and water vapor are given by

$$F = \rho \overline{u'w'}$$
$$H = \rho c_p \overline{\theta'w'}$$
$$E = \rho \overline{q'w'} \quad (4.20)$$

Note that the momentum flux is always downward (negative) as a result of increasing wind speed with height in the lower part of the PBL, but heat and moisture fluxes can be upward or downward, depending on the variations of mean potential temperature and specific humidity with height above the surface. Typically, the vertical fluxes are large near the surface and vary more or less linearly with height in the convective mixed layer. At night, the fluxes are much weaker and decrease more rapidly with height.

4.6 GRADIENT-TRANSPORT THEORIES

Almost all turbulence theories and models are semi-empirical in nature, because of the fundamental problems and difficulties encountered in purely mathematical and statistical descriptions of turbulence (see Chapter 5). In this section, we will give an overview of the simpler gradient-transport theories, which are still widely used in applications to air pollution meteorology and dispersion. More sophisticated theories and models will be discussed later.

4.6.1 Eddy Diffusivity Hypothesis

The kinetic theory of gases, as well as experimental observations of molecular exchange processes in laminar viscous flows, has shown simple proportionality between the average molecular fluxes of momentum, heat, and mass and the gradients of bulk velocity, temperature, and concentration, respectively, in the direction of fluxes. These flux-gradient relations have come to be known as Newton's law of viscosity, Fourier's law of heat conduction, and Fick's law of mass diffusion:

$$F = -\rho \nu \frac{\partial u}{\partial z}$$
$$H = -\rho c_p \kappa \frac{\partial \theta}{\partial z}$$
$$M = -D \frac{\partial c}{\partial z} \quad (4.21)$$

Here, ν is the kinematic viscosity, κ is the thermal diffusivity, and D is the mass diffusivity—all molecular properties of the fluid and the diffusing material.

In direct analogy with molecular exchanges, turbulent transports or fluxes are also assumed (as originally proposed by J. Boussinesq in 1877) to be related to the mean gradients as

$$\overline{u'w'} = -K_m \frac{\partial \overline{u}}{\partial z}$$
$$\overline{\theta'w'} = -K_h \frac{\partial \overline{\theta}}{\partial z}$$
$$\overline{c'w'} = -K_z \frac{\partial \overline{c}}{\partial z} \quad (4.22)$$

in which K_m, K_h, and K_z are called the eddy diffusivities or exchange coefficients of momentum, heat, and mass, respectively (K_m is also called eddy or turbulent viscosity in analogy with molecular viscosity).

The gradient-transport (K-theory) relations are not the expressions of any sound physical laws in the same sense as their molecular counterparts are. These are not based on any rigorous theory, but only on an intuitive analogy between molecular and turbulent exchange processes. This analogy is found to be rather weak and qualitative only. Unlike molecular viscosity and other diffusivities, eddy diffusivities cannot be regarded as fluid properties. The latter are actually properties of flow (turbulence), which can vary widely from one flow to another and also from one region to another in the same flow. For example, in the PBL, eddy diffusivities may be different in different directions (horizontal diffusivities, K_x and K_y, are usually larger than the vertical diffusivity, K_z), and eddy viscosity may differ from the eddy diffusivities of heat and mass. Even the basic premise of the down-gradient transport implied in the theory becomes questionable under certain conditions, particularly when convective motions dominate transport and diffusion processes. For example, in the convective mixed layer, the vertical fluxes of momentum, heat, and mass are not always down the mean gradients of velocity, potential temperature, and concentration, implying negative or infinitely large eddy diffusivities. The gradient-transport theory is not valid in such cases. In neutral and stably stratified flows with stronger gradients, there are no such conceptual difficulties, but specification of eddy diffusivities is mostly ad-hoc and empirical.

4.6.2 The Mixing-length Hypothesis

In an attempt to specify eddy viscosity as a function of geometry and flow parameters, L. Prandtl in 1925 further extended the analogy between molecular and turbulent transfer processes to the hypothetical mechanisms of such transfers. He hypothesized large eddies, parcels, or blobs (analogous to molecules) of fluid breaking away from the main body of the fluid, traveling a certain variable distance called the mixing length (analogous to molecular free path length), and then suddenly mixing with the new environment and transferring their momentum, heat, etc.) to the environment. Fluctuations in properties such as velocity and temperature, were assumed to occur as a result of eddy mixing. Prandtl's mixing-length theory leads to the following types of expressions for the vertical eddy diffusivities (see e.g., Sutton, 1953; Arya, 1988):

$$K_m = l_m^2 \left| \frac{\partial \overline{V}}{\partial z} \right|$$

$$K_h = l_m l_h \left| \frac{\partial \overline{V}}{\partial z} \right| \quad (4.23)$$

in which l_m and l_h are the average mixing lengths for the transfer of momentum and heat, respectively. Alternatively, the mixing-length theory also yields

$$K_m = c_m \, l_m \, \sigma_w$$
$$K_h = c_h \, l_h \, \sigma_w \quad (4.24)$$

where c_m and c_h are empirical constants. Further generalizing Eqs. (4.24) to turbulent diffusivities of nonbuoyant, passive material in different directions,

$$K_x = c \, l_x \, \sigma_u$$
$$K_y = c \, l_y \, \sigma_v$$
$$K_z = c \, l_z \, \sigma_w \quad (4.25)$$

in which l_x, l_y, and l_z are the characteristic length scales of diffusion/dispersion in x, y, and z directions, respectively. These are also called large-eddy scales or simply, scales of turbulence.

Although Prandtl's original mixing-length theory and his proposed hypothetical mechanisms for turbulent transfer processes are not given much credence now in the light of overwhelming experimental evidence to the contrary, the expressions (4.23) through (4.25) for eddy diffusivities based on that theory are considered to be plausible, empirical relations. In fact, Eqs. (4.24) and (4.25) can be written from dimensional considerations alone, recognizing that eddy diffusivity must be a product of the appropriate length and velocity scales, requiring no hypothetical (e.g, mixing length) mechanism for the turbulent transfer.

The mixing-length theory also has all the limitations of the K-theory mentioned earlier. In addition, Prandtl's hypothesized mechanism of momentum and material transfer does not quite fit with what we now know about these transfer processes. His extended analogies between molecules and "blobs" of fluid, between free path length and mixing length, and between mean molecular velocity and mean flow velocity are not found to be very satisfactory. In spite of these limitations, gradient-transport theories are extensively used for solving practical problems in nature and technology. They provide the simplest and most direct relationships between turbulent fluxes of momentum, heat, and mass and mean gradients of velocity, temperature, and mass concentration.

The use of gradient-transport relations in the equations of mean motion and of other mean variables makes possible their closed-form, analytical or numerical solutions. We will demonstrate this later in chapters 6 and 11 by giving some solutions of the mean diffusion equation.

4.6.3 Specification of Eddy Diffusivities and Mixing Lengths

There is not much theoretical guidance on the specification of eddy diffusivities in gradient-transport relations (4.22). Their empirical estimates require simultaneous measurements of turbulent fluxes and mean gradients in a variety of meteorological conditions. Such estimates become highly uncertain in the regions and directions of vanishingly small gradients, which may not be possible to resolve under variable conditions of the atmosphere. Some direct estimates of vertical eddy diffusivities have been reported, mostly for the lower part of the PBL. The values of K_m reported in the literature range from 0.1 to 2000 m^2 s^{-1}, with typical values of 1, 10, and 100 m^2 s^{-1} for stable, neutral, and unstable boundary layers, respectively. Note that these are many orders of magnitude larger than the value of the molecular kinematic viscosity $\nu \simeq 1.5 \times 10^{-5}$ m^2 s^{-1} for air. Horizontal diffusivities have never been directly measured, but these are expected to increase with the characteristic length scale of diffusive/dispersive motions. In particular, K_x and K_y are found to increase initially with the time of travel after release. Thus, for travel (diffusion) times greater than a few hours, the apparent horizontal diffusivities may become orders of magnitude larger than the vertical diffusivity, as the vertical scale of atmospheric diffusion is always restricted by the presence of the ground surface, stable stratification, and elevated inversions.

The fact that l_x, l_y, and so on, are some physically measurable (see Chapter 5) length scales of turbulence, makes the relations (4.24) and (4.25) very useful in practical applications. Empirically estimating eddy diffusivities using these relations, and their subsequent use in turbulent transport and diffusion models, might be preferable to their estimates from direct measurements of fluxes and gradients. The former procedure avoids the problem of negative or infinitely large diffusivities encountered in the latter approach, when fluxes might be counter to the local mean gradients (e.g., in the convective mixed layer). Of course, a gradient-transport theory would not predict countergradient fluxes, unless the original flux-gradient relations (4.22) were modified to allow for the same. Such a modification has been proposed for the heat flux (Deardorff, 1966):

$$\overline{\theta' w'} = -K_h \left(\frac{\partial \overline{\theta}}{\partial z} - \gamma_c \right) \quad (4.26)$$

in which $\gamma_c \simeq 6.5 \times 10^{-4}$ K m^{-1} is an empirical constant.

Many specifications or parameterizations of mixing lengths and eddy diffusivities have been proposed in the literature (for recent reviews of the same, see Stull, 1988; Sorbjan, 1989). In atmospheric diffusion applications, the assumptions of constant mixing length and constant eddy diffusivities are quite common when the region of interest is the free atmosphere. The assumption of constant mixing length is found to be reasonable in the outer part of the PBL also where mixing length is assumed proportional to the PBL depth ($l_m = l_o \propto h$). It is preferable to the assumption of constant eddy diffusivity (K is expected to vanish at the top of the PBL as turbulence dies down there). In the surface layer, however, mixing lengths and eddy diffusivities are expected to increase with height in response to the increasing scales of turbulence. Their magnitudes also depend on the surface roughness, the PBL depth, stability, and possibly other parameters. Particularly in the neutral surface layer, both the mixing length and eddy viscosity are found to vary linearly with height, that is,

$$l_m = kz; \quad K_m = kzu_* \quad (4.27)$$

where $k \simeq 0.40$ is the von Karman constant and $u_* \equiv (\tau_o/\rho)^{1/2}$ is the friction velocity that is related to the surface drag or stress τ_o. Either of the above relations (4.27) leads to the well-known logarithmic velocity profile in the constant stress, neutral surface layer (Arya, 1988). Further discussion of wind profiles in the atmosphere surface layer will be given later.

For the entire PBL, the mixing length is often prescribed by an interpolation formula

$$\frac{1}{l_m} = \frac{1}{kz} + \frac{1}{l_b} + \frac{1}{l_o} \quad (4.28)$$

which gives the desired linear behavior close to the surface, the approach to the constant value $l_o \propto h$ in the upper part of the PBL, and a limiting value l_b (the buoyancy length scale) in the presence of stable stratification ($l_b \sim \sigma_w/N$, where $N = s^{1/2}$ is the Brunt–Vaisala frequency).

4.7 SIMILARITY THEORIES

For a variety of flow situations, particularly turbulent flows in boundary layers, mixing layers, and others, our knowledge of the governing physics is not complete enough to derive flux-profile relations based on first principles. It is necessary to have some experimental input to an approximate or incomplete theory in order to develop useful empirical relationships between variables of interest (e.g., turbulent fluxes and mean gradients). A similarity theory, based on dimensional analysis, provides a means of grouping the variables into some dimensionless similarity parameters and organizing the experimental data in the most efficient manner to derive universal similarity relationships (see e.g., Arya, 1988; Stull, 1988).

4.7.1 Neutral Surface-layer Similarity Theory

We first consider the neutral surface layer over a flat and homogeneous surface, and exclude from our consideration any viscous sublayer that may exist over a smooth surface, or the roughness canopy layer in which the flow is not likely to be horizontally homogeneous because it is disturbed by individual roughness elements. In the fully turbulent and homogeneous surface layer, the momentum flux is observed to be nearly constant with height and is related to the surface stress τ_o. An appropriate similarity hypothesis for this flow is that the mean velocity gradient and turbulent quantities depend only on the height above the surface and the kinematic momentum flux or surface stress τ_o/ρ. This implies that the influence of other possible variables, such as the surface roughness, geostrophic winds, and the PBL depth, is indirectly taken into account through τ_o. The plausible similarity hypothesis for the mean velocity gradient

$$\frac{\partial \overline{u}}{\partial z} = f\left(z, \frac{\tau_o}{\rho}\right) \quad (4.29)$$

leads to z and $u_* \equiv (\tau_o/\rho)^{1/2}$ as the only appropriate length and velocity scales, respectively, and also to a constant dimensionless wind shear

$$\frac{z}{u_*} \frac{\partial \overline{u}}{\partial z} = \text{const.} = \frac{1}{k} \quad (4.30)$$

which follows from dimensional analysis.

The above similarity prediction has been verified by many observations of velocity profiles in the wall layers of laboratory channel, pipe, and boundary layer flows, as well as in the near-neutral atmospheric surface layer. The dimensionless wind shear is the inverse of the von Karman constant k, which is an empirical constant with a value of about 0.40. In spite of an uncertainty of the order of 5 percent in empirical estimates of k, k is considered to be a universal constant in the sense that the same value is obtained for different surface or wall layers. Note that Eq. (4.30) also follows from the mixing-length and eddy viscosity theories with l_m and K_m given by Eq. (4.27). The similarity theory prediction is more elegant and implies no particular mechanisms of momentum exchange.

The integration of Eq. (4.30) with respect to z gives the well-known logarithmic velocity profile law,

$$\frac{\overline{u}}{u_*} = k^{-1} \ln \frac{z}{z_o}, \quad \text{for } z \gg z_o \quad (4.31)$$

in which z_o is the surface roughness parameter (length) defined such that at $z = z_o$, $\overline{u} = 0$, theoretically (actually, the velocity will become zero only at the air–surface interface). Equation (4.31) is valid only for $z > h_o \gg z_o$ where h_o is the average height of dominant roughness elements (e.g., sand grains, grass, trees, or buildings). Figure 4.9 shows some of the observed wind profiles over a grass surface during the Wangara Experiment; the best-fitted straight lines through the data points confirm the validity of the log law (4.31). It also illustrates how u_* and z_o can be determined from the slope and intercept of the best-fitted line for each run. The surface roughness length (parameter) is found to be strongly dependent on the average height of dom-

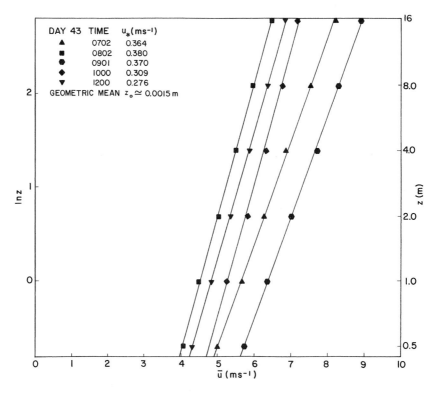

Figure 4.9 Comparison of the observed mean wind profiles in the neutral surface layer of Day 43 of the Wangara Experiment with the log law [Eq. (4.31)] (solid lines).

inant roughness elements and, to a lesser extent, on their areal density and aerodynamic characteristics.

For very rough and uneven natural surfaces, such as croplands, forests, and urban areas, the most appropriate datum (effective surface) for measuring heights in the surface layer is not necessarily the ground level, but it may lie somewhere between the ground level and the top of the roughness canopy. To account for this displaced reference datum, another roughness parameter, called the zero-plane displacement (d_o), is introduced such that the effective height is $z = z' - d_o$, where z' is the height above the ground level. Both z_o and d_o are empirically determined from best fitting of Eq. (4.31), with $z = z' - d_o$, to the measured wind profiles under near-neutral conditions. Micrometeorologists have estimated z_o and d_o for a wide variety of natural surfaces (see Arya, 1988). Although z_o varies over five orders of magnitude (from about 10^{-4} m for smooth ice and water surfaces to several meters for forests and urban areas), the ratio z_o/h_o falls within a much narrower range between 0.03 and 0.30; it varies, to some extent, with the type, height, and areal density of roughness elements. The ratio d_o/h_o depends more strongly on the roughness-element density; it increases with increasing density of roughness elements. For most agricultural crops, for example, $z_o/h_o \simeq 0.15$ and $d_o/h_o \simeq 0.80$. Systematic studies of the dependence of the above ratios on roughness density have been made in laboratory wind tunnels and channels only for certain types of roughness elements arranged in simple geometric patterns.

The application of the same similarity hypothesis to turbulent quantities leads to the simple prediction that the ratios σ_u/u_*, σ_v/u_*, and σ_w/u_* must be constants in the neutral surface layer. Again, many observations from laboratory and atmospheric surface layers have confirmed this expectation; the best estimated values of the normalized standard deviations are

$$\frac{\sigma_u}{u_*} \simeq 2.5; \quad \frac{\sigma_v}{u_*} \simeq 1.9; \quad \frac{\sigma_w}{u_*} \simeq 1.3 \qquad (4.32)$$

4.7.2 Monin–Obukhov Similarity Theory

For the more frequently encountered stratified surface layer of the atmosphere, the Monin–Obukhov similarity theory has been found to be more appropriate and widely accepted. It is based on the similarity hypothesis proposed by Monin and Obukhov (1954), which stipulates that mean gradients and turbulence characteristics of a stratified surface layer depend only on the height z, the kinematic surface stress τ_0/ρ, the kinematic heat flux $H_o/\rho c_p$, and the buoyancy variable g/T_o. The addition of heat flux and buoyancy variables to the original list for the neutral surface layer leads to the following independent scales, which are used to form dimensionless groups or parameters of the M–O similarity theory:

Friction velocity: $\quad u_* = \left(\dfrac{\tau_o}{\rho}\right)^{1/2}$

Friction temperature: $\quad \theta_* = -\dfrac{H_o}{\rho c_p u_*}$

Height above surface: $\quad z$

Buoyancy length:

$$L = -\frac{(\tau_o/\rho)^{3/2}}{k(g/T_o)(H_o/\rho c_p)} = \frac{u_*^2}{k(g/T_o)\theta_*} \quad (4.33)$$

The buoyancy length scale, also known as the Obukhov (or Monin–Obukhov) length, is a measure of the depth of the near-surface layer in which shear effects are likely to be significant under any stability condition. The ratio of the two length scales, z/L, is the Monin–Obukhov similarity or stability parameter, which is a measure of the relative importance of buoyancy versus shear effects, similar to the Richardson number (Ri) introduced earlier. In fact, Ri and z/L are uniquely related to each other, even though they have different distributions with respect to height, particularly in stably stratified conditions. The buoyancy effects of water vapor can be considered easily by replacing temperature and heat flux in the definition of L by virtual temperature and virtual heat flux, respectively. Similarly, the virtual potential temperature gradient should be considered in the definition of Ri.

The application of the M–O similarity hypothesis to mean velocity and potential temperature gradients and dimensional analysis lead to the M–O similarity relations

$$\left(\frac{kz}{u_*}\right)\left(\frac{\partial \bar{u}}{\partial z}\right) = \phi_m(\zeta)$$

$$\left(\frac{kz}{\theta_*}\right)\left(\frac{\partial \bar{\theta}}{\partial z}\right) = \phi_h(\zeta) \quad (4.34)$$

which imply that dimensionless wind shear and temperature gradients are some unique functions of $\zeta = z/L$. The integration of Eq. (4.34) with respect to z gives the normalized profile relations

$$\frac{\bar{u}}{u_*} = \frac{1}{k}\left(\ln\frac{z}{z_o} - \Psi_m(\zeta)\right)$$

$$\frac{\bar{\theta} - \bar{\theta}_o}{\theta_*} = \frac{\alpha}{k}\left(\ln\frac{z}{z_o} - \Psi_h(\zeta)\right) \quad (4.35)$$

in which $\bar{\theta}_o$ is the potential temperature at $z = z_o$, α is a constant between 0.9 and 1, and $\Psi_m(\zeta)$ and $\Psi_h(\zeta)$ are uniquely related to $\phi_m(\zeta)$ and $\phi_h(\zeta)$, respectively, as

$$\Psi_m(z/L) = \int_{z_o/L}^{z/L}[1 - \phi_m(\zeta)]\frac{d\zeta}{\zeta}$$

$$\Psi_h(z/L) = \int_{z_o/L}^{z/L}[1 - \phi_h(\zeta)]\frac{d\zeta}{\zeta} \quad (4.36)$$

The profile relations (4.35) may be considered as simple modifications of the log-law in which both the stability-dependent Ψ-functions are found to decrease with increasing stability (Panofsky and Dutton, 1984; Arya, 1988). Deviations from the log-law increase with increasing instability or stability. Under stable conditions, Ψ functions are negative and vary linearly with z/L. Consequently, the wind and temperature profiles are log-linear and tend to become linear as stability (z/L) increases. The wind profile curvature increases as stability decreases to near neutral and changes to unstable.

Other variables and parameters involving mean gradients can also be expressed in terms of the basic M–O similarity functions $\phi_m(\zeta)$ and $\phi_h(\zeta)$, for example,

$$\text{Ri} = \frac{\zeta \, \phi_h(\zeta)}{\phi_m^2(\zeta)}$$

$$\frac{K_m}{kzu_*} = \frac{1}{\phi_m(\zeta)}$$

$$\frac{K_h}{kzu_*} = \frac{1}{\phi_h(\zeta)} \quad (4.37)$$

The empirical forms of the M–O similarity functions have been determined from micrometeorological experiments at various flat and homogeneous sites (see Arya, 1988; Garratt, 1992). There is no general consensus on the best formulas to be used. The most widely used and simplest forms are the Businger–Dyer relations:

$$\phi_h = \phi_m^2 = (1 - 15\zeta)^{-1/2}, \quad \text{for } -5 < \zeta < 0$$

$$\phi_h = \phi_m = 1 + 5\zeta, \quad \text{for } 1 > \zeta \geq 0 \quad (4.38)$$

The corresponding relations between ζ and Ri are

$$\zeta = \text{Ri}, \quad \text{for Ri} < 0$$

$$\zeta = \frac{\text{Ri}}{1 - 5\text{Ri}}, \quad \text{for } 0 \leq \text{Ri} < 0.2 \quad (4.39)$$

The second of Eq. (4.39) implies a critical value of Ri $\simeq 0.2$, although the above empirical forms of the similarity functions are not expected to remain valid in extremely stable conditions approaching the critical condition (Ri) for the maintenance of turbulence (Arya, 1972).

For turbulence quantities, the Monin–Obukhov similarity theory predicts that the normalized standard deviations, σ_u/u_*, σ_v/u_*, σ_w/u_*, and σ_θ/θ_*, must be universal functions of the M–O stability parameter z/L only. Measurements of turbulence in the atmospheric surface layer are not entirely consistent, however, with

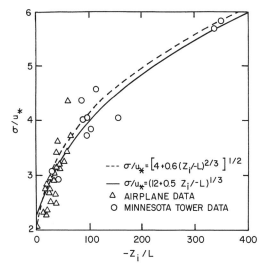

Figure 4.10 Normalized standard deviations of horizontal velocity fluctuations in the surface layer as functions of z_i/L compared with the mixed-layer similarity relations. After Panofsky et al., 1977.

the above theoretical prediction. In particular, the statistics of horizontal velocity fluctuations in unstable and convective conditions do not follow the M–O similarity scaling; these show a strong dependence on the PBL thickness (h), which was ignored in the Monin–Obukhov similarity hypothesis. The statistics of vertical velocity, temperature, and specific humidity fluctuations are consistent with the theory and so are the other turbulence statistics (e.g., σ_u/u_* and σ_v/u_*) in neutral and stably stratified conditions. With increasing stability, turbulence becomes very weak and σ_u/u_* and σ_v/u_* (etc.) show rather large scatter with no definite dependence on z/L. Therefore, the empirical values of turbulence parameters given in Eq. (4.32) are often used for all the neutral and stable conditions ($z/L \geq 0$). For unstable and convective conditions, on the other hand, the following empirical relations based on the data of Figure 4.10 and Figure 4.11 can be recommended:

$$\frac{\sigma_{u,v}}{u_*} = \left(12 - 0.5 \frac{h}{L}\right)^{1/3}, \quad \text{for } \frac{h}{L} < 0$$

$$\frac{\sigma_w}{u_*} = 1.3\left(1 - 3\frac{z}{L}\right)^{1/3}, \quad \text{for } \frac{z}{L} \leq 0 \quad (4.40)$$

The dependence of $\sigma_{u,v}/u_*$ on h/L and the exponent one-third are consistent with the prediction of the mixed-layer similarity theory for the CBL, which will be discussed in the next section.

4.7.3 Local Free-convection Similarity Theory

Another similarity hypothesis of validity limited to only vertical velocity and temperature in the free convective surface layer is Obukhov's local free convection similarity hypothesis in which u_* is considered irrelevant and only z, g/T_o, and $H_o/\rho c_p$ are considered appropriate independent variables. These lead to the following scales for local free convection:

Length: z

Velocity: $u_f = \left(\dfrac{g}{T_o}\dfrac{H_o}{\rho c_p} z\right)^{1/3}$

Temperature: $\theta_f = \left(\dfrac{H_o}{\rho c_p}\right)^{2/3}\left(\dfrac{g}{T_o} z\right)^{-1/3}$ (4.41)

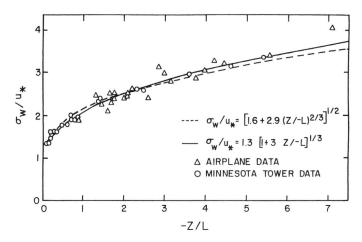

Figure 4.11 Normalized standard deviations of vertical velocity fluctuations in the surface layer as a function of z/L compared with surface-layer similarity relations. After Panofsky et al., 1977.

The important predictions of the local free-convection similarity theory are

$$\frac{\sigma_w}{u_f} = C_w$$

$$\frac{\sigma_\theta}{\theta_f} = C_\theta \qquad (4.42)$$

in which $C_w \simeq 1.4$ and $C_\theta \simeq 1.3$ are empirical constants. This free-convection similarity theory implies that σ_w increases with one-third power of z and is consistent with Eq. (4.40) for large values of $-z/L$. For estimating σ_w, Eq. (4.42) has the advantage over Eq. (4.40) in that the former does not require knowledge of u_* and L and is found to be valid for a wide range of unstable and convective conditions. One can consider the local free-convection similarity theory as a special case of the Monin–Obukhov similarity theory in the limit of $u_* \to 0$ or $z/L \to -\infty$ (more appropriately, $-z/L \gg 1$). The conditions of true free convection in the sense of $u_* \to 0$, and strong surface heating, occur very rarely, if at all, in the atmospheric surface layer. Observations indicate that a quasi-free convection regime often occurs in unstable conditions, in which the statistics of vertical velocity and temperature fluctuations are found to be generally consistent with the local free-convection similarity scaling as shown in Figure 4.12. In particular, the similarity relations (4.42) appear to be valid for $-z/L > 0.5$.

The simpler law of wall (surface) similarity, implying the similarity relations (4.30) through (4.32) for the neutral surface layer, is also a special case of the Monin–Obukhov similarity theory in the limit of $|z/L| \to 0$ (more appropriately, $|z/L| \ll 1$, or $|Ri| \ll 1$). The strict condition of neutral stability in the sense of $H_o = 0$ is essentially transient, which occurs momentarily during morning and evening transitions of the diurnal heating and cooling cycle. The mean profiles and turbulence structure at these moments change so rapidly with time that the assumption of stationarity is not likely to be valid. It is doubtful that any similarity theory would be valid during such rapidly changing transition conditions. Quasi-stationary, near-neutral stability conditions in the sense of $|z/L|$ or $|Ri| \ll 1$ are encountered in the atmospheric surface layer only during strong winds and overcast skies. During these near-neutral as well as moderately stable conditions, one would expect essentially shear-generated turbulence, which can be scaled by the friction velocity u_* only.

4.7.4 Mixed-layer and PBL Similarity Theories

Above the constant-flux surface layer, mean flow and turbulence structure of a stationary and horizontally homogeneous PBL are expected to depend on the surface-layer variables (g/T_o, u_*, and $Q_o = H_o/\rho c_p$), the

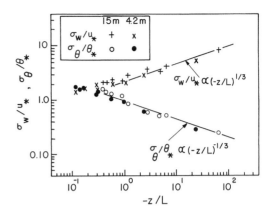

Figure 4.12 Normalized standard deviations of vertical velocity and temperature fluctuations as function of $-z/L$, compared with local free-convection similarity relations. From Businger, 1973, after Monji 1972.

local height z, the PBL depth h, the Coriolis parameter f, and possibly other variables (e.g., geostrophic winds or mean winds at the top of the PBL, thermal winds or geostrophic shears, the rate of entrainment, presence of clouds, etc.). Consideration of all these independent variables would necessarily lead to a large number of dimensionless parameters in a generalized similarity formulation. Since an empirical determination of any similarity function of more than one or two parameters is not yet possible, special similarity theories that might be valid only for selected variables, under certain stability conditions, and in certain regions of flow, are of particular interest.

The most widely used, in parameterizations of turbulence and diffusion in the convective boundary layer, is the mixed-layer similarity theory. It is based on Deardorff's (1970b) similarity hypothesis that z, g/T_o, Q_o, and h are the most appropriate independent variables, which gives the following scales for the convectively mixed layer:

Length: h

Velocity: $W_* = \left(\frac{g}{T_o} Q_o h\right)^{1/3}$

Temperature: $T_* = \frac{Q_o}{W_*}$ (4.43)

In the above definition of convective velocity, W_*, temperature, and heat flux should be replaced by virtual temperature and virtual heat flux, respectively, whenever there is a substantial water vapor flux contributing to buoyancy effects.

The mixed-layer similarity theory predicts that the normalized turbulent quantities σ_u/W_*, σ_v/W_*, σ_w/W_*, σ_θ/T_*, and so on must be unique functions of z/h only. Observations of turbulence in the CBL (see Fig. 4.13) further indicate that there is hardly any dependence of

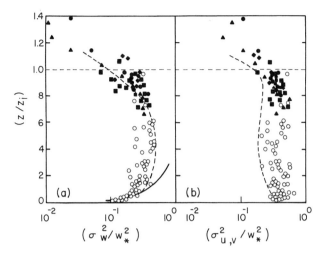

Figure 4.13 Normalized variances of horizontal and vertical velocity as functions of z/z_i. From Caughey and Palmer, 1979. Solid line represents the local free convection prediction and dashed lines the convection tank results of Willis and Deardorff (1974).

horizontal velocity fluctuations on z/h and, for all practical purposes,

$$\frac{\sigma_u}{W_*} \simeq \frac{\sigma_v}{W_*} \simeq 0.60 \tag{4.44}$$

Note that Eq. (4.44) is also found to be valid in the convective surface layer and is consistent with the first part of empirical relations (4.40) for large values of $-h/L$. It can be argued that the horizontal velocity fluctuations are essentially governed by large convective cells (updrafts and downdrafts) with dimensions of the order of h, even close to the surface. Therefore, the appropriate height scale for horizontal turbulent motions is h rather than z, even in the surface layer. This explains the earlier noted failure of the Monin–Obukhov similarity theory in describing horizontal turbulence.

Figure 4.13 also shows that, in contrast to σ_u and σ_v, the vertical turbulent velocity σ_w has a stronger dependence on the normalized height z/h; σ_w/W_* increases with height, attains a broad maximum in the middle of the boundary layer, and then decreases sharply in the upper part of the mixed layer. The increase near the surface is consistent with the prediction of the local free convection similarity theory. Within the bulk of the mixed layer, however, the average value of $\sigma_w/W_* \simeq 0.6$, is similar to that of horizontal components. This may be used in simpler parameterizations of turbulence in the convective mixed layer.

The mixed-layer similarity theory cannot deal with mean wind and potential temperature profiles, which are, in fact, assumed to be uniform in the convectively well-mixed layer. In reality, significant wind shears and low-level wind maxima sometimes develop even in the convective boundary layer, particularly in the presence of strong geostrophic shears (thermal winds). In the majority of convective cases, however, wind speed and direction shears are small enough to be neglected. Therefore, the simplifying assumption of constant wind speed and direction made in most operational dispersion models is probably best justified for convective conditions.

4.8 BOUNDARY-LAYER PARAMETERIZATION FOR DISPERSION APPLICATIONS

Transport and diffusion of material released in the atmospheric boundary layer depend on the vertical distributions of mean wind speed, wind direction, and turbulence in the PBL. These are strongly dependent on the atmospheric stability, which also influences the rise of a buoyant plume or puff. The PBL depth or the mixing height usually determines the vertical extent of the mixing of the material from surface and elevated releases in the PBL. Long-range transport of material is also mostly confined to the maximum depth of the PBL on a diurnal basis. The above meteorological (primarily, boundary layer) parameters must be specified as an input to a dispersion model, unless the latter is tied to an appropriate boundary-layer model.

The quality of a dispersion model strongly depends on the quality of its meteorological input. Since all the desired meteorological parameters are not usually measured at the location and time of release, whose dispersion is to be modeled, some method of specifying or parameterizing the boundary-layer meteorology must be used in dispersion applications. The type of information available might be standard surface meteorological data and upper-air sounding data from the nearest meteorological station (usually an airport), on-site meteorological data from one or two levels of a meteorological tower, and, on rare occasions, turbulence and mixing height measurements at the site. The type of meteorological input desired also depends on the sophistication of the dispersion model. Most Gaussian-type models, currently used in regulatory applications, require only the input of mean wind speed and direction at the effective release height, some measure (index) of stability or turbulence type, and mixing height at each time step (usually, 1 hour). More sophisticated analytical and numerical models can use vertical profiles of mean winds, turbulence, characteristic length scales or eddy diffusivities, and mixing height as functions of time. Some of the theoretical and semiempirical bases for specifying these have already been discussed in earlier sections. Here we review a few simple methods of estimating or specifying boundary layer parameters for dispersion modeling (van Ulden and Holtslag, 1985). We start with three primary parameters—the surface stress or friction velocity, the surface heat flux, and the PBL depth—which determine a host of secondary parameters and scales used for specifying (parameterizing) mean winds and turbulence.

4.8.1 Estimation of Surface Stress and Heat Flux

The various methods of determining the surface fluxes of momentum and heat are described in the micrometeorological literature (for recent reviews of the same, see Arya, 1988; Stull, 1988). In the absence of surface drag and turbulence measurements, which are made only during special research expeditions, the Monin–Obukhov similarity relations (4.34) through (4.39) are used to estimate u_* and θ_* and, hence, the surface fluxes of momentum and heat. Different approaches that can be used are bulk transfer, gradient, and profile methods, depending on the type of measurements available. Here we discuss a version of the profile method that requires the absolutely minimum measurements of mean wind speed at one level (z_r) and the mean temperature difference between two levels (z_1 and z_2) in the surface layer. It is assumed that estimates of the surface roughness parameters z_o and d_o are also available for the particular site.

From the Monin–Obukhov similarity profile relations (4.35), u_* and θ_* can be obtained as

$$u_* = \frac{k\bar{u}_r}{[\ln(z_r/z_o) - \Psi_m(z_r/L)]}$$

$$\theta_* = \frac{k\Delta\bar{\theta}}{\alpha[\ln(z_2/z_1) - \Psi_h(z_2/L) + \Psi_h(z_1/L)]} \quad (4.45)$$

in which $\alpha \simeq 1$ and Ψ_m and Ψ_h are stability-dependent similarity functions whose approximate empirical forms, according to Eqs. (4.36) and (4.38), are

$$\Psi_m = \Psi_h = -\frac{5z}{L}, \quad \text{for } \frac{z}{L} \geq 0$$

$$\Psi_m = 2\ln\left(\frac{1+x}{2}\right) + \ln\left(\frac{1+x^2}{2}\right)$$
$$- 2\tan^{-1} x + \frac{\pi}{2}, \text{ for } \frac{z}{L} < 0$$

$$\Psi_h = 2\ln\left(\frac{1+x^2}{2}\right), \quad \text{for } \frac{z}{L} < 0 \quad (4.46)$$

where $x = (1 - 15\, z/L)^{1/4}$.

When measurements are available for the wind speed $\bar{u}_r$ at the reference height z_r and the potential temperature difference $\Delta\bar{\theta}$ between the height levels z_2 and z_1, one can solve Eq. (4.45) for u_* and θ_* by iteration, since L is defined by the same scales through Eq. (4.33). The iteration procedure can be simplified or avoided altogether if one makes use of the functional relationship between the bulk Richardson number Ri_b and z_r/L

$$\text{Ri}_b \equiv \frac{g\Delta\bar{\theta}\, z_r}{T_o\, \bar{u}_r^2}$$

$$= \frac{z_r}{L}\frac{[\ln(z_2/z_1) - \Psi_h(z_2/L) + \Psi_h(z_1/L)]}{[\ln(z_r/z_o) - \Psi_m(z_r/L)]^2} \quad (4.47)$$

for specified values of z_o, z_1, z_2 and z_r.

An alternative gradient method (Arya, 1988) can be used if the wind speed difference $\Delta\bar{u}$ between the two height levels z_2 and z_1 is also available, in addition to the temperature difference. From the basic M–O relations (4.34), using the often-recommended logarithmic finite-difference approximations for the gradients (Arya, 1988, 1991),

$$u_* = \frac{k\Delta\bar{u}}{[\phi_m(z_m/L) \ln(z_2/z_1)]}$$

$$\theta_* = \frac{k\Delta\bar{\theta}}{[\phi_h(z_m/L) \ln(z_2/z_1)]} \quad (4.48)$$

Here $z_m = (z_1 z_2)^{1/2}$ is the geometric mean of the two heights, where the gradient Richardson number can be estimated by

$$\text{Ri}_m = \frac{g}{T_o}\frac{\Delta\bar{\theta}z_m}{(\Delta\bar{u})^2}\ln\left(\frac{z_2}{z_1}\right) \quad (4.49)$$

which is related to the M–O similarity parameter as

$$\frac{z_m}{L} = \text{Ri}_m, \quad \text{for } \text{Ri}_m < 0$$

$$\frac{z_m}{L} = \frac{\text{Ri}_m}{(1 - 5\,\text{Ri}_m)}, \text{ for } 0 \leq \text{Ri}_m < 0.2 \quad (4.50)$$

An advantage of the gradient method is that it does not require the knowledge of the roughness length parameter z_o (it can, however, be determined from the same measurements, if necessary). The primary disadvantage over the previous method is the measurement of wind speed at an additional level.

From the above discussed profile and gradient methods, the surface heat flux can be estimated as

$H_o = -\rho c_p u_* \theta_*$. Another method of estimating the same is the energy budget method based on Eq. (4.1) or (4.2) in which other energy fluxes are measured, modeled, or parameterized (see e.g., Van Ulden and Holtslag, 1985). This often requires the use of empirical parametric relations developed from data collected at similar sites. The primary advantage is that the method is also applicable in weak winds, during morning and evening transition periods, and in nonstationary conditions in which the usual similarity relations may become invalid. Whenever the latent heat flux is a significant component of the surface energy budget, an estimation of the same, or the rate of evaporation, is needed for the determination of the sensible heat flux. Alternatively, one can estimate the Bowen ratio, $B = H_o/L_e E_o$, from the gradient-transport relations for H_o and E_o, assuming the same eddy diffusivity for heat and water vapor,

$$B = \frac{c_p}{L_e}\frac{\partial\bar{\theta}/\partial z}{\partial\bar{q}/\partial z} \simeq \frac{c_p\Delta\bar{\theta}}{L_e\Delta\bar{q}} \quad (4.51)$$

Here $\Delta\bar{\theta}$ and $\Delta\bar{q}$ are the differences in potential temperature and specific humidity, respectively, between the same two heights, which can easily be determined from the measurements of dry and wet bulb temperatures at the two heights in the surface layer.

Another method of estimating H_o, particularly in daytime unstable conditions, is based on Eq. (4.5). This requires monitoring of air temperature at a height where the rate of warming becomes independent of height (e.g., in the convective mixed layer) and, hence, can be considered representative of the whole PBL and the measurement or estimation of the PBL depth.

4.8.2 Mixing Height or the PBL Depth

The mixing height or the PBL depth h is the most important parameter, which not only determines the limit on the vertical diffusion of plumes or puffs of materials released in the PBL but also determines a host of other parameters (e.g., z/h and h/L) and scales (e.g., W_*, h/W_*, l_x, l_y, etc.) related to turbulence and diffusion. On-site continuous monitoring of the mixing height, using remote sensing devices such as acoustic sounder and lidar, is always desirable for dispersion modeling. In the absence of direct measurements, however, one has to estimate h using some diagnostic or prognostic relations.

From simple theoretical (similarity) considerations, the depth of a neutral PBL in middle and high-latitude regions is given by

$$h \simeq 0.3 \frac{u_*}{|f|} \quad (4.52)$$

The applicability of Eq. (4.52) is restricted, however, to steady-state near-neutral conditions with $|h/L| < 1$ and absence of low-level elevated inversions. Whenever an inversion is present at a height less than that given by Eq. (4.52), the height of inversion base z_i would determine the mixing height.

For the stably stratified nocturnal boundary layer with moderate and strong winds, the mixing height is also given by a simple diagnostic formula

$$h \simeq 0.4\left(u_*\frac{L}{|f|}\right)^{1/2}, \text{ for } \frac{h}{L} > 0 \quad (4.53)$$

which is strictly valid for the steady-state, equilibrium conditions. Comparisons with direct observations of h indicate that Eq. (4.53) provides a fair estimate of the mixing height even during the more typical, slowly evolving, nonstationary conditions prevailing in the nocturnal boundary layer (Nieuwstadt, 1984b). At high wind speeds and low heat flux, the Obukhov length L may become too large, so that it would be advisable to limit h by its neutral value given by Eq. (4.52). An elevated low-level inversion can also limit the PBL height to z_i if the latter is less than that given by Eq. (4.53).

No diagnostic relation is found to be suitable for the daytime unstable and convective boundary layers. Instead, a prognostic rate equation for the mixing height,

$$\frac{\partial h}{\partial t} = \overline{w}_h + w_e \quad (4.54)$$

is considered to be more appropriate for the diurnally evolving PBL. Here $\overline{w}_h$ is the mean vertical velocity at the top of the PBL and w_e is the entrainment velocity. Whereas $\overline{w}_h$ must be estimated from mesoscale or large scale divergence, the entrainment velocity is usually parameterized in terms of the friction velocity u_*, the convective velocity W_*, the environmental lapse rate just above the PBL, and possibly other parameters (see e.g., Deardorff, 1974). A large number of mixing-height models have been proposed in the literature (for a recent review, see Stull, 1988). Since the accuracy of estimating h is more often limited by the lack of information on the mean vertical velocity $\overline{w}_h$, even a crude but simple model might be considered adequate for practical applications.

4.8.3 Mean Wind Profile

Generally, the mean wind speed increases with height, at least in the lower half of the PBL, while the mean wind direction also shifts with increasing height. The wind speed profile in the surface layer is given by the Monin–Obukhov similarity relation (4.35), which can also be written in the form

$$\frac{\bar{u}(z)}{\bar{u}(z_r)} = \frac{\ln(z/z_o) - \Psi_m(z/L)}{\ln(z_r/z_o) - \Psi_m(z_r/L)} \quad (4.55)$$

where z_r is the reference height at which measurements are available. The specification of Ψ_m through Eq. (4.46) requires knowledge of L, which can be estimated using the profile or gradient method discussed earlier. The applicability of Eq. (4.55) is, however, limited to only the lowest 10 to 15 percent of the PBL in which wind direction does not change significantly with height.

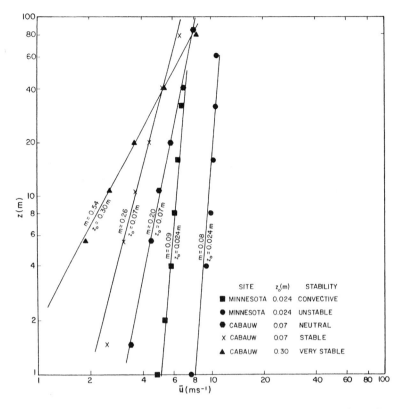

Figure 4.14 Comparison of observed wind speed profiles at different sites (z_o represents the surface roughness) under different stability conditions with the power-law profile. Minnesota data from Izumi and Caughey, 1976.

Above the surface layer, the variations in wind speed and wind direction with height are found to be highly dependent on both the stability and the baroclinicity of the PBL. The theoretical basis for generalizing the wind profile is much more tenuous for the whole PBL than it is for the surface layer only. Observed wind profiles (see section 4.4) under different stability conditions suggest that mean wind speed can be assumed to be more or less uniform, independent of height, in unstable and convective conditions. The mean mixed-layer wind speed is approximately given by (Garratt et al., 1982)

$$\frac{\overline{V}_m}{u_*} = \frac{1}{k}\left[\ln\frac{h}{z_o} - \frac{1}{2}\ln\left|\frac{h}{L}\right| - 2.3\right],$$

$$\text{for } \frac{h}{L} < 0 \qquad (4.56)$$

In very stable conditions, on the other hand, there is considerable observational evidence for an approximately linear wind speed profile up to the height of the low-level jet, which often coincides with the top of the PBL. That is,

$$\overline{V} = \overline{V}_s + \frac{z - h_s}{h - h_s}(\overline{V}_h - \overline{V}_s), \text{ for } \frac{h}{L} > 0 \qquad (4.57)$$

in which the subscripts s and h refer to the top of the surface layer and the PBL, respectively, and $\overline{V}_s$ can be obtained from the surface-layer profile relations ($\overline{V}_s = \overline{u}_s$ in the surface-layer coordinate system).

An alternative empirical representation of the mean wind distribution frequently used in air pollution dispersion applications is the power-law profile

$$\frac{\overline{V}}{\overline{V}_r} = \left(\frac{z}{z_r}\right)^m \qquad (4.58)$$

where $\overline{V}_r$ is the wind speed at the reference height z_r and m is an exponent less than or equal to unity. The power-law profile does not have a sound theoretical basis, but it can provide a reasonable fit to the observed wind speed profiles in the lower part of the PBL over a wide range of surface roughness and stability conditions (see Fig. 4.14). The only parameter of the above profile law is the exponent m, which is found to depend on the surface roughness, stability, and possibly the height range over which Eq. (4.58) is fitted. Comparison of observed wind-speed profiles from different sites and different stability conditions with Eq. (4.58) clearly indicates that m increases with increasing surface roughness and, for the same site, it increases with increasing stability. Under near-neutral conditions, the empirically estimated values of m range from 0.15 for smooth water, snow, and ice surfaces to about 0.40 for well-developed urban areas. Over a moderately rough surface, m may range from 0.10 in very unstable or convective conditions to near unity (corresponding to the linear wind profile) under extremely stable conditions.

Note that the power-law wind profile would be appropriate only for the lower part of the PBL in which wind speed increases monotonically with height. At the upper levels, wind speed may be prescribed differently (e.g., constant or linearly varying), depending on stability and other conditions.

Finally, we consider the change of wind direction with height, which is an important factor in air pollution modeling, because it affects both the direction of pollutant transport and the lateral dispersion. From observations of wind profiles, it appears that the wind direction shear is often small (less than 10°/km) in a convective boundary layer, increases with increasing stability, and becomes very large (up to 45°/100 m) in extremely stable conditions. The balance between pressure gradient, Coriolis, and friction forces on air parcels in the PBL (see section 4.4) dictates that, in the northern hemisphere, wind direction must increase with height, particularly above the surface layer. On the basis of some observed data on mean wind direction from a 200-m mast at Cabauw in the Netherlands, Van Ulden and Holtslag (1985) have proposed the following empirical relation for the turning angle (α) of the wind relative to the near-surface wind

$$\frac{\alpha(z)}{\alpha(h)} = d_1[1 - \exp(-d_2 z/h)] \quad (4.59)$$

where $d_1 \simeq 1.58$ and $d_2 \simeq 1.0$ are empirical coefficients. There is no generally applicable empirical or theoretical relationship for the total turning of the wind $\alpha(h)$ across the PBL depth. It is expected to depend on the surface roughness, stability, baroclinicity, latitude, and possibly other factors. For a given site, $\alpha(h)$ is primarily dependent on stability and baroclinicity; the theoretical range of 0° to 45° is strictly valid only for barotropic conditions (constant geostrophic winds).

4.8.4 Stability

For the purpose of diffusion parameterization and modeling, it is often required to specify the stability and/or turbulence conditions in which diffusion of pollutants would occur. In simpler dispersion models requiring no direct input of turbulence, stability is usually considered as a surrogate for turbulence, and diffusion is parameterized as a function of stability. In more sophisticated models, turbulence may be specified explicitly, but usually as a function of stability.

We have already introduced a number of stability or buoyancy parameters in this and previous chapters. Some of them are based on the vertical gradient of temperature or potential temperature (e.g., $\partial\overline{\theta}/\partial z$ and

Table 4.1 General Characterizations of Static Stability

Lapse Rate	$\dfrac{\partial \overline{\theta}_v}{\partial z}$ or s	$\dfrac{\partial \overline{T}_v}{\partial z}$	Static Stability
Subadiabatic	>0	>$-\Gamma$	Stable
Adiabatic	=0	=$-\Gamma$	Neutral
Superadiabatic	<0	<$-\Gamma$	Unstable

s) and are not considered to be good measures of turbulence, because they contain no information on winds or wind shears. These can be used only for a broad characterization of static stability, as in Table 4.1. Others are dimensionless parameters that represent the relative magnitudes of buoyancy and shear effects on turbulence. Important parameters in this category are the various forms of Richardson number, for example,

$$\text{Gradient: Ri} = \frac{g}{T_o} \frac{\partial\overline{\theta}/\partial z}{|\partial\overline{V}/\partial z|^2}$$

$$\text{Bulk: Ri}_b = \frac{g}{T_o} \frac{\Delta\overline{\theta}\, z_r}{\overline{V}_r^2}$$

$$\text{Mixed: Ri}_B = \frac{g}{T_o} \frac{(\partial\overline{\theta}/\partial z)z_r^2}{\overline{V}_r^2} \quad (4.60)$$

and the Obukhov length (L), or the Monin–Obukhov stability parameter

$$\frac{z}{L} = -\frac{zk}{u_*^3} \frac{g}{T_o} \frac{H_o}{\rho c_p} = \frac{zk}{u_*^2} \frac{g\theta_*}{T_o} \quad (4.61)$$

The above parameters are all local, because they depend on height above the surface, local gradient, or wind speed at a reference height z_r. These are considered to be good measures of turbulence in the surface layer where vertical gradients of potential temperature and velocity are most significant. In this layer, the various Richardson numbers are uniquely related to z/L (Ri$_b$ and Ri$_B$ also depend on z/z_o) through the Monin–Obukhov similarity theory. Their magnitudes increase with height monotronically (e.g., $|z/L|$ increases linearly, because L is a fixed buoyancy length scale), indicating increasing influence of buoyancy with height.

Above the surface layer, z/L is not a relevant parameter, but the gradient Richardson number is still an appropriate local stability parameter that determines the existence or not of local turbulence, as well as the intensity of turbulence in a stably stratified layer. None of the above parameters are adequate measures of turbulence in unstable and convective mixed layers. Their use could be very misleading, because, locally, $\partial\overline{\theta}/\partial z$ may be zero, slightly negative, or positive in a mixed layer with vigorous convective turbulence and upward heat flux. Therefore, the use of any gradient-based local parameter must be avoided in unstable and convective boundary layers.

The most appropriate parameter of the overall stability of the PBL is the ratio

$$\frac{h}{L} = -\frac{hk}{u_*^3}\frac{g}{T_o}\frac{H_o}{\rho c_p} = \frac{hk}{u_*^2}\frac{g\theta_*}{T_o} \quad (4.62)$$

which is based on the PBL similarity theory and is most recommended by boundary layer and air pollution meteorologists. In the absence of any information on the surface fluxes of momentum and heat (alternatively, u_* and θ_*), a bulk Richardson number of the PBL

$$\text{Ri}_h = \frac{g}{T_o}\frac{\Delta\bar{\theta}h}{\overline{V}_h^2} \quad (4.63)$$

in which $\Delta\bar{\theta}$ is the potential-temperature difference across the PBL depth, may be used as a substitute for h/L. Note that Ri_h does not have the limitations of the local bulk or the gradient Richardson number and is well defined for stable, unstable, or convective boundary layer. Both h/L and Ri_h can be considered as good measures of turbulence in the PBL, as well as of the relative effects of buoyancy and shear on turbulent mixing and diffusion. For example, it is easy to see that the ratio of convective and friction velocity scales is uniquely related to h/L in unstable and convective conditions, that is,

$$\frac{W_*}{u_*} = k^{1/3}\left(-\frac{h}{L}\right)^{1/3} \quad (4.64)$$

which can also be used to estimate h/L.

4.8.5 Turbulence

In the absence of on-site turbulence measurements, standard deviations of velocity fluctuations can be specified on the basis of similarity theory relations discussed in section 4.7 and the available experimental data from other sites. For the PBL as a whole, the following empirical similarity relations can be recommended for use in dispersion modeling applications:

$$\frac{\sigma_u}{u_*} = 2.5\left(1 - \frac{z}{h}\right)^a$$

$$\frac{\sigma_v}{u_*} = 1.9\left(1 - \frac{z}{h}\right)^a$$

$$\frac{\sigma_w}{u_*} = 1.3\left(1 - \frac{z}{h}\right)^a \quad (4.65)$$

with the value $a = 0.5$ to 1 based on the PBL turbulence data (Pasquill and Smith, 1983; Nieuwstadt, 1984a&b; Lenschow et al., 1988a&b) under neutral and stable conditions, and

$$\frac{\sigma_u}{w_*} = \frac{\sigma_v}{w_*} = \frac{\sigma_w}{w_*} \simeq 0.60 \quad (4.66)$$

for unstable and convective conditions. Here we have ignored the expected weak dependence of normalized σ's on the PBL stability parameter h/L, particularly in slightly stable and slightly unstable conditions, and also the height dependence of σ_w/W_* near the surface and near the top of the PBL.

The other turbulence variables used in diffusion modeling are the eddy diffusivities K_x, K_y, and K_z, which are best prescribed using similarity relations (4.25). The characteristic length scales of turbulence and their parameterization will be discussed in the next chapter. For obtaining simple, analytical solutions of the diffusion equation to be discussed later, eddy diffusivities are either assumed to be constants or specified as functions of height through power-law relations of the type

$$\frac{K}{K_r} = \left(\frac{z}{z_r}\right)^n \quad (4.67)$$

in which n is an exponent that may depend on the surface roughness and stability.

There is an approximate inverse relationship between n and m, but not necessarily the strict conjugate relationship $n = 1 - m$, which is often suggested in the literature. The conjugate relationship is obtained from the condition that the momentum flux must be independent of height in the surface layer, that is, $K\,\partial\bar{u}/\partial z = $ constant. The constant-flux condition is found to be inconsistent, however, with power-law profiles of both $\bar{u}$ and K. For example, in the constant-flux, neutral-surface layer, the eddy viscosity varies linearly with height ($n = 1$) while the velocity profile is logarithmic, which can be approximated by a power-law profile only with a finite exponent $m > 0$, so that the conjugate relationship between the two exponents is not satisfied ($m + n > 1$). The same is true for extremely stable conditions ($m \simeq 1$), as well as extremely unstable or free convective conditions ($n = 4/3$), so that, in general, $m + n > 1$.

Probably, the simplest and most frequently used characterization of stability and turbulence for diffusion modeling and parameterization in regulatory applications is the discrete stability classification or turbulence typing scheme originally proposed by Pasquill (1961) and later modified by D. B. Turner (1970). This scheme is given in Table 4.2.

"Strong" solar radiation corresponds to a solar altitude angle greater than 60° with clear skies, while "slight" insolation corresponds to a solar altitude less than 35° with clear skies. Cloudiness decreases incoming solar radiation and should be considered along with solar altitude in determining the strength of insolation. Objective methods of determining Pasquill's stability/turbulence classes from standard hourly meteorological observations have been suggested.

The primary advantages of Pasquill's stability classification scheme are its simplicity and its requirement of only routinely available information from surface meteorological stations, such as the near-surface (10 m) wind speed, solar radiation, and cloudiness. The discrete stability classes A through F are found to be almost linearly related to the lateral and vertical turbulence intensities. This can be seen from Table 4.3

Table 4.2 Meteorological Conditions Defining Pasquill's Stability/Turbulence Types*

Surface (10 m) Wind Speed (m s^{-1})	Daytime			Nighttime	
	Incoming Solar Radiation			Cloudiness	
	Strong	Moderate	Slight	≥4/8	≤3/8
<2	A	A–B	B	—	—
2–3	A–B	B	C	E	F
3–5	B	B–C	C	D	E
5–6	C	C–D	D	D	D
>6	C	D	D	D	D

*A, extremely unstable; B, moderately unstable; C, slightly unstable; D, neutral (applicable to overcast conditions day or night); E, slightly stable; F, moderately stable.

giving the approximate correspondence between Pasquill's stability types and standard deviations of horizontal and vertical wind direction fluctuations, which are related to turbulence intensities through Eqs. (4.16). It is appropriate only for steady-state conditions, a standard measurement height of 10 m, and for a flat rural terrain of roughness length $z_o \simeq 0.15$ m. The correspondence would break down during nighttime weak winds with highly variable wind direction.

There are several limitations of Pasquill's stability and turbulence typing scheme. The most severe is the fact that, unlike other continuously varying, quantitative stability parameters such as Ri, z/L, and h/L, the scheme uses only six discrete classes, each of which covers a rather wide range of stability and turbulence conditions. For example, lateral and vertical turbulence intensities (or σ_θ and σ_ϕ) nearly double in going from the lower end to the upper end of the range of values for the stability class E. It is even worse for the stability class F, which has apparently no lower limit for turbulence intensity. The approximate inverse relationship between the lateral turbulence intensity and stability breaks down in extremely stable conditions (a stability category higher than F and sometimes referred to as G), which are frequently encountered at night during weak and near-calm winds. Wind direction becomes highly variable and σ_θ values may correspond to anything between categories A and F. The other limitation is that the correspondence between Pasquill's stability classes and turbulence is not unique but strongly depends on the surface roughness, the PBL depth, and possibly other parameters. Any cor-

Table 4.3 Approximate Correspondence Between Pasquill's Stability Classes and Turbulence Parameters σ_θ and σ_ϕ

Pasquill's Stability Class	σ_θ (deg.)	σ_ϕ (deg.)
A	≥22.5	≥11.5
B	17.5–22.5	10.0–11.5
C	12.5–17.5	7.8–10.0
D	7.5–12.5	5.0–7.8
E	3.8–7.5	2.4–5.0
F	<3.8	<2.4

Source: Irwin, 1980.

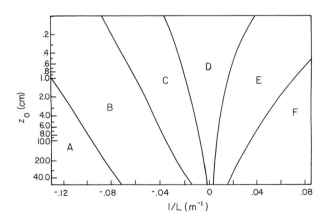

Figure 4.15 Dependence of Pasquill's stability classes on surface roughness and inverse Obukhov length. From Golder, 1972.

respondence between these stability categories and other surface-layer stability parameters, such as the Obukhov length and the Richardson number at a reference height, must also depend on the surface roughness. This is clearly shown in Figure 4.15, which is based on experimental data from five sites with different surface roughness characteristics.

One might expect considerable uncertainty in any empirical relationship between Pasquill's stability classes and the Obukhov length, because the former is based on the near-surface wind speed, insolation, and cloudiness whereas the latter is defined in terms of the surface fluxes of momentum and heat. Still, it is clear from Figure 4.15 that each stability class represents a range of L or $1/L$ values that depends on the roughness length. Pasquill's original stability categories and the associated dispersion parameterization scheme were proposed for a rather small roughness ($z_o \simeq 0.03$ m). The same classes, when applied to significantly rougher or smoother surfaces, could represent widely different turbulence and diffusion characteristics. Consequently, turbulence and diffusion parameters cannot be uniquely specified in terms of these qualitative stability categories alone. This is often ignored in regulatory guidelines and applications of dispersion modeling.

PROBLEMS AND EXERCISES

1. Discuss the scope of micrometeorology and its importance and applications to air pollution meteorology and atmospheric dispersion.
2. a. Discuss the important factors that determine the PBL depth or mixing height in the lower atmosphere.
   b. Give a schematic of the typical diurnal variation of the mixing height over land surfaces under clear skies, indicating the times of sunrise and sunset.
3. Discuss the primary exchange processes between the earth's surface and the atmosphere and their importance to the dispersion of air pollutants in the atmosphere.
4. a. Net radiation measurements at a flat and homogeneous site on a clear sunny day can be represented by a half-sine wave with an amplitude R_{Nmax} and a half-period P of the wave. If the ground heat flux at the site can be parameterized by $H_G = C_G R_N$ and the Bowen ratio B can be assumed to remain constant throughout the daytime, derive an expression for the surface heat flux from considerations of the surface energy budget.
   b. Using the measured or estimated values of $R_{Nmax} = 500$ W m^{-2}, $P = 13.5$ hr, $C_G = 0.04$, and $B = 0.33$, estimate the maximum values of the surface heat flux and the latent heat flux.
5. a. Discuss different types of measurements that can be used in conjunction with Eq. (4.5) for estimating the surface heat flux in the daytime convective boundary layer.
   b. Temperature measurements at the top of a 100-m stack indicate hourly mean temperatures of 23.0°C and 24.5°C at 1200 and 1300 hours, respectively, while a temperature sounding in the middle of the period indicated a mixing height of 850 m. Estimate the surface heat flux as well as the convective velocity and time scales at the time of these measurements.
6. a. Schematically show the diurnal evolution of the potential temperature profiles in the summertime unstable or convective boundary layer over land, indicating the tops of the surface layer and the mixed layer in each profile.
   b. Compare and contrast the stable nocturnal boundary layer from the mid-day convective boundary layer, considering their relevant characteristics for pollutant transport and diffusion.
7. a. Distinguish between local and nonlocal definitions of static stability. Which of these is more appropriate for characterizing turbulence and diffusion in the PBL and why?
   b. Why is a dynamic stability parameter, such as the gradient Richardson number, a better indicator of turbulence and mixing in a stably stratified layer than the static stability parameter?
8. The following observations of mean winds and temperatures were made at night from the 200-m mast at Cabauw in the Netherlands:

Height (m)	Wind Speed (ms^{-1})	Wind Direction (deg.)	Temperature (°C)
10	3.1	80	14.67
20	3.9	84	14.81
40	5.9	90	15.17
80	9.1	98	16.33
120	11.1	112	18.73
160	10.6	122	19.71
200	9.8	127	19.77

   a. Plot on a graph the wind speed, wind direction, and potential temperature profiles.
   b. Calculate and plot the component velocity profiles, with the x axis along the near-surface wind.
   c. Calculate and plot the static stability parameter s and the gradient Richardson number as functions of height.
   d. Estimate the mixing height based on the critical Richardson number of 0.25 and compare it with the observed value of 140 m.
9. a. Show that the standard deviation of the vertical wind direction (elevation angle) is approximately equal to the vertical turbulence intensity.
   b. How would you expect the lateral and vertical turbulence intensities to vary qualitatively with height, stability, and wind speed in the nocturnal boundary layer?
10. a. Derive the logarithmic wind profile law for the neutral-surface layer, using the eddy viscosity and mixing length hypotheses.
    b. What are the basic limitations of gradient transport theories when applied to the atmospheric boundary layer?
11. a. How would you specify the horizontal and the vertical eddy diffusivities in a gradient-transport model of a

contaminant diffusing in neutral and stable atmospheric boundary layers.

b. Calculate and plot the mixing length l_m as a function of height z, using the interpolation formula (4.28) for the various combinations of $l_b = 10, 30,$ and 100 m, and $l_o = 10$ and 100 m.

12. The following wind profile observations were made on a short grass surface, under near-neutral conditions, during the Wangara Experiment:

z (m):	0.5	1	2	4	8	16
$\bar{u}$ (ms^{-1}):	4.08	4.51	5.04	5.52	5.97	6.48

a. Using the graphical procedure, determine the roughness length and the friction velocity.

b. Estimate σ_u, σ_v, and σ_w in the surface layer.

c. Estimate the turbulent intensities i_v and i_w at the heights of 10, 30, and 100 m.

d. Estimate the vertical eddy diffusivity and mixing length at the same heights (10, 30, and 100 m).

13. a. List all the assumptions implied in the Monin–Obukhov similarity hypothesis, with plausible justifications for the same.

b. What is the physical significance of the Obukhov length (L)?

14. a. What are the limitations of using L or z/L as a stability parameter?

b. How would you estimate L in the absence of direct measurements of momentum and heat fluxes?

15. a. Using the Monin–Obukhov similarity relations (4.37)–(4.39), express eddy diffusivities of heat and momentum as functions of Ri and discuss the variation of their ratio K_h/K_m with stability.

b. Calculate and plot K_h/u_* as a function of height ($z = 5$–50 m) in the surface layer and the Obukhov length ($L = -100, -50, -10, 50,$ and 100 m).

16. a. Compare and contrast the local free-convection similarity hypothesis with the Monin–Obukhov similarity theory. Under what situations would the former be more appropriate?

b. How does mixed-layer similarity scaling differ from the local free-convection scaling? Show that in the free convective surface layer, $\sigma_w/W_* = C_w (z/h)^{1/3}$.

17. The following observations were made at a suburban site during the afternoon convective conditions:

Mean wind speed at 10-m height $= 2.5$ m s^{-1}
Mean temperature at 10-m height $= 17.0°C$
Sensible heat flux near surface $= 300$ W m^{-2}
Mixing height from sounding $= 1500$ m

Using the appropriate local free-convection or mixed-layer similarity relations, estimate the following: (a) u_f and W_*; (b) σ_v and σ_w at 10 m and 100 m heights; (c) average values of σ_v and σ_w for the mixed layer.

18. The following measurements were made of the hourly averaged wind speed and temperature at a homogeneous rural site:

	Daytime Run		Nighttime Run	
z (m)	$\bar{u}$ (ms^{-1})	$\bar{T}$ (°C)	$\bar{u}$ (ms^{-1})	$\bar{T}$ (°C)
2	3.34	29.04	1.57	23.63
8	3.98	28.10	2.94	24.94

For each of these runs, estimate the following, using the gradient method: (a) velocity and temperature gradients at 4 m; (b) gradient Richardson number at 4 m; (c) Obukhov length; (d) friction velocity and surface heat flux; (e) surface roughness length.

19. For the nighttime data minus the wind speed at 2 m of Prob. 18 and an estimated roughness length of 0.035 m for the site, estimate the friction velocity, surface heat flux, and Obukhov length using the profile method and compare results with estimates from the gradient method.

20. a. Calculate and plot the normalized power-law profiles for the exponent values of 0.1, 0.2, 0.3, 0.4, and 0.5, covering a normalized height (z/z_r) range from 0 to 50.

b. Discuss the observed or expected dependence of the power-law exponent on surface roughness and stability.

21. In certain atmospheric diffusion models, mean velocity and eddy diffusivity profiles in the lower part of the PBL are represented by the power-law relations (4.58) and (4.67) in which both the exponents are positive and finite.

a. Show that the condition of constant momentum flux in the surface layer implies a conjugate relationship $m + n = 1$.

b. Argue why it may not be desirable to require that $m + n = 1$.

c. By comparing Eqs. (4.58) and (4.67) with the corresponding Monin–Obukhov similarity relations, express m and n in terms of the M–O similarity functions.

d. Derive more specific expressions for m and n as functions of the PBL stability parameter h/L by matching the power-law and the similarity relations at the top of the surface layer ($z = h_s = 0.1$ h).

5

Statistical Description of Atmospheric Turbulence

5.1 REYNOLDS AVERAGING

We have already emphasized the importance of statistical approach in describing turbulence. The simplest statistical measures—averages, variances, and covariances—have been discussed in Chapter 4. These measures can tell much about the mean values of meteorological variables, relative magnitudes (standard deviations or intensities) of fluctuations around their mean values, and correlations between different fluctuating variables at the same time and the same point in space. But they tell nothing about the distribution of fluctuations around their means, about the correlations between fluctuations at different times and at different points in space, or about the frequency composition of time series of fluctuations. Information about these other characteristics of turbulence is provided by more sophisticated statistical measures, such as probability, correlation, and spectrum functions, which will be discussed in this chapter. All of these statistics are some kind of averages (time, space, or ensemble), which are required to follow the Reynolds averaging rules or conditions introduced in Chapter 2 and expressed again as follows:

$$\overline{f + g} = \bar{f} + \bar{g}$$
$$\overline{cf} = c\bar{f}, \text{ or } \overline{\bar{g}f} = \bar{g}\bar{f}$$
$$\overline{\left(\frac{\partial f}{\partial s}\right)} = \frac{\partial \bar{f}}{\partial s}$$
$$\overline{\int f \, ds} = \int \bar{f} \, ds \quad (5.1)$$

where f and g are fluctuating variables or some functions of variables, c is a constant, and $s = x, y, z,$ or t. The above conditions express the distributive property of summation and the commutivity of the operations of averaging (or integration) and differentiation. All averages behave like constants and are not affected by further averaging operations.

In the following statistical description of turbulence, we will consider only the fluctuating turbulent variables with zero means. Further restriction to stationary variables (time series) will be made for the sake of simplicity and also because only under stationary situations might time averaging be appropriate and time averages be equivalent to ensemble averages.

5.2 PROBABILITY FUNCTIONS

The average distribution of values of a fluctuating variable $u'(t)$ around its mean value of $\bar{u}' = 0$ can be described by a probability density or distribution function. The probability density function describes the probability that $u'(t)$ will assume a value within some defined range, say between u' and $u' + \Delta u'$. This probability may be obtained by determining and summing all the time intervals during which the variable falls within the specified range, that is,

$$\text{Prob}[u' < u'(t) \leq u' + \Delta u'] = \lim_{T \to \infty} \frac{T_u}{T} \quad (5.2)$$

where $T_u = \Sigma_i \Delta t_i$, with $i = 1, 2, \ldots$, are the time intervals and the summation is over all values of i. This is illustrated in Figure 5.1 in which the various time intervals corresponding to a specified range of $u'(t)$ between u' and $u' + \Delta u'$ are marked by dashed lines.

For small $\Delta u'$, one may expect the right-hand side of Eq. (5.2) to be proportional to $\Delta u'$ and define the probability density function $p(u')$ as a proportionality coefficient, that is,

$$\text{Prob}[u' < u'(t) \leq u' + \Delta u'] \simeq p(u')\Delta u' \quad (5.3)$$

106 STATISTICAL DESCRIPTION OF ATMOSPHERIC TURBULENCE

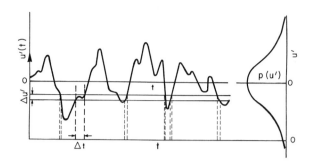

Figure 5.1 Schematic of determining the probability density function of a stationary turbulence variable. Modified after Tennekes and Lumley, 1972.

More precisely, $p(u')$ can be defined as

$$p(u') = \lim_{\Delta u' \to 0} \frac{\text{prob}[u' < u'(t) \leq u' + \Delta u']}{\Delta u'}$$

$$= \lim_{\Delta u' \to 0} \lim_{T \to \infty} \frac{T_u}{T \Delta u'} \quad (5.4)$$

The probability density function, by definition, is a real-valued, positive function, whose integral with respect to the variable over the whole possible range of values must be equal to unity, that is,

$$\int_{-\infty}^{\infty} p(u')du' = 1 \quad (5.5)$$

The probability that the instantaneous value $u'(t)$ is less than or equal to some specified value u' is given by the cumulative probability distribution function $P(u')$, which is related to $p(u')$ as

$$P(u') = \text{prob}[u'(t) \leq u'] = \int_{-\infty}^{u'} p(u)du \quad (5.6)$$

The cumulative probability distribution function is a real-valued, positive, monotonically increasing function of the variable, which is bounded by zero and one. When $P(u')$ is continuous and its derivative exists,

$$p(u') = \frac{\partial P(u')}{\partial u'} \quad (5.7)$$

Figure 5.2 shows the well-known Gaussian probability density function and the corresponding distribution function. These are also typical probability functions for randomly fluctuating atmospheric turbulence variables, such as velocity, temperature, and concentration fluctuations, in moderate and strong winds.

The knowledge of the probability density function is important in that the average of any function of a variable can be expressed in terms of the former as

$$\overline{f(u')} = \int_{-\infty}^{\infty} f(u')p(u')du' \quad (5.8)$$

A particular application of this relationship is the determination of moments, which are defined as averages of the integer powers of the variable. For example,

$$\overline{u'^n} = \int_{-\infty}^{\infty} u'^n p(u')du' \quad (5.9)$$

defines the nth moment. The first moment is the mean that, for a turbulence variable, must be equal to zero, that is,

$$\overline{u'} = \int_{-\infty}^{\infty} u'p(u')du' = 0 \quad (5.10)$$

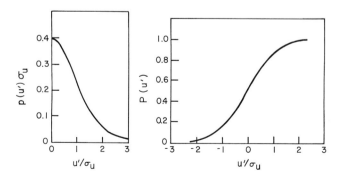

Figure 5.2 One-dimensional Gaussian probability density function $p(u')$ and the corresponding distribution function $P(u')$ of a stationary variable. After Lumley and Panofsky, 1964.

The second moment is called the variance and is defined by

$$\overline{u'^2} = \sigma_u^2 = \int_{-\infty}^{\infty} u'^2 p(u') du' \qquad (5.11)$$

The standard deviation σ_u is a measure of the width of the probability density function $p(u')$; it can also be considered a measure of the dispersion or spread of $u'(t)$ values. For example, about 99.6 percent of the area under the Gaussian probability density function shown in Figure 5.2 falls within the range of values $-3\sigma_u \leq u' \leq 3\sigma_u$. The second moment is not affected by any lack of symmetry in the probability density function.

The third moment u'^3 depends only on the lack of symmetry in $p(u')$; all odd moments are zero for a symmetrical probability density function. A dimensionless measure of the asymmetry of the probability density function is given by the skewness (S)

$$S = \frac{\overline{u'^3}}{\sigma_u^3} \qquad (5.12)$$

A positive value of skewness indicates that large positive values of the variable are more frequent than the large negative values. This often occurs in the convective boundary layer in which both the vertical velocity and temperature fluctuations show positive skewness ($S > 0$). Figure 5.3 shows the computed probability density of vertical velocity at different levels in the convective mixed layer, based on a three-dimensional numerical turbulence model (large eddy simulation). Here both $p(w')$ and w' are normalized by convective velocity W_*. Note that the vertical velocity has a zero mean but large negative mode (most probable value). The area under $p(w')$ on the negative side is about 50 percent larger than that on the positive side, indicating that downdrafts occupy more area than updrafts in a convective boundary layer. The mass continuity equation then requires that ascending motions (updrafts) be stronger than the descending motions (downdrafts). This is evident from the positive skewness of probability density functions shown in

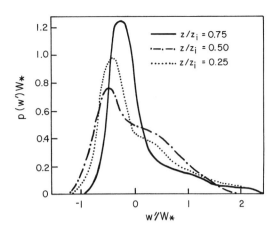

Figure 5.3 Probability density function of vertical velocity at three normalized heights in the convective boundary layer derived from a large-eddy simulation (LES). After Lamb, 1982.

Figure 5.3. Similar results have been obtained from field observations as well as from laboratory simulations of turbulence in the convective mixed layer (see Fig. 5.4).

The fourth moment $\overline{u'^4}$ is considered to be a measure of the peakiness or flatness of the probability density function. When normalized by the standard deviation, it defines the kurtosis or flatness factor

$$K = \frac{\overline{u'^4}}{\sigma_u^4} \qquad (5.13)$$

It can be easily shown that $K = 3$ for the Gaussian probability function, which is used as a standard for comparison. A smaller kurtosis is indicative of a flatter density function around the mode, but smaller tails at the extremities. A large kurtosis, on the other hand, indicates peaky function near the mode, but more extended tails on both sides, compared with the Gaussian form.

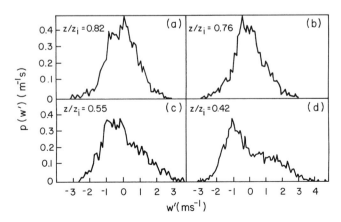

Figure 5.4 Measured probability density functions of vertical velocity at four normalized heights in the convective boundary layer. After Caughey et al., 1983.

It can be shown that all the moments taken together are equivalent to the probability density function. The main contribution to the fourth and higher moments, however, comes from the few extreme values of the variable, so that it is imperative to have very long, but statistically stationary, time series of turbulence variables for a satisfactory determination of these higher moments. The conflicting requirements of very long sampling times and stationarity are rarely met in the evolving atmospheric boundary layer under the prevailing diurnal and larger-scale influences. Therefore, it is not surprising to find almost total absence of the higher-order statistics in the reported measurements of atmospheric turbulence (for recent reviews, see Monin and Yaglom, 1971; Nieuwstadt and van Dop, 1982; Panofsky and Dutton, 1984; Stull, 1988). In terms of the probability density function, the tail ends are not satisfactorily determined due to the limitations of finite sampling time and stationarity. For this reason, probability functions or higher than third-order moments are seldom used in describing atmospheric turbulence, except, perhaps, in certain numerical simulations of atmospheric turbulence and diffusion.

5.3 AUTOCORRELATION FUNCTIONS

An autocorrelation function of time describes the relationship between any variable at a particular time to the same variable at a later or earlier time. In general, the autocorrelation function of $u'(t)$ is defined as

$$R'_{uu}(t_1, t_2) = \overline{u'(t_1)u'(t_2)} \tag{5.14}$$

where the overbar implies an ensemble average. For a stationary variable, the ensemble average can be replaced by the time average over a sufficiently long sampling or averaging time ($T \to \infty$) and the autocorrelation function depends only on the time difference $\tau = t_2 - t_1$, rather than on the absolute times t_1 and t_2. Thus, for a stationary variable, the autocorrelation function can be defined as

$$R'_{uu}(\tau) = \overline{u'(t)u'(t + \tau)}$$
$$= \lim_{T \to \infty} \frac{1}{T} \int_0^T u'(t)u'(t + \tau) dt \tag{5.15}$$

in which τ is called the time lag. In air pollution and diffusion literature, the normalized form of the autocorrelation function or the autocorrelation coefficient

$$R_{uu}(\tau) = \frac{\overline{u'(t)u'(t + \tau)}}{u'^2} \tag{5.16}$$

is more commonly used. It is easy to show from the definition that $R(\tau)$ is an even function, with its maximum value of unity at $\tau = 0$, that is,

$$R(\tau) = R(-\tau)$$
$$R(0) = 1$$
$$R(\tau) \leq R(0) = 1 \tag{5.17}$$

Here we have dropped the double subscript for the particular variable; $R(\tau)$ expresses the correlation between the values of the variable at two different times separated by τ. The correlation is expected to be perfect ($R = 1$) when there is no time lag involved, it is expected to decrease with increasing time lag, and become insignificant for sufficiently long time lags. In particular for randomly varying turbulent fluctuations with zero mean,

$$R(\tau) \to 0, \text{ as } \tau \to \pm \infty \tag{5.18}$$

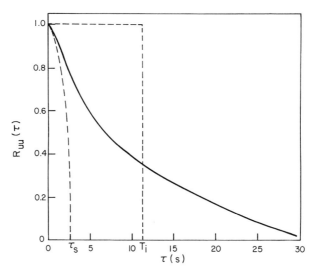

Figure 5.5 Schematic of the autocorrelation function of a stationary turbulence variable indicating the integral scale and Taylor's microscale of turbulence.

A schematic of the autocorrelation function of turbulent variables, possessing the above-mentioned properties, is shown in Figure 5.5. Several characteristic scales of turbulence can be defined from the shape of $R(\tau)$. The most widely used in diffusion and other applications is the integral time scale

$$T_i = \int_0^\infty R(\tau)d\tau \tag{5.19}$$

which is a crude measure of how long the variable remains significantly correlated with its previous values. By definition, T_i represents the area under $R(\tau)$ versus τ curve. It is often used to represent the characteristic time scale of large eddies in the flow that are considered to be responsible for significant correlations between the values of turbulence variables at different times. If one visualizes a large eddy or coherent structure passing over an observation point, one would expect to find significant correlations between the values of any variable during the passage of the eddy. Actually, $R(\tau)$ represents an average correlation coefficient, accounting for the influence of many large-eddy structures over the sampling duration $(T \gg T_i)$, while T_i can be considered as an average large-eddy time scale. In general, $R(\tau)$ and T_i will be different for different variables, such as the fluctuating velocities u', v', and w'. The variable subscripts will be used to distinguish them, whenever necessary.

Another time scale defined from the shape of the autocorrelation function is the Taylor microscale τ_s of turbulence

$$\tau_s = \left[-2 \Big/ \frac{d^2R(0)}{d\tau^2} \right]^{1/2} \tag{5.20}$$

which is determined by the second derivative or the curvature of the $R(\tau)$ versus τ curve at $\tau = 0$. A more physical interpretation of τ_s can be given if one approximates the autocorrelation function for small time lags $(\tau \to 0)$ by a Taylor series

$$R(\tau) = R(0) + \frac{dR(0)}{d\tau}\tau + \frac{d^2R(0)}{d\tau^2}\frac{\tau^2}{2!}$$
$$+ \cdots \simeq 1 + \frac{d^2R(0)}{d\tau^2}\frac{\tau^2}{2} \tag{5.21}$$

Note that τ_s is the intercept on the τ axis of the parabola (5.21) having the same curvature as $R(\tau)$ at $\tau = 0$. This is also schematically shown in Figure 5.5, although in reality $\tau_s \ll T_i$. The condition of stationarity requires that $dR(0)/d\tau = 0$ and $d^2R(0)/d\tau^2 < 0$, so that the autocorrelation function must have a cusp near $\tau = 0$. In reality, however, the cusp might be too small to be detected from the measurements of $R(\tau)$, because τ_s is usually very small (less than 0.1 s). The curvature $d^2R(0)/d\tau^2$ may be indirectly estimated, however, from the relationship

$$\overline{u'^2}\frac{d^2R_{uu}(0)}{d\tau^2} = \overline{u'(t)\frac{\partial^2 u'(t)}{\partial t^2}}$$
$$= -\overline{\left(\frac{\partial u'}{\partial t}\right)^2} \tag{5.22}$$

so that

$$\tau_s = \left[2\overline{u'^2} \Big/ \overline{\left(\frac{\partial u'}{\partial t}\right)^2} \right]^{1/2} \tag{5.23}$$

which requires an estimation of the variance of the time derivative of the variable of interest. The microscale is a characteristic scale of small (but not the smallest) eddies in the flow. Its estimation requires turbulence measurements with extremely small, fast-response sensors, such as hot-wire or hot-film anemometers.

Small-scale eddies are extremely important for the dissipation of turbulent kinetic energy (TKE) and, hence, in the overall TKE budget. But they do not contribute much to the diffusion and dispersion of contaminants. Therefore, the actual curvature of the autocorrelation function near $\tau = 0$ and the microscale of turbulence are not considered important in diffusion applications. The overall shape of $R(\tau)$ and the integral scale T_i are, however, very important. The autocorrelation function is often approximated by an exponential form

$$R(\tau) = \exp\left(-\frac{|\tau|}{T_i}\right) \tag{5.24}$$

which is not even consistent with the existence of microscale. However, Eq. (5.24) has other desirable properties of well-defined integral scale and $R(\tau) \to 0$ as $\tau \to \pm \infty$.

An empirically estimated autocorrelation function of an observed variable (time series) in the atmosphere may not satisfy the above requirements, if the variable has a trend or a periodicity superimposed on otherwise irregular or random fluctuations. It is easy to show from the definition of $R(\tau)$, that the autocorrelation function of a linear trend with time is a constant $(R(\tau) = 1)$ and the autocorrelation function of a sine wave of frequency ω is a cosine wave $(R(\tau) \propto \cos \omega \tau)$ of the same frequency. Furthermore, if a trend or periodicity is superimposed on a random variable, the autocorrelation of the combination will not go to zero as $\tau \to \infty$, but it will approach to a constant value or a periodic function of τ. In either case, one cannot define the integral scale. These inconsistencies arise because the requirements of stationarity and ergodicity are not satisfied by the observed data (time series). It is extremely important to remove any apparent or hidden trends and periodicities (e.g., the diurnal period) from observed time series of meteorological variables, such as velocity fluctuations, before computing their autocorrelation functions and associated integral scales. One should also ensure the condition that the mean value of the variable over the sampling duration is zero. Correlation functions are particularly sensitive to low-frequency errors and therefore depend on the method of separation of the fluctuating motion from the mean motion, as well as on the averaging time or sampling duration.

Analogous to the time autocorrelation function $R(\tau)$, spatial correlation functions are used to correlate values of any variable at different points in space, but at a fixed time, in a homogeneous field, for example,

$$R_{uu}(\underline{r}) = \frac{\overline{u'(\underline{x},t)u'(\underline{x}+\underline{r},t)}}{\overline{u'^2}} \quad (5.25)$$

In place of the simplifying condition of stationarity, the condition of spatial homogeneity must be satisfied in order for the autocorrelation function of the separation vector to be independent of spatial coordinates. Even then, $R(\underline{r})$ depends on both the magnitude and the direction of the separation vector. Alternatively, $R(\underline{r})$ can be represented by its three scalar components $R(r_1)$, $R(r_2)$, and $R(r_3)$. Each component has the same basic properties as the time autocorrelation function. For any variable, one can also define spatial microscales and integral scales of turbulence in different directions. Thus, spatial correlation functions, in principle, can depict the average, three-dimensional large-eddy structure. The spatial statistics (correlations and scales) is extremely difficult to obtain in practice, however, due to the limitations of sensor deployment in three-dimensional space and the lack of complete spatial homogeneity of mean flow and turbulence in the atmospheric boundary layer. Mean wind shear and buoyancy generally cause inhomogeneity of turbulence in the vertical direction. Therefore, only the correlations involving horizontal separations at fixed heights, over uniform and homogeneous surfaces (more appropriately, in horizontally homogeneous flows) are likely to be meaningful. A few of these have been reported in the literature (Lumley and Panofsky, 1964; Shiotani and Iwantani, 1976; King, 1988).

Measurements show an elongation of correlation patterns in the mean wind direction, indicating the largest integral-length scale in that direction. The correlation functions involving spatial separations along the mean wind direction are, therefore, considered to be special; these are also simply and directly related to the time autocorrelation functions through Taylor's frozen turbulence hypothesis, to be discussed later.

The integral length scale in the mean wind direction

$$L_1 = \int_0^\infty R(r_1) dr_1 \quad (5.26)$$

is considered to be the characteristic large-eddy-length scale in that direction. It is simply related to the integral time scale as $L_1 = \overline{u}T_i$, assuming advection of spatial eddy structure by mean flow.

5.4 SPECTRUM FUNCTIONS

Any variable as a function of time may be thought of as a superposition or combination of many sine waves of different frequencies. Spectrum functions are used to describe the frequency decomposition of data (time series) and to indicate the relative contribution of each frequency (or a range of frequencies) to the variance. The variance associated with the sinusoidal wave components within the range of frequencies ω and $\omega + \Delta\omega$ may be obtained by filtering the time series with an ideal band-pass filter having sharp-cutoff characteristics, and computing the average of the squared output from the filter (Bendat and Piersol, 1966), that is,

$$\overline{u'^2}(\omega,\Delta\omega) = \lim_{T\to\infty} \frac{1}{T} \int_0^T u'^2(t,\omega,\Delta\omega) dt \quad (5.27)$$

where $u'(t,\omega,\Delta\omega)$ denotes the filtered portion of $u'(t)$ in the frequency range from ω to $\omega + \Delta\omega$. For small $\Delta\omega$, the fractional variance $\overline{u'^2}(\omega,\Delta\omega)$ may be expected to increase in proportion to $\Delta\omega$, and a spectral density function $S'_{uu}(\omega)$ can be defined such that

$$\overline{u'^2}(\omega,\Delta\omega) \simeq S'_{uu}(\omega)\Delta\omega \quad (5.28)$$

or, more precisely,

$$S'_{uu}(\omega) = \lim_{\Delta\omega\to 0} \lim_{T\to\infty} \frac{1}{\Delta\omega T} \int_0^T u'^2(t,\omega,\Delta\omega) dt \quad (5.29)$$

The spectral density function, also commonly known as the power spectrum or the variance spectrum, is a real and positive function whose integral over all possible frequencies must be equal to the total variance, that is,

$$\int_0^\infty S'_{uu}(\omega) d\omega = \overline{u'^2} \quad (5.30)$$

In air pollution and diffusion literature, a normalized spectrum function $S_{uu}(\omega) = S'_{uu}(\omega)/\overline{u'^2}$ is more commonly used, such that

$$\int_0^\infty S(\omega) d\omega = 1 \quad (5.31)$$

where we have dropped the variable subscripts in the more general representation. One can also define the spectrum as a function of ordinary or linear frequency n (expressed in units of cycles per second, or hertz (Hz)) instead of the circular frequency ω (radians per second). The two spectrum functions are simply related as

$$S(n) = 2\pi S(\omega) \quad (5.32)$$

for corresponding values of $\omega = 2\pi n$. The above relationship follows from the more basic relation

$$S(n)dn = S(\omega)d\omega \quad (5.33)$$

which expresses the equality of fractional variances associated with the narrow range of frequencies ($d\omega = 2\pi dn$) in the two spectra centered around the frequency $\omega = 2\pi n$.

Here we have given an intuitively simple and physical definition of the spectrum function, rather than more rigorous mathematical representations involving ordinary Fourier series or integrals, or more complex Fourier–Stieltjes integrals (Lumley and Panofsky, 1964; Panofsky and Dutton, 1984). Spectral analysis is a powerful tool developed in mathematical statistics but extensively used in many other fields of science, engineering, and social sciences. Many textbooks and monographs have been written on this (see e.g., Blackman and Tukey, 1958; Priestley, 1981; Bendat and Piersol, 1986).

5.4.1 Relations between Correlation and Spectrum Functions

Mathematically, the autocorrelation and spectrum functions are related by the following Fourier transform relations:

$$S(\omega) = \frac{1}{\pi} \int_{-\infty}^{\infty} R(\tau) e^{-i\omega\tau} d\tau$$

$$= \frac{2}{\pi} \int_{0}^{\infty} R(\tau) \cos(\omega\tau) d\tau$$

$$R(\tau) = \int_{0}^{\infty} S(\omega) \cos(\omega\tau) d\omega \quad (5.34)$$

Here we are using the physically measurable one-sided spectrum function, instead of the two-sided theoretical spectrum function used in some other textbooks and monographs. In terms of the linear frequency, we have

$$S(n) = 4 \int_{0}^{\infty} R(\tau) \cos(2\pi n\tau) d\tau$$

$$R(\tau) = \int_{0}^{\infty} S(n) \cos(2\pi n\tau) dn \quad (5.35)$$

which gives $S(0) = 4 T_i$, indicating a proportionality relationship between the theoretically expected value of $S(n)$ at $n = 0$ and the integral time scale. In practice, it is not possible to directly estimate the spectral density at zero frequency, because it would require an infinite time series record.

The above mathematical relations between correlation and spectrum functions imply that the latter contains no information that is not contained also in the former. However, the spectral representation is more frequently used in statistical descriptions of turbulence, because it clearly depicts the relative contributions of different frequencies or eddy sizes (as indicated by wavelengths) to the total variance or turbulent kinetic energy. A characteristic large-eddy scale (time or length) can also be defined from the spectrum function. The frequency n_m, representing the maximum relative contribution $nS(n)$ to the variance, is used to define and estimate the large-eddy scale. The integral scale based on the correlation function might be expected to be proportional to the large-eddy time scale n_m^{-1} based on the spectrum function. Both types of scales are found to be useful in diffusion theories and models to be discussed later.

The measured spectra of atmospheric turbulence indicate a rather wide range of low frequencies (extending over several decades) contributing to the variance or energy. There is an equally wide range of high frequencies with little or no contribution to the variance or energy, which are primarily responsible for the molecular dissipation of variance or energy. Thus the range of frequencies of interest in atmospheric spectra is usually very wide. In the graphical representation of such spectra, it is most appropriate to use a logarithmic scale for the frequency along the abscissa. The ordinate represents either the spectral density $S(n)$ on a log scale or the fractional variance $nS(n)$ on a linear scale. The latter has the advantage of preserving the area under the curve

$$\int_{0}^{\infty} nS(n) d(\ln n) = \int_{0}^{\infty} S(n) dn = 1 \quad (5.36)$$

and is used for estimating the frequency n_m corresponding to the peak in $nS(n)$. Therefore, atmospheric spectra are most commonly represented as $nS(n)$ versus n, using linear and log scales, respectively. Log–log plots are also used sometimes to show a particular power-law behavior of the spectrum predicted by theory.

5.4.2 Wave-number Spectra

Analogous to the frequency spectra of time series, one can introduce the wave-number spectra of spatial series, which represent the relative contributions of elementary sine waves of different wave numbers (number of waves per unit length, or radians per unit length) to the total variance or energy. Since wave number is a vector quantity, one can define three-dimensional spectra of vector wave number, or conceptually simpler one-dimensional spectra of component wave numbers. Wave-number spectra are more commonly used in theory than they are in practical measurements. Simultaneous measurements in space at spatial resolutions fine enough to estimate wave-number spectra are rarely possible in the atmosphere or even in the laboratory. An instrumented aircraft rapidly moving through the atmosphere provides probably the closest approximation to a spatial series (it actually is a space-time series).

Wave-number spectra are almost always estimated from measurements with time at fixed points in space or on moving platforms, such as an aircraft. We illustrate in Figure 5.6 the simple relationships between temporal and spatial sinusoidal waves that are advected at mean speed $\bar{u}$ in the x direction. Here, P denotes the wave period, λ the wavelength, κ_1 the circular wave number (rad. m^{-1}), and κ_1' the linear wave number (m^{-1}).

The spectrum functions $S(\kappa_1)$ and $S(\kappa_1')$ are related to the spatial autocorrelation function $R(r_1)$ as

$$S(\kappa_1) = \frac{2}{\pi} \int_{0}^{\infty} R(r_1) \cos(\kappa_1 r_1) dr_1$$

$$S(\kappa_1') = 4 \int_{0}^{\infty} R(r_1) \cos(2\pi \kappa_1' r_1) dr_1 \quad (5.37)$$

which are analogous to Eqs. (5.34) and (5.35). The two spectrum functions differ by a factor of 2π, for

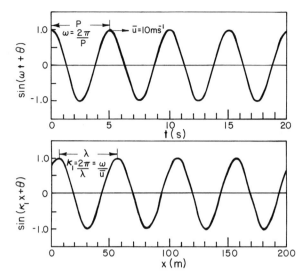

Figure 5.6 Schematic of a sinusoidal wave of frequency ω propagating at a mean velocity $\bar{u}$ in the x direction, illustrating the relationship between temporal and spatial wave characteristics.

corresponding values of $\kappa_1 = 2\pi\kappa_1'$. The relationships between wave-number spectra and corresponding frequency spectra simply follow from the relations between wave numbers and frequencies shown in Figure 5.6 and the condition

$$S(\kappa_1)d\kappa_1 = S(\omega)d\omega = S(n)dn \qquad (5.38)$$

These relationships are

$$S(\kappa_1) = \bar{u}S(\omega), \quad \text{for } \kappa_1 = \omega/\bar{u}$$

$$S(\kappa_1') = \bar{u}S(n), \quad \text{for } \kappa_1' = n/\bar{u} \qquad (5.39)$$

which are used to estimate the one-dimensional wave-number spectra from the measured frequency spectra.

5.5 TAYLOR'S HYPOTHESIS

In the previous section we showed how certain wave-number spectra can be related to frequency spectra if each elementary wave component in space can be assumed to be advected in the mean flow direction at a fixed mean velocity (wave speed = $\bar{u}$). Since spectral representation of a field of turbulence in space implies that the random field can be decomposed in terms of an infinite number of elementary sine waves, each moving at the same mean flow velocity, the whole turbulence structure may be assumed to be advected by the mean flow. This idea was originally proposed by Taylor (1938) as a plausible assumption or hypothesis:

> If the velocity of air stream which carries the eddies is very much greater than the turbulent velocity, one may assume that the sequence of changes in u' at the fixed point are simply due to the passage of an unchanging pattern of turbulent motion over the point, i.e., one may assume that $u'(t) = u'(x/\bar{u})$ where x is measured upstream at time $t = 0$ from the fixed point where u' is measured.

This hypothesis was originally suggested for a homogeneous field of low-intensity turbulence in a uniform mean flow. Taylor used it to derive the appropriate relationships between spatial and temporal statistics (e.g., correlations and spectra). We have already discussed the spectral relations. The spatial and temporal correlation functions are also simply related as

$$R(r_1) = R(\tau), \quad \text{for } r_1 = \bar{u}\tau \qquad (5.40)$$

which imply simple relations between space and time scales

$$L_1 = \bar{u}T_i; \quad \lambda_1 = \bar{u}\tau_s \qquad (5.41)$$

where λ_1 is called Taylor's microscale. Here the subscript 1 denotes the longitudinal or the mean flow direction.

Taylor's hypothesis implies a direct correspondence between spatial changes in a turbulence variable in the mean flow direction and its temporal changes at a fixed point, for example,

$$u'(x,y,z,t) = u'(x - \bar{u}t, y, z, 0) \qquad (5.42)$$

$$\frac{\partial u'}{\partial t} = -\bar{u}\frac{\partial u'}{\partial x} \qquad (5.43)$$

which, of course, implies an unchanging or frozen pattern of turbulence. In reality, turbulence changes continuously with time, as well as in space. Therefore the assumption of frozen turbulence appears to be questionable when applied to instantaneous values of variables or their derivatives as in Eqs. (5.42) and (5.43).

The corresponding relations between average statistics derived from Eq. (5.42) or (5.43) are likely to be more robust, for example,

$$\overline{\left(\frac{\partial u'}{\partial t}\right)^2} = \bar{u}^2 \overline{\left(\frac{\partial u'}{\partial x}\right)^2} \qquad (5.44)$$

and other relations between spatial and temporal statistics. The hypothesis is mostly used in connection with the average statistics. Earlier, it provided a practical and convenient means of testing the results of the statistical theory of homogeneous and isotropic turbulence (Batchelor, 1953; Hinze, 1975; Monin and Yaglom, 1975), using grid turbulence produced in wind tunnels and water channels. Subsequently, the hypothesis has been extended and generalized to all turbulent variables and also to nonhomogeneous shear flows. The basic assumption implied in the hypothesis, however, remains the same, that changes with time at a fixed point are entirely due to advection by mean flow.

Taylor's hypothesis has been critically examined theoretically, as well as experimentally, by many investigators. Two types of limitations are found on the validity of the hypothesis in atmospheric flows. The first arises from finite turbulence intensities, which have the apparent effect of increasing the mean advection velocity. The ratio of the apparent or effective advection velocity ($\bar{u}_e$) to the actual mean velocity ($\bar{u}$) is found to depend on the turbulent quantity whose temporal and spatial characteristics (averages) are to be compared or related (see e.g., Wyngaard and Clifford, 1977). For the second-order statistics, such as autocorrelations and spectra, the ratio $\bar{u}_e/\bar{u}$ is close to unity and Taylor's hypothesis can be used without any corrections, if turbulence intensities are less than about 20 percent. The validity of the frozen-turbulence hypothesis becomes questionable, however, in free-convection conditions when standard deviations of velocity fluctuations are of the same order or larger than the mean velocity.

The second limitation of Taylor's hypothesis arises from the mean flow shear. In atmospheric applications, we are primarily concerned with the vertical wind shear that is often responsible for producing turbulence. In the presence of shear, large eddies may not be advected at the same local mean velocity that small eddies are, because large velocity differences are expected to exist across large eddies. A wellknown theoretical criterion for the validity of Taylor's hypothesis in nonuniform (shear) flows is

$$\omega = \kappa_1 \bar{u} \gg \left|\frac{\partial \bar{u}}{\partial z}\right| \qquad (5.45)$$

which says that only eddies with frequencies much greater than the magnitude of mean wind shear may be assumed to be transported by the local mean velocity, as implied by Taylor's hypothesis. The above criterion was experimentally verified for the atmospheric boundary layer by Powell and Elderkin (1974), who suggested a more definite empirical criterion, $\omega \geq 3.6 \ |\partial \bar{u}/\partial z|$, for the validity of the frozen-turbulence hypothesis under stable and near-neutral conditions. Taylor's hypothesis has also been verified through comparisons of the frequency spectra measured on an instrumented tower and the wave-number spectra measured by an aircraft at the same height (see e.g., Kaimal et al., 1982).

Taylor's hypothesis has widespread applications to the dispersion of pollutants. Conversion between the average travel time and the mean travel distance of particles released in the atmosphere is routinely made using it. Like turbulent eddies, clouds or puffs of material are assumed to be transported by local mean flow as they diffuse and grow in size under the influence of turbulence. In the presence of strong wind shear, different parts of puffs would move at different speeds and in different directions, so that the validity of the hypothesis then becomes questionable.

5.6 STATISTICAL THEORY OF TURBULENCE

5.6.1 Homogeneous and Isotropic Turbulence

A sophisticated mathematical and statistical theory of turbulence has been developed for the simplest case of homogeneous and isotropic turbulence (Batchelor, 1953; Monin and Yaglom, 1975). This idealized field of turbulence has no boundaries, no mean flow and shear, and no buoyancy effects that may cause deviations from homogeneity and isotropy. Consequently, there are no momentum, heat, and mass fluxes in any direction and the variance of velocity fluctuations is the same in all directions. In the absence of primary production mechanisms (shear and buoyancy), an isotropic turbulence cannot be stationary, but decays with time.

Considerations of kinematics with incompressibility and symmetry conditions lead to general expressions of two-point (spatial) velocity correlations of second, third, and higher order, in terms of a few simple scalar correlation coefficients (see e.g., Hinze, 1975). Similar expressions have been derived for the various spatial spectrum functions. Further considerations of dynamics, using the equations of motion, yield dynamical equations for spatial correlations and spectra. The latter are found to contain an unknown third-order correlation coefficient or its wave-number spectrum. For example, the dynamical equation for the energy spectrum $E(\kappa)$, defined such that $\int_0^\infty E(\kappa)d\kappa = \bar{e}$, is given by

$$\frac{\partial E(\kappa)}{\partial t} = T(\kappa) - 2\nu\kappa^2 E(\kappa) \qquad (5.46)$$

where $T(\kappa)$ is the unknown energy transfer spectrum function that represents the transfer of energy from other wave numbers to the particular wave number κ. There is no net transfer when one considers all the wave numbers, so that

$$\int_0^\infty T(\kappa)d\kappa = 0 \qquad (5.47)$$

The second term on the right-hand side of Eq. (5.46) is the local (at the particular wave number κ) dissipation of energy by viscosity; its integral over all wave numbers must equal the total rate of dissipation of turbulence kinetic energy, that is,

$$2\nu \int_0^\infty \kappa^2 E(\kappa)d\kappa = \varepsilon \qquad (5.48)$$

In the integrated form, Eq. (5.46) becomes the equation for TKE

$$\frac{\partial \overline{e}}{\partial t} = -\varepsilon \qquad (5.49)$$

that states that the rate of energy decay of an isotropic turbulence is equal to the rate of energy dissipation.

5.6.2 Nonhomogeneous Turbulence

Atmospheric turbulence is not homogeneous and isotropic, due to the presence of mean wind shears and thermal stratification, which are actually responsible for the generation and maintenance of turbulence. This can be easily seen from the simplified turbulence kinetic energy equation for the horizontally homogeneous flow (Stull, 1988)

$$\frac{\partial \overline{e}}{\partial t} = -\overline{u'w'}\frac{\partial \overline{u}}{\partial z} - \overline{v'w'}\frac{\partial \overline{v}}{\partial z} + \frac{g}{T_o}\overline{w'\theta'}$$

$$- \frac{\partial}{\partial z}\left(\overline{e'w'} + \frac{1}{\rho_o}\overline{p'w'}\right) - \varepsilon \qquad (5.50)$$

where $e' = (u'^2 + v'^2 + w'^2)/2$ is the fluctuating energy. Note that the consideration of vertical inhomogeneity alone leads to the addition of what are called the shear production terms, the buoyancy term, and the turbulence transport terms to the much simpler TKE budget equation (5.49) for the isotropic turbulence.

The shear production terms in Eq. (5.50) represent an interaction between the appropriate components of momentum flux and mean wind shear. These represent the conversion of kinetic energy of mean motion into that of turbulent motion. These are most important source terms near the surface, but decrease rapidly with height because both the mean wind shear and the momentum flux normally decrease with height above the surface.

The buoyancy term in Eq. (5.50) represents the conversion of potential energy into turbulent kinetic energy, or vice versa. This term can be positive (source) or negative (sink), depending on the direction of the heat flux. It is the source or production term in unstable and convective conditions and the destruction term in stable stratification. The ratio of negative buoyancy to shear production terms is called the flux Richardson number

$$\mathrm{Rf} = \frac{g}{T_o}\frac{\overline{w'\theta'}}{[\overline{u'w'}(\partial \overline{u}/\partial z) + \overline{v'w'}(\partial \overline{v}/\partial z)]} \qquad (5.51)$$

which is another measure of the relative effects of buoyancy and shear on turbulence. Writing turbulent fluxes in terms of mean gradients, Eq. (5.51) may be written as

$$\mathrm{Rf} = \frac{K_h}{K_m}\mathrm{Ri} \qquad (5.52)$$

Recognizing that the ratio K_h/K_m is some unique function of stability (Ri or Rf), Ri and Rf are equivalent parameters as measures of buoyancy effects on turbulence. The gradient Richardson number is simpler to define and estimate than the flux Richardson number. The former is well defined for turbulent as well as non-turbulent flows, whereas Rf can be defined only for stratified turbulent flows. The criterion for the critical condition for maintenance of turbulence in stably stratified flows is more easily derived in terms of Rf (Richardson, 1920; Arya, 1972). The best empirical and theoretical estimates of the critical flux Richardson number range from 0.15 to 0.25, with a middle value of 0.20 often used for the atmospheric surface layer. The TKE equation (5.50) gives only an upper bound of unity for Rf, which may not be reached due to finite energy dissipation observed under conditions approaching the critical, when a sudden breakdown of turbulence may occur. There is a more rigorous theoretical criterion of Ri $<$ Ri$_c$ = 0.25 for the instability of a stratified laminar or inviscid flow and possible transition to turbulence.

The turbulence kinetic energy budget is commonly used for describing the turbulence structure of the atmospheric PBL. The various source and sink terms in Eq. (5.50) are measured or estimated and their balance is represented as a function of dimensionless height z/h. These energy budget estimates show that the time rate of change term is usually negligible, while the shear production, buoyancy, and dissipation terms are most important throughout the PBL. The turbulent transport term is relatively small in stable conditions, but becomes very important in unstable and convective boundary layers. For a more detailed discussion of the TKE budget and the importance of the various terms in the same, the reader may refer to Stull (1988).

The presence of vertical inhomogeneity considerably complicates the spectral energy budget also. One can, however, identify similar shear production, buoyancy, and turbulent transport terms in the wave number space as in the TKE budget Eq. (5.50). Figure 5.7 shows a schematic of spectral energy budget in a stably stratified homogeneous shear flow (Lumley and Panofsky, 1964). The energy spectrum is also shown in the same figure, even though it has different dimensions from those of the energy budget terms. Note

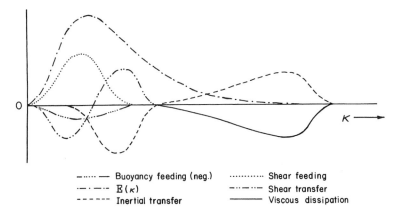

Figure 5.7 A schematic of spectral energy budget in a stably stratified homogeneous shear flow. After Lumley and Panofsky, 1964.

that the peaks in the production and dissipation terms are separated by four decades on the frequency or wave-number scale. Practical and theoretical implications of this will be discussed in the following.

5.6.3 The Energy Cascade Hypothesis

There has been considerable interest in knowing how turbulence kinetic energy is transferred between different wave numbers or eddy sizes. Even before the concept and measurements of an energy spectrum were introduced to the field of turbulence by Taylor (1935) and others, Richardson (1922) suggested a hypothetical mechanism of energy transfer between turbulent eddies in the form of a parody:

> Big whorls have little whorls,
> Which feed on the velocity;
> And little whorls have lesser whorls,
> And so on to viscosity.

It was recognized even earlier that in large-Reynolds-number turbulent flows, almost all of the energy is supplied by mean flow to large eddies, while almost all of the energy is dissipated by very small eddies. Richardson suggested that the transfer of energy from large (energy-containing) eddies to very small (energy-dissipating) eddies occurs through a cascade-like process involving a whole wide range of intermediate-size eddies. This energy cascade hypothesis provided the foundation for the more formal statistical and local similarity theories of turbulence proposed by Kolmogorov (1941) and others.

The measured spectral energy budget in Figure 5.7 clearly indicates a range of low frequencies or wave numbers that contribute to the shear and buoyant production of energy and a range of very high frequencies or wave numbers that are responsible for the dissipation of energy. The energy transfer term, estimated as an imbalance between the production and dissipation terms, is negative, indicating a sink of energy, at low wave numbers and a source term at high wave numbers. Consequently, energy must be transferred continuously down the scale from the largest energy-containing eddies to the smallest dissipating eddies. The actual mechanism of energy transfer at each step of the "cascade" cannot be easily isolated and examined. Smaller eddies are assumed to be created through instability and breakdown of slightly larger eddies, and energy transfer from the latter to the former occurs during this process. The largest eddies are produced by shear and buoyancy forces driving the mean flow. These are characterized by a length scale l, which is of the same order as the characteristic length scale of the mean flow, and a velocity scale u of the order of $e^{1/2}$. If the large-eddy Reynolds number $Re_l = ul/\nu$ is sufficiently high, these large eddies become unstable and produce eddies of a somewhat smaller size, which themselves become unstable and produce eddies of still smaller size, and so on further down the scale. This cascade process is terminated when the Reynolds number based on the smallest eddy scales becomes small enough (order of unity) for these smallest eddies to become stable under the influence of viscosity. This qualitative concept of energy cascade provided an underpinning for subsequent theoretical ideas, as well as experimental investigations of turbulence.

5.6.4 Kolmogorov's Local Similarity Theory

Kolmogorov (1941) postulated that at sufficiently large Reynolds numbers, small-scale motions in all turbulent flows should have some universal characteristics. He proposed a similarity theory to describe these characteristics. The basic foundation of his widely accepted theory lies in the energy cascade hypothesis and the presumed mechanism of energy transfer through successive instabilities and breakdown of eddies or disturbances in the flow.

Kolmogorov argued that if the characteristic Reynolds number of the mean flow or that of the most energetic large-scale eddies or disturbances is sufficiently large, there will be many steps in the energy cascade process before it terminates at the smallest scales allowable by the fluid viscosity. The ratio l/η between the largest (most energetic) and the smallest (most dissipating) scales determines the range of eddy sizes involved; it is expected to increase with increasing Reynolds number (Re_l). The large-scale motions (eddies) that receive energy directly from mean shear or buoyancy are expected to be inhomogeneous and anisotropic. But the small eddies that are formed after many successive breakdowns of large and intermediate-size eddies are likely to become homogeneous and isotropic, because they are far enough removed from the original sources of inhomogeneity (shear and buoyancy) and have no memory of the large-scale processes. Following this reasoning, Kolmogorov (1941) proposed his local isotropy hypothesis, which states that at sufficiently large Reynolds numbers, small-scale structure is locally isotropic whether large-scale motions are isotropic or not. Here "local" refers to the range of small-scale turbulent motions or disturbances that may be considered isotropic.

The concept of local isotropy proved to be very useful in that it applied to all turbulent flows. It made possible for the previously developed statistical theory of homogeneous and isotropic turbulence to be applied to small-scale motions in all the practical turbulent flows. The condition of sufficiently large Reynolds number is particularly well satisfied in the atmosphere. Consequently, the local isotropy hypothesis is found to be valid over a wide range of scales and eddies in the atmosphere.

Equilibrium Range Hypothesis For locally isotropic turbulence, Kolmogorov (1941) proposed two similarity hypotheses. His first hypothesis, also called the equilibrium range hypothesis, states that at sufficiently high Reynolds numbers there is a range of small scales or high wave numbers for which turbulence structure is in statistical equilibrium and is uniquely determined by the rate of energy dissipation (ε) and kinematic viscosity (ν). Here the statistical equilibrium refers to the stationarity of small-scale turbulence, even when the large-scale turbulence may not be stationary. The existence of stationarity or equilibrium condition for small-scale turbulence is justified, because the characteristic time scale of small-scale motions is much smaller than that for the most energetic large-scale turbulent motions, or for the time scale associated with changes in the latter.

According to the above similarity hypothesis, ν and ε are the only parameters that are appropriate for the dynamics of turbulence in the locally isotropic equilibrium range. These parameters lead to the following Kolmogorov's microscales:

Length: $\eta = \nu^{3/4}\varepsilon^{-1/4}$

Velocity: $\upsilon = \nu^{1/4}\varepsilon^{1/4}$ (5.53)

which should be used to normalize turbulent statistics and represent them as universal functions of appropriate dimensionless parameters. For example, the energy spectrum in the equilibrium range can be assumed to be a function of κ, ε, and ν, that is,

$$E(\kappa) = f(\kappa, \varepsilon, \nu) \quad (5.54)$$

or, in the dimensionless form,

$$\frac{E(\kappa)}{\upsilon^2 \eta} = E_*(\kappa \eta) \quad (5.55)$$

where E_* is a universal function of $\kappa\, \eta$. Thus, Kolmogorov's first similarity hypothesis predicts a universal spectrum for small-scale turbulence, irrespective of its origin, in large Reynolds number flows. Similar predictions are made for probability, correlation, and structure functions (see e.g., Monin and Yaglom, 1975).

Approximate relationships between large-eddy scales and Kolmogorov's microscales can be obtained from the definitions of the latter and the commonly used parameterization for the rate of energy dissipation $\varepsilon \sim u^3/l$. These are

$$\frac{l}{\eta} \sim Re_l^{3/4}$$

$$\frac{u}{\upsilon} \sim Re_l^{1/4} \quad (5.56)$$

where $Re_l = ul/\nu$. The above relations show that the ratio of macroscale to microscale of turbulence increases with increasing Reynolds number; for the commonly encountered range of $Re_l = 10^5$ to 10^8 in the atmosphere, the ratio $l/\eta \simeq 10^4$ to 10^6. The microscale η is of the order of millimeters and varies over a narrow range (0.5–5 mm). On the other hand, the large-eddy scale in the lower atmosphere varies over a much wider range ($l \sim 10$–10^3 m).

Inertial Subrange Hypothesis Kolmogorov's second similarity hypothesis is concerned with the inertial subrange of eddies or wave numbers in which the energy is transferred from larger eddies to smaller eddies at a fixed rate determined by the rate of energy dissipation. It states that at sufficiently high Reynolds numbers, there is an inertial subrange of wave numbers in which turbulent kinetic energy is neither produced nor dissipated and the turbulence structure depends only on ε. It is based on experimental observation as well as on the theoretical expectation that the dissipation of energy by viscosity is confined to the high wave-number end of the spectrum called the dissipation subrange. This is quite obvious from the spectral energy budget shown in Figure 5.7, in which we have also identified the various ranges and subranges of wave numbers. Note that both the production and dissipation of energy become insignificant in the inertial subrange. An important consequence of the above similarity hypothesis is that the energy spectrum in the inertial subrange is given by

$$E(\kappa) = \alpha \varepsilon^{2/3} \kappa^{-5/3} \quad (5.57)$$

in which α is a universal Kolmogorov's constant. The corresponding forms of the one-dimensional variance spectra as functions of the longitudinal wave number in the inertial subrange are

$$S'_{uu}(\kappa_1) = \alpha_1 \varepsilon^{2/3} \kappa_1^{-5/3}$$
$$S'_{vv}(\kappa_1) = S'_{ww}(\kappa_1) = \alpha_2 \varepsilon^{2/3} \kappa_1^{-5/3} \quad (5.58)$$

in which the various constants are related through the following invariant relations for the isotropic turbulence:

$$\alpha_1 = \frac{3}{4}\alpha_2 = \frac{18}{55}\alpha \quad (5.59)$$

Many observations of turbulence spectra in the atmosphere, as well as in laboratory flows at high Reynolds numbers have confirmed the predictions of Kolmogorov's local similarity theory (see Monin and Yaglom, 1975). In particular, the atmospheric turbulence spectra display an extensive inertial subrange with the $\kappa^{-5/3}$ spectral behavior, as shown in the following section.

5.7 OBSERVED SPECTRA AND SCALES

Observations of atmospheric turbulence have been made using instrumented towers, tethered balloons/kytoons, aircraft, and more recently, remote sensing devices. The most frequently derived statistics are the variances and covariances, as well as their frequency spectra. There are only a few reported measurements of correlation functions, which are of course related to spectra. Measurements of probability functions and higher-order moments of atmospheric turbulence have been rarely reported (see Fig. 5.4). Here we give a brief overview of the observed correlations and spectra of atmospheric turbulence and the associated large-eddy scales. For more comprehensive and detailed reviews of the same, the reader should refer to Monin and Yaglom (1975), Panofsky and Dutton (1984), and Stull (1988).

5.7.1 The Surface Layer

A comparison of time and space correlations for the longitudinal, lateral, and vertical velocity components over a moderately rough ($z_o \simeq 0.03$ m) and flat terrain under near-neutral stability conditions is given in Figure 5.8. These are based on turbulence measurements in the surface layer, using a row of instrumented towers logarithmically spaced at 4, 8, 16, 32, 64, and 128 m separations along the mean wind direction. The time correlations based on the slow response propellor anemometers (Gill) and fast response sonic anemometers are somewhat different, especially at small time lags. The abscissa scale along the bottom of each figure is for space lag and that along the top is for time lag such that $\tau = r_1/\bar{u}$. Note that there is a good correspondence between the sonic time and space correlation functions, verifying the validity of

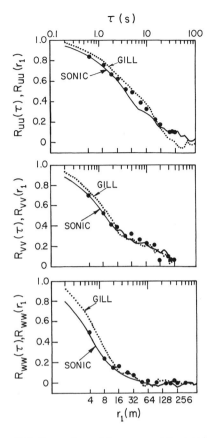

Figure 5.8 Comparison of the measured time and space autocorrelation functions for each of the three turbulence velocity components in the surface layer. Solid and dashed curves represent time correlations measured by sonic and Gill anemometers, respectively, and large dots represent space correlations. After Powell and Elderkin, 1974.

Taylor's hypothesis. The correlation data also indicate much larger scales for the horizontal components than for the vertical component.

The variance spectra of three components corresponding to the temporal correlation functions of Figure 5.8 (sonic) are shown in Figure 5.9. These are normalized using the usual surface-layer similarity scaling, but are otherwise raw (unsmoothed and uncorrected) spectra. The spectra show the theoretically predicted inertial subrange behavior at high frequencies where $nS(n) \propto n^{-2/3}$. At the very low frequency end also the spectrum drops off, with a peak or broad maximum occurring at some low frequency. The frequency n_m, or $f_m = n_m(z/\bar{u})$, corresponding to the peak or maximum in the spectrum $nS(n)$ is often used to define the characteristic wavelength $\lambda_m = \bar{u}/n_m$ of large eddies that contribute most to the variance. Note that the peak wavelengths differ considerably for the three components and, in general, $\lambda_{mu} > \lambda_{mv} > \lambda_{mw}$. These

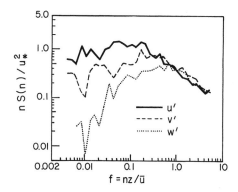

Figure 5.9 Normalized spectra of velocity variances in the near-neutral surface layer. After Powell and Elderkin, 1974.

f_m on z/L is shown in Figure 5.11. Note that the normalized peak spectral wavelength λ_m/z is the inverse of f_m and decreases with increasing stability (z/L). On the stable side, the following approximate relations between the peak frequencies and wavelengths corresponding to different velocity spectra are suggested:

$$f_{mu} \simeq 0.4 f_{mv} \simeq 0.2 f_{mw}$$
$$\lambda_{mu} \simeq 2.5 \lambda_{mv} \simeq 5 \lambda_{mw} \qquad (5.60)$$

In unstable and convective conditions, on the other hand, λ_{mu} and λ_{mv} are expected to be approximately equal and proportional to the PBL height ($\lambda_{mu} \simeq \lambda_{mv} \simeq 1.5\,h$), whereas λ_{mw} is proportional to the local height.

5.7.2 Convective Boundary Layer

The most comprehensive and systematic study of turbulence structure in the CBL has been reported by Kaimal et al. (1976). It was based on turbulence measurements from a 32-m tower and a tethered balloon during the 1973 Minnesota boundary-layer experiment. The normalized spectra, using the mixed-layer

are typically four to six times larger than the corresponding integral-length scales based on the autocorrelation functions.

Many studies have been made of the effects of stability on the spectra and scales of turbulence in the atmospheric surface layer. Probably the most comprehensive and systematic study was that of Kaimal et al. (1972) during the 1968 Kansas experiment. In that experiment, turbulence measurements were made with fast response sensors over a wide range of stabilities ($-2.1 < z/L < 3.3$) at three different levels in the surface layer. The spectra were computed using the fast Fourier transform technique over a fairly wide range of frequencies 3×10^{-4} to 10 Hz.

All the Kansas turbulence spectra showed the inertial subrange from which a best estimate of the Kolmogorov constant $\alpha_1 \simeq 0.50$ was made. The lower frequency end of the inertial subrange depends on the particular velocity component, measurement height, mean wind speed, and the Obukhov length. In terms of the normalized frequency $f = nz/\bar{u}$, the inertial subrange begins at much lower values of f in convective conditions than in very stable conditions. This is clearly shown in the appropriately normalized spectra of Figure 5.10, which were averaged and smoothed over specified ranges of z/L. The particular normalization by $u_*^2 \phi_\varepsilon^{2/3} = (kz\varepsilon)^{2/3}$ where $\phi_\varepsilon = kz\varepsilon/u_*^3$ is the dimensionless dissipation, collapses all the spectra in the inertial subrange. On the other hand, the low-frequency part of spectra become separated according to the stability parameter z/L. The systematic dependence of the normalized spectra on z/L in the energy-containing range is consistent with the Monin–Obukhov similarity theory. The horizontal velocity spectra in unstable and convective conditions do not show any systematic dependence on z/L. The PBL height h, rather than the local height, is expected to be the appropriate length scale for the same.

The above spectra show that the normalized frequency corresponding to the peak in the spectrum increases with stability. The systematic dependence of

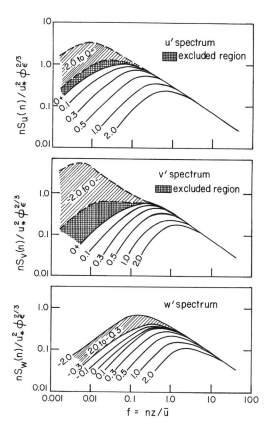

Figure 5.10 Normalized spectra of velocity variances in the stratified surface layer. After Kaimal and Finnegan, 1994.

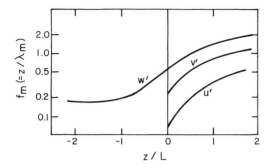

Figure 5.11 Normalized frequency at spectral maxima for the three turbulence velocity components as functions of the stability parameter z/L. After Kaimal and Finnegan, 1994.

similarity scaling, are shown in Figure 5.12. The normalization by the quantity $W_*^2 \psi_\varepsilon^{2/3} = (h\varepsilon)^{2/3}$, where $\psi_\varepsilon = \varepsilon h/W_*^3$ is the normalized dissipation, collapses all the spectra in the inertial subrange. In the energy-containing range, the spectra can be separated according to the normalized height z/h, irrespective of the PBL stability parameter h/L. A striking feature of these observed spectra is the lack of significant dependence on height in the mixed layer. Even in the surface layer, horizontal velocity spectra show only a weak dependence on z/h for $z/h < 0.02$, while the vertical velocity spectrum shows much stronger dependence. The peak wavelength λ_{mw} for the vertical component also shows a strong dependence on height (see Fig. 5.12), whereas $\lambda_{mu} \simeq \lambda_{mv} \simeq 1.5\,h$, independent of height in the CBL. The three scales become approximately equal in the middle part ($0.4 < z/h < 0.8$) of the CBL, but λ_{mw} decreases sharply near the top of the mixed layer.

5.7.3 Stable Boundary Layer

Limited observations of turbulence at nighttime during the 1973 Minnesota experiment, covering a slight to moderate stability range ($3.5 < h/L < 8$), have been used to compute the spectra and scales of turbulence in the SBL (Caughey et al., 1979). These were taken during the first few hours following the evening transition when wind speed, the PBL height, and turbulence decreased rapidly with time. In spite of these nonstationary conditions, which are typical of the early phase of the nocturnal boundary layer over land, the normalized variances and covariances were found to be unique functions of z/h, with no systematic dependence on the stability parameter h/L. Similarly, the normalized spectra $nS(n)$, when plotted against normalized frequency n/n_m or f/f_m, showed some weak dependence on z/h, but no dependence on h/L (see Fig. 5.13). This type of normalization constrains both the area under the curve and the position of maximum in the spectrum. It may, therefore, mask any dependence

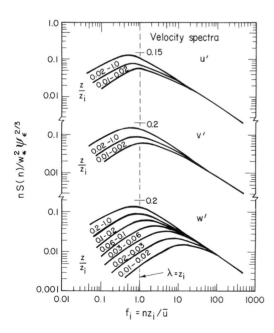

Figure 5.12 Normalized spectra of velocity variances in the convective boundary layer. After Kaimal et al., 1976.

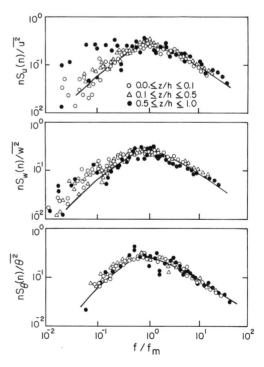

Figure 5.13 Normalized spectra of velocity variances in the moderately stable boundary layer. After Caughey et al., 1979.

120 STATISTICAL DESCRIPTION OF ATMOSPHERIC TURBULENCE

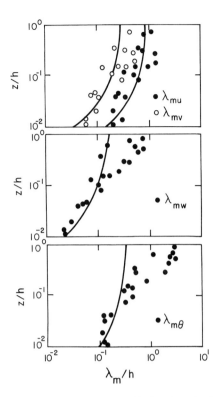

Figure 5.14 Normalized large-eddy length scales based on the frequency at spectral maxima as functions of dimensionless height in the stable boundary layer with $h/L = 5$. After Caughey et al., 1979.

on height and stability that might exist on spectral densities. The normalized large-eddy-length scales based on the spectral peak frequency, however, did show systematic dependence on z/h (see Fig. 5.14); they all increase with height, approaching near unity at the top of the SBL.

Turbulence spectra from the Minnesota experiment shown here for both the CBL and the SBL are based on the sampling period of 75 min. Also, the fluctuation data were filtered with a high-pass digital filter to minimize or eliminate any long-term trends. The filter effectively attenuated frequencies below 0.001 Hz (Kaimal et al., 1976). For the stable boundary layer, the variances and covariances were estimated by integrating the spectral and cospectral curves over the bandwidth 0.001 to 5 Hz. This technique reduces contributions from long-term trends and wave motions that would otherwise appear as a secondary peak at the low-frequency end (Caughey et al., 1979). Without appropriate high-pass filtering of observed turbulence data to remove significant contributions of long-term trends and wave motions, turbulence spectra in the PBL cannot be generalized, even with use of the appropriate similarity scaling. Contaminants in the atmosphere are dispersed by three-dimensional, nearly random turbulent motions, as well as by more organized wave motions and nearly two-dimensional mesoscale and larger-scale motions. Therefore, one needs to know the variances and spectra of velocity fluctuations across the whole range of scales of interest, including those caused by terrain inhomogeneities (Lenschow et al., 1988a, 1988b).

5.8 EFFECTS OF SMOOTHING AND FINITE SAMPLING

Practical measurements of atmospheric turbulence and other meteorological variables are invariably subjected to some smoothing due to limited response of the sensor, speed of recording equipment, or finite sampling interval (time between consecutive samples of a digitized record). Certain data analysis procedures also require some smoothing of the time series in order to reduce the undesirable influence of high-frequency fluctuations and to minimize uncertainties or errors in the estimates of turbulence statistics. Measurements are also sampled over finite duration, which may depend on the measurement height, wind speed, the PBL height, stability conditions, and requirements of stationarity or homogeneity of the data. While smoothing of a time series due to limited instrument response or finite sampling rate cuts off high-frequency fluctuations from the record, finite sampling duration limits the lower end of the frequencies that can be resolved from the finite record. The combination of the two effects amounts to a band-pass filtering of the original time series. It is of great interest to know the quantitative effects of smoothing and finite sampling on turbulence statistics, such as variances and spectra. Here, we essentially follow the derivations of Pasquill (1974) with some differences in notation.

5.8.1 Effect of Smoothing

Consider a time series $u'(t)$ that is subjected to smoothing through finite sampling or averaging (e.g., running mean) with a characteristic averaging time or sampling interval s. The effect of smoothing the original time series $u'(t)$ is to replace it with a new series

$$u'_s(t) = \frac{1}{s} \int_{t-s/2}^{t+s/2} u'(t')dt' \qquad (5.61)$$

where the subscript s refers to the particular smoothing or averaging time.

To see the effect of smoothing on variance and spectra, we assume the original series to be composed of a spectrum of sinusoidal components with an infinite range of frequencies. Representing a particular frequency component by

$$u'(t,\omega) = A(\omega)\sin(\omega t) \qquad (5.62)$$

the corresponding component of the smoothed series, using Eq. (5.61), would be

$$u'_s(t,\omega) = A(\omega)\frac{\sin(\omega s/2)}{(\omega s/2)}\sin(\omega t) \quad (5.63)$$

or, in terms of the ordinary frequency $n = \omega/2\pi$,

$$u'_s(t,n) = u'(t,n)\frac{\sin(\pi ns)}{(\pi ns)} \quad (5.64)$$

Equation (5.64) indicates that the amplitude of a particular sinusoidal component of frequency n will be reduced by a factor $\sin(\pi ns)/\pi ns$ by the operation of smoothing or averaging over an interval s. Assuming an infinite sampling duration, the variance associated with that component will be reduced by a factor $\sin^2(\pi ns)/(\pi ns)^2$. Using the definition of the spectral density, we can express the reduced variance associated with the frequency n in the form

$$\overline{u'^2_s}(n) = \overline{u'^2}(n)\frac{\sin^2(\pi ns)}{(\pi ns)^2}$$

$$= S'_{uu}(n)dn\frac{\sin^2(\pi ns)}{(\pi ns)^2} \quad (5.65)$$

Equation (5.65) can be integrated with respect to n over the whole spectrum to yield the overall variance of the smoothed time series

$$\overline{u'^2_s} = \int_0^\infty S'_{uu}(n)\frac{\sin^2(\pi ns)}{(\pi ns)^2}dn \quad (5.66)$$

or, in terms of the normalized spectrum

$$\overline{u'^2_s} = \overline{u'^2}\int_0^\infty \frac{S_{uu}(n)\sin^2(\pi ns)}{(\pi ns)^2}dn \quad (5.67)$$

Thus, the spectrum is also reduced by the factor $\sin^2(\pi ns)/(\pi ns)^2$, which is shown in Figure 5.15 as a function of the product ns. As ns is increased, the ratio of the spectrum of smoothed series to that of the original series decreases gradually at first and sharply for $ns > 0.1$. There is little contribution from frequencies larger than about $0.5\ s^{-1}$, while the spectrum is completely cut off at $n = s^{-1}$. The overall effect on the total variance depends on the integral of the reduced spectrum with the high-frequency part filtered out or cut off. It is obvious that as s is increased, more and more of the high-frequency end of the spectrum will be cut off and less will be the ratio of the variances of smoothed and original series.

5.8.2 Effect of Finite Sampling

The effect of finite sampling duration T on the computed variance can also be easily derived. Again considering the time series of the variable $u(t)$, we denote the temporal mean and variance of a particular sample of finite duration T by $\overline{u}_T$ and u'^2_T, respectively, which are expected to differ from the ensemble mean $\overline{u}$ and the ensemble variance u'^2. A truly infinite and stationary time series may be considered to consist of

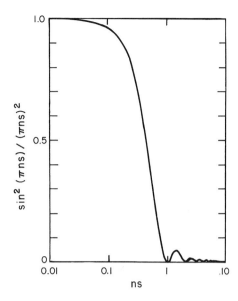

Figure 5.15 Effect of smoothing by averaging over a time s on the variance of a sinusoidal component of frequency n. After Pasquill and Smith, 1983.

an infinite number of finite duration samples of the same variable, of which only one particular sample may be available (recorded) for the statistical analysis. Thus $\overline{u}_T$ and u'^2_T represent the sample mean and sample variance, which are likely to differ from the true (ensemble) mean and variance in a random manner for different samples.

Expressing the turbulent fluctuation from the true mean in the form

$$u' = u - \overline{u} = (u - \overline{u}_T) + (\overline{u}_T - \overline{u}) \quad (5.68)$$

squaring and averaging over all the samples (infinite record), one obtains

$$\overline{u'^2} = \overline{u'^2_T} + \overline{(\overline{u}_T - \overline{u})^2} \quad (5.69)$$

in which we have assumed that turbulent fluctuations $u - \overline{u}_T$ from the sample mean are not likely to be correlated with the deviations $\overline{u}_T - \overline{u}$ of the sample means from the true mean. According to Eq. (5.69), the true variance of an infinitely long time series is simply the sum of the average of sample variances and the variance of the sample means. The latter is equivalent to obtaining the variance of the smoother time series using a smoothing interval T, that is,

$$\overline{(\overline{u}_T - \overline{u})^2} = \int_0^\infty S'_{uu}(n)\frac{\sin^2(\pi nT)}{(\pi nT)^2}dn \quad (5.70)$$

From Eqs. (5.69) and (5.70), the average of sample variances is given by

$$\overline{u'^2_T} = \int_0^\infty S'_{uu}(n)\left[1 - \frac{\sin^2(\pi nT)}{(\pi nT)^2}\right]dn \quad (5.71)$$

or, in terms of the normalized spectrum,

$$\overline{u_T'^2} = \overline{u'^2} \int_0^\infty S_{uu}(n)\left[1 - \frac{\sin^2(\pi nT)}{(\pi nT)^2}\right]dn \quad (5.72)$$

Thus, on the average, the sample variance is reduced by the factor represented by the integral in Eq. (5.72). The spectrum is reduced by the factor $1 - \sin^2(\pi nT)/(\pi nT)^2$, which is a function of the normalized frequency nT and has a shape complementary to that shown in Figure 5.15. Therefore, the effect of finite sampling duration is equivalent to applying a filter that removes contribution from all frequencies much smaller than $1/T$; effectively, the contribution of all frequencies less than about $1/2T$ is eliminated.

The combined effect of smoothing and finite sampling duration is equivalent to applying both the reduction factors or weighting functions at the same time, with the resulting expression (Pasquill, 1974)

$$\overline{u_{T,s}'^2} = \overline{u'^2} \int_0^\infty S_{uu}(n)\left[\frac{\sin^2(\pi ns)}{(\pi ns)^2}\right]$$

$$\left[1 - \frac{\sin^2(\pi nT)}{(\pi nT)^2}\right]dn \quad (5.73)$$

which amounts to a band-pass filtering of the original spectrum with the weighting function

$$W(n,s,T) = \frac{\sin^2(\pi ns)}{(\pi ns)^2}\left[1 - \frac{\sin^2(\pi nT)}{(\pi nT)^2}\right] \quad (5.74)$$

The reduced spectrum is the product of the original spectrum and the above weighting function. Thus, the effect of smoothing and finite sampling duration is essentially to cut off the spectrum at both the high-frequency $(n > 0.5\ s^{-1})$ and the low-frequency $(n < 0.5\ T^{-1})$ ends.

Note that Eq. (5.73) is an expression of the expected or average value of the sample variance; it may differ from the particular sample variance $u_T'^2$, depending on the sample duration. For a sufficiently long T, the difference between the two would be small, and one may use Eq. (5.73) or its abbreviated form

$$\overline{u_{T,s}'^2} = \overline{u'^2} \int_0^\infty S_{uu}(n)W(n,s,T)dn \quad (5.75)$$

for the particular sample variance as well.

The weighting function and, hence, the integral in Eq. (5.75) approach unity only when $s \to 0$ and $T \to \infty$, in which case the sample variance will be equal to the true ensemble variance. For most practical purposes, however, the sample variance would closely approach the true variance if $T \gg n_m^{-1}$ and $s \ll n_m^{-1}$, where n_m^{-1} is the characteristic time scale of large energy-containing eddies. With the modern fast response turbulence sensors and recording and digitizing equipment available, the condition of $s \ll n_m^{-1}$ is not difficult to achieve. Sufficiently long sampling duration $(T \gg n_m^{-1})$, which might be necessary to avoid a significant reduction in variance due to the low-frequency cutoff of the spectrum, may, however,

be more difficult to achieve, especially with the simultaneous requirement of the stationarity of mean flow and turbulence structure. In particular, measurements of horizontal velocity variances in the atmosphere show them to be increasing functions of sampling duration. Important implications of this observation and the theoretical result (5.72) to atmospheric diffusion will be discussed later.

5.9 LAGRANGIAN DESCRIPTION OF TURBULENCE

So far we have discussed the fluid motion at fixed points in space in either a fixed or a moving frame of reference. This type of motion or velocity field is easily measured by instruments mounted on masts, towers, and moving platforms, such as aircraft, and is called the Eulerian motion (denoted here by u, v, and w components). Such measurements always involve a continually changing sample of fluid or fluid particles. The Eulerian motion may not be the most appropriate for describing the dispersion of contaminants (gases or particles) in a turbulent flow. For this a Lagrangian description of motion following some specified (tagged) fluid particles is found to be more appropriate. The Lagrangian motion associated with a tagged fluid particle is much more difficult to measure in a turbulent flow, because it requires faithful tracing of the highly tortuous path that the fluid particle might take with increasing time, as it is acted upon by different eddy motions. Neutral balloons and constant-level tetroons (tetrahedral-shaped balloons) have been used in the few reported measurements of Lagrangian motion in the atmosphere.

The Lagrangian motion is simply and directly related to the position coordinates of the tagged fluid particle as functions of time and the initial coordinates of the particle (X_o, Y_o, Z_o) at $t = 0$, that is,

$$U(X_o,Y_o,Z_o,t) = \frac{\partial X(X_o,Y_o,Z_o,t)}{\partial t} \quad (5.76)$$

$$X(X_o,Y_o,Z_o,t) = X_o + \int_0^t U(t')dt' \quad (5.77)$$

and similar expressions for other components. Thus, for a particular (tagged) fluid particle, the Lagrangian velocity components are functions of time only. One can define the instantaneous turbulent fluctuations by subtracting the appropriate mean values, that is,

$$U'(t) = U(t) - \overline{U} \quad (5.78)$$

in which $\overline{U}$ may also be a function of time if the Lagrangian motion is not stationary. For a stationary Lagrangian field of turbulence, however, we can define the various moments, probability, time correlation, and frequency spectrum functions in the same manner as we did for the stationary Eulerian field. We will use the subscripts L and E to denote and distinguish between Lagrangian and Eulerian statistics. In particular, we introduce the Lagrangian autocorrelation function

$$R_L(\xi) = \frac{\overline{U'(t)U'(t+\xi)}}{\overline{U'^2}} \quad (5.79)$$

and the corresponding integral time scale

$$T_{iL} = \int_0^\infty R_L(\xi)d\xi \quad (5.80)$$

The Lagrangian spectrum function is related to the autocorrelation function through expressions similar to Eq. (5.35)

$$S_L(n) = 4\int_0^\infty R_L(\xi)\cos(2\pi n\xi)d\xi$$

$$R_L(\xi) = \int_0^\infty S_L(n)\cos(2\pi n\xi)dn \quad (5.81)$$

Unlike the Eulerian velocity correlations and spectra, the Lagrangian velocity correlations and spectra are defined only in terms of lag time and frequency. Spatial correlations and wave number spectra are not appropriate or even relevant for describing Lagrangian turbulence. The statistical theory of turbulence (see section 5.6) deals primarily with the spatial statistics and may not be very useful in describing Lagrangian statistics. Kolmogorov's local similarity hypotheses can be applied, however, to indicate the relevance of microscales defined in Eq. (5.53). At sufficiently large Reynolds numbers, one-dimensional frequency spectrum is expected to have the following inertial subrange form derived from the appropriate similarity hypothesis and dimensional analysis:

$$S'_L(n) = B\varepsilon n^{-2} \quad (5.82)$$

This may be compared with the corresponding Eulerian velocity spectrum in the inertial subrange, obtained from Eq. (5.58)

$$S'_E(n) = C\,\bar{u}^{2/3}\varepsilon^{2/3}n^{-5/3} \quad (5.83)$$

The difference in the power-law forms of the two spectra in the inertial subrange arises entirely from considering the Lagrangian spectrum as a function of frequency and the original Eulerian spectrum as a function of wave number and using the appropriate similarity hypotheses for the two spectra. Limited experimental measurements of Lagrangian velocity spectra have verified the above noted difference and indicated a value of $B \simeq 0.2$ for the lateral and vertical velocity components (Hanna, 1981). For the same components, the empirically estimated value of C is $\simeq 0.20$, so that $B \simeq C$.

5.9.1 Relationships Between Lagrangian and Eulerian Statistics

Since Lagrangian turbulent motions are extremely difficult to measure, it has always been of great interest to develop theoretical and empirical relationships between the Lagrangian and the Eulerian statistics, so that estimates of the former can be derived from knowledge of the latter. Instantaneously, the Lagrangian velocity of a tagged fluid particle is equal to the Eulerian velocity at the location of the particle at time t, that is,

$$U(\underline{X}_o,t) = u(\underline{X},t)$$

$$V(\underline{X}_o,t) = v(\underline{X},t)$$

$$W(\underline{X}_o,t) = w(\underline{X},t) \quad (5.84)$$

where the particle position $\underline{X}$ is a function of the initial position $\underline{X}_o$ and t. Therefore, specifying the Eulerian velocity everywhere and for all times is fully equivalent to specifying the Lagrangian velocity field everywhere and for all times. Still, the problem of relating the Lagrangian statistics to the Eulerian statistics of turbulence, in general, is an extremely complex one. Some simple relationships can be given only for the idealized case of stationary and homogeneous turbulence.

One can show that if the Eulerian field is stationary and homogeneous, the Lagrangian field must also be stationary and homogeneous. Furthermore, if the velocity field is also incompressible (nondivergent), then the probability density of the Lagrangian velocity of a tagged fluid particle at any time is the same as the probability density of the Eulerian velocity at a fixed point (Lumley and Panofsky, 1964). This implies that all the moments of Lagrangian velocities must be equal to the corresponding moments of Eulerian velocities, for example, $\overline{U'^2} = \overline{u'^2}, \overline{W'^3} = \overline{w'^3}$, and so on.

This direct correspondence between Lagrangian and Eulerian one-point or one-time statistics does not extend, however, to the joint (two points or two times) statistics. For diffusion applications, we are primarily interested in relating the Lagrangian autocorrelation function and the integral scale to their Eulerian counterparts. It has not been possible to derive such relationships rigorously from theory alone. But, several attempts using heuristic arguments, theoretical conjectures, and semiempirical forms of correlation or spectrum functions have been made to derive an expression for the ratio $\beta = T_{iL}/T_{iE}$ of Lagrangian and Eulerian integral time scales (see e.g., Pasquill and Smith, 1983).

The simplest approach is based on Taylor's frozen turbulence hypothesis according to which the spatial pattern of turbulence (eddy structure) is advected by mean flow virtually unchanged. The characteristic time scale for the Eulerian velocity measurements at a fixed point is the integral time scale T_{iE}, which is related to the large-eddy (integral) length scale L_1 in the mean flow direction as $T_{iE} = L_1/u$. Within a particular large-eddy structure, however, fluid particles may be considered to move in the same direction at a characteristic speed of σ_u, so that the characteristic integral time scale for Lagrangian measurements should be proportional to L_1/σ_u. According to this argument, the ratio of the two time scales

$$\beta = \frac{T_{iL}}{T_{iE}} \propto \frac{1}{i} \quad (5.85)$$

where $i = \sigma/\bar{u}$ is the intensity of turbulence in the appropriate direction.

Similar inverse-proportionality relationships between β and i have been derived by several investigators, using more sophisticated theoretical conjectures and semiempirical information on the shape of autocorrelation and spectrum functions. The estimated values of the coefficient of proportionality or βi range from 0.35 to 0.8 (Pasquill and Smith, 1983). Empirical estimates of the same constant based on simultaneous measurements of Lagrangian and Eulerian statistics show even a larger scatter. The most recent and probably the best experimental estimate of the constant is $\beta i \simeq 0.7$ (Hanna, 1981). However, this was limited to the daytime convective boundary layer in which β was found to lie within a narrow range of 1.4 to 2.1. Hanna's measurements also support the theoretically predicted behavior of Lagrangian velocity spectra, consistent with Eq. (5.82), in the inertial subrange. In the energy-containing range, the peak in $nS_L(n)$ occurs at a lower frequency compared with the peak in $nS_E(n)$. This is again consistent with the expectation that the Lagrangian integral time scale is considerably larger than the Eulerian integral time scale.

5.10 PARAMETERIZATION OF TURBULENCE FOR DIFFUSION APPLICATIONS

Here we review some of the simpler parametric relations that are frequently used in the literature for purposes of specifying turbulence statistics in applications of the various diffusion theories and models. Parametric relations for turbulent velocity variances based on similarity theories and empirical atmospheric data have been given in the previous chapter.

5.10.1 Probability Functions

In numerical random-walk or Monte-Carlo simulations of diffusion, probability density functions of velocity components are required to be specified. The most commonly used is the Gaussian function,

$$p(u') = \frac{1}{(2\pi)^{1/2}\sigma_u} \exp\left(-\frac{u'^2}{2\sigma_u^2}\right) \quad (5.86)$$

which is a symmetrical bell-shaped function with well-known properties (skewness $S = 0$, kurtosis $K = 3$). The Gaussian function provides a good approximation for the probability density of observed turbulence under near-neutral, stable, and slightly unstable conditions.

Deviations from the Gaussian distribution are found to increase with increasing instability and convective activity. In particular, the probability function of vertical velocity fluctuations becomes highly skewed in a convective boundary layer. A superposition of two Gaussian functions is often used to approximate the expected (observed or computed) skewed distribution, for example,

$$p(w) = \frac{c_1}{(2\pi)^{1/2}\sigma_1} \exp\left[\frac{-(w-\bar{w}_1)^2}{2\sigma_1^2}\right]$$
$$+ \frac{c_2}{(2\pi)^{1/2}\sigma_2} \exp\left[\frac{-(w-\bar{w}_2)^2}{2\sigma_2^2}\right] \quad (5.87)$$

where $c_1 + c_2 = 1$. The two Gaussian distributions denoted by subscripts 1 and 2 represent the probability densities of downdraft and updraft velocities, respectively (Weil, 1988b). One may choose the appropriate values of the parameters c_1, σ_1, $\bar{w}_1$, and so on, to obtain the desired probability function $p(w)$.

5.10.2 Autocorrelation Functions

The most widely used approximation or parameterization of the autocorrelation function of turbulent velocity is the exponential function

$$R_E(\tau) = \exp\left(\frac{-|\tau|}{T_{iE}}\right)$$
$$R_L(\xi) = \exp\left(\frac{-|\xi|}{T_{iL}}\right) \quad (5.88)$$

whose integral with respect to the time lag from 0 to ∞ is the integral scale. This approximation cannot be right for very small time lags, because it is not consistent with the existence of Taylor's microscale. However, it is considered to be adequate for diffusion applications in which small-scale eddies are not expected to be important. An exponential correlation function implies an n^{-2} behavior of the spectrum at large frequencies. Since such a behavior is also theoretically predicted for the spectrum of Lagrangian turbulence in the inertial subrange, an exponential form (5.88) might be quite appropriate for $R_L(\xi)$. The same approximation would be inferior, however, for $R_E(\tau)$, because it is not entirely consistent with Kolmogorov's $-5/3$ law for the spectrum of Eulerian velocity in the inertial subrange.

Other, somewhat less satisfactory parameterizations of $R_L(\xi)$ have also been used in the literature, for example,

$$R_L(\xi) = \left(1 + \frac{|\xi|}{T_{iL}}\right)^{-2} \quad (5.89)$$

$$R_L(\xi) = \begin{cases} 1 - \dfrac{|\xi|}{2T_{iL}}, & \text{for } \xi \leq 2T_{iL} \\ 0, & \text{for } \xi > 2T_{iL} \end{cases} \quad (5.90)$$

$$R_L(\xi) = \begin{cases} 1, & \text{for } \xi \leq T_{iL} \\ 0, & \text{for } \xi > T_L \end{cases} \quad (5.91)$$

Even the crudest approximation might be adequate for some diffusion applications, if the Lagrangian time scale can be accurately specified (Pasquill, 1974). In practice, due to the lack of Lagrangian turbulence measurements over a wide range of stability conditions, T_{iL} is specified in terms of T_{iE} using the semiempirical relation

$$T_{iL} = \beta T_{iE} \simeq 0.7 \frac{T_{iE}}{i} \quad (5.92)$$

In the absence of direct measurements of T_{iE}, Hanna (1981) has suggested the following parametric relations for a convective mixed layer:

$$T_{iL} \simeq \frac{0.25\beta h}{\bar{u}} \simeq \frac{0.17\, h}{\sigma_{u,v,w}} \quad (5.93)$$

In the surface layer of the CBL, Eq. (5.93) should still be applicable for horizontal velocity components, while a height-dependent relation is proposed for the vertical velocity

$$T_{iL} \simeq 0.60\, \beta \frac{z}{u_*} \simeq 0.42 \frac{z}{u_*} \frac{\bar{u}}{\sigma_w},\ \text{for } z \leq 0.1\, h \quad (5.94)$$

which would require the knowledge of u_* and β or vertical turbulence intensity.

5.10.3 Spectrum Functions

Corresponding to the most frequently used exponential form of the autocorrelation function given by Eq. (5.88), the one-dimensional frequency spectrum is given as

$$S(n) = 4T_i(1 + 4\pi^2 n^2 T_i^2)^{-1} \quad (5.95)$$

which has an n^{-2} behavior at high frequencies ($n > 1/T_i$).

As discussed earlier, Eq. (5.95) is more appropriate for the spectrum of Lagrangian than Eulerian motion. The normalized form of the Lagrangian spectrum representing the fractional variance is given by

$$nS_L(n) = \frac{4nT_{iL}}{1 + 4\pi^2 n^2 T_{iL}^2} \quad (5.96)$$

which has a maximum at the frequency $n_{mL} = 1/2\,\pi T_{iL}$. If we define a normalized frequency as $N_L = n/n_{mL}$, then Eq. (5.96) can also be expressed in the form

$$nS_L(n) = \frac{2\pi^{-1} N_L}{1 + N_L^2} \quad (5.97)$$

An alternative form of the Lagrangian spectrum with the same high-frequency behavior and the appropriate integral constraint is (Pasquill, 1974)

$$nS_L(n) = \frac{N_L}{(1 + N_L)^2} \quad (5.98)$$

Note that the spectrum represented by Eq. (5.98) is somewhat flatter with a smaller pleak than the spectrum of Eq. (5.97).

For the normalized frequency spectrum of the Eulerian motion, a simple form that is consistent with Kolmogorov's law for the inertial subrange has been suggested (Pasquill, 1974):

$$nS_E(n) = \frac{N_E}{(1 + 1.5 N_E)^{5/3}} \quad (5.99)$$

An alternative form with the same inertial subrange behavior but with a larger maximum value is

$$nS_E(n) = \frac{0.644 N_E}{1 + 1.5 N_E^{5/3}} \quad (5.100)$$

All of the above suggested normalized spectra satisfy the integral constraint

$$\int_0^\infty S(n)dn = \int_0^\infty S(N)dN = 1 \quad (5.101)$$

as well as the condition that $nS(n)$ or $NS(N)$ must be maximum at $N = 1$. The same conditions can also be satisfied by a variety of other analytical forms of the normalized spectra (see e.g., Sorbjan, 1989), but those given above are the most frequently used parameterizations for the spectra with well-defined peaks. The observed atmospheric spectra of horizontal velocity fluctuations under unstable and convective conditions are often characterized by a rather broad maximum or double maxima. The above suggested parameterizations may not be adequate for representing such spectra. Instead, a more complex parameterization or spectral model might be necessary (see e.g, Hojstrup, 1982).

PROBLEMS AND EXERCISES

1. Consider a stationary random variable u' with zero mean and a Gaussian probability density function

$$p(u') = \frac{1}{\sqrt{2\pi}\sigma_u} \exp\left(\frac{-u'^2}{2\sigma_u^2}\right)$$

   a. Show that the skewness $S = 0$ and the kurtosis $K = 3$.
   b. What is the corresponding probability distribution function $P(u')$?
   c. Plot $p(u')$ and $P(u')$ as functions of u'/σ_u.
   d. Derive an approximate expression for the probability of u' becoming much larger than σ_u. Use this expression to estimate the probability of $u' > 3\,\sigma_u$ and compare it with the exact value from probability tables.

2. Consider a sawtooth wave with a period T and amplitude a. Numerically or analytically estimate and sketch the following: (a) probability density function; (b) cumulative probability (distribution) function; (c) autocorrelation function and the associated integral and microscales.

3. Determine and sketch the probability, autocorrelation, and spectrum functions for the periodic sine wave of frequency ω and amplitude a.

4. Show that, if a measured turbulence variable $u_m'(t)$ has a periodic component $s'(t)$ superimposed on its random part $u'(t)$, $R_{u_m u_m}(\tau) = R_{uu}(\tau) + R_{ss}(\tau)$. What is the practical significance of this result?

5. Schematically show the autocorrelation and spectrum functions of a measured turbulence variable with a superimposed (a) linear trend, and (b) sine wave.

6. The autocorrelation function of fluctuating velocity is often approximated by an exponential function $R(\tau) = \exp(-|\tau|/T_i)$.
   a. Show that T_i is the integral scale and Taylor's microscale cannot be defined.

b. Determine the corresponding frequency spectrum function $S(\omega)$ or $S(n)$.

c. Determine the maximum value of $\omega S(\omega)$ or $nS(n)$, as well as the frequency at which the maximum occurs.

7. Derive the relationship between the frequency spectrum $S(n)$ and the one-dimensional wave-number spectrum $S(\kappa_1)$.

8. Discuss the usefulness and limitations of Taylor's frozen-turbulence hypothesis to problems of atmospheric turbulence and diffusion.

9. Simplify the TKE budget equation (5.50), normalize it by appropriate length and velocity scales, and schematically show the important budget terms for the following conditions and regions of the PBL: (a) neutral surface layer with negligible turbulent transport; (b) stably stratified boundary layer; (c) convective boundary layer with negligible shear.

10. Using the TKE budget equation, derive Richardson's criterion for the maintenance of turbulence in a stably stratified boundary layer. Discuss later modification of this criterion and the implied critical Richardson number.

11. Discuss the energy cascade hypothesis and its implications for the small-scale turbulence structure.

12. What is the basis of Kolmogorov's local similarity theory? Discuss the concept of local isotropy and its practical significance and applications.

13. What is the basis for Kolmogorov's microscales? Derive the approximate relations (5.56) for the ratios of large-eddy scales and microscales of turbulence.

14. Using Kolmogorov's inertial-subrange hypothesis and appropriate dimensional analysis, show that the one-dimensional spectrum of longitudinal velocity fluctuations is given by

$$S_{uu}(\kappa_1) = \alpha_1 \varepsilon^{2/3} \kappa_1^{-5/3}$$

$$S_{uu}(n) = C \bar{u}^{2/3} \varepsilon^{2/3} n^{-5/3}$$

Also, estimate the value of C if $\alpha_1 = 0.5$.

15. The observed spectrum of vertical velocity fluctuations in the PBL can be approximately represented by

$$nS_{ww}(n) = NS_{ww}(N) = \frac{aN}{(1 + bN)^{5/3}}$$

where $N = n/n_m = n\lambda_m/\bar{u}$ is the normalized frequency and λ_m is the wavelength corresponding to the peak in the spectrum.

a. Determine the values of a and b from the known properties of the above spectrum.

b. How are the effects of height and stability considered in the above representation?

16. The following measurements were made from a micrometeorological tower and a tethered balloon during the Minnesota experiment:

Surface heat flux = 0.20 K m s^{-1}
Height of inversion base = 1250 m
Mixed-layer wind speed = 11.8 m s^{-1}
Surface temperature = 300 K

a. Using the mixed-layer similarity relations, estimate the variance $\overline{u'^2}$ and the frequency n_m corresponding to the peak in u' spectrum in the mixed layer.

b. If velocity fluctuations were measured with a fast response sonic anemometer and digitally recorded at the sampling rates of 1 and 10 Hz for the sample durations of 1500 and 150 s, respectively, calculate and plot the expected "measured" spectra for the two combinations of sampling rate and sample duration, and compare them with the normalized "true" spectrum $nS(n) = N(1 + 1.5 N)^{-5/3}$, where $N = n/n_m$.

17. Using Kolmogorov's similarity hypothesis, derive an expression for the Lagrangian velocity spectrum in the inertial subrange and explain why it differs from the Eulerian velocity spectrum.

18. Using the appropriate expressions (5.82) and (5.83) for the Lagrangian and Eulerian velocity spectra in the inertial subrange and their parametric relations (5.98) and (5.99) for the whole range of frequencies, derive an expression for the ratio $\beta = T_{iL}/T_{iE}$, and estimate the proportionality coefficient in the expected relation $\beta \propto 1/i$.

6

Gradient Transport Theories

6.1 EULERIAN APPROACH TO DESCRIBING DIFFUSION

Of the two basic approaches or frameworks for representing flow and diffusion in fluids (including the atmospheric environment), Eulerian and Lagrangian, the former is better suited for measurements and observations at fixed points in space in either a fixed or a moving frame of reference. Measurements from instruments or samplers located at fixed locations on the ground, masts, stacks, and towers are some of the examples of Eulerian measurements in a fixed frame of reference, while those from moving vehicles, boats, and aircraft are examples of Eulerian measurements in moving reference frames. The latter can be easily converted into or represented in a fixed frame of reference, provided the motion of the moving platform is known. The Eulerian approach is also more suitable for describing the flow using theoretical and numerical models, because the equations of Eulerian motion are much simpler than those of the Lagrangian motion (see e.g., Kao, 1984). Both the Eulerian and the Lagrangian approaches can be used for describing turbulent diffusion in relatively simple flows (e.g., stationary and homogeneous flows), but the former is found to be more suitable for describing diffusion in more complex (nonstationary and nonhomogeneous) flows.

6.1.1 Concentration and Flux of Contaminant

The primary variable describing the presence of a contaminant in a fluid is its mass or volumetric concentration. The mass concentration is usually defined as the mass of the contaminant per unit volume of the fluid. The most commonly used unit of mass concentration in the atmosphere is microgram per cubic meter. The volumetric concentration in terms of parts per million (ppm), parts per billion (ppb), and so on, are also used to express small amounts of trace gases, including gaseous pollutants, in the atmosphere. These are not really concentrations, but dimensionless mole fractions of gaseous contaminants and air. The appropriate conversion relations between mass and volumetric concentrations have been given in Chapter 1.

The above definition of mass or volume concentration is somewhat ambiguous, because in practice concentration might depend on the volume of fluid sampled. To avoid this ambiguity, a point concentration is defined by reducing the sample volume around a point to an arbitrarily small value, but a value still large enough to allow the treatment of fluid as a continuum. In theory we can deal with instantaneous point concentrations as well as with volume-averaged or time-averaged concentrations, whereas practical measurements almost always involve concentrations averaged over some finite sample volume and finite averaging time which depend on the sensor size and response characteristics. Very rarely, if at all, measured concentrations would be representative of theoretical or computed concentrations and vice versa. This is one of the fundamental difficulties in the validation of diffusion theories and models against concentration measurements under similar conditions.

The mass or volume flux of a contaminant in a fluid flow is the amount (mass or volume) of contaminant flowing per unit area, per unit time, in a given direction. Thus, like heat and momentum fluxes, the mass flux ($\underline{M}$) is a vector quantity. It comprises two parts—transport by flow ($\underline{T}$) and a molecular diffusion ($\underline{F}$)—each involving a different flux-producing mechanism. It is easy to show that the flow transport is simply the product of concentration and velocity vector ($\underline{T} = c\,\underline{V}$), which would vanish only in the absence of bulk flow. The diffusive flux, on the other hand, is independent of bulk flow and is caused by molecular mo-

Table 6.1 Values of Molecular Diffusivity of Gaseous Pollutants in the Atmosphere at 1000 mb

Diffusing Substance	Molecular Diffusivity (m² s⁻¹)	
	0°C	25°C
Ammonia (NH_3)	1.98×10^{-5}	2.29×10^{-5}
Carbon dioxide (CO_2)	1.42×10^{-5}	1.64×10^{-5}
Carbon disulfide (CS_2)	0.94×10^{-5}	1.07×10^{-5}
Hydrocarbons	$0.5–1.3 \times 10^{-5}$	$0.5–1.6 \times 10^{-5}$

Source: Handbook of Tables for Applied Engineering Science, R.C. Weast, Editor, CRC Press, 1970.

tion in the presence of concentration gradients. The latter will vanish only when the contaminant is thoroughly well mixed, so that there is no bulk concentration gradient.

6.1.2 Fick's Law

Analogous to Newton's law of viscosity and Fourier's law of heat conduction, Fick's law states that the diffusive mass flux is proportional to the concentration gradient, the direction of flux being opposite to the gradient, that is,

$$\underline{F} = -D\underline{\Delta}c \quad (6.1)$$

where $\underline{\Delta}$ is the gradient operator and the proportionality coefficient D is the molecular diffusivity whose value depends on both the contaminant and the fluid in which it is diffusing. Fick's law was originally proposed as an empirical, phenomenological expression of proportionality between the diffusive flux and the concentration gradient. More rigorous theoretical derivations of Eq. (6.1) and similar laws for the diffusion of heat and momentum have been based on the kinetic theory of gases, which expresses molecular diffusivities of heat, momentum, and mass in terms of the mean molecular speed (V_m) and the mean free-path length (Λ_m) (molecular diffusivity $\propto V_m \Lambda_m$).

In the kinetic theory, gas molecules are treated as elastic spheres moving in random directions at high speeds and colliding with each other. The molecules of nitrogen, oxygen, and argon, the three principal constituents of air, have diameters of about 10^{-10} m, the mean free-path length (average distance traveled between collisions) is about 10^{-7} m, and the mean molecular speed is about 400 m s⁻¹ at the normal temperature and pressure. The number of collisions experienced by each molecule is of the order of 10^9/s. Molecular diffusivity arises as a direct consequence of this ceaseless agitation and collision of molecules. Typical values of molecular diffusivity of some of the gaseous pollutants in the atmosphere at the normal pressure are given in Table 6.1. Note that, typically, molecular diffusivity of gaseous contaminants in air is of the order of 10^{-5} m² s⁻¹ and it is only weakly dependent on temperature and pressure.

6.2 MASS CONSERVATION AND DIFFUSION EQUATIONS

Fick's law provides a relationship between the spatial distribution of concentration and the diffusive mass flux of the particular chemical species of interest. A second mathematical relationship between the same variables may be obtained by the application of the principle of the conservation of mass in an elementary volume, resulting in a mathematical closure of the problem. For the sake of simplicity, we will derive an instantaneous diffusion equation for the concentration of a neutrally buoyant, nonreacting chemical species released in the atmosphere and dispersed by both the molecular and the bulk air motions.

Consider the conservation of mass of a chemical species within an elementary control volume $\Delta x \Delta y \Delta z$, with infinitesimal dimensions Δx, Δy, and Δz, centered around the point (x,y,z), as shown in Figure 6.1. The mass $c\Delta x \Delta y \Delta z$ or concentration c within the elementary volume, in the absence of any sources and sinks

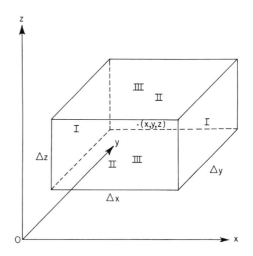

Figure 6.1 Schematic of an elementary control volume with the pairs of faces I–I, II–II, and III–III shown normal to x, y, and z directions, respectively, and the point (x,y,z) located at the center of the control volume.

within the volume itself, can change only in the following ways:

1. Advection or transport of species per unit time to or from the element due to air motions. Considering the net transport flux through each pair of opposite faces, the total advection (A) or transport of species mass into the element is given as

$$A = \left[cu + \frac{\partial(cu)}{\partial x} \frac{\Delta x}{2} - cu + \frac{\partial(cu)}{\partial x} \frac{\Delta x}{2} \right] \Delta y \Delta z$$

$$+ \left[cv + \frac{\partial(cv)}{\partial y} \frac{\Delta y}{2} - cv + \frac{\partial(cv)}{\partial y} \frac{\Delta y}{2} \right] \Delta x \Delta z$$

$$+ \left[cw + \frac{\partial(cw)}{\partial z} \frac{\Delta z}{2} - cw + \frac{\partial(cw)}{\partial z} \frac{\Delta z}{2} \right] \Delta x \Delta y \quad (6.2)$$

or

$$A = - \left[\frac{\partial(cu)}{\partial x} + \frac{\partial(cv)}{\partial y} + \frac{\partial(cw)}{\partial z} \right] \Delta x \Delta y \Delta z$$

2. Diffusion of species mass per unit time due to molecular agitation. Again, considering each pair of opposite faces of the element, the total diffusion of mass to the element is given by

$$D_m = - \left(\frac{\partial F_x}{\partial x} + \frac{\partial F_y}{\partial y} + \frac{\partial F_z}{\partial z} \right) \Delta x \Delta y \Delta z \quad (6.3)$$

As a consequence of net advection and diffusion, the species mass ($M = c\Delta x \Delta y \Delta z$) within the element changes with time. The conservation of mass requires that

$$\frac{\partial M}{\partial t} = A + D_m \quad (6.4)$$

or, after substituting from (6.2) and (6.3) and dividing by the volume $\Delta x \Delta y \Delta z$,

$$\frac{\partial c}{\partial t} + \frac{\partial(cu)}{\partial x} + \frac{\partial(cv)}{\partial y} + \frac{\partial(cw)}{\partial z}$$

$$= - \left(\frac{\partial F_x}{\partial x} + \frac{\partial F_y}{\partial y} + \frac{\partial F_z}{\partial z} \right) \quad (6.5)$$

According to Fick's law, however,

$$F_x = -\frac{D \partial c}{\partial x}; \quad F_y = -\frac{D \partial c}{\partial y}; \quad F_z = -\frac{D \partial c}{\partial z} \quad (6.6)$$

so that Eq. (6.5) can be written as

$$\frac{\partial c}{\partial t} + \frac{\partial(cu)}{\partial x} + \frac{\partial(cv)}{\partial y} + \frac{\partial(cw)}{\partial z}$$

$$= D \left(\frac{\partial^2 c}{\partial x^2} + \frac{\partial^2 c}{\partial y^2} + \frac{\partial^2 c}{\partial z^2} \right) \quad (6.7)$$

which is the appropriate instantaneous diffusion equation for a nonbuoyant, nonreactive (conservable) species, in the absence of any sources and sinks at the point (x,y,z) where concentration is desired.

After considering the changes in mass concentration due to possible chemical transformations and other sources and sinks (e.g., precipitation scavenging), Eq. (6.7) can be generalized to

$$\frac{\partial c}{\partial t} + \frac{\partial(cu)}{\partial x} + \frac{\partial(cv)}{\partial y} + \frac{\partial(cw)}{\partial z}$$

$$= D \left(\frac{\partial^2 c}{\partial x^2} + \frac{\partial^2 c}{\partial y^2} + \frac{\partial^2 c}{\partial z^2} \right) + R + S \quad (6.8)$$

where R and S represent the rates of change of concentration with time due to chemical reactions and due to the presence of sources/sinks. Using the incompressible form of the continuity equation, Eq. (6.8) can also be expressed as

$$\frac{\partial c}{\partial t} + u\frac{\partial c}{\partial x} + v\frac{\partial c}{\partial y} + w\frac{\partial c}{\partial z}$$

$$= D \left(\frac{\partial^2 c}{\partial x^2} + \frac{\partial^2 c}{\partial y^2} + \frac{\partial^2 c}{\partial z^2} \right) + R + S \quad (6.9)$$

or, using the simplifying notations for the total derivative and the Laplacian,

$$\frac{Dc}{Dt} = D\nabla^2 c + R + S \quad (6.10)$$

One can write similar instantaneous diffusion equations for all the chemical species of interest that might react with each other and with the primary atmospheric constituents, such as oxygen, water vapor, and so on. The chemical reaction term in any particular species equation would depend on the concentrations of all other species that might react with the former. Thus, in general, it might be necessary to obtain simultaneous solutions of the coupled partial differential equations for the conservation of mass of many different species

$$\frac{Dc_i}{Dt} = D_i \nabla^2 c_i + R_i(c_1, c_2, \cdots c_N; T) + S_i \quad (6.11)$$

where $i = 1, 2, \ldots, N$. In addition, the coupled equations for the conservation of momentum and thermodynamic energy may have to be solved, separately, to get the velocity and temperature fields that are required as inputs to mass-conservation (diffusion) equations. It is usually assumed, with good justification, that concentrations of pollutant species are too small to affect the dynamics and thermodynamics of the atmosphere. Therefore, velocity and temperature fields can be considered independent of concentration fields, with the possible exception of water vapor, which is not considered a pollutant anyway, except in the case of huge quantities of water vapor released into the atmosphere from large cooling towers that may cause significant adverse impacts, such as fog formation and visibility reduction, on a local scale.

Table 6.2 Idealized Source Characteristics Used for Representing Pollutant Sources in Diffusion Models/Theories

Type of Release	Source Location	Source Geometry
Instantaneous	Ground surface	Point or volume source
Continuous	Elevated in PBL	Line source
Intermittent	Free atmospheric	Area source

6.2.1 Characterization of Sources

Knowledge of source characteristics, such as type of release, source location and geometry, and source strength, is a necessary starting point for diffusion estimates. Idealized source characterizations frequently used in the literature, based on the above considerations, are given in Table 6.2.

Note that an ideal source may be characterized by any combination of the type of release, source location, and source geometry. For example, one can specify an instantaneous, ground-level, volume source; a continuous, elevated, line source; and so on. Often, the distinction between a continuous point source and a finite line, area, or volume source depends on the distance from the source where diffusion estimates are desired. At large enough distances where the pollutant plume becomes much thicker than the maximum linear dimension of the source, a finite line, area, or volume source may simply be treated as a point source of equivalent emission rate. For example, even a moderate-size city may be considered as a point source of emissions in evaluating its impact on regional air quality far away from the city.

Perhaps the most important characteristic of a pollutant source is its emission rate or the amount of material released. The characterization of source strength depends on the type of release, as well as on the source geometry. Strengths of instantaneous sources are normally expressed by the amount (mass or volume) of material released by a point source (strength Q_{ip}), or that released per unit length of a line source (strength Q_{il}), or per unit area of an area source (strength Q_{ia}). Similarly, strengths of continuous sources can be expressed by the emission rates Q_{cp}, Q_{cl}, and Q_{ca} for point, line, and area sources, respectively. Here, the two subscripts to the source strength symbol Q are used to indicate the type of release and the source geometry. Different dimensions of strengths of different sources must be considered when relating them to concentrations.

6.2.2 Integral Mass Conservation Equations

The instantaneous diffusion equation derived earlier is an expression for the conservation of mass within an arbitrarily small volume around a point in space. Its solution for the specified sources and the flow field would give, at least in principle, concentrations at all points and times. For a physically realizable solution, however, the resulting concentration field must also satisfy an overall integral constraint that is also based on the principle of mass conservation and accounts for the strengths of various sources affecting the field. The integral mass conservation equation or the mass continuity condition to be satisfied depends on the contaminant sources, as well as on the physical boundaries of the domain that might constrain the diffusing material. The simplest conditions are obtained for the simplest situations in which no chemical transformations and removal processes are taking place. These would be applicable, for example, for short-range dispersion of relatively inert pollutants.

For an instantaneous release, the total mass in the diffusing puff or cloud at any time after release must remain equal to the source strength, that is,

$$\int_{-\infty}^{\infty}\int_{-\infty}^{\infty}\int_{-\infty}^{\infty} c\, dx\, dy\, dz = Q_{ip} \qquad (6.12)$$

for an instantaneous point source. Similar integral mass conservation equations can also be written for the infinite line and area sources. The limits of integrations are arbitrarily set at $\pm\infty$ for convenience only, because one may not know the exact outline of the diffusing cloud as a function of time and concentrations are expected to become vanishingly small at large distances from the center of the cloud. In describing diffusion of a puff or a contaminant cloud, it is most appropriate and convenient to use a frame of reference moving with the puff or cloud at the mean transport velocity.

For representing diffusion in a continuous source plume, on the other hand, a fixed frame of reference, relative to the source, is more appropriate because the material is continuously emitted and stretches out far downwind from the source. For this, the integral mass conservation condition is an expression of the requirement that the total rate of mass passing through a cross section normal to the plume axis must equal the source strength (emission rate), that is,

$$\int_{-\infty}^{\infty}\int_{-\infty}^{\infty} cu\, dy\, dz \\ -\int_{-\infty}^{\infty}\int_{-\infty}^{\infty} D\, \frac{\partial c}{\partial x}\, dy\, dz = Q_{cp} \qquad (6.13)$$

for a continuous point source. Here, we have assumed steady state and oriented the x axis with the plume axis along the direction of mean flow. The second term in Eq. (6.13) represents the mass flow rate due to molecular diffusion, which can often be neglected in comparison to the advection term. The approximate mass continuity equation, ignoring diffusion in the direction of flow, is

$$\int_{-\infty}^{\infty}\int_{-\infty}^{\infty} cu\, dy\, dz = Q_{cp} \qquad (6.14)$$

Note that Eq. (6.13) or (6.14) would not be valid in calm or near-calm wind conditions in which diffusion of material may not be in the form of an extended plume in a particular direction.

For sources near the ground, when diffusion is more likely to be confined to the PBL, the more appropriate limits of integration with respect to the vertical coordinate z are from 0 to h. The above integral mass continuity equations can be modified accordingly for sources in the PBL.

6.2.3 The Principle of Superposition

Atmospheric pollution is a consequence of transport and diffusion of pollutants from a variety of sources. Major sources may be point sources (e.g., industrial stacks and volcanoes), line sources (e.g., highways and airport runways), or area sources (e.g., house-top chimneys and city traffic). Air quality of an urban or rural area depends on the sum total of all the sources in that area and the surrounding region. Therefore, urban or regional air quality models must deal with multiple sources. Even when the impact of a particular (existing or proposed) source is required to be evaluated, one cannot ignore the effects of other existing and proposed sources on local and regional air quality. A basic assumption often implied in air quality and dispersion modeling is that concentration fields of multiple sources are additive. This assumption is sometimes stated as a more fundamental "principle of superposition" of concentration fields. However, the principle of superposition strictly implies the independence of concentration fields due to different sources, which cannot be expected to hold if pollutant species from different sources are likely to react with each other. In modeling chemical transformations and dispersion of highly reactive species, it would be more appropriate to consider simultaneously all the species diffusion equations, complete with reaction terms, as well as all the possible sources of those reactive species. This is normally done in regional and larger-scale air quality models and sometimes also in urban air quality models that include chemical reactions and transformations.

In most local and urban (short-range) dispersion models, impacts of particular sources on air quality can be evaluated by using the principle of superposition. A good justification for this lies in the nature of the diffusion equation. Since velocity and temperature fields can be taken (specified) essentially independent of concentration fields, considering concentrations of the order of 0.1 percent or less for major pollutants of concern, the diffusion equation is essentially a linear equation. The solutions of linear diffusion equation can be linearly combined and the combination is also a solution to the equation. This is the mathematical justification for the principle of superposition. The principle allows for the individual sources to be treated in an isolated manner and for the composite concentration field due to all the sources to be determined by simply summing their individual contributions.

The principle of superposition can also be used to determine concentration distribution due to a volume, area, or line source from knowledge of the concentration field of a point source, because the above source configurations can be assumed to consist of an infinite number of point sources arranged in an appropriate fashion. Thus, the point source may be considered the fundamental building block of dispersion theory and modeling. It is not surprising that most atmospheric diffusion experiments have been conducted using this basic source configuration, particularly the continuous point source. One may argue that, since a continuous point source may be considered as a sequence in time of an infinite number of instantaneous puffs, the instantaneous point source should be a more fundamental source configuration. However, the idea of an instantaneous point source suffers from some conceptual difficulties and limitations, which will be discussed in Chapter 7.

The set of diffusion and integral mass conservation equations for species concentrations form the basis of the various local, urban, and regional air quality models. The meteorological fields in such models are either specified explicitly or are computed from other mesoscale or regional flow models. In numerical atmospheric flow and air quality models of this type, the conservation equations are further subjected to appropriate time and space averaging, before their solutions are obtained numerically. Important consequences of such averaging will be discussed later. First, we consider some classical analytical solutions of the diffusion equation for the simpler cases of molecular diffusion, because those solutions are also found relevant to certain atmospheric diffusion problems.

6.3 MOLECULAR DIFFUSION

For the sake of simplicity, we consider here the molecular diffusion of a nonbuoyant, passive (nonreactive) contaminant in an infinite, homogeneous medium at rest, or moving at a constant uniform velocity u in the x direction. Then, the diffusion equation (6.9) is reduced to

$$\frac{\partial c}{\partial t} + u\frac{\partial c}{\partial x} = D\left(\frac{\partial^2 c}{\partial x^2} + \frac{\partial^2 c}{\partial y^2} + \frac{\partial^2 c}{\partial z^2}\right) \quad (6.15)$$

whose solution has been obtained for different source configurations and boundary conditions (see e.g., Sutton, 1953; Carslaw and Jaeger, 1959). For representing diffusion from idealized sources, far from any constraining boundaries, the simplest condition of vanishing concentration at very large (infinite) distances from the source is specified.

6.3.1 Instantaneous Point Source

The concept of an instantaneous point release is a purely mathematical or theoretical one. All the material is assumed to be concentrated at a point and released instantaneously. To be consistent with our basic assumption of fluid as a continuum, the point source must actually have finite dimensions much larger than the mean molecular path length, so that molecular diffusion can spread the material in the form of an expanding puff or cloud following the instantaneous release.

For the simpler case of a medium at rest or moving at a uniform velocity u, it is more convenient to look at the expanding puff as a function of time in a reference frame moving with the uniform fluid velocity. In this moving framework, $u = 0$, and the diffusion equation (6.15) is further simplified to

$$\frac{\partial c}{\partial t} = D\left(\frac{\partial^2 c}{\partial x^2} + \frac{\partial^2 c}{\partial y^2} + \frac{\partial^2 c}{\partial z^2}\right) = D\nabla^2 c \quad (6.16)$$

Considering the spherical symmetry of the expanding puff, it would be more convenient to use the spherical coordinate system with its origin located at the initial point of release. In a symmetrical puff, the concentration is only a function of radial distance (r) and time. The solution of the diffusion Eq. (6.16) satisfying the appropriate initial and boundary conditions and the integral mass continuity condition (6.12) is

$$c(r,t) = \frac{Q_{ip}}{8(\pi D t)^{3/2}} \exp\left(-\frac{r^2}{4Dt}\right) \quad (6.17a)$$

or, in a Cartesian coordinate system,

$$c(x,y,z,t) = \frac{Q_{ip}}{8(\pi D t)^{3/2}} \exp\left(-\frac{x^2 + y^2 + z^2}{4Dt}\right) \quad (6.17b)$$

A physical interpretation and several implications of this solution can be given. It describes the concentration field as a function of space and time due to the diffusion of an instantaneous point release at the origin at $t = 0$ (the implied infinite concentration at the time of release is not realistic, but merely an artifact of the mathematical "point" source). The expanding puff and surfaces of constant concentration at any time are spherical in shape, with the maximum concentration at the center (c_{max}) decreasing monotonically with time ($c_{max} \propto t^{-3/2}$). Concentration at any radial distance from the center initially increases with time, attains a maximum value, and then decreases with further increase in time. At any particular instant (t), the concentration distribution in the puff is the symmetrical, bell-shaped Gaussian curve. The second moment of this distribution in any direction (say, x) through the center, normalized by the integral of concentration along the same direction,

$$\sigma^2 = \int_{-\infty}^{\infty} cx^2 \, dx / \int_{-\infty}^{\infty} c \, dx = 2Dt \quad (6.18)$$

represents the mean-square distance to which the material has diffused. The square root of this, that is, $\sigma = \sqrt{2Dt}$, defines a convenient scale of the width of the distribution, and is known as the standard deviation of the distribution. In diffusion literature, σ is more commonly referred to as the diffusion or dispersion parameter.

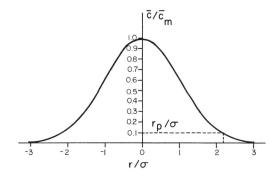

Figure 6.2 The theoretical Gaussian concentration distribution in an ensemble-averaged puff and the definition of the 10-percent puff radius.

In terms of σ, Eq. (6.17b) can be written as

$$c(x,y,z,t) = \frac{Q_{ip}}{(2\pi)^{3/2}\sigma^3} \exp\left(-\frac{x^2 + y^2 + z^2}{2\sigma^2}\right) \quad (6.19)$$

which is the more recognizable form of the three-dimensional Gaussian distribution. Note that the time dependency of concentration in Eq. (6.19) comes through σ, which is an increasing function of time ($\sigma \propto t^{1/2}$). The diffusion parameter σ is also a measure of the puff size. The radial distance at which concentration falls to 10 percent of the maximum value at the center is often used to define 10 percent puff radius r_p, which is related to σ as

$$r_p \approx 2.15\sigma \quad (6.20)$$

This is illustrated in Figure 6.2 together with the Gaussian form of the concentration distribution.

A more general expression of the concentration at a point (x,y,z) due to an instantaneous point release at (x',y',z') at time t' can be obtained through a simple coordinate transformation

$$c(x,y,z,t) = \frac{Q_{ip}}{(2\pi)^{3/2}\sigma^3} \exp\left[-\frac{(x-x')^2}{2\sigma^2} - \frac{(y-y')^2}{2\sigma^2} - \frac{(z-z')^2}{2\sigma^2}\right] \quad (6.21)$$

where

$$\sigma = [2D(t-t')]^{1/2} \quad (6.22)$$

Equation (6.21) would be more useful for determining concentrations due to multiple point, as well as area and volume sources.

6.3.2 Instantaneous Line Source

The solution of the diffusion equation for an instantaneous line source of strength Q_{il}, with dimensions of mass per unit length [ML^{-1}], along the y axis, can be obtained by considering the concentration due to the instantaneous point sources of strength $Q_{il}dy'$

located at y' and integrating it with respect to y' from $-\infty$ to ∞. The result is

$$c(x,z,t) = \frac{Q_{il}}{2\pi\sigma^2} \exp\left(-\frac{x^2+z^2}{2\sigma^2}\right) \quad (6.23)$$

indicating that the diffusing cloud has a cylindrical shape with the maximum concentration at its axis, which is assumed here to coincide with the line source at the time of release. The diameter of the cloud expands with time at the same rate ($\sigma \propto t^{1/2}$) as for the point source, but the center line (axial) concentration decreases less rapidly ($c_{\max} \propto t^{-1}$) because diffusion is now limited to two, rather than three, dimensions. Note that Eq. (6.23) is indeed the solution of the two-dimensional diffusion equation

$$\frac{\partial c}{\partial t} = D\left(\frac{\partial^2 c}{\partial x^2} + \frac{\partial^2 c}{\partial z^2}\right) \quad (6.24)$$

and satisfies the mass continuity condition

$$\int_{-\infty}^{\infty}\int_{-\infty}^{\infty} c\, dx\, dz = Q_{il} \quad (6.25)$$

6.3.3 Instantaneous Area Source

Following the same procedure as for the line source, one can obtain the solution of the diffusion equation for an area source of strength Q_{ia}, which is located, for example, in the x–y plane:

$$c(z,t) = \frac{Q_{ia}}{(2\pi)^{1/2}\sigma} \exp\left(-\frac{z^2}{2\sigma^2}\right) \quad (6.26)$$

Equation (6.26) again implies the Gaussian distribution in the z direction. The diffusing cloud is homogeneous in the plane of the source and expands only in the direction normal to the source plane. Note that Eq. (6.26) is the exact solution of the one-dimensional diffusion equation

$$\frac{\partial c}{\partial t} = D\frac{\partial^2 c}{\partial z^2} \quad (6.27)$$

and satisfies the mass conservation or continuity condition

$$\int_{-\infty}^{\infty} c\, dz = Q_{ia} \quad (6.28)$$

In all the above cases of molecular diffusion from point, line, and area sources, the expanding puff or cloud grows in proportion to $\sigma = (2Dt)^{1/2}$ in the direction of diffusion. Using a typical value of $D = 1.5 \times 10^{-5}$ m^2 s^{-1} for the molecular diffusivity of air, the following values of σ are obtained at various diffusion times:

t (s):	10^2	10^4	10^6	10^8
σ (m):	0.05	0.55	5.5	55

These indicate that the process of molecular diffusion in the atmosphere is very slow indeed, such that a puff or cloud of material released instantaneously in a non-turbulent free atmosphere would grow to only a few meters thick in one day.

Actual observations of purely molecular diffusion in the atmosphere are practically nonexistent. But the process is operative all the time and is believed to be responsible for the slow but steady diffusion of inert trace gases such as CO_2, methane, chlorofluorocarbons (CFCs), and others from the PBL to the upper atmosphere. Most instantaneous releases in the atmosphere are in the form of injections from volcanic eruptions, bomb blasts, shell bursts, container explosions, and exhausts from rockets and other flying machines. The initial diffusion of material from such explosive events near the surface or in the PBL is quite rapid, because it is caused by both the ever-present PBL turbulence and violent motions produced by explosion. Even for releases in the free atmosphere, the initial growth of puff or cloud is due to turbulence produced by explosion or wake of flying object and, hence, is quite rapid. After the resulting cloud or exhaust wake grows to certain finite dimensions and turbulence responsible for the initial phase of rapid diffusion dies down, subsequent expansion of the cloud or contrail may be essentially due to molecular diffusion. Thus, a mathematical description of molecular diffusion of a finite-size contaminant cloud should be more useful in applications to the atmosphere.

6.3.4 Finite Thickness Area Source

All real sources have finite dimensions and finite initial concentrations. We consider here an instantaneous area source in the x–y plane, with an initial thickness d in the vertical and a uniform initial concentration c_o at $t = 0$ within the source region ($c = 0$ at $t = 0$ outside the source region). The vertical diffusion of this finite-size cloud can be determined by breaking it into thin sheets (plane sources) of infinitesimal thicknesses dz', each having a concentrated source strength of $Q_{ia} = c_o\, dz'$. Then, integrating the elementary area source solution over the finite initial source depth, the result is (Csanady, 1973)

$$c(z,t) = \frac{c_o}{2}\left[\text{erf}\left(\frac{0.5d+z}{\sqrt{2}\sigma}\right) + \text{erf}\left(\frac{0.5d-z}{\sqrt{2}\sigma}\right)\right] \quad (6.29)$$

where the error function is defined as

$$\text{erf}(x) = \frac{2}{\sqrt{\pi}} \int_0^x \exp(-x'^2)dx' \quad (6.30)$$

and can be obtained from standard mathematical tables.

The behavior of the solution (6.29) is illustrated in Figure 6.3 in which c/c_o is plotted as a function of $2z/d$ and the parameter $\sqrt{8}\sigma/d$, which is proportional to $t^{1/2}$. Note that the maximum concentration at the center of the cloud ($z = 0$) is given by

$$c_{\max}(t) = c_o\, \text{erf}\left(\frac{d}{\sqrt{8}\sigma}\right) \quad (6.31)$$

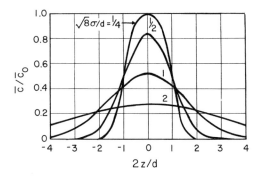

Figure 6.3 Normalized concentration distribution in the instantaneous cloud due to a finite thickness area source as a function of the dimensionless distance from the source center and the dimensionless time after release. After Csanady, 1973.

From the asymptotic behavior of the error function, one can infer that the maximum concentration within the cloud does not change significantly from its initial value c_o, until $\sigma \gtrsim 0.2\,d$. In the later stages ($\sigma/d \gg 1$) of diffusion, the concentration distribution within the cloud approaches the Gaussian distribution for an instantaneous concentrated plane source with a source strength $Q_{ia} = c_o d$. Thus, the ideal plane source model should be adequate for describing diffusion from an initial finite-size cloud at later stages of diffusion when the cloud has grown much bigger than its initial size.

The above results can also be easily extended to other finite volume sources (e.g., cylindrical and spherical). An instantaneous point source may serve as an adequate model of a spherical volume source only after the cloud has expanded to much larger dimensions compared with its initial diameter. In early stages of diffusion, the exact solution involves error function terms similar to those in Eq. (6.29), in all three dimensions.

6.3.5 Continuous Point Source Moving with the Medium

When a contaminant is emitted at a point in the fluid environment at some fixed rate for some finite time or indefinitely, its diffusion can be described using an idealized continuous point source model. We first consider the simpler case of a point source emitting continuously while moving with the medium. This will also include the extreme case of a fixed source in a medium at rest ($u = 0$). In any case, we again use a coordinate system moving with the uniform velocity of the environment (relative velocity $u = 0$), with the origin located at the source position.

A continuous source may be assumed to be a sequence of instantaneous puffs released one after another at very small intervals dt'. Let the rate of emission at any time t' be $Q(t')$, so that $Q(t')dt'$ is the amount of material emitted during any short interval dt'. At any time t after the emission started, the puff size or standard deviation σ, given by Eq. (6.22), differs from puff to puff. Using the method of superposition, the concentration at a point (x,y,z) at time t is given by

$$c(x,y,z,t) = \frac{1}{(2\pi)^{3/2}} \int_0^t \frac{Q(t')dt'}{\sigma^3} \exp\left[-\frac{r^2}{2\sigma^2}\right]$$

$$= \frac{1}{8(\pi D)^{3/2}} \int_0^t \frac{Q(t')dt'}{(t-t')^{3/2}}$$

$$\exp\left[-\frac{r^2}{4D(t-t')}\right] \quad (6.32)$$

which may be integrated numerically for any arbitrary (given) emission rate as a function of time.

For a constant rate of emission Q, Eq. (6.32) can be simplified to (Sutton, 1953)

$$c(c,y,z,t) = \frac{Q}{4\pi D r}\left[1 - \mathrm{erf}\left(\frac{r}{\sqrt{4Dt}}\right)\right] \quad (6.33)$$

which predicts an ever-expanding cloud in which concentrations increase with time. If the emission continues long enough for the argument of the above error function to become much less than unity ($r/\sqrt{4Dt} \ll 1$), however, the concentration field becomes independent of time, that is,

$$c(r) = \frac{Q}{4\pi D r} \quad (6.34)$$

Equation (6.34) is an interesting result, which shows that concentration due to a point source emitting continuously in medium at rest (relative to the source) varies inversely proportional to the radial distance from the source. One can obtain the same result more simply from the relationship between the mass flux and the concentration gradient and the mass continuity condition

$$Q = -4\pi r^2 D \frac{\partial c}{\partial r} \quad (6.35)$$

which states that, in a steady state, the total amount of material passing through any spherical surface, with the source at the center, must be equal to the emission rate. The solution of Eq. (6.35), with the appropriate boundary condition that $c \to 0$ as $r \to \infty$, yields Eq. (6.34).

6.3.6 Continuous Point Source in a Uniform Flow

The diffusion problem of greatest interest in the atmospheric environment is that of continuous sources in a uniform flow, as well as in more complex stratified shear flows. In such cases, the source is usually fixed so that it is more convenient to use a fixed frame of reference. For the sake of simplicity, we consider a continuous point source emitting at a fixed rate Q for a long enough time that a steady-state concentration field is associated with the plume emanating from the source. The steady-state diffusion equation for a

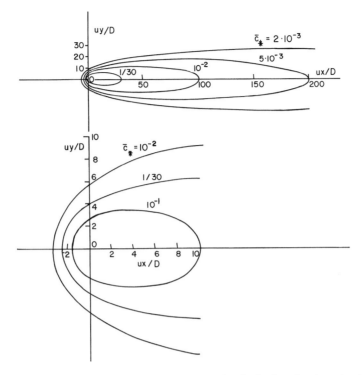

Figure 6.4 Contours of constant normalized concentration in the longitudinal section (x–y plane) of the continuous source plume in uniform flow in the x direction. After Csanady, 1973.

point source in uniform wind (with the x axis aligned with the wind) is

$$u\frac{\partial c}{\partial x} = D\left(\frac{\partial^2 c}{\partial x^2} + \frac{\partial^2 c}{\partial y^2} + \frac{\partial^2 c}{\partial z^2}\right) \quad (6.36)$$

The appropriate boundary conditions are

$$c \to 0, \text{ as } x,y,z \to \pm\infty \quad (6.37)$$

The exact solution of Eq. (6.36) satisfying the above boundary conditions and the appropriate mass continuity condition, is

$$c(x,y,z) = \frac{Q}{4\pi Dr}\exp\left[-u\left(\frac{r-x}{2D}\right)\right] \quad (6.38)$$

which also reduces to Eq. (6.34) for the case of no wind ($u = 0$). Equation (6.38) can also be obtained from the integration of the instantaneous point-source solution with respect to time (see e.g., Sutton, 1953). The resulting concentration distribution (6.38) is shown in Figure 6.4, using normalized coordinates and dimensionless concentration $c_* = (4\pi D^2/Qu)c$.

Unlike the symmetrical, three-dimensional cloud of material resulting from an instantaneous point source or a continuous point source in a medium at rest (the calm or no-wind condition), the cloud resulting from a continuous point source in a uniform flow is considerably elongated in the direction of flow. Except for the immediate neighborhood of the source and in very light winds, the resulting "plume" may be considered thin or slender. Since the interest is usually centered on concentrations at points not too far removed from the plume axis ($y = z = 0$), the following "slender-plume approximation" is made for the terms involving r in Eq. (6.38):

$$r = (x^2 + y^2 + z^2)^{1/2} \simeq x\left(1 + \frac{y^2 + z^2}{2x^2}\right),$$
$$\text{for } y^2 + z^2 \ll x^2 \quad (6.39)$$

so that, to the lowest order of approximations,

$$r \simeq x$$
$$r - x \simeq \left(\frac{y^2 + z^2}{2x}\right) \quad (6.40)$$

Substituting from Eq. (6.40) into Eq. (6.38), one obtains the Gaussian plume formula

$$c(x,y,z) = \frac{Q}{2\pi u\sigma^2}\exp\left(-\frac{y^2 + z^2}{2\sigma^2}\right) \quad (6.41)$$

in which we have introduced the plume diffusion parameter

$$\sigma = \left(\frac{2Dx}{u}\right)^{1/2} \quad (6.42)$$

which is similar to the puff diffusion parameter introduced earlier if one recognizes the obvious relationship

$t = x/u$ between the travel or diffusion time of a puff and the distance downwind from the continuous source.

The plume diffusion parameter σ is also the standard deviation of the concentration distribution normal to the plume axis, which is predicted to be Gaussian in the slender-plume approximation. It is directly proportional to the plume thickness. For example, the 10-percent half-plume width is given by $y_p = z_p \simeq 2.15\,\sigma$, which is identical to Eq. (6.20) for the 10-percent puff radius. According to Eq. (6.42), plume width or $\sigma \propto x^{1/2} u^{-1/2}$, so that plume grows monotonically with distance downwind of the source and, at a given distance, plume becomes more and more slender with increasing flow speed.

Another implication of the slender-plume approximation is that the concentration distribution in the plume cross section at a distance x is identical to that in a puff at $t = x/u$. In fact, Eq. (6.41) is an exact solution of the approximate diffusion equation

$$u\frac{\partial c}{\partial x} = D\left(\frac{\partial^2 c}{\partial y^2} + \frac{\partial^2 c}{\partial z^2}\right) \tag{6.43}$$

in which diffusion in the direction of flow has been neglected in comparison with advective transport. Thus, the slender-plume approximation is equivalent to neglecting diffusion in the direction of flow.

The appropriate mass continuity condition for the plume diffusion is given by Eq. (6.14), which is indeed satisfied by Eq. (6.41). The condition for the validity of the slender-plume approximation is $ux/D \gg 1$, or $\sigma/x \ll 1$. Note that, despite the above approximation, the plume centerline concentration given by Eq. (6.41) is identical to that given by the exact solution (6.38), that is,

$$c(x,0,0) = \frac{Q}{2\pi u \sigma^2} = \frac{Q}{4\pi D x} \tag{6.44}$$

which shows inverse relationship of plume centerline concentration with distance from the source.

6.3.7 Continuous Cross-wind Line Source

The exact solution of the two-dimensional diffusion equation for a continuous, cross-wind line source (say, in y direction) is rather complicated, involving Bessel functions (see e.g., Roberts, 1923). However, at a sufficiently large distance downwind from the source and in the presence of some flow in one (x) direction, one can again use the slender-plume approximation to simplify that solution to the following form:

$$c(x,z) = \frac{Q_{cl}}{(2\pi)^{1/2}\,u\sigma}\exp\left(-\frac{z^2}{2\sigma^2}\right) \tag{6.45}$$

Note that Eq. (6.45) is also the exact solution of the approximate diffusion equation, ignoring diffusion in the direction of flow,

$$u\frac{\partial c}{\partial x} = D\frac{\partial^2 c}{\partial z^2} \tag{6.46}$$

and satisfies the approximate mass continuity equation

$$\int_{-\infty}^{\infty} cu\,dz = Q_{cl} \tag{6.47}$$

According to Eq. (6.45), the concentration distribution in the plume is again Gaussian, with the centerline concentration decreasing with distance as $c(x,0) \propto x^{-1/2}$.

6.3.8 Continuous Area Source

For a continuous infinite area source, for example, in the x–y plane, one cannot expect a steady-state solution (concentration field) similar to those for the continuous point and line sources. Both the cloud thickness and the concentration may be expected to increase with time, especially in the absence of any sinks in the atmosphere. The relevant diffusion equation is

$$\frac{\partial c}{\partial t} = D\frac{\partial^2 c}{\partial z^2} \tag{6.48}$$

which is identical to Eq. (6.27) for the instantaneous area source. However, the mass conservation condition to be satisfied for the continuous area source is

$$\int_{-\infty}^{\infty} c\,dz = Q_{ca}t \tag{6.49}$$

which requires concentration to be an increasing function of time.

The solution of Eq. (6.48) for an area source with a prescribed constant flux or emission rate Q_{ca} at $z = 0$, in a semiinfinite medium ($z > 0$), is given by (Sutton, 1953)

$$c(z,t) = \frac{Q_{ca}}{D}\left[\left(\frac{Dt}{\pi}\right)^{1/2}\exp\left(-\frac{z^2}{4Dt}\right) \right.$$
$$\left. - \frac{z}{2}\left\{1 - \mathrm{erf}\left(\frac{z}{\sqrt{4Dt}}\right)\right\}\right] \tag{6.50}$$

Note that Eq. (6.50) implies that the maximum concentration at $z = 0$ increases in proportion to $t^{1/2}$.

The concept of a continuous infinite area source is quite useful in micrometeorological applications, because the earth's surface often acts as an extensive source of water vapor in the atmosphere as well as a source or sink of heat. However, the area source strengths (surface fluxes) do not remain constant for any extensive period, but show strong diurnal variations and even reversal of signs (e.g., in heat flux). Other source and sink mechanisms associated with phase changes of H_2O also become important over an extended period of time. Most important, diffusion processes in the lower part of the atmosphere, particularly in the PBL, are not molecular but almost always turbulent. For this reason, direct application of the above theoretical/analytical results would be severely limited unless molecular diffusivity D could be replaced by an effective turbulent diffusivity. This in fact is done in the generalized Fickian diffusion theory to be discussed later. First, we consider some of the fundamental problems and difficulties associated with turbulent diffusion.

6.4 TURBULENT DIFFUSION

In previous chapters we have given an overview of turbulent exchange processes in the atmospheric boundary layer and discussed some of the observed statistical properties as well as prevailing theoretical concepts of turbulence. Since turbulent diffusion is a consequence of highly irregular (nearly random), three-dimensional turbulent motions characterized by a rather wide range of scales, its description faces all the problems and difficulties that are associated with turbulence. The lack of a rigorous and general theory of turbulence has obviously hindered the development of a similar theory of turbulent diffusion. In this chapter we introduce the simpler gradient-transport theory based on an analogy with molecular diffusion. Other more sophisticated diffusion theories and models will be described in later chapters.

6.4.1 Mean Diffusion Equation

The instantaneous diffusion equation derived in section 6.2 is also applicable in a turbulent flow. However, due to the random variability in time and space of concentration and velocity components, the diffusion equation must be solved simultaneously with the Navier–Stokes equations of motion. These equations are highly nonlinear and their complete numerical solution has not yet become possible for large Reynolds number flows. The fastest and the biggest supercomputers of today cannot resolve all the scales of turbulent motion that contribute to atmospheric diffusion. They can resolve, however, larger eddies and bring out the important role played by these eddies in the diffusion process. Such large-eddy simulations (LES), using the state-of-the-art computers and numerical algorithms, are probably the most sophisticated, but also the most expensive, numerical models of turbulence and diffusion (Deardorff, 1972; Lamb, 1982; Nieuwstadt and de Valk, 1987). In the foreseeable future, however, large-eddy simulations are more likely to remain expensive tools for basic research rather than off-the-shelf operational models for practical applications to air quality. The latter are mostly based on analytical solutions (or their empirical modifications) of the mean diffusion equation that is obtained by Reynolds averaging of the instantaneous equation

$$\frac{\partial \bar{c}}{\partial t} + \bar{u}\frac{\partial \bar{c}}{\partial x} + \bar{v}\frac{\partial \bar{c}}{\partial y} + \bar{w}\frac{\partial \bar{c}}{\partial z}$$
$$= D\nabla^2\bar{c} - \left(\frac{\partial \overline{c'u'}}{\partial x} + \frac{\partial \overline{c'v'}}{\partial y} + \frac{\partial \overline{c'w'}}{\partial z}\right) \quad (6.51)$$

Note that in the above mean diffusion equation all the averages are ensemble averages, which may not be equivalent to time averages if the diffusion process is time dependent, even when the flow is stationary. An important feature of Eq. (6.51) is the appearance of three covariances (turbulent fluxes) as additional unknowns. Even when the mean flow field is specified, Eq. (6.51) has essentially four unknowns and cannot be solved. Similarly, each of the mean flow equations has three unknown covariances. This problem of more unknowns than the number of mean flow and diffusion equations is called the closure problem of turbulence, which arises from the nonlinearity of the original instantaneous equations. Otherwise, the Reynolds-averaged equations deal with well-behaved, smoothly varying variables and can be further simplified depending on the particular flow and source conditions. The molecular diffusion terms can be neglected, because turbulence dominates the diffusion process. In particular, the mean diffusion equation for a uniform mean flow is reduced to

$$\frac{\partial \bar{c}}{\partial t} + \bar{u}\frac{\partial \bar{c}}{\partial x}$$
$$= -\left(\frac{\partial \overline{c'u'}}{\partial x} + \frac{\partial \overline{c'v'}}{\partial y} + \frac{\partial \overline{c'w'}}{\partial z}\right) \quad (6.52)$$

which can be solved only if the terms on the right-hand side can be parameterized or specified in terms of the mean concentration field and some bulk turbulence properties (e.g., eddy diffusivities), or additional equations are developed for the turbulent covariances, which have their own closure problems.

6.4.2 First-order Closure Based on the K-Theory

The simplest and probably the most widely used closure approach in diffusion applications is the first-order closure based on the gradient-transport hypothesis discussed in Chapter 4. According to this hypothesis, turbulent mass fluxes can be expressed in terms of mean concentration gradients as

$$\overline{c'u'} = -K_x\frac{\partial \bar{c}}{\partial x}$$
$$\overline{c'v'} = -K_y\frac{\partial \bar{c}}{\partial y}$$
$$\overline{c'w'} = -K_z\frac{\partial \bar{c}}{\partial z} \quad (6.53)$$

Substituting from Eq. (6.53) into Eq. (6.52) yields

$$\frac{\partial \bar{c}}{\partial t} + \bar{u}\frac{\partial \bar{c}}{\partial x} = \frac{\partial}{\partial x}\left(K_x\frac{\partial \bar{c}}{\partial x}\right)$$
$$+ \frac{\partial}{\partial y}\left(K_y\frac{\partial \bar{c}}{\partial y}\right) + \frac{\partial}{\partial z}\left(K_z\frac{\partial \bar{c}}{\partial z}\right) \quad (6.54)$$

Turbulent or eddy diffusivities (K_x, K_y, and K_z) have to be specified for the complete closure and solution of Eq. (6.54) for any desired initial, boundary, and source conditions. The most satisfactory, yet simple, specification for eddy diffusivities is given by Eq. (4.28) in which the proportionality coefficient may depend on the definition of large-eddy length scales (e.g., those based on the autocorrelation function or the spectrum function). A value of $c \simeq 0.15$ has been suggested (Hanna, 1982) when length scales correspond to the spectral peak wavelengths (λ_{mu}, λ_{mv}, and λ_{mw}). In the absence of turbulence measurements, large-eddy length scales and turbulent velocities can be specified using the empirical similarity relations given in chapters 4 and 5. In this way, the approximate (parameterized) mean diffusion equation (6.54) becomes closed and can be solved for the appropriate initial, boundary, and source conditions. We have already discussed the mass continuity and other conditions associated with idealized sources in an infinite medium. Those conditions have to be modified when one deals with sources near the earth's surface or in the atmospheric PBL capped by an inversion from above.

6.4.3 Surface Boundary Conditions

The presence of the underlying earth's (land or water) surface and its interaction with diffusing clouds or plumes of contaminants released from near-surface sources are usually accounted for through the lower (surface) boundary condition. The concentration field in a diffusing cloud is considerably altered by the presence of the surface as a physical barrier to downward diffusion. If the diffusing cloud contains particulate matter, one would expect considerable deposition of material on the surface through gravitational settling or falling out of larger particles due to gravity. Even in clouds of very small particles and gases, for which gravitational settling is not significant, some deposition is likely to occur due to turbulent transport of material downward and its partial absorption by the ground surface including soil, water, and vegetation. A quantitative measure of the surface deposition is the downward mass flux at the surface, which is usually parameterized as the product of the average surface concentration $\bar{c}_o$ and the deposition velocity v_d, that is,

$$F_{zo} = -\bar{c}_o v_d \quad (6.55)$$

Three different types of the surface boundary conditions can be used, depending on the extent of the interaction between the surface and the diffusing material (contaminant):

1. Perfectly reflecting surface. If the contaminant does not interact (chemically or physically) with the surface at all, it is simply reflected away from the surface. Consequently, there is no downward mass flux and, hence, no deposition at the surface. Mathematically, the reflecting boundary condition is expressed as

$$F_{zo} = -K_z \frac{\partial \bar{c}}{\partial z} = 0, \text{ at } z = 0 \quad (6.56)$$

In most situations involving dispersion of gaseous pollutants, the bare ground surface may be regarded, without serious error, as a perfect reflector. The reflecting boundary condition may also be appropriate for describing the short-range dispersion of gases over many other surfaces, unless gases are highly reactive with surface materials (e.g., vegetation and water). It is considered to be a conservative boundary condition that would have an overall effect of overestimating the ground-level concentrations from near-surface sources.

2. Perfectly absorbing surface. If the contaminant reacts with the surface instantly and is completely absorbed on contact, the surface would be considered as a perfect absorber. Mathematically, the perfectly absorbing boundary condition is specified by the requirement that the concentration must vanish at the surface, that is,

$$\bar{c}_o = 0, \text{ at } z = 0 \quad (6.57)$$

An example of the perfectly absorbing surface might be the diffusion of particulates over a water surface. Liquid droplets over solid surfaces would also tend to stick to the surface. In most other cases, however, the assumption of zero ground-level concentration would not be realistic and might lead to considerable underestimation of concentrations near the ground. Since air pollution control regulations often specify limits on ground-level concentrations of specific contaminants (see e.g., The National Ambient Air Quality Standards), a perfectly absorbing boundary condition, implying $\bar{c}_o = 0$, would not be permissible in a regulatory air-quality model.

3. Partly reflecting/absorbing surface. Most natural surfaces are neither perfect reflectors nor perfect absorbers of gaseous or particulate contaminants diffusing toward them. There is usually a finite amount of deposition of the contaminant at the surface, which becomes especially important in problems of long-range dispersion and deposition. Mathematically, the partly reflecting and partly absorbing surface boundary condition is simply expressed as

$$K_z \frac{\partial \bar{c}}{\partial z} = \bar{c}_o v_d, \text{ at } z = 0 \quad (6.58)$$

in which the deposition velocity is to be specified. There is considerable uncertainty, however, in the specification of v_d. The physics of the dry deposition process as well as the experimental data on v_d will be discussed in Chapter 10. Here, suffice to say that our current theoretical and empirical understanding of the deposition process is not sufficient to specify deposition velocities for the commonly encountered air pollutants, natural surfaces, and meteorological conditions with a reasonable degree of confidence; the specification of v_d often involves a large uncertainty.

6.4.4 Upper Boundary Conditions

In theoretical considerations of diffusion from sources in the PBL, one must consider the effect of any capping inversion at the top of the PBL where upward turbulent diffusion may cease. In the absence of other vertical transport mechanisms, such as breaking gravity waves and deep convective clouds, the contaminant must become concentrated near the top of the PBL and then diffuse downward. For example, it is quite common to see a thick-haze layer of pollutants over and around large urban areas. The visibility improves dramatically as one gets out of the haze layer at the top of the PBL on an aircraft or a balloon. In most fair-weather conditions, pollutants released from sources within the PBL are usually confined to the PBL, because they are transported long distances by boundary layer winds. For all practical purposes, the inversion at the top of the PBL acts as an impervious surface for which the reflecting boundary condition

$$F_{zh} = -K_z \frac{\partial \overline{c}}{\partial z} = 0, \text{ at } z = h \quad (6.59)$$

should be appropriate.

There are, of course, exceptional situations when highly buoyant plumes from tall stacks may penetrate through weaker low-level inversions and towering cumulus clouds or thunderstorms vent out the pollutants from the PBL to the free atmosphere. Such situations have to be dealt with separately and not necessarily through a simple upper boundary condition.

6.4.5 Other Boundary and Continuity Conditions

So long as a contaminant cloud or a plume from an elevated release in the PBL is not affected either by the ground surface or by a capping inversion, one can use the more general and obvious boundary condition that concentration must vanish at infinite distances from the cloud or plume center at any time t after release or distance from the source. The same condition would also be appropriate at lateral boundaries of the puff or the plume, even when the contaminant may spread over the whole depth of the PBL. Such a boundary condition is automatically satisfied by the parabolic-type diffusion equation resulting from the gradient transport (K) closure.

Integral mass conservation equations for the various idealized source configurations can be obtained by averaging the similar equations in instantaneous forms given earlier. For example, for a continuous point source, Eq. (6.14) yields, upon averaging,

$$\int_{-\infty}^{\infty} \int_{-\infty}^{\infty} (\overline{c}\,\overline{u} + \overline{c'u'}) dy\, dz = Q \quad (6.60)$$

If plume is essentially confined to the PBL and turbulent diffusion in the direction of flow can be neglected in comparison with the transport by mean flow, Eq. (6.60) can be simplified to

$$\int_0^h \int_{-\infty}^{\infty} \overline{c}\,\overline{u}\, dy\, dz = Q \quad (6.61)$$

which is the more commonly used, but approximate, form of the mass continuity equation. The implied assumption here is that $|\overline{c'u'}| \ll \overline{c}\,\overline{u}$, or $|\partial \overline{c}/\partial x| \ll \overline{c}\,\overline{u}/K_x$. It can easily be shown, by using an appropriate power-law relationship $\overline{c} \propto x^{-\beta}$, that the above condition (assumption) would be satisfied only if $\overline{u}x/K_x \gg 1$. Thus, the slender-plume approximation for turbulent diffusion can be justified only at large enough wind speeds and distances from the source that their product $\overline{u}x \gg K_x$. This requirement is much more stringent than that for molecular diffusion, since K_x is several orders of magnitude larger than D.

6.5 CONSTANT K (FICKIAN DIFFUSION)-THEORY

In stationary and homogeneous turbulent flows, eddy diffusivities are not expected to depend on time and space, although, in general, these may be different in different directions. For the specified constant K_x, K_y, and K_z, Eq. (6.54) is simplified to

$$\frac{\partial \overline{c}}{\partial t} + \overline{u} \frac{\partial \overline{c}}{\partial x} = K_x \frac{\partial^2 \overline{c}}{\partial x^2}$$
$$+ K_y \frac{\partial^2 \overline{c}}{\partial y^2} + K_z \frac{\partial^2 \overline{c}}{\partial z^2} \quad (6.62)$$

which is quite similar to molecular diffusion equation (6.12). In fact, the two are identical for the special case of isotropic turbulence when $K_x = K_y = K_z = K$. The solutions given in section 6.3 for molecular diffusion in an infinite medium should be directly applicable to this particular case of turbulent diffusion after D is replaced by K. Those solutions of the diffusion equation for an isotropic diffusivity can easily be generalized to nonisotropic conditions, typical of the atmosphere, by following the same methods and procedures. Some of the solutions will be given in this section, with appropriate modifications for sources near the earth's surface (e.g., surface and elevated sources in the PBL).

6.5.1 The Method of Images

The reflecting boundary condition at the surface or at the top of the PBL requires that $\partial \overline{c}/\partial z = 0$ at the impervious (reflecting) surface. The mathematical device by which such a boundary condition is automatically satisfied, while still using the simpler solutions of the diffusion equation in an infinite medium, is the introduction of image sources of strengths equal to the real sources but located at the mirror images of true sources in reflecting surfaces. The actual medium confined by reflecting boundaries with real sources is replaced by an infinite medium containing both the real and image sources. Then, the solution is obtained through a superposition of similar solutions due to each of the real and image sources. The solution would, of course, be meaningful (valid) only in the original confined space, and not outside the real boundaries. The method of images is only a procedure for obtaining solutions to the linear diffusion equation in a medium confined by reflecting boundaries from its known solution in an infinite medium with no boundaries. Alternatively, one can obtain the solution with any appropriate boundary conditions without the use of this method. We will illustrate the use of the method of images for the reflecting boundary conditions at the earth's surface and, in some cases, at the top of the PBL as well.

6.5.2 Instantaneous Releases

For an instantaneous point release in an infinite and isotropic medium, the solution of the Fickian diffusion equation (6.62) in a moving frame of reference with its origin at the cloud center, is

$$\bar{c}(x,y,z,t) = \frac{Q_{ip}}{(2\pi)^{3/2}\sigma_x\sigma_y\sigma_z}$$
$$\exp\left(-\frac{x^2}{2\sigma_x^2} - \frac{y^2}{2\sigma_y^2} - \frac{z^2}{2\sigma_z^2}\right) \quad (6.63)$$

where

$$\sigma_x = (2K_x t)^{1/2}; \quad \sigma_y = (2K_y t)^{1/2};$$
$$\sigma_z = (2K_z t)^{1/2} \quad (6.64)$$

Note that Eq. (6.63) is a generalization of Eq. (6.19) in which diffusion parameters are related to turbulent (eddy) diffusivities. Since eddy diffusivities are normally four to eight orders of magnitude larger than the molecular diffusivity in the atmosphere, the diffusing cloud at a certain time after release would be two to four orders of magnitude larger in the turbulent atmosphere. On the average (taken over a large number of ensembles or releases), the contaminant cloud is predicted to have the form of an ellipsoid. The surface of constant concentration has the same form; it grows initially with time, attains its maximum outline at some time (t_m) after release, and then decreases with further increase in time, until it disappears at a later time (et_m). A simple theory for describing the visible outline of a puff of smoke in a turbulent atmosphere was first proposed by Roberts (1923), who also presented some observations of antiaircraft shellbursts. An illustration of his computed outline of the ensemble-averaged puff in an isotropic turbulence medium as a function of time after release is given in Figure 6.5.

For the instantaneous point release at a height H above the reflecting ground surface, we can use the method of images and consider an image source at the mirror-image point of the real source. Assuming that the origin of the moving frame of reference is at the ground level just below the point of release, the coordinates of the real and the image sources are $(0,0,H)$ and $(0,0,-H)$, respectively. Then, the concentration field due to both the real and the image sources in an infinite medium (the original boundary is removed, while considering the image source) is given by

$$c(x,y,z,t) = \frac{Q_{ip}}{(2\pi)^{3/2}\sigma_x\sigma_y\sigma_z}$$
$$\exp\left(-\frac{x^2}{2\sigma_x^2} - \frac{y^2}{2\sigma_y^2}\right)\left\{\exp\left[-\frac{(z-H)^2}{2\sigma_z^2}\right]\right.$$
$$\left. + \exp\left[-\frac{(z+H)^2}{2\sigma_z^2}\right]\right\} \quad (6.65)$$

in which only positive values of z need to be considered. The reader can verify that Eq. (6.65) is indeed a solution of the Fickian diffusion equation (6.62), which satisfies the reflecting boundary condition (6.56) as well as the mass continuity condition

$$\int_0^\infty \int_{-\infty}^\infty \int_{-\infty}^\infty \bar{c}\, dx\, dy\, dz = Q_{ip} \quad (6.66)$$

Recognizing that our primary interest is in the ground-level concentration (g.l.c.) directly below the puff center, that is, at the point $(0,0,0)$,

$$\bar{c}_o = \frac{2Q_{ip}}{(2\pi)^{3/2}\sigma_x\sigma_y\sigma_z}\exp\left(-\frac{H^2}{2\sigma_z^2}\right) \quad (6.67)$$

which can be shown to initially increase with time, attain a maximum value $\bar{c}_{omax}$ at time t_{max} after release, and decrease monotonically with time thereafter (see Fig. 6.6). In particular, for the case when $\sigma_x = \sigma_y = \alpha\sigma_z$, where α is a constant, one can show that the maximum g.l.c.

$$\bar{c}_{omax} \simeq \frac{0.147}{\alpha^2}\frac{Q_{ip}}{H^3} \quad (6.68)$$

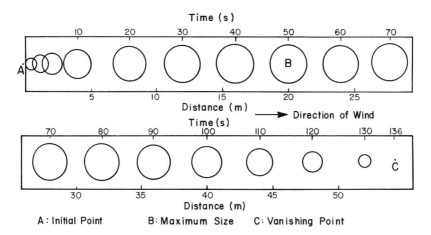

Figure 6.5 Computed expansion and contraction of an ensemble-averaged puff of smoke in an isotropic turbulence medium as a function of time after release for $K = 1$ m^2 s^{-1} and $\bar{u} = 4$ m s^{-1}. After Roberts, 1923.

141 CONSTANT K (FICKIAN DIFFUSION)-THEORY

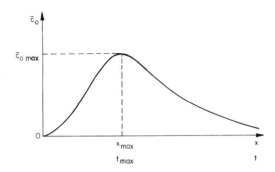

Figure 6.6 Schematic of the ground-level concentration as a function of time after an elevated release or distance from a continuous source.

occurs at a time

$$t_{omax} = \frac{1}{6}\frac{H^2}{K_z} \quad (6.69)$$

when $\sigma_z = H/\sqrt{3}$. One should note the extreme sensitivity of the maximum g.l.c. to the release height.

Similar results can be obtained for the instantaneous line and area releases at some height H above the ground level. It is found that $\bar{c}_{omax} \propto H^{-2}$ for the line source in x or y direction, and $\bar{c}_{omax} \propto H^{-1}$ for the area source in the x–y plane. For any ground-level release ($H = 0$), the maximum g.l.c. would monotonically decrease with time; it would be twice the maximum concentration due to a similar release in an infinite medium.

6.5.3 Continuous Sources

In the absence of any mean wind, for example, under true free convection conditions, it might be reasonable to assume that $K_x = K_y = K_z = K$. Then diffusion from a continuous point source is described by Eq. (6.34) with D replaced by K. When the source is located at the ground level (e.g., an idling vehicle), considering the effect of the reflecting ground surface one obtains

$$\bar{c}(r) = \frac{Q}{2\pi K r} \quad (6.70)$$

A more general solution of the steady mean diffusion equation for a continuous point source in an infinite, anisotropic, turbulent atmosphere was first derived by Roberts (1923) as

$$\bar{c}(x,y,z) = \frac{Q \exp\left\{-\dfrac{\bar{u}}{2\,K_x^{1/2}}\left[\left(\dfrac{x^2}{K_x} + \dfrac{y^2}{K_y} + \dfrac{z^2}{K_z}\right)^{1/2} - \dfrac{x}{K_x^{1/2}}\right]\right\}}{4\pi(K_x K_y K_z)^{1/2}\left(\dfrac{x^2}{K_x} + \dfrac{y^2}{K_y} + \dfrac{z^2}{K_z}\right)^{1/2}} \quad (6.71)$$

which is applicable even under situations of very weak or no wind. Introducing the usual definitions of plume diffusion parameters

$$\sigma_x = \left(\frac{2K_x x}{\bar{u}}\right)^{1/2};\; \sigma_y = \left(\frac{2K_y x}{\bar{u}}\right)^{1/2};$$

$$\sigma_z = \left(\frac{2K_z x}{\bar{u}}\right)^{1/2} \quad (6.72)$$

the above solution can also be expressed as

$$\bar{c}(x,y,z) = \frac{Q(x_*/r_*)}{2\pi\bar{u}\sigma_y\sigma_z}\exp(x_*^2 - x_* r_*) \quad (6.73)$$

where

$$x_* = \frac{x}{\sigma_x};\; y_* = \frac{y}{\sigma_y};\; z_* = \frac{z}{\sigma_z}$$

$$r_* = (x_*^2 + y_*^2 + z_*^2)^{1/2} \quad (6.74)$$

Further simplification of the above solution can be made by using the slender-plume approximation, provided

$$y_*^2 + z_*^2 \ll x_*^2 \quad (6.75)$$

in the plume. For the part of the plume satisfying the above condition, Eq. (6.73) reduces to the Gaussian plume formula

$$\bar{c}(x,y,z) \simeq \frac{Q}{2\pi\bar{u}\sigma_y\sigma_z}\exp\left(-\frac{y^2}{2\sigma_y^2} - \frac{z^2}{2\sigma_z^2}\right) \quad (6.76)$$

For an elevated continuous point source of height H above a reflecting ground surface, the method of images yields

$$\bar{c}(x,y,z) \simeq \frac{Q}{2\pi\bar{u}\sigma_y\sigma_z}\exp\left(-\frac{y^2}{2\sigma_y^2}\right)$$

$$\left\{\exp\left[-\frac{(z-H)^2}{2\sigma_z^2}\right] + \exp\left[-\frac{(z+H)^2}{2\sigma_z^2}\right]\right\} \quad (6.77)$$

which is also known as the reflected Gaussian plume formula. From Eq. (6.77), the g.l.c. at the plume centerline is given by

$$\bar{c}_o(x) = \frac{Q}{\pi\bar{u}\sigma_y\sigma_z}\exp\left(-\frac{H^2}{2\sigma_z^2}\right) \quad (6.78)$$

which indicates that the g.l.c. increases with distance from the source, attains a maximum value $\bar{c}_{omax}$ at a distance where $\sigma_z = H/\sqrt{2}$, and decrease monotonically with distance thereafter (see Fig. 6.6). The Fickian diffusion theory predicts that the maximum g.l.c.

$$\bar{c}_{omax} = \frac{2}{e\pi}\frac{\sigma_z}{\sigma_y}\frac{Q}{\bar{u}H^2} \simeq 0.234\frac{\sigma_z}{\sigma_y}\frac{Q}{\bar{u}H^2} \quad (6.79)$$

is inversely proportional to the square of the source height. Later, in Chapter 9, this will be shown to be a more general result of the Gaussian plume model, without any stipulation of the Fickian theory.

For a continuous point source at the ground ($H = 0$), the approximate Gaussian plume formula predicts the g.l.c. as

$$\bar{c}(x,y,0) = \frac{Q}{\pi \bar{u} \sigma_y \sigma_z} \exp\left(-\frac{y^2}{2\sigma_y^2}\right) \quad (6.80)$$

which may be compared with the g.l.c. obtained from the exact solution (6.73), after accounting for the reflecting ground surface,

$$\bar{c}(x,y,0) = \frac{Q(x/r)}{\pi \bar{u} \sigma_y \sigma_z} \exp\left[-\left(\frac{r}{x} - 1\right)\frac{x^2}{2\sigma_y^2}\right] \quad (6.81)$$

Here, we have also made a reasonable assumption that $\sigma_x = \sigma_y$, implying horizontal homogeneity. The ratio of the g.l.c. given by the approximate Eq. (6.80) to that by the exact Eq. (6.81) is

$$F = \frac{r}{x}\exp\left[-\frac{(r-x)^2}{2\sigma_y^2}\right] \quad (6.82)$$

which is a measure of the underestimation or overestimation of the g.l.c. by the Gaussian plume formula. It is obvious that, at the plume centerline ($y = 0$ and $r = x$), $F = 1$ for a finite σ_y, so that the use of the Gaussian plume formula or the slender-plume approximation does not introduce any error in the determination of centerline g.l.c. The off-centerline ground-level concentrations would be in error depending on the factor F. We have not considered here the practical difficulties associated with the estimation of σ_y in very light or near-calm winds.

The solution for a continuous cross-wind line source can be obtained by suitably integrating the point source solution (see e.g., Roberts, 1923). For a source in an infinite medium, the simplified solution based on the slender-plume approximation is given by Eq. (6.45) in which σ should be replaced by σ_z. For an elevated line source above the reflecting ground surface, the concentration field is given by the reflected Gaussian formula

$$\bar{c}(x,z) = \frac{Q_{cl}}{(2\pi)^{1/2} \bar{u} \sigma_z}\left\{\exp\left[-\frac{(z-H)^2}{2\sigma_z^2}\right] + \exp\left[-\frac{(z+H)^2}{2\sigma_z^2}\right]\right\} \quad (6.83)$$

which reduces to the simpler form

$$\bar{c}(x,z) = \left(\frac{2}{\pi}\right)^{1/2} \frac{Q_{cl}}{\bar{u}\sigma_z}\exp\left(-\frac{z^2}{2\sigma_z^2}\right) \quad (6.84)$$

for line sources near the ground, such as busy roads and highways. For an elevated source, the g.l.c. is given by

$$\bar{c}_o(x) = \left(\frac{2}{\pi}\right)^{1/2} \frac{Q_{cl}}{\bar{u}\sigma_z}\exp\left(-\frac{H^2}{2\sigma_z^2}\right) \quad (6.85)$$

from which one can easily show that the maximum g.l.c.

$$\bar{c}_{o\max} = \left(\frac{2}{\pi e}\right)^{1/2} \frac{Q_{cl}}{\bar{u} H} \simeq 0.484 \left(\frac{Q_{cl}}{\bar{u} H}\right) \quad (6.86)$$

occurs at a distance where $\sigma_z = H$.

No steady-state solution (concentration field) can be obtained for an infinite continuous area source. In practical applications, one often deals with finite area sources (e.g., an urban area, a landfill, a large spill, etc.) that can be treated as combinations of multiple point sources spread over a network of grids. In this manner, spatially variable emission rates of the area sources can also be considered.

6.5.4 The Effect of Capping Inversion

We recognize that the PBL is often capped by an inversion, which simply reflects back the material reaching the inversion base. We can consider the effect of capping inversion simply by treating both the ground and inversion base as reflecting surfaces and considering image sources located at all the mirror-image points. Such a parallel-mirror geometry leads to an infinite number of images, although contribution from image sources is expected to decrease as their distance from the ground surface increases. Considering such multiple reflections, the concentration field due to an elevated continuous point source in the PBL (Fig. 6.7) is given by

$$\bar{c}(x,y,z) = \frac{Q}{2\pi \bar{u}\sigma_y \sigma_z}\exp\left(-\frac{y^2}{2\sigma_y^2}\right) \sum_N$$
$$\left\{\exp\left[-\frac{(z - H + 2Nh)^2}{2\sigma_z^2}\right] + \exp\left[-\frac{(z + H + 2Nh)^2}{2\sigma_z^2}\right]\right\} \quad (6.87)$$

in which h is the inversion height or the PBL depth and summation is implied over all the integer values of $N = 0, \pm 1, \pm 2$, and so on. All the image sources with heights greater than h are represented by positive values of N, while image sources with heights less than $-h$ are represented by negative values of N. The above series is found to converge rapidly and, for all practical purposes, terms involving N up to ± 5 might be sufficient. Two further approximations may be considered:

1. For small distances ($x < x_h$ in Fig. 6.7) from the source up to which the plume is not significantly affected by the inversion, that is, where the half-plume thickness $2.15\,\sigma_z \lesssim (h - H)$, the presence of inversion can be ignored altogether and one can use the reflected Gaussian formulae given earlier.
2. For very large distances from the source, such that the material has gone through multiple reflections from both the ground surface and the inversion, one may assume that concentration becomes uniform, independent of height and distance from the source. This assumption, in conjunction with the assumption of Gaussian distribution

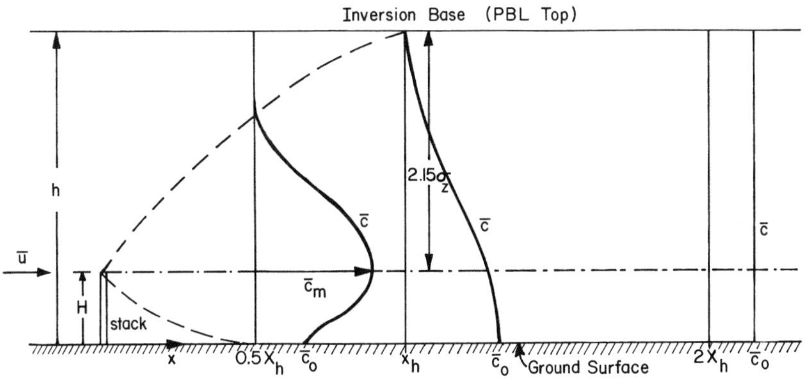

Figure 6.7 Schematic of a continuous-source plume trapped between the ground surface and the inversion at the top of the planetary boundary layer and hypothetical vertical concentration profiles at different distances from the source relative to x_h.

in the lateral direction and the mass continuity condition (6.61), yields

$$\overline{c}(x,y) = \frac{Q}{(2\pi)^{1/2} \, \overline{u}\sigma_y h} \exp\left(-\frac{y^2}{2\sigma_y^2}\right) \quad (6.88)$$

$$\overline{c}_o(x) = \frac{Q}{(2\pi)^{1/2} \, \overline{u}\sigma_y h} \quad (6.89)$$

where $\overline{c}_o$ denotes the g.l.c. at the plume centerline. Note that at very large distances ($x \gg x_h$) from the source, concentrations in the plume and, hence, $\overline{c}_o$ become independent of the release height. Turner (1970) suggested using Eq. (6.88) for any release height between 0 and h and for $x > 2x_h$.

In the same way, one can consider the confining effects of both the ground surface and the capping inversion on diffusion from other sources. The overall effect is increased ground-level concentrations, because the vertical plume thickness is limited by the inversion or the PBL height h. The product ($\overline{u}h$) of the average wind speed across the plume and the mixing height is sometimes referred to as the ventilation factor, because g.l.c. is inversely proportional to this factor.

6.6 VARIABLE K-THEORY

The basic assumption of constant-eddy diffusivity implied in the Fickian diffusion theory can be justified only in regions of more or less uniform flow and homogeneous turbulence. It may not be a satisfactory assumption, however, in regions of strong shear and nonhomogeneous turbulence, such as the atmospheric surface layer or the stably stratified nocturnal boundary layer. In such regions or shear layers, it is imperative to consider the effects of both the mean wind shear and the vertical inhomogeneity of turbulence. The latter is considered through an appropriate specification of eddy diffusivities as functions of height, stability, and other boundary-layer parameters. It forms the basis for the variable K-theory. In this section, we will discuss only some of the better known analytical solutions of the diffusion equation for some idealized $\overline{u}$ and K profiles. A more general numerical approach to solving the same equation for any desired $\overline{u}$ and K profiles and boundary conditions will be discussed later in Chapter 11.

6.6.1 Linear K-Profile

In the neutral atmospheric surface layer, the vertical-eddy diffusivity is found to be linearly increasing with height, that is,

$$K_z = ku_* z \quad (6.90a)$$

or, alternatively,

$$K_z = c\overline{u}_r z \quad (6.90b)$$

where $c = ku_*/\overline{u}_r = k\, C_D^{1/2}$ is a dimensionless coefficient related to the surface drag coefficient C_D, and $\overline{u}_r$ is the wind speed at the reference height.

With the above specification for eddy diffusivity, the diffusion equation can be solved for different types of sources. The simplest case is that of an instantaneous plane (infinite area) source at the ground, for which the solution is obtained as

$$\overline{c}(z,t) = \frac{Q_{ia}}{ku_* t} \exp\left(-\frac{z}{ku_* t}\right) \quad (6.91)$$

which satisfies the reflecting boundary condition at the surface.

Note that the above solution is independent of the wind profile, because the cloud diffuses only in the vertical direction. A finite area release at the ground will be subjected also to height-dependent advective transport by winds and horizontal diffusion by turbulence. The resulting cloud would be greatly stretched out in the direction of wind. For the infinite ground-level area source, however, Eq. (6.91) predicts an exponential concentration profile at any time after release, with the maximum concentration at the ground level decreasing inversely proportional to time. This can be compared with the earlier prediction of the Gaussian concentration distribution, with the maximum g.l.c. decreasing in proportion to $t^{1/2}$, by the Fickian (constant K) diffusion theory. With variable K_z, the puff or cloud grows in proportion to t, compared with $t^{1/2}$ for the Fickian theory. The vertical concentration distribution is exponential in the former, whereas it is Gaussian in the latter. The linear K_z-profile is, of course, more realistic for the neutral surface layer, while the constant K assumption might be more suitable for the middle part of the PBL. Therefore, the validity of the above solution for the variable K_z is limited to small diffusion times when the cloud is still confined to the surface layer. This limiting time is estimated to be about $0.1\ h/u_*$. At larger diffusion times ($t > 0.1\ h/u_*$), the assumption of a linear K_z profile (6.90) across the entire depth of the cloud would become highly questionable or invalid.

A similar solution is obtained for an infinite, continuous, cross-wind line source, provided diffusion in the x direction can be neglected in comparison with the advection by mean flow (the slender-plume approximation), and the variation of mean wind with height is also ignored, that is, a uniform wind profile is assumed. The solution is (Bosanquet and Pearson, 1936; Calder, 1952):

$$\bar{c}(x,z) = \frac{Q_{cl}}{ku_*x} \exp\left(-\frac{\bar{u}z}{ku_*x}\right) \quad (6.92)$$

which may be compared with Eq. (6.84) for a homogeneous layer with constant-eddy diffusivity. The comparison shows that the plume would grow much more rapidly with distance ($z_p \propto x$) in a vertically inhomogeneous surface layer than it does in a homogeneous layer ($z_p \propto x^{1/2}$). The g.l.c. decreases with distance in proportion to x^{-1} in the former (linear K_z) and in proportion to $x^{-1/2}$ for the latter (constant K_z). We have not considered, so far, the possible influence of mean wind shear (height-dependent velocity) on the vertical dispersion in the former.

6.6.2 Power-Law $\bar{u}$ and K Profiles

It is well known that both the wind speed and vertical-eddy diffusivity increase with height in the lower part of the atmospheric boundary layer. The Monin–Obukhov similarity theory probably best describes their variations with height, stability, and other parameters in the surface layer, as discussed in Chapter 4. The validity of the empirical surface layer similarity relations is, however, limited to rather small ranges of height (say, $10\ z_o < z < 0.1\ h$) and the stability parameter (say, $-5 < z/L < 2$). Moreover, the M–O similarity representation of $\bar{u}$ and K profiles does not normally yield analytical solutions of the diffusion equation. Simpler power-law profile representations, also discussed in Chapter 4, are found to be more convenient in diffusion applications, because they lead to analytical solutions and also because they can approximately represent the actual profiles to much larger heights, often covering more than half the PBL.

The height-dependent wind speed and the vertical eddy diffusivity can be expressed as

$$\bar{u} = az^m$$
$$K_z = bz^n \quad (6.93)$$

in which $a = \bar{u}_r/z_r^m$ and $b = K_{zr}/z_r^n$ are the dimensional coefficients, which are given in terms of the wind speed $\bar{u}_r$ and the eddy diffusivity K_{zr} at the reference height z_r, and m and n are the profile exponents.

6.6.3 Continuous Line Source

The solution of the two-dimensional diffusion equation for a cross-wind, infinite continuous line source at a reflecting ground surface is given by (Sutton, 1953; Huang, 1979)

$$\bar{c}(x,z) = AQ_{cl}x^{-\beta} \exp\left(-\frac{z^\alpha}{Bx}\right) \quad (6.94)$$

where

$$\alpha = m - n + 2 > 0$$
$$\beta = \frac{(m+1)}{\alpha} \quad (6.95)$$

are functions only of the power-law exponents m and n, whereas

$$A = \frac{\alpha^{1-2\beta}}{(a^{1-\beta}\ b^\beta\ \Gamma_\beta)}$$
$$B = \frac{b\alpha^2}{a} \quad (6.96)$$

also involve the coefficients a and b.

Note that Eq. (6.94) is a more general solution that reduces to the Gaussian form (6.84) for $m = n = 0$ (constant $\bar{u}$ and K_z) and to the exponential form (6.92) for $m = 0, n = 1$, and $b = ku_*$ (linearly increasing K_z). According to Eq. (6.94), the vertical concentration profile is expected to lie somewhere between the above two forms, that is,

$$\bar{c}(x,z) = \bar{c}_o(x)\exp\left(-\frac{z^\alpha}{Bx}\right) \quad (6.97)$$

where

$$\bar{c}_o(x) = AQ_{cl}x^{-\beta} \quad (6.98)$$

is the ground-level concentration.

The exponents α and β in Eqs. (6.97) and (6.98) depend on m and n, which, in turn, depend on the surface roughness and stability. Their estimated ranges of values are

$$1 \leq \alpha \leq 3$$
$$1 \geq \beta \geq 0.5 \quad (6.99)$$

Note that α and β have an approximately inverse relationship with each other, whereas the profile exponents m and n have an approximately conjugate relationship ($n \simeq 1 - m$). The latter relationship has been assumed by some investigators for obtaining solutions of the diffusion equation (Sutton, 1953; Smith, 1957). As pointed out earlier in Chapter 4, however, the conjugate relationship is not always strictly valid; it would be more appropriate to consider m and n as independent parameters or profile exponents. A very stable boundary layer would be characterized by a large m, a small n, and, hence, a large α and small β. Therefore, the vertical concentration profile might be nearly Gaussian, while the g.l.c. decreases only slowly with distance ($\bar{c}_o \propto x^{-1/2}$). A very unstable boundary layer, on the other hand, would be characterized by a small m, a large n and, hence, a small α and large β. Consequently, the theory predicts that the vertical concentration profile would be more nearly exponential, while the g.l.c. decreases rapidly with distance ($\bar{c}_o \propto x^{-1}$). Most atmospheric situations probably lie somewhere between these two extremes. Thus, in general, the variable K-theory predicts non-Gaussian vertical concentration profiles in the lower part of the PBL due to continuous sources at the surface.

The solution to the diffusion equation for an elevated, continuous, cross-wind line source is more complex, involving Bessel functions (Smith, 1957). However, the ground-level concentration, in which we are primarily interested, has a simpler expression:

$$\bar{c}_o(x,H) = AQ_{cl}x^{-\beta} \exp\left(-\frac{H^\alpha}{Bx}\right) \quad (6.100)$$

which reduces to Eq. (6.98) for a ground source. From Eq. (6.100), one can easily determine the location and magnitude of the maximum g.l.c. as

$$x_{max} = \frac{H^\alpha}{B\beta} \quad (6.101)$$

$$\bar{c}_{o\,max} = A(B\beta)^\beta \exp(-\beta) Q_{cl} H^{-(m+1)} \quad (6.102)$$

which predict a stronger effect of the effective release height in reducing the maximum g.l.c., compared with the case of constant $\bar{u}$ ($m = 0$).

6.6.4 Continuous Point Source

The above results for the continuous cross-wind line source can also be used for describing diffusion from continuous point sources, if we recognize that the cross-wind integrated equation of diffusion, neglecting diffusion in the x direction, is

$$\bar{u}\frac{\partial \bar{c}_y}{\partial x} = \frac{\partial}{\partial z}\left(K_z \frac{\partial \bar{c}_y}{\partial z}\right) \quad (6.103)$$

where

$$\bar{c}_y = \int_{-\infty}^{\infty} \bar{c}\, dy \quad (6.104)$$

is the cross-wind-integrated concentration. Since Eq. (6.103) is identical to the diffusion equation for the cross-wind line source, with $\bar{c}_y$ replaced by $\bar{c}$, the solutions of the latter discussed above also represent the cross-wind integrated concentration due to a point source. The $\bar{c}_y$ can be related to the point concentration $\bar{c}$ by assuming a Gaussian distribution in the lateral direction, that is,

$$\bar{c}(y) = \frac{\bar{c}_y}{(2\pi)^{1/2}\sigma_y}\exp\left(-\frac{y^2}{2\sigma_y^2}\right) \quad (6.105)$$

Thus, for the power-law profiles given by Eq. (6.93), the concentration field due to a point source at the surface ($H = 0$) can be obtained from Eq. (6.94) (written for $\bar{c}_y$) and Eq. (6.105) as

$$\bar{c}(x,y,z) = \frac{AQx^{-\beta}}{(2\pi)^{1/2}\sigma_y}\exp\left(-\frac{y^2}{2\sigma_y^2}\right)\exp\left(-\frac{z^\alpha}{Bx}\right) \quad (6.106)$$

At the plume centerline ($y = 0$), the vertical concentration distribution due to a point source at the ground is given by

$$\frac{\bar{c}(x,0,z)}{\bar{c}_o(x,0,0)} = \exp\left(-\frac{z^\alpha}{Bx}\right) \quad (6.107)$$

where $\bar{c}_o$ is the g.l.c. at the plume centerline. Defining the 10-percent plume height z_p, where concentration falls to 10 percent of the g.l.c., the above distribution can also be expressed in the form

$$\frac{\bar{c}}{\bar{c}_o} = \exp\left[-(\ln 10)\left(\frac{z}{z_p}\right)^\alpha\right] \quad (6.108)$$

where

$$z_p = [(\ln 10)Bx]^{1/\alpha} \quad (6.109)$$

The normalized concentration profiles corresponding to different values of α in Eq. (6.108) are shown in Figure 6.8. Note that both the end points of these pro-

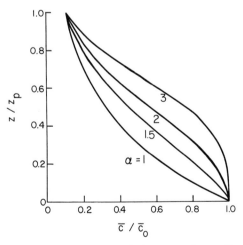

Figure 6.8 Normalized vertical concentration profiles in the plume, with the power-law profiles of wind speed and eddy diffusivity in the PBL, as functions of the profile-shape parameter α for a continuous point or line source at the surface.

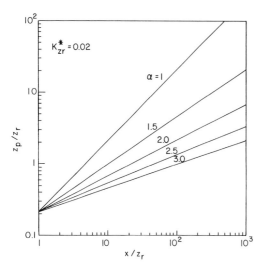

Figure 6.9 Theoretically predicted variation of the normalized plume height with the dimensionless distance from the source as a function of the profile shape parameter α for $K_{zr}^* = 0.02$.

files are fixed by our normalization procedure, which resulted in a single profile shape parameter α.

The plume height, according to Eq. (6.109), is expected to depend on the distance from the source, the profile shape parameter α, and another profile parameter

$$B = \frac{b}{a}\alpha^2 = \left(\frac{K_{zr}}{\bar{u}_r z_r}\right)(z_r^{\alpha-1}\,\alpha^2) \qquad (6.110)$$

which, for a given reference height z_r (say, 10 m) and α, essentially depends on the dimensionless eddy diffusivity $K_{zr}^* = K_{zr}/\bar{u}_r z_r$ at the reference level. From equations (6.109) and (6.110), the normalized plume height z_p/z_r can be represented as a function of the normalized distance x/z_r from the source as

$$\frac{z_p}{z_r} = (\ln 10\, K_{zr}^*)^{1/2}\left(\frac{x}{z_r}\right)^{1/\alpha} \qquad (6.111)$$

The above theoretically predicted variation of plume height with distance from the source is shown in Figure 6.9 for the specified values or ranges of α and K_{zr}^*.

Both the plume height parameters α and K_{zr}^* are expected to depend on surface roughness and stability. The profile exponent α varies from near unity for smoother surfaces and very unstable or convective conditions to 3 for very rough surfaces and extremely stable conditions. It determines the rate of growth of the plume height with distance, as $z_p/z_r \propto (x/z_r)^{1/\alpha}$. The proportionality coefficient in the above plume height relationship is determined by K_{zr}^*, which can be expressed in terms of the Monin–Obukhov similarity relations given in Chapter 4, provided z_r falls in the surface layer. It is easy to show that

$$K_{zr}^* = \frac{K_{zr}}{\bar{u}_r z_r} = \frac{k\, C_{Dr}^{1/2}}{\phi_h(z_r/L)} \qquad (6.112)$$

where $\phi_h(z_r/L)$ is the M–O similarity function evaluated at $\zeta = z_r/L$. Equation (6.112) indicates that K_{zr}^* is a strong function of stability and also a weaker function of surface roughness ($C_{Dr}^{1/2} = 1/\ln(z_r/z_o)$) for neutral stability. It is expected to have a very wide range of values, of which we have specified in Figure 6.9 only a typical one for the neutral stability condition. Note that the Fickian diffusion theory, implying the large value of $\alpha = 2$, considerably underpredicts the rate of growth of the plume with distance from a ground source in all conditions, except for the extremely stable ones.

The ground-level concentration at the plume centerline is given by

$$\bar{c}_o(x) = \frac{AQx^{-\beta}}{(2\pi)^{1/2}\sigma_y} \qquad (6.113)$$

in which σ_y is an increasing function of x. The behavior of the g.l.c. with distance from the source depends on σ_y as a function of x; if the latter can be approximated by a power law of the form $\sigma_y \propto x^p$, then the g.l.c. is predicted to vary as $\bar{c}_o \propto x^{-(\beta+p)}$. The combined exponent $(\beta + p)$ is estimated to range from 1 to 2, with the lower value applicable to very stable conditions and large distances from the source and the upper value applicable to extremely unstable or convective conditions and small distances. The possible dependence of p on distance from the source will be discussed later. The above theoretically predicted dependence of the decay of g.l.c. with distance and atmospheric stability is consistent with our physical intuition of turbulence and diffusion processes, as well as with the available experimental data to be discussed later. In contrast to this, the Fickian (constant K)-theory predicts that, for a continuous point source at the ground, $\bar{c}_o \propto x^{-1}$ irrespective of atmospheric stability and distance from the source.

The g.l.c. along the plume centerline due to an elevated, continuous, point source can be obtained from Eq. (6.100) (written for $\bar{c}_y$) and Eq. (6.105) as

$$\bar{c}_o(x,H) = \frac{AQx^{-\beta}}{(2\pi)^{1/2}\sigma_y}\exp\left(-\frac{H^\alpha}{Bx}\right) \qquad (6.114)$$

Again, assuming $\sigma_y \propto x^p$, one can determine the location and the magnitude of the maximum g.l.c. as

$$x_{\max} = \frac{H^\alpha}{B(\beta+p)} \qquad (6.115)$$

$$\bar{c}_{o\max} \propto H^{-\alpha(\beta+p)} \qquad (6.116)$$

Instead of following the simpler procedure of solving the two-dimensional diffusion equation for $\bar{c}_y$, some investigators have attempted to solve the three-dimensional diffusion equation for $\bar{c}$ after assuming (Yeh and Huang, 1975; Huang, 1979)

$$K_y = \frac{1}{2}\bar{u}\frac{d\sigma_y^2}{dx} \qquad (6.117)$$

This also results in the Gaussian distribution of concentration in the lateral direction and there is no difference in the solution for $\bar{c}$ from that given by the simpler procedure. For an elevated point source, the exact solution of the diffusion equation is given by

$$\bar{c}(x,y,z) = \frac{Q}{(2\pi)^{1/2}\sigma_y} \exp\left(-\frac{y^2}{2\sigma_y^2}\right) \frac{(zH)^\delta}{b\,\alpha\,x}$$

$$\exp\left(-\frac{z^\alpha + H^\alpha}{Bx}\right) I_{-\gamma}\left[\frac{(zH)^{\alpha/2}}{Bx}\right] \quad (6.118)$$

where

$$\gamma = \frac{(1-n)}{\alpha}; \quad \delta = \frac{(1-n)}{2} \quad (6.119)$$

are related to the profile exponents and $I_{-\gamma}$ is the modified Bessel function of the first kind of the order $-\gamma$. The above solution is valid only for $\gamma < 1$, a condition easily met in the PBL.

Using the series representation for $I_{-\gamma}$ and approximating it for a small release height ($H \to 0$), Eq. (6.118) can be simplified to

$$\bar{c}(x,y,z) = \frac{QAx^{-\beta}}{(2\pi)^{1/2}\sigma_y}$$

$$\exp\left(-\frac{y^2}{2\sigma_y^2}\right)\exp\left(-\frac{z^\alpha + H^\alpha}{Bx}\right) \quad (6.120)$$

This approximate solution should be valid for continuous point sources in the surface layer ($H < 0.1\,h$). Both the exact and approximate solutions yield the same equation (6.114) for the g.l.c., implying that the approximation made in obtaining the simpler expression (6.120) is particularly good for near-surface concentrations.

6.6.5 Effect of Capping Inversion

The analytical solutions of the diffusion equation, discussed so far in this section, did not consider the confining effect of a capping inversion at the top of the PBL. Rather, they imply unrestricted diffusion of the resulting puff or plume in the vertical direction. This does not normally happen in the real atmosphere, where an inversion layer of vanishing turbulence at the top of the PBL almost always restricts vertical diffusion from near-surface sources to within the PBL, with the possible exception of highly buoyant plumes penetrating through weaker low-level inversions. Therefore, the validity of the preceding solutions must be limited to rather short travel times or distances, until the upper edge of the resulting puff or plume just reaches the inversion base or the top of the PBL. Since the g.l.c. due to an elevated source in the lower part of the PBL will probably reach its maximum value by this time or distance, the above analytical solutions should be quite useful for estimating the maximum g.l.c. and its time or location. The confining effect of the capping inversion at the top of the PBL must be considered, however, for more accurate representations of dispersion from elevated releases, particularly in the upper-half of the PBL, as well as for long-range or far-field estimates of concentrations from near-surface sources. This can easily be done by using a re-flecting boundary condition or the zero mass flux condition (Eq. 6.59) at the top of the PBL.

Several analytical solutions of the diffusion equation satisfying the reflecting boundary condition at $z = h$ have been reported in the literature (Rounds, 1955; Smith, 1957; Demuth, 1978; Nieuwstadt, 1980a). They differ mostly in the manner $\bar{u}$ and K_z are specified. In most analytical solutions, K_z is assumed to be a monotonic function of height, such as the linear or power-law profile. Such specifications cannot be very realistic in the upper part of the PBL, where K_z is expected to decrease with height and become zero at the top of the PBL where turbulence vanishes. A more realistic, yet simple, specification of K_z, which satisfies the upper and lower boundary conditions, is

$$K_z = c_1 u_* z\left(1 - \frac{z}{h}\right) \quad (6.121)$$

Here, c_1 is an empirical constant that determines the maximum value of K_z in the middle of the PBL, as $K_{z\max} = 0.25\,c_1 u_* h$. For the neutral PBL, $c_1 = k = 0.4$ from surface layer similarity considerations. Perhaps the same value might be used for near-neutral and stably stratified boundary layers. The validity of the gradient diffusion approach becomes highly questionable under very unstable and convective conditions. If it is to be used at all, the friction velocity u_* must be replaced by the convective velocity W_* in Eq. (6.121).

The solution of the time-dependent, one-dimensional diffusion equation, with K_z given by Eq. (6.121), has been obtained for an instantaneous, infinite area release at any height $H < h$ in the PBL as (Smith, 1957; Nieuwstadt, 1980a)

$$\bar{c}(z,t) = \frac{Q_{ia}}{h}\left\{1 + \sum_{n=1}^{\infty}(2n+1)P_n\left(2\frac{H}{h} - 1\right)\right.$$

$$\left. P_n\left(2\frac{z}{h} - 1\right)\exp[-n(n+1)\frac{c_1 u_* t}{h}]\right\} \quad (6.122)$$

where $P_n(s)$ are the Legendre polynomials, which follow the recurrence relation (Abramowitz and Stegan, 1965)

$$(n+1)P_{n+1}(s) = (2n+1)sP_n(s) - nP_{n-1}(s) \quad (6.123)$$

with $P_0(s) = 1$ and $P_1(s) = s$.

The negative exponential function in Eq. (6.122) ensures a fast convergence of the infinite series, so that only a few terms of the series may suffice for not-too-small values of $c_1 u_* t/h \geq 0.1$. For large values of $c_1 u_* t/h \gg 1$, the concentration approaches the constant value of Q_{ia}/h, independent of height.

The normalized ground-level concentration $\bar{c}_o h/Q_{ia}$ is plotted in Figure 6.10 as a function of normalized time $c_1 u_* t/h$ for several different source heights H/h. It shows that the g.l.c. will attain a maximum value greater than Q_{ia}/h only for release heights in the lower half of the PBL ($H/h < 0.5$). For the case of a ground source ($H = 0$) and $t \to 0$, Eq. (6.122) can be approximated by the simpler solution of Eq. (6.91) for $K_z = ku_* z$ with $k = c_1$. This indicates that, initially, the

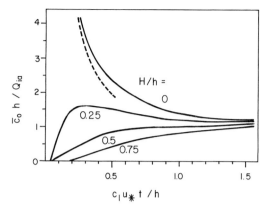

Figure 6.10 Normalized ground-level concentration ($\bar{c}_o h/Q_a$) due to an elevated instantaneous area source in the planetary boundary layer as a function of the dimensionless time after release ($c_1 u_* t/h$) and the normalized release height (H/h) according to the analytical solution (6.122); the dashed curve – – – represents the simpler solution (6.91). From Nieuwstadt, 1980a.

upper boundary condition does not have any significant effect on diffusion from a ground source, which is essentially governed by the K_z-distribution near the surface. The simpler solution (6.91) is also shown in Figure 6.10; it deviates from the exact solution (6.122) with increasing value of cu_*t/h.

The above results can also be used to describe the vertical diffusion from continuous sources in a uniform wind $\bar{u}$ by using the transformation $x = \bar{u}t$. For example, for a cross-wind, continuous-line source in the y direction, the solution of the approximate diffusion equation obtained after neglecting diffusion in the longitudinal direction is

$$\bar{c}(z,x) = \frac{Q_{cl}}{\bar{u}h}$$
$$\left\{ 1 + \sum_{n=1}^{\infty} Q_n \exp\left[-n(n+1)\frac{c_1 u_* x}{\bar{u}h} \right] \right\} \quad (6.124)$$

where

$$Q_n = (2n+1)P_n\left(2\frac{H}{h} - 1\right)P_n\left(2\frac{z}{h} - 1\right) \quad (6.125)$$

For this case, a plot of $\bar{c}\,\bar{u}\,h/Q_{cl}$ versus $c_1 u_* x/\bar{u}h$ will be identical to Figure 6.10.

Similarly, the cross-wind-integrated concentration due to a continuous point source is given by

$$\bar{c}_y(z,x)$$
$$= \frac{Q}{\bar{u}h}\left\{ 1 + \sum_{n+1}^{\infty} Q_n \exp\left[-n(n+1)\frac{c_1 u_* x}{\bar{u}h} \right] \right\} \quad (6.126)$$

Thus, Figure 6.10 could also represent $\bar{c}_y\,\bar{u}h/Q$ as a function of $c_1 u_* x/\bar{u}h$ and H/h. The local (point) concentrations in the plume are related to $\bar{c}_y$ through Eq. (6.105).

The above expressions for $\bar{c}$ and $\bar{c}_y$ for the continuous line and point sources, respectively, are strictly based on the assumptions of uniform winds and no diffusion in the mean wind direction (the slender-plume approximation). The assumption of uniform wind field, also implied in the Gaussian plume-diffusion model, may not be justified in stable conditions in which wind shear effects on dispersion may be as important as the realistic K_z profile. Demuth (1978) has given an analytical solution of the diffusion equation for the power-law wind speed and eddy-diffusivity profiles. Demuth's solution is rather complicated and will not be given here. The monotonic increase of K_z with height, even in the upper part of the PBL, remains the most serious limitation of that solution.

6.7 LIMITATIONS OF GRADIENT TRANSPORT THEORIES

The gradient transport hypothesis provides the simplest way of tackling the fundamental closure problem of turbulence and diffusion in that the equations of mean motion and diffusion can be closed and then solved by properly specifying the relevant eddy diffusivities. Gradient transport theories are based on the hypothetical relationships between turbulent fluxes and spatial gradients of appropriate mean properties. Eddy diffusivities appear as proportionality coefficients in the gradient transport relations, which are analogous to the molecular transfer relations. Because of the simplicity of the gradient transport hypothesis and the first-order closure of the relevant conservation equations it provides, K-theories have great appeal to micrometeorologists and air pollution meteorologists who deal with practical problems of atmospheric turbulence and diffusion. They should also be aware of the several limitations of these theories, so that theoretical results are judiciously applied.

The most serious limitation arises from the lack of general validity of the basic hypothesis of down-gradient transfer in a turbulent atmosphere. Whereas, under normal conditions, one might expect heat to flow from a warmer region to a colder region, that is, down the local mean potential temperature gradient, and similarly expect a contaminant to be transported down the mean concentration gradient, there are also exceptions to these intuitive expectations. For example, it is well documented that under unstable and convective conditions, vertical heat flux becomes independent of the near-zero potential temperature gradient in the mixed layer, implying infinitely large or even negative K_h. Heat is transferred by large-scale (of the order of the PBL depth) convective-eddy motions that manifest themselves in the forms of updrafts and downdrafts. The same eddies may also transport material up and down, irrespective of mean concentration gradients in the mixed layer. Thus, the gradient transport hypothesis becomes highly questionable or invalid for turbulent transport and diffusion processes in a convective mixed layer, where buoyant motions dominate over mechanical or shear-generated turbulence.

The gradient transport hypothesis has been proposed on the basis of an apparent or suggested analogy between molecular and turbulent transfer processes. Originally, Boussinesq (1877) suggested gradient transport relations for turbulent fluxes that were analogous to the flux-gradient relations for molecular transfers in laminar flows (e.g., Newton's law for momentum transfer, Fourier's law for heat transfer, and Fick's law for mass transfer). Thus eddy diffusivities of momentum, heat, and mass were considered analogous to molecular viscosity and thermal and mass diffusivities. Later, Prandtl (1925) extended the analogy further to even physical mechanisms of transfer in laminar and turbulent flows. He suggested analogies between molecules and "blobs" of fluid, between molecular and turbulent velocities, and between molecular free path length and turbulent mixing length. On the basis of these analogies, he derived plausible relations for turbulent momentum flux and eddy viscosity, which were analogous to the more rigorous and well-tested kinetic theory relations for molecular flux and viscosity. Prandtl's mixing length theory was later extended to describe turbulent transports (fluxes) of passive scalars (Sutton, 1953). The most useful result of the mixing length theory for diffusion applications is the expression of eddy diffusivity being proportional to the product of a mean mixing length and the appropriate turbulent velocity. This is very similar to the expression (4.28) based on purely dimensional considerations, without the involvement of any questionable analogies and hypotheses.

Upon more extensive experimental and theoretical studies of turbulent transport processes by many investigators, the above mentioned analogies between molecular and turbulent transfer processes have been found to be rather weak and qualitative only. For example, unlike their molecular counterparts, eddy diffusivities and mixing lengths are not some fluid properties but are properties of flow (primarily, turbulence). The latter are in no way related to the corresponding (analogous) molecular properties. Prandtl's analogy between molecules and blobs or eddies is also not good. Whereas in a laminar flow the velocities of molecules are very large compared with the velocity of bulk flow, and free path lengths traveled by molecules before their collisions are extremely small compared with the length scale (L) of bulk flow, the reverse is true for analogous variables in turbulent flows ($u' < \bar{u}$ and $l' \sim L$). Thus, as an eddy or a blob of fluid moves a distance l' in the vertical direction, it will travel a rather long distance in the mean flow direction and may find itself in entirely new surroundings.

The three-dimensional nature of mixing and transfer processes is somehow ignored in the simpler gradient transport theory. It can be considered only

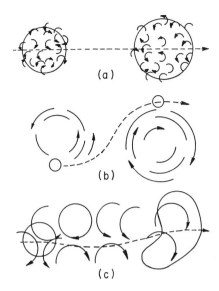

Figure 6.11 Schematics of idealized cloud dispersion patterns under the influence of different eddy sizes relative to that of the cloud. (a) A large cloud in a field of smaller eddies. (b) A small cloud in a uniform field of large eddies. (c) A cloud in a field of the eddies of the same size as the cloud. After Slade, 1968.

through the use of generalized three-dimensional gradient-transport relations, in which eddy diffusivity becomes a second or higher-order tensor (see e.g., Monin and Yaglom, 1971; Kao, 1984). The simpler K-theory assumes eddy diffusivity to be a vector with no more than three independent components (e.g., K_x, K_y, and K_z).

Another limitation of the K-theory is that there is no generally applicable method of specifying eddy diffusivities under the wide range of conditions encountered in the atmosphere. Experimental data on K_x, K_y, and K_z are lacking due to the lack of direct measurements of material fluxes and the corresponding mean concentration gradients. Data from diffusion experiments have been mostly analyzed in terms of concentration distributions and dispersion parameters, from which eddy diffusivities can be indirectly estimated by using Eq. (6.72) or more general relationships between K_x and σ_x to be given later. Such data are usually confined to the near-surface layer of the atmosphere in which eddy diffusivities are expected to vary with height, but the available information is not sufficient to determine the spatially varying Ks. Consequently, empirical diffusion data have not provided a practical method of specifying eddy diffusivities for use in the K-theory. More useful empirical relations based on dimensional and similarity considerations have been discussed in Chapter 4.

Yet another limitation of the K-theory is that only the eddies or scales of motion that are smaller than the diffusing cloud or plume can possibly be considered in the gradient-transport process (see e.g., Fig. 6.11a). Larger eddies often cause systematic distortion or large-scale bodily movement of a cloud (see Fig. 6.11b,c) and looping or meandering of a plume. These effects cannot be represented as a simple gradient diffusion process. Figure 6.11 gives an idealized schematic of cloud dispersion patterns under the influence of eddies of smaller, larger, and same size as the cloud. In the real atmosphere, eddies of different sizes are always present, so that a material cloud will always grow in time due to diffusion by smaller eddies, while it will be distorted and tossed around by larger eddies. The influence of the latter cannot be considered in the K-theory, except in a superficial manner by specifying a larger eddy diffusivity consistent with the large-eddy size.

Lastly, the Gaussian or modified exponential distributions predicted by the K-theory are unrealistic in the sense that they have infinitely long tails, that is, concentration goes to zero only at infinite distances from the center of a puff or a plume. This implies an unlimited speed of propagation of material outward from a source. This is a direct consequence of the parabolic type of mean diffusion equation that results from the simple closure relations based on the gradient transport hypothesis. This point has been discussed in more detail by Monin and Yaglom (1971), who have also proposed an alternative hyperbolic form of the diffusion equation, incorporating a finite speed of propagation. The solution of the modified (hyperbolic) diffusion equation does not differ from that of the conventional (parabolic) equation, except near the edges of a puff or plume where concentration suddenly drops to zero according to the former solution, while it is about 10 percent of the maximum value and decreases very gradually according to the latter solution (Monin and Yaglom, 1971; Pasquill and Smith, 1983). Thus, simpler K-theory solutions implying infinite tails for concentration distributions may not be applicable at large distances (say, more than 2σ) from the puff or plume center. The so-called 10-percent width may be close to the actual width of an average puff or plume to which the material is confined.

6.8 EXPERIMENTAL VERIFICATION OF K-THEORIES

Experimental verification of any gradient transport (K) theory may be made at different stages in the development of the theory. The first stage is testing the basic hypothesis on the transport of material down the mean concentration gradient and verifying the assumed spatial distribution of eddy diffusivity in different directions. Verifying these underpinnings of the theory requires simultaneous measurements of material fluxes and mean concentration distributions or gradients in different directions. Due to difficulties encountered in the direct measurements of turbulent fluxes, however, an experimental verification of the gradient transport hypothesis for mass diffusion has not been made, even in well-controlled diffusion experiments conducted in laboratory facilities, such as wind tunnels, water channels, and convection chambers. More successful attempts at verifying the same hypothesis for turbulent transports of momentum and heat (with or without the buoyancy effects) have confirmed the validity of the hypothesis in surface or wall layers, jets, and plumes where gradients are large (see e.g., Monin and Yaglom, 1971; Raupach and Legg, 1983; Panofsky and Dutton, 1984; Stull, 1988). The hypothesis also works well in the upper parts of neutral and stably stratified boundary layers (Arya, 1975; Lenschow et al., 1988a&b). Under very unstable and convective conditions, however, turbulent transports are often dominated by buoyancy-generated convective motions (circulations), and the simple concept of down-gradient transfer of heat and momentum has been shown to become generally invalid. One might expect the K-theory relations for the turbulent mass fluxes to also become questionable or invalid, since material is transported by the same large-scale convective motions (eddies), particularly in the convective mixed layer. Therefore, the usefulness and applicability of the K-theory are probably limited to atmospheric conditions and regions not dominated by convective motions or buoyancy.

Experimental verifications of diffusion theories and models are mostly limited to the comparisons of theoretically predicted and measured ground-level concentrations under different meteorological conditions. In some cases, predicted and measured concentration distributions in lateral and vertical directions across the plume at various distances from the source are also compared. Such detailed concentration profile measurements are rarely made in field experiments due to the expense and logistical difficulties of monitoring above the ground level and at large distances from the source. Much more information is available on the cross-wind distribution at the ground level at relatively short distances from continuous point sources (see e.g., Pasquill and Smith, 1983). Individual cross-wind (lateral) concentration profiles rarely show smooth Gaussian form predicted by theory, but averages of several profiles taken during repeated releases, under the same meteorological conditions, show a much closer fit to the Gaussian shape. The deviations of the observed profiles from the theoretical Gaussian form may be attributed to inadequate averaging time, inhomogeneity of the underlying terrain, and nonstationary meteorological conditions. The lateral concentration profiles measured under more ideal (homogeneous and stationary) conditions simulated in the laboratory show almost perfect Gaussian forms (see Fig. 6.12).

The existence of the Gaussian distribution in the lateral direction, in which K_y may be considered to be constant, does not necessarily prove the validity of the K-theory, because the Gaussian distribution is also implied by other statistical dispersion theories and mod-

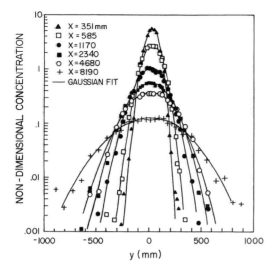

Figure 6.12 Measured lateral concentration profiles at the surface at various distances from a continuous surface release in a wind-tunnel simulated boundary layer, compared with the Gaussian distribution. After Pendergrass and Arya, 1984.

and wind speed. The constant-eddy diffusivity assumed in the Fickian theory predicts again the Gaussian concentration distribution in the vertical with $\sigma_z \propto x^{1/2}$. These predictions are not verified by experimental observations of vertical profiles and σ_z versus x (see e.g., Pasquill and Smith, 1983).

The shape of vertical concentration distribution in plumes resulting from near-surface releases has been extensively studied. Both field and laboratory measurements indicate that the vertical distribution can be represented by the modified exponential form (6.107) in which the profile shape exponent α has been found to lie between 1 and 3, depending mainly on the atmospheric stability and, to a lesser extent, on the surface roughness also. An analysis of the Project Prairie Grass (PPG) data shows a clear trend in α increasing with stability from a value of unity (implying the exponential profile shape) under convective conditions to 2 (implying the Gaussian profile shape) under moderately stable conditions (see Nieuwstadt and van Ulden, 1978; Pasquill and Smith, 1983). There is also some indication of $\alpha > 2$ under extremely stable conditions. This trend of α increasing with stability has been confirmed by wind tunnel experiments in stably stratified boundary layers, which have also indicated large values of α (some exceeding 3) for extremely stable conditions (Ogawa et al., 1985). Under such an extreme stability, turbulence intensities are reduced below 2 percent and the plume is seen to collapse vertically and spread instead in a very thin horizontal layer.

A more recent analysis of the PPG data for near-neutral conditions has indicated an average value of α between 1.4 and 1.5, which is consistent with the average value estimated from the measured concentrations profiles in wind tunnel-simulated neutral boundary layers (Brown et al., 1993). Actually, wind tunnel results show α to be a slightly decreasing function of x/z_o, suggesting a weak dependence of α on the distance from the source, as pointed out by Hunt and Weber (1979), as well as on the surface roughness. The effect of surface roughness can be seen by comparing the estimated values of 1.15 for an aerodynamically smooth surface (Ogawa et al., 1985) and 1.55 for moderately rough surfaces (Brown et al., 1993).

The above mentioned studies of the vertical distribution of concentration from near-surface releases in the boundary layer clearly show that the Gaussian shape is an exception rather than the rule and that a non-Gaussian model would be more appropriate for describing the vertical dispersion. The variable K-theory model based on the power-law profiles for wind speed and K_z has been suggested as an alternative to the Gaussian model. The analytical solutions to the diffusion equation, given earlier, imply that $\alpha = m - n + 2$, where m and n are the power-law profile exponents for $\bar{u}$ and K_z, respectively. Several investigators have compared the model estimated values of α with those based on the observed concentration profiles under different stability conditions (Ogawa et al.

els. A more severe test is the comparison of the theoretically predicted and measured variations of the dispersion parameter σ_y with distance from the source. For a constant K_y, the K-theory predicts $\sigma_y \propto x^{0.5}$. However, there is overwhelming observational evidence from many field and laboratory experiments indicating a variable exponent, p, in the dispersion relationship, $\sigma_y \propto x^p$, whose value ranges between 0.5 and 1, depending on the distance from the source and atmospheric stability conditions. In general, the exponent decreases with increasing distance from the source, as well as with increasing stability (see e.g., Pasquill and Smith, 1983). Only at very large distances, compared with the characteristic large-eddy length scale, the lateral dispersion might become consistent with the prediction of the K-theory. At shorter distances, the K-theory cannot properly account for the dispersion caused by eddies that are larger than the plume width. Large eddies can easily transport material over large distances, irrespective of local gradient of mean concentration. Therefore, the gradient transport theory should not be used for describing lateral dispersion at short range (up to 10 km) from any source. The theory is extensively used, however, in regional and long-range dispersion and transport models. Its experimental verification on those scales is far more difficult, if not impossible, than on local scales at which individual plumes can be identified with particular sources.

The K-theory is considered to be particularly useful for describing vertical diffusion from near-surface sources. In this context, the vertical inhomogeneity of the surface layer and the lower part of the PBL can be considered through the use of the height-dependent K_z

1985; Brown et al., 1993). For neutral stability, the model appears to underestimate α by about 20 percent if one assumes $n = 1$ (i.e., linear variation of K_z with height). Much better agreement between the model estimated and observed values of α is obtained, however, if one uses the conjugate power-law relationship $n = 1 - m$, which implies a nonlinear K_z profile and relates the concentration-profile shape parameter to the wind-profile exponent ($\alpha = 2m + 1$). Thus, there may be some merit to the use of the conjugate power law for specifying K_z in the absence of more direct experimental information on the same.

Further comparisons of the model predicted and observed concentration profiles in both the longitudinal and the vertical directions show that the non-Gaussian model, with appropriately specified power-law profiles of $\bar{u}$ and K_z, performs reasonably well and much better than the Gaussian diffusion model, particularly for near-surface releases (see Brown et al., 1993). Still, the model is far from perfect and consistently overestimates (from 10–50%) the ground-level concentrations at large distances (beyond the location of the maximum g.l.c.) from elevated sources. This may be partly due to the loss of material due to deposition to the surface during diffusion experiments from which observations were taken. The model also overestimates the vertical diffusion due to the assumed monotonic increase of vertical diffusivity with height up to the top of the boundary layer. A more realistic variation of K_z with height, with its maximum value in the middle of the PBL, might be expected to better model the vertical diffusion from elevated releases.

6.9 APPLICATIONS OF K-THEORIES TO ATMOSPHERIC DISPERSION

Gradient transport theories have been applied to a variety of atmospheric dispersion problems on local, regional, and global scales. In short-range dispersion problems, the validity of the K-theory is limited to vertical diffusion from near-surface releases. The variable K-theory, with the vertical-eddy diffusivity and wind speed profiles specified by simple power laws or by more complex similarity relations, can be used to describe the diffusion of material in the vertical direction. Analytical solutions to the diffusion equation are known only for the simpler specifications of $\bar{u}$ and K_z, such as the power-law profiles, while numerical solutions would be necessary for more realistic specifications. Non-Gaussian dispersion models based on analytical solutions may be adequate for most practical dispersion applications (Berlyand, 1975; Huang, 1979; Tirabassi et al., 1986; Ide et al., 1988).

The variable K-theory is also extensively used in micrometeorological applications, where exchanges of momentum, heat, and water vapor are related to the mean profiles of wind speed, potential temperature, and specific humidity, respectively, through gradient transport relations, especially in the surface layer. The validity of the K-theory becomes questionable above the surface layer when turbulent transport and mixing are dominated by convective motions.

The K-theory is more commonly applied in regional dispersion and deposition models, particularly in Eulerian, three-dimensional grid models (Zwerver and van Ham, 1985; Chang et al., 1987; Carmichael et al., 1991). Such models, to be discussed in more detail later, are based on mass conservation equations for many chemical species, each containing appropriate transport (advection), diffusion, chemical reaction, scavenging, and deposition terms. The turbulent diffusion in horizontal and vertical directions is parameterized through the usual gradient transport relations in which eddy diffusivities are prescribed as functions of height, wind speed or wind shear, and stability. Some models also include subgrid scale diffusion and all contain some "numerical or pseudo diffusion" introduced by numerical integration of advection equations.

Other regional and long-range transport and diffusion models have been formulated using the Lagrangian frame of reference in which pollutant particles or puffs are tracked along trajectories determined by the sum of the mean transport velocity and the diffusive velocity at each time step. The diffusive velocity in a particular direction is defined as the ratio of mass flux to mean concentration, with the former being parameterized in terms of eddy diffusivity and concentration gradient in that direction (Sklarew et al., 1972; Diehl et al., 1982). Some of the regional models have both Eulerian and Lagrangian features and are therefore called hybrid models (Zwerver and van Ham, 1985). The gradient transport approach is used in most of the regional and long-range transport models. For simplifying the numerical solutions, horizontal diffusion may be neglected in comparison with horizontal transport in some models, while complete vertical mixing may be assumed to occur in the mixing (boundary) layer with the mixing height varying in time and space. Further descriptions of regional dispersion and air quality models based on the K-theory will be given in Chapter 12.

PROBLEMS AND EXERCISES

1. a. Distinguish between advective and diffusive fluxes of material.
   b. State Fick's law of diffusion and discuss its validity, usefulness, and applications to atmospheric diffusion.

2. a. Show that, for a contaminant diffusing in an incompressible flow,

$$\frac{\partial(cu)}{\partial x} + \frac{\partial(cv)}{\partial y} + \frac{\partial(cw)}{\partial z} = u\frac{\partial c}{\partial x} + v\frac{\partial c}{\partial y} + w\frac{\partial c}{\partial z}$$

b. Derive the instantaneous diffusion equation for a contaminant species, considering the possible changes in its concentration with respect to time due to chemical reactions with other contaminants and natural atmospheric constituents and also due to cloud scavenging processes.

3. Considering the conservation of mass of a passive (nonreactive) contaminant, write down the appropriate units or dimensions of source strength, as well as the integral mass conservation or continuity equations for the following source configurations: (a) instantaneous infinite line source in the free atmosphere; (b) instantaneous point source in the PBL; (c) continuous point source in the PBL; (d) continuous finite-area source at the ground (x–y plane); (e) continuous cross-wind line source in the PBL.

4. a. Discuss the theoretical basis, usefulness, and practical applications of the principle of superposition of concentrations arising from different sources.

b. Under what conditions would the above principle become invalid?

5. By way of direct substitution, verify that Eq. (6.19) is indeed a solution of the diffusion Eq. (6.16) with $\sigma = (2Dt)^{1/2}$, and shows that the former also satisfies the appropriate boundary conditions as well as the integral mass conservation condition for an instantaneous point release.

6. Consider the steady molecular diffusion from a continuous point source emitting in an infinite calm environment.

a. Mathematically express the mass conservation condition that the radial mass flow rate at any radial distance from the source must be equal to the source emission rate.

b. Using Fick's law, express the above condition as a differential equation for concentration and solve it for the appropriate boundary conditions. Verify that concentration decreases inversely proportional to the radial distance from the source, as given by Eq. (6.34).

7. a. How is molecular diffusion from a continuous point source in an infinite medium modified by the presence of uniform flow in one (say, x) direction?

b. Show that the slender-plume approximation, resulting in the Gaussian plume formula (6.41), is equivalent to neglecting diffusion in the x direction.

8. a. Starting from the instantaneous diffusion equation (6.7), derive the mean diffusion equation (6.51) using Reynolds' averaging rules or conditions.

b. Point out the basic differences between the equations for c and $\bar{c}$.

9. a. Simplify the mean diffusion equation for a horizontally homogeneous turbulent flow outside any thin viscous sublayers. Note that, in general, concentration field may not be homogeneous.

b. Give the gradient transport (K-theory) relations commonly used for parameterizing the turbulent fluxes and, hence, closing the mean diffusion equation. How would you specify eddy diffusivities in the above relations for modeling diffusion in the atmospheric boundary layer?

10. a. Discuss the possible confining effects of the earth's surface and the capping inversion at the top of the PBL on diffusion from sources within the PBL.

b. What mathematical boundary conditions may be used in obtaining the solutions to the diffusion equation, which would account for the presence of both the surface and the capping inversion?

11. a. Starting from the instantaneous, integral mass conservation equation (6.13), obtain the corresponding Reynolds-averaged equation.

b. Simplify the above equation to the more commonly used mass continuity equation (6.61) and discuss the possible justifications for the neglected terms. Show that a necessary condition for neglecting turbulent transport in comparison with the mean transport in the direction of flow is $\bar{u}x/K_x \gg 1$.

12. An ensemble-averaged smoke puff due to an antiaircraft shell burst in the atmosphere can be described by the solution (6.63) for the instantaneous point release in an infinite medium.

a. Give a general expression for the surface of constant concentration as a function of time and simplify it for the case of homogeneous turbulence with $K_x = K_y = K_z = K$.

b. Show that the surface of constant concentration $\bar{c}$ reaches a maximum diameter of

$$d_{max} = \left(\frac{6}{\pi e}\right)^{1/2} \left(\frac{Q_{ip}}{c}\right)^{1/3}$$

at time t_{max} and it vanishes at time $t_o = et_{max}$. Also, give an expression for t_{max}.

c. Show that at any time $t \leq t_o$, the diameter is given by

$$d = \left[24Kt \ln\left(\frac{t_o}{t}\right)\right]^{1/2}$$

13. Starting from the Fickian diffusion equation and its solution for an instantaneous point source in an infinite medium, obtain the simplified diffusion equations and concentration distributions, using the principle of superposition, for the following source configurations: (a) An infinite cross-wind line source in the y direction; (b) An infinite area source in the x–y plane.

14. a. Using the method of images, derive an expression for the ground-level concentrations due to an elevated cross-wind line source.

b. Hence, obtain the maximum g.l.c. and its time of occurrence after release as functions of the release height.

15. A highly volatile liquid is spilled on the ground in the form of a thin sheet over a large area, and quickly vaporizes and diffuses vertically upward. If the cloud is observed to be 200 m thick (10% concentration thickness) 10 minutes after release,

a. Calculate the effective vertical diffusivity and state all the assumptions you make.

b. Calculate and plot the vertical concentration profiles at 5, 10, and 15 minutes after release.

16. The exhaust of an idling vehicle during calm wind conditions or in a large garage may be regarded as a con-

tinuous point source on the ground diffusing in still but turbulent air. If the average turbulent diffusivities are related as $K_x = K_y = 4 K_z$,

    a. Obtain the appropriate expression of concentration distribution in the exhaust cloud, using Robert's (1923) solution (6.71).

    b. Plot and compare the horizontal and vertical distributions of $\bar{c}/Q$ when $K_z = 0.25$ m² s⁻¹.

17. By direct substitution, verify that the Gaussian plume formula (6.76) is a solution for the approximate diffusion equation

$$\bar{u}\frac{\partial \bar{c}}{\partial x} = K_y\frac{\partial^2 \bar{c}}{\partial y^2} + K_z\frac{\partial^2 \bar{c}}{\partial z^2}$$

and it satisfies the following conservation condition:

$$\int_{-\infty}^{\infty}\int_{-\infty}^{\infty} \bar{c}\bar{u}\, dy\, dz = Q$$

18. Consider ground-level concentrations due to a continuous point source at the surface, under light winds with $\bar{u} = 1$ m s⁻¹, and vigorous convective mixing with $K_x = K_y = 50$ m² s⁻¹ and $K_z = 25$ m² s⁻¹, in the lower part of the convective boundary layer.

    a. Compute and plot the lateral profiles of $\bar{c}/Q$ in the half plume ($y \geq 0$) at downwind distances of 100, 400, and 1600 m from the source, using the exact solution to the diffusion equation.

    b. Compare the above profiles with those obtained from the approximate Gaussian solution, and discuss the conditions for approximate validity of the latter for estimating ground-level concentrations.

19. Using the Gaussian plume formula for a continuous point source with an effective release height H above the reflecting ground surface, show that the maximum g.l.c. occurs at a distance where $\sigma_z = H/\sqrt{2}$, and is given by Eq. (6.79). Assume that the ratio σ_y/σ_z is a constant, independent of x.

20. Using the Gaussian plume formula for an elevated, cross-wind line source above a perfectly reflecting surface, show that the maximum g.l.c. occurs at a distance where $\sigma_z = H$ and is given by Eq. (6.86). What are the implications of these results?

21. Assuming that, at a sufficiently large distance from a ground or an elevated continuous point source in the PBL, the mean concentration distribution in the vertical becomes uniform ($\bar{c}$ independent of height), while it stays Gaussian in the lateral direction, derive an expression of $\bar{c}(x,y)$ from the integral mass conservation equation

$$\int_0^h \int_{-\infty}^{\infty} \bar{c}\bar{u}\, dy\, dz = Q$$

for the above plume trapping conditions.

22. a. What are the different types of inhomogeneities of flow created by the presence of earth's surface?

    b. How would you account for the vertical inhomogeneities of flow on diffusion in the atmospheric boundary layer?

23. A two-lane highway carries thirty vehicles per minute with an average speed of 80 km/hr in each direction. The emission rate of CO from an average vehicle is 0.4 kg hr⁻¹ traveled. If the prevailing wind is normal to the highway, calculate, plot, and compare the 10-percent plume thickness and the g.l.c. of CO as functions of downwind distance from 50 to 1000 m, using the following K-theory models: (a) the Gaussian plume model based on the Fickian diffusion theory with a mean wind speed of 5 m s⁻¹ and vertical diffusivity of 1 m² s⁻¹, both referred to the 10-m reference height; (b) the exponential plume model based on the linear K_z-profile with $K_{10} = 1$ m² s⁻¹ and a uniform wind speed of 5 m s⁻¹; (c) The variable K-theory with the power-law wind speed and eddy diffusivity profiles given as

$$\bar{u} = \bar{u}_{10}\left(\frac{z}{10}\right)^{0.2}$$

$$K_z = k\, u_* z$$

for neutral stability. Use the same reference values of $\bar{u}_{10}$ and K_{10} as in (a).

Discuss the results of different models and draw appropriate conclusions from their comparison.

24. Considering atmospheric dispersion of gaseous pollutants from a point or a line source at the ground, discuss the possible influence of atmospheric stability on the following, using an appropriate K-theory model: (a) the g.l.c. at the plume centerline; (b) the vertical plume thickness z_p.

25. a. Briefly state the various limitations of K-theories, in general, and the Fickian diffusion theory, in particular.

    b. Under which atmospheric conditions or dispersion applications could you justifiably use the gradient transport theory?

7

Statistical Theories of Diffusion

7.1 LAGRANGIAN APPROACH TO DESCRIBING DIFFUSION

In the previous chapter we discussed molecular and turbulent diffusion in an Eulerian frame of reference. That approach allows concentrations and mass fluxes at fixed points in space to be described by appropriate diffusion and mass conservation equations. Another, perhaps more fundamental, approach to describing diffusion is the statistical one, which uses the Lagrangian viewpoint or frame of reference. In this, one follows the paths of certain tagged molecules or particles, which are subjected to random movements in the fluid environment, and deduces their dispersion characteristics, such as concentration and spatial distribution, from the statistics of particle displacements. Since the statistical properties of particle motions and displacements are used for describing the diffusion of particles in a fluid, the theory based on the Lagrangian approach has been called the "statistical" one, in contrast to the "phenomenological" gradient transport theory of the previous chapter.

The statistical approach, when applied to molecular or Brownian diffusion (Einstein, 1905; Chandrasekhar, 1943), also leads to the same results as given by the phenomenological theory and verified by molecular diffusion experiments in laminar flows. The new approach does provide considerably more physical insight into the diffusion phenomenon and some useful information on diffusivities associated with the Brownian diffusion of small particles in a fluid environment (e.g., aerosols in air). But it yields few new results that are not given by the older gradient transport approach. Therefore, the statistical theory of Brownian diffusion will not be covered here; the reader may refer to the original references given above or the more recent reviews by Csanady (1973) and Seinfeld (1986).

Taylor (1921) extended the statistical approach to turbulent diffusion by continuous movements, and contrasted it with molecular diffusion by discontinuous random movements. The Lagrangian description of turbulence following the motion of tagged fluid particles has been introduced in Chapter 5. It is clear from that discussion that, in spite of the high degree of irregularity or randomness associated with turbulent motions, they are essentially continuous and have finite temporal correlations over time periods (lag times) of the order of the integral time scale. This is markedly different from the random but discontinuous molecular motions. The statistical theory, when applied to turbulent diffusion, yields some novel results, fundamentally different from those of the K-theory.

7.2 STATISTICAL THEORY OF ABSOLUTE DIFFUSION

7.2.1 Taylor's Theorems

Taylor (1921) considered the one-dimensional diffusion of tagged fluid particles released from the same point (one at a time and independent of each other) in a relatively simple, stationary, and homogeneous turbulent flow with zero or uniform mean velocity. The spatial homogeneity of turbulence is implied at least in the direction of diffusion, if not in all directions (a fully homogeneous flow cannot remain stationary, because it may not have any mean shear or unstable stratification in order to generate and maintain turbulence). The effect of any uniform mean motion is simply to transport particles at a fixed speed in the direction of flow. For the sake of simplicity, this uniform translation can be taken out of consideration using a frame of reference moving with the mean flow. Let us con-

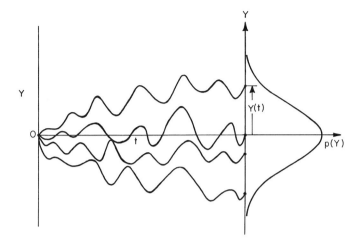

Figure 7.1 Schematics of the lateral displacements of tagged fluid particles released independently of each other from a point (the origin) in a stationary and homogeneous field of turbulence and the hypothetical probability density function of $Y(t)$.

sider the Lagrangian motion of a fluid particle initially located or released at the origin in the above frame of reference. Although the Lagrangian motion is three dimensional, following Taylor (1921), we can focus on particle movements in one direction (say, in the Y direction) only. The particle displacement Y from its initial position at the origin is only a function of time t after release and is schematically shown in Figure 7.1. This is related to the Lagrangian turbulent velocity $V'(t)$ as

$$V'(t) = \frac{dY(t)}{dt}$$

$$Y(t) = \int_0^t V'(t') dt' \qquad (7.1)$$

Following the paths of an infinitely large number of marked fluid particles released from the same point (the origin), independent of each other, we can look at the ensemble-averaged statistics of particle displacements. Obviously, in the frame of reference used here, their mean values at any time after release must be identically equal to zero, that is, $\overline{X} = 0$, $\overline{Y} = 0$, and so on. However, the range of particle displacements in any direction increases with time, as larger and larger deviations from the origin become more probable. This is essentially the nature of the dispersion process. The simplest meaningful statistical measures of dispersion we can compute are the mean-square particle displacements $\overline{X^2}(t)$, $\overline{Y^2}(t)$, and so on, which, following the above reasoning, must be increasing functions of time. This is in contrast to the variances of Lagrangian velocity fluctuations, which must be independent of time in a stationary and homogeneous field of turbulence and equal to the corresponding Eulerian velocity variances.

The rate of change of the mean-square particle displacement $\overline{Y^2}(t)$ with time can be related to the statistics of turbulence, using Eq. (7.1) and Reynolds' averaging rules, as

$$\frac{d\overline{Y^2}}{dt} = \overline{2Y(t)\frac{dY}{dt}} = 2\overline{\left[\int_0^t V'(t')dt'\right]V'(t)}$$

$$= 2\int_0^t \overline{V'(t)V'(t')}dt' \qquad (7.2)$$

In a stationary field of turbulence, the average of the velocity product in Eq. (7.2) will depend on the time difference, $\xi = t' - t$, only and may be replaced by $\overline{V'^2}R_L(\xi)$, so that

$$\frac{d\overline{Y^2}}{dt} = 2\overline{v'^2}\int_0^t R_L(\xi)d\xi \qquad (7.3)$$

where $R_L(\xi)$ is the autocorrelation coefficient of Lagrangian velocity. This is one of the fundamental results obtained by Taylor (1921), who also expressed it in terms of the covariance between particle velocity and displacement

$$\overline{V'(t)Y(t)} = \frac{1}{2}\frac{d\overline{Y^2}}{dt} = \overline{v'^2}\int_0^t R_L(\xi)d\xi \qquad (7.4)$$

Integration of Eq. (7.3) yields an alternative "law of diffusion"

$$\overline{Y^2}(t) = 2\overline{v'^2}\int_0^t \int_0^{t'} R_L(\xi)d\xi dt' \qquad (7.5)$$

Similar expressions can be written for $\overline{X^2}(t)$ and $\overline{Z^2}(t)$.

Taylor (1921) called Eq. (7.5) "rather remarkable because it reduces the problem of diffusion, in a simplified type of turbulent motion, to the consideration of a single quantity, namely, the correlation coefficient between the velocity of a particle at one instant and that at a time ξ later." Subsequently, Eq. (7.3) or (7.5) has been referred to as Taylor's theorem. Knowing the general behavior of $R_L(\xi)$, further deductions can be made from Taylor's theorem for small and large diffusion times.

For small diffusion time $t \ll T_{iL}$, $R_L(\xi)$ does not differ appreciably from unity, so that Eqs. (7.3) and (7.5) can be approximated to

$$\frac{d\overline{Y^2}}{dt} \simeq 2\overline{v'^2}t \quad (7.6)$$

$$\overline{Y^2} \simeq \overline{v'^2}t^2 \quad (7.7)$$

indicating that, initially, mean-square particle displacement increases in proportion to t^2.

For large diffusion time $t \gg T_{iL}$, one would expect the integral in Eq. (7.3) to approach a constant value, equal to the Lagrangian integral time scale T_{iL}, so that Eqs. (7.3) and (7.5) may be approximated to

$$\frac{d\overline{Y^2}}{dt} \simeq 2\overline{v'^2}T_{iL} \quad (7.8)$$

$$\overline{Y^2} \simeq 2\overline{v'^2}T_{iL}t \quad (7.9)$$

indicating that the mean-square diffusion (displacement) eventually becomes proportional to t.

The above results were originally derived by Taylor (1921) and are frequently referred to as Taylor's theorems for small and large diffusion times. These can also be written in terms of the standard deviation of particle displacement or the so-called one-particle diffusion parameter, $\sigma_y = (\overline{Y^2})^{1/2}$, as

$$\sigma_y \simeq \sigma_v t, \quad \text{for } t \ll T_{iL} \quad (7.10)$$
$$\sigma_y \simeq \sqrt{2}\sigma_v(T_{iL}t)^{1/2}, \quad \text{for } t \gg T_{iL} \quad (7.11)$$

An important implication of Taylor's theorems is that turbulent diffusion initially proceeds rather rapidly at a rate proportional to t, but eventually settles down to a much slower rate proportional to $t^{1/2}$ at large diffusion times compared with T_{iL}. The initial, more rapid, stage of turbulent diffusion by continuous movements is in marked contrast to the molecular or Brownian diffusion, which is predicted to proceed at the slower rate proportional to $t^{1/2}$ for all diffusion times. Only in the later stage of diffusion, that is, when $t \gg T_{iL}$, the turbulent diffusion becomes analogous to molecular or Brownian diffusion.

Taylor (1921) also noted that, when $t \gg T_{iL}$, the covariance between the particle displacement and its velocity, $\overline{YV'}$, becomes constant in spite of the fact that $\overline{Y^2}$ continually increases. It obviously follows from Eqs. (7.4) and (7.8)

$$\overline{YV'} = \frac{1}{2}\frac{d\overline{Y^2}}{dt} = \overline{v'^2}T_{iL}, \quad \text{for } t \gg T_{iL} \quad (7.12)$$

which implies that particle displacement must always be positively correlated with particle velocity, but the correlation coefficient must decrease with increasing diffusion time and, hence, with increasing $\overline{Y^2}$. Later on, the above covariance, or $\frac{1}{2}d\overline{Y^2}/dt$, will be shown to be analogous (equivalent) to eddy diffusivity K_y, so that Eq. (7.12) can also be considered as an expression for eddy diffusivity as a product of the variance of velocity and the integral time scale. One would expect a stationary and homogeneous field of turbulence to be characterized by a constant eddy diffusivity. Taylor's results indicate that this is true only in the limit of large diffusion time. For small diffusion times, the very concept of eddy diffusivity is found to be inconsistent with the predictions of the statistical theory, unless one is willing to allow for the eddy diffusivity to be a function of diffusion time. Taylor (1959) later called this an "illogical conception."

7.2.2 Form of $R_L(\xi)$

The exact limits of the applicability of Taylor's relations for small and large diffusion times cannot be determined theoretically. Experimentally also, Lagrangian statistics, such as $R_L(\xi)$, are extremely difficult to obtain with precision. It was argued in Chapter 5 that a simple exponential form of $R_L(\xi)$ given by Eq. (5.88) is probably the most appropriate analytical expression that is also consistent with the theoretically predicted behavior of Lagrangian spectrum in the inertial subrange. In fact, Taylor (1921) assumed the same exponential form for $R_L(\xi)$ in deriving the following general expression for the diffusion parameter:

$$\sigma_y = \sigma_v T_{iL}\left\{2\frac{t}{T_{iL}} - 2\left[1 - \exp\left(-\frac{t}{T_{iL}}\right)\right]\right\}^{1/2} \quad (7.13)$$

which reduces to Eqs. (7.10) and (7.11) in the limits of small and large diffusion times.

Note that the exponential form of the autocorrelation coefficient

$$R_L(\xi) = \exp\left(-\frac{|\xi|}{T_{iL}}\right) \quad (7.14)$$

does not satisfy the physical constraint that the second derivative of $R_L(\xi)$ at $\xi = 0$ should be finite (implying finite accelerations and a finite microscale). An alternative form of the autocorrelation function, which satisfies the above constraint, is

$$R_L(\xi) = \exp\left(-\frac{\pi}{4}\frac{\xi^2}{T_{iL}^2}\right) \quad (7.15)$$

which implies the following expression for the dispersion parameter:

$$\sigma_y = \sigma_v T_{iL}\left[2\frac{t}{T_{iL}}\mathrm{erf}\left(\frac{\sqrt{\pi}}{2}\frac{t}{T_{iL}}\right) + \frac{4}{\pi}\exp\left(-\frac{\pi}{4}\frac{t^2}{T_{iL}^2}\right) - \frac{4}{\pi}\right]^{1/2} \quad (7.16)$$

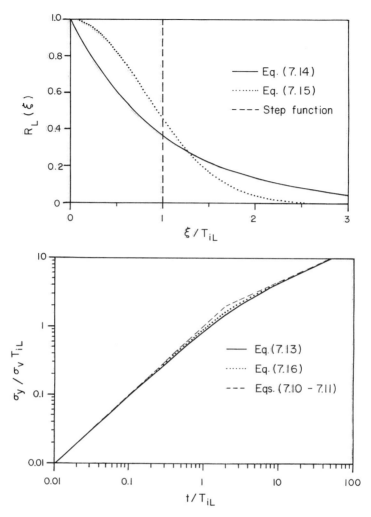

Figure 7.2 Comparison of (a) several hypothetical but plausible forms of the Lagrangian autocorrelation function and (b) the predicted dispersion parameter as a function of the dimensionless diffusion time.

A comparison of the above predictions of the dimensionless dispersion parameter $\sigma_y/\sigma_v T_{iL}$, as a function of t/T_{iL}, is made in Figure 7.2 for different forms of $R_L(\xi)$. Similar comparisons have been made by Pasquill (1974) and Pasquill and Smith (1983) for other plausible and some rather crude forms of $R_L(\xi)$.

An inevitable conclusion drawn from such comparisons is that dispersion in a homogeneous turbulent flow is relatively insensitive to the exact form of $R_L(\xi)$, provided $R_L(\xi)$ remains positive everywhere, that is, for all values of ξ. Thus, a good estimation of dispersion in a particular direction can be made from the estimates or measurements of the turbulence intensity and the Lagrangian integral time scale of turbulence in that direction. Even a rough estimate of T_{iL} may suffice, because dispersion remains essentially independent of T_{iL} for small diffusion times and increases in proportion to $\sqrt{T_{iL}}$ for large diffusion times. The linear law of dispersion, corresponding to Taylor's theorem for small diffusion times, can be used, without much error, up to $t \simeq 0.5T_{iL}$, and the 1/2-power law, corresponding to Taylor's theorem for large diffusion times, can be applied for $t \geq 5T_{iL}$. For intermediate diffusion times, $0.5T_{iL} < t < 5T_{iL}$, an appropriate interpolation formula, such as Eq. (7.13) or Eq. (7.16), can be used.

7.2.3 Further Development of the Theory

Taylor's (1921) results have undergone a considerable amount of elaboration and generalization by other investigators, notably by Kampe de Feriet (1939), Batchelor (1949), Ogura (1952, 1959), Hay and Pasquill (1959), F.B. Smith (1968), and others.

A useful simplification of Taylor's theorem is obtained by performing the integration in Eq. (7.5) with respect to t' by parts, using Leibinitz's rule,

$$\overline{Y^2}(t) = 2\overline{v'^2}\int_0^t (t-\xi)R_L(\xi)d\xi \qquad (7.17)$$

This result can also be written in terms of the normalized Lagrangian spectrum function, $S_L(n)$, by using the Fourier transform relations (5.81),

$$\overline{Y^2}(t) = 2\overline{v'^2}t^2 \int_0^\infty S_L(n) \frac{\sin^2(\pi nt)}{(\pi nt)^2}dn \qquad (7.18)$$

The mathematical steps involved in the derivations of Eqs. (7.17) and (7.18) are left as an exercise for the reader. Similar expressions can be obtained for $\overline{X^2}(t)$ and $\overline{Z^2}(t)$, provided turbulence is homogeneous in those directions also. For a completely homogeneous field of turbulence, Batchelor (1949) derived three-dimensional versions of these results. He also expressed covariances between particle displacements in different directions, in terms of complex cross-correlation and cross-spectrum functions of Lagrangian motion. For the sake of simplicity, we discuss here only one-dimensional relations, which describe dispersion in one direction at a time and are often used in applications to atmospheric dispersion modeling.

Equation (7.18) is an alternative form of Taylor's theorem, which clearly shows that, with increasing diffusion time, small (high-frequency) eddies contribute less and less to the mean-square particle displacement and large (low-frequency) eddies come to dominate the diffusion process. While small-scale motions merely oscillate particles in small neighborhoods around their mean positions, large eddies tend to displace them in a sustained way. The relative importance of different scales of motion (eddy sizes) is quantitatively expressed by Eq. (7.18) in which the factor $\sin^2(\pi nt)/(\pi nt)^2$ acts like a low-pass filter that effectively cuts off the high frequency part of the spectrum. The properties of such a filter or smoothing function have been discussed earlier in section 5.8, particularly in Figure 5.16. It is obvious that there is little or no contribution to the integral in Eq. (7.18) from frequencies larger than about $0.5/t$. Thus, with increasing t, increasingly large eddies become ineffective in diffusing particles, whereas initially (at $t = 0$), all scales of motion are effective in dispersion, in proportion to their contribution to the velocity variance.

This selective action of different size eddies on particle dispersion, as a function of travel time, is depicted in Figure 7.3 Here, the integrand in Eq. (7.18), which represents the spectrum of a smoothed time series of Lagrangian velocity with a smoothing interval equal to the diffusion time t, is shown as a function of the normalized frequency $2\pi n T_{iL}$ and diffusion time t/T_{iL}. Note that, initially, the area under the unsmoothed spectrum is unity, but it reduces with increasing diffusion time as more and more of the high-frequency part is cut off. Here the original spectrum, $nS_L(n)$, was specified by Eq. (5.97), which corresponds to the exponential form of autocorrelation coefficient.

Following the simplifying notation of the effect of smoothing of a time series, discussed in section 5.8, we can also express Eq. (7.18) as

$$\overline{Y^2}(t) = \overline{V_t'^2}t^2 \qquad (7.19)$$

where $\overline{V_t'^2}$ represents the variance of the smoothed Lagrangian velocity with a smoothing interval of t. The area under the smoothed spectrum of Figure 7.3, corresponding to a particular value of t/T_{iL}, actually represents the ratio $\overline{V_t'^2}/\overline{V'^2}$ of smoothed to actual variance. In the limit of large diffusion times $t/T_{iL} \gg 1$, the above ratio is expected to decrease inversely proportional to t. Equation (7.19) can also be interpreted

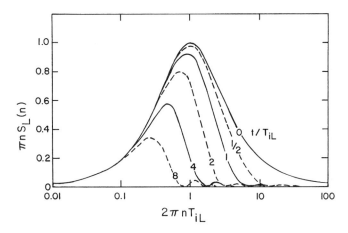

Figure 7.3 A spectral representation of the effect of time of travel on the dispersion of particles released from a fixed point in homogeneous turbulence. The ordinate represents the weighted or smoothed spectrum $\pi n S_L(n)[\sin^2(\pi nt)/(\pi nt)^2]$ as a function of the normalized frequency $2\pi n T_{iL}$ and the dimensionless travel time t/T_{iL}, such that area under the curve represents the integral in Eq. (7.18) at different travel times. Modified after Pasquill, 1974.

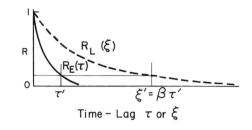
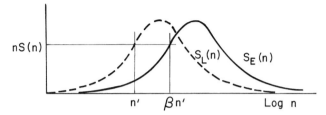

Figure 7.4 Schematics of the hypothetical similarity between Eulerian and Lagrangian autocorrelation functions (a) and spectra (b). After Pasquill, 1974.

as a simple but general relationship for the dispersion parameter σ_y as a product of diffusion (travel) time t and the root-mean-square of particle velocity averaged over t (Ogura, 1952).

7.2.4 The Hay–Pasquill Approach

The main difficulty in using the otherwise simple dispersion relation (7.19) is the measurement of Lagrangian velocity whose variance is to be determined after appropriate smoothing or averaging. Alternatively, the use of Eq. (7.18) or the original Taylor's relations requires an estimate or measurement of total variance, which is the same for Eulerian and Lagrangian velocities and can easily be obtained from the former. But the additional specification of the form of $R_L(\xi)$ or $S_L(n)$, as well as the appropriate value of the Lagrangian integral time scale T_{iL} in the above relations, introduces the same degree of difficulty associated with Lagrangian motion as in the determination of $\overline{V_t'^2}$. To avoid this difficulty, Hay and Pasquill (1959) sought to express the above dispersion relations entirely in terms of fixed-point (Eulerian) statistics. The basis of their modification is the simple hypothesis that the forms of $R_L(\xi)$ and $R_E(\tau)$ are similar, that is,

$$R_L(\xi) = R_E(\tau), \text{ for } \xi = \beta\tau \qquad (7.20)$$

where $\beta = T_{iL}/T_{iE}$ is the ratio of Lagrangian and Eulerian integral scales. The assumption of similarity between Lagrangian and Eulerian autocorrelation coefficients also implies the similarity between their corresponding spectral functions, that is,

$$nS_L(n) = \beta nS_E(\beta n) \qquad (7.21)$$

which follows from the appropriate Fourier-transform relations (5.35) between autocorrelation and spectrum functions. The hypothetical similarity relations (7.20) and (7.21) are schematically shown in Figure 7.4.

Substituting from Eq. (7.21) into (7.18) one obtains

$$\overline{Y^2}(t) = \overline{v'^2}t^2 \int_0^\infty S_E(n) \frac{\sin^2(\pi nt')}{(\pi nt')^2} dn \qquad (7.22)$$

in which $t' = t/\beta$ may be considered as a modified diffusion or travel time. Note that Eq. (7.22) is similar to Eq. (7.18) but involves the more easily measured spectrum of Eulerian velocity. Utilizing the earlier expression (5.67) for the reduced variance of a smoothed time series, Eq. (7.22) can also be written as

$$\overline{Y^2}(t) = \overline{v_t'^2}t^2 \qquad (7.23)$$

where $t' = t/\beta$ is now the appropriate averaging time or smoothing interval. Although Eq. (7.23) is quite similar to Eq. (7.19), it involves the more easily obtainable variance of Eulerian velocity after appropriate smoothing.

The approximate dispersion relations (7.22) and (7.23) based on Hay and Pasquill's (1959) hypothesis are the Eulerian analogs of the exact dispersion relations (7.18) and (7.19) obtained from Taylor's (1921) theorem. The hypothesis regarding the similar forms of Eulerian and Lagrangian autocorrelations or spectra has not been thoroughly tested so far. Limited comparisons of Lagrangian and Eulerian spectra (see e.g., Hanna, 1981), as well as theoretical considerations have pointed out significant differences in their forms for the inertial subrange. However, it can be argued that turbulent diffusion is mostly caused by large-scale motions and those in the inertial subrange have little or no effect. The implied similarity of $nS_L(n)$ and $nS_E(n)$ around their spectral peaks still needs to be demonstrated. It has been argued, however, that the aforementioned dispersion relations, including those based on the Hay–Pasquill hypothesis, are not very sensitive to the exact forms of autocorrelations or spectra (Pasquill and Smith, 1983). Far more important to know or specify is the ratio of the integral time scales $\beta = T_{iL}/T_{iE}$. Hay and Pasquill (1959) used Eq. (7.23), in conjunction with available dispersion and velocity measurements, to determine β indirectly. For a wide range of conditions and over an enormous range of scales of turbulence encountered in the atmosphere as well as in laboratory wind tunnels and water channels, they showed that the computed values of β lie in the range of 1 to 10. Other theoretical considerations and experimental observations have established an inverse relationship between β and turbulence intensity, as discussed in section 5.9.

The statistical theory described in this section and the various dispersion relations based on the same express the ensemble-averaged dispersion parameters as functions of diffusion time and ensemble-averaged statistics of turbulence. Strictly, the theory deals with the statistics of one-particle displacements relative to fixed coordinate axes passing through the release point as the origin. For this reason, this theory is also called the one-particle diffusion theory or the absolute diffusion theory, in contrast to the relative diffusion theory that deals with relative displacements or separations between two or more particles released simultaneously. Since very simple and ideal conditions of flow and particle release are implied in the theory, we should discuss the relevance and applicability of theoretical results to practical diffusion problems.

7.3 PLUME DIFFUSION FROM CONTINUOUS SOURCES

A continuous line, area, or volume source can be considered as a superposition of an infinite number of continuous point sources. Thus, the continuous point source can be used as the fundamental source geometry for describing diffusion from other continuous sources. An ideal point source is assumed to release neutrally buoyant, passive, tagged fluid particles from a given point in succession. In a stationary and homogeneous field of turbulence and uniform mean flow, such a maintained source would be equivalent to a very large (infinite) number of realizations of one-particle (absolute) diffusion. The statistics of displacements of many (approaching to infinity) particles released in succession should be identical to those of the displacements of a single particle observed a large number of times. Here, the particle displacements are to be taken from their expected positions as a result of being carried by the mean flow alone. Thus, in the usual coordinate system with the origin at the source position and the x-axis along the mean flow direction, the mean particle position from which displacements are considered is $(x,0,0)$, with $x = \bar{u}t$.

Using the above simple transformation ($t = x/\bar{u}$) between the travel time and the average travel distance from the source, the dispersion relations of the previous section can all be expressed in terms of the average travel distance from the source. For example, Eqs. (7.10) and (7.11) can be written as

$$\sigma_y \simeq \frac{\sigma_v}{\bar{u}}x = i_v x, \quad \text{for } x \ll L_y \quad (7.24)$$

$$\sigma_y \simeq \sqrt{2}i_v L_y^{1/2} x^{1/2}, \quad \text{for } x \gg L_y \quad (7.25)$$

where i_v is the turbulence intensity and $L_y = \bar{u}T_{iL}$ is a large-eddy length scale corresponding to the Lagrangian integral time scale. Similarly, the more general Hay–Pasquill relation (7.23) can also be expressed as

$$\sigma_y = i'_v x \quad (7.26)$$

where i'_v is the reduced turbulence intensity obtained after smoothing the turbulent velocity signal with a smoothing interval of $t' = x/\bar{u}\beta$. Note that i'_v is expected to be a decreasing function of x, with its maximum value i_v near the source and becoming proportional to $x^{-1/2}$ far from the source.

7.3.1 Effects of Finite Sampling and Release Times

In all the statistical theory relations presented in section 7.2, the averages denoted by an overbar, such as $\overline{Y^2}$, $\overline{V'^2}$, and so on, are ensemble averages that would be extremely difficult, if not impossible, to obtain from observations or experiments on turbulence and diffusion. The application of theoretical relations to the practical problem of diffusion from a continuous point source in a stationary and homogeneous turbulent medium requires that the sampling duration for averaging as well as the release time be large enough (approaching infinity) to fully account for the effects of large eddies on variances of velocity and particle displacements. These restrictions follow from the conditions for the equivalence of time and ensemble averages, discussed in section 4.5. In practice, however, such ideally long sample and release times are not usually obtained in the atmosphere, because lower atmospheric conditions may not remain stationary for more than a few hours (in some cases, a few minutes). Sampling time in the atmosphere is often limited by diurnal variations in the PBL, evolution of mesoscale and large-scale weather systems, duration of steady release or emission rate, and possibly other factors. Sampling duration should be much larger than the time of travel of marked particles to the sample location, but less than the duration of steady release.

The effect of finite sampling duration on spectrum and variance of a stationary variable (time series) has been discussed in section 5.8. In particular, Eq. (5.72) gives the reduction factors for both the spectrum and the variance of $u'(t)$ as functions of the finite sampling duration T. Further recognizing the direct correspondence (proportionality) between the variances of particle displacement and velocity, one can easily write the finite-sampling analogs of the statistical theory relations given in section 7.2. For example, from Eq. (7.5) and Eq. (7.18), respectively,

$$\overline{Y_T^2} = 2\overline{v_T'^2}\int_0^t\int_0^{t'} R_L(\xi)d\xi dt' \qquad (7.27)$$

$$\overline{Y_T^2} = \overline{v_T'^2}t^2\int_0^t S_L(n)\frac{\sin^2(\pi nt)}{(\pi nt)}dn = \overline{V_{T,t}'^2}\,t^2 \qquad (7.28)$$

where the first subscript refers to the finite sampling duration and the second subscript indicates the finite smoothing or averaging interval. The above relations still involve the difficult to obtain Lagrangian statistics. The Hay–Pasquill approach, described earlier, yields the following dispersion relation entirely in terms of the fixed-point Eulerian velocity:

$$\overline{Y_T^2} = \overline{v'^2}t^2\int_0^\infty S_E(n)\frac{\sin^2(\pi nt')}{(\pi nt')^2}$$
$$\left[1 - \frac{\sin^2(\pi nT)}{(\pi nT)^2}\right]dn \qquad (7.29)$$

which, following Eq. (5.73), can also be written as

$$\overline{Y_T^2} = \overline{v_{T,t'}'^2}\,t^2 \qquad (7.30)$$

Here, $t' = t/\beta$ represents the appropriate interval for smoothing or averaging the finite-duration time series $v'(t)$ before determining its variance.

One can use either Eq. (7.29) or Eq. (7.30), depending on the type of turbulence information or data available. If the total variance and spectrum of velocity are estimated or specified from appropriate theoretical or empirical similarity relations, then Eq. (7.29) can be used to estimate dispersion as a function of travel time. On the other hand, if on-site turbulence measurements are available for estimating the reduced variance $\overline{v_{T,t'}'^2}$ as a function of travel time, then Eq. (7.30) provides the dispersion estimates more readily. The latter method is superior, because it is based on actual measurements of turbulence and is essentially devoid of questionable assumptions of turbulence theories or models. In order for the proper determination of the reduced variance $\overline{v_{T,t'}'^2}$, it would be necessary for the sampling duration to be much larger than the smoothing or averaging interval, that is, $T \gg t' = t/\beta$. Thus, the finite sampling duration does restrict the maximum travel time or distance from a continuous source to which Eq. (7.30) could be usefully applied. Otherwise, the above dispersion relation is a very simple and practically useful result of the statistical theory in that the mean-square dispersion of particles over a useful range of distance is determined completely by appropriately smoothing the fluctuating velocity at the point of release and then determining the variance over the desired sampling period.

7.3.2 Relationship Between Concentration Field and Dispersion Parameters

The statistical theory predicts only the second moments of particle displacements as functions of diffusion time; first moments or mean displacements are zero by virtue of the reference frame moving with the steady and uniform mean flow. To describe the mean concentration field due to a continuous point source in a homogeneous field of turbulence, we need to know the entire spatial probability density function $p(X,Y,Z; t)$, not only its second moments. Here $p(X,Y,Z; t)$ denotes the probability density of particle displacement vector $\underline{X} = (X,Y,Z)$ after a travel time t, such that $p(\underline{X}; t)d\underline{X}$ is the small probability that the displacement vector will end in an arbitrarily small volume element $d\underline{X} = dX\,dY\,dZ$ surrounding the point $\underline{X}$, at time t.

In a homogeneous field of turbulence, the probability density function $p(\underline{X}; t)$ is expected to be Gaussian in all three spatial dimensions and, hence, can be fully specified in terms of its second moments $\overline{X^2}$, $\overline{Y^2}$, and so on, or the so-called dispersion parameters $\sigma_x = (\overline{X^2})^{1/2}$, etcetera. The Gaussian distribution of particle displacements has been confirmed by statistical random-walk (discontinuous motion) models and also by experiments in grid turbulence, although a rigorous theoretical proof of its validity for homogeneous turbulent flows (continuous motion) is still lacking (Csanady, 1973).

For a continuous point source, the average concentration at a point (x,y,z) is equivalent (proportional) to the probability of finding a particle at (x,y,z) at any time. Considering contributions from all particle travel times, one can write

$$\overline{c}(x,y,z) = Q\int_0^\infty p(x,y,z; t)dt \qquad (7.31)$$

where Q is the emission rate of point source. With the Gaussian form of probability density functions of particle displacements in the y and z directions, one would expect from Eq. (7.31) the concentration distributions in y and z directions to be Gaussian also. It is easy to see that the widths of two distributions must be identical, so that the particle dispersion parameters σ_y and σ_z also represent the standard deviations of concentration distributions across the plume in y and z directions, respectively, that is,

$$\sigma_y = \left[\int_{-\infty}^\infty y^2\overline{c}(x,y,z)dy / \int_{-\infty}^\infty \overline{c}(x,y,z)dy\right]^{1/2}$$
$$\sigma_z = \left[\int_{-\infty}^\infty z^2\overline{c}(x,y,z)dz / \int_{-\infty}^\infty \overline{c}(x,y,z)dz\right]^{1/2} \qquad (7.32)$$

Concentration distributions for different source configurations and different boundary conditions have already been given in Chapter 6 in terms of the Gaussian plume-dispersion parameters. These were obtained as solutions of the average diffusion equation for constant-eddy diffusivities. The same concentration distributions can also be obtained by assuming the usual Gaussian forms for these distributions in an infinite, homogeneous medium and by satisfying the appropriate mass continuity equation. The influence of confining boundaries, such as the reflecting ground surface and capping inversion, is easily considered by applying the method of images and the principle of superposition, as discussed in Chapter 6. Thus, the Gaussian plume formulas given in the previous chapter are not necessarily tied to the Fickian diffusion theory, but they have a more general validity if dispersion parameters are assumed to be related to certain statistical properties of turbulence and diffusion time or travel distance.

By identifying the plume-dispersion parameters with the particle-dispersion parameters in a stationary and homogeneous turbulent medium, the statistical theory provides a sound theoretical basis for specifying these parameters. The Fickian gradient diffusion theory, discussed earlier in Chapter 6, provided another, not-so-sound theoretical basis of specifying dispersion parameters in terms of constant-eddy diffusivities. The predictions of the Fickian diffusion theory are reasonably consistent with those of the more rigorous statistical theory only in the limit of large diffusion times or travel distances from the source. The initial stage of the more rapid, linear growth of the plume with distance from the sources is not predicted by the Fickian theory.

7.3.3 Apparent Eddy Diffusivity

Does the process of diffusion by continuous movements (turbulent motions) imply an apparent turbulent diffusivity? This question has been examined in detail by Batchelor (1949) and Csanady (1973). Starting from the Gaussian expression for concentration distribution in a slender plume, resulting from a continuous point source in an infinite, homogeneous medium, that is,

$$\bar{c}(x,y,z) = \frac{Q}{2\pi \bar{u} \sigma_y \sigma_z} \exp\left[-\frac{y^2}{2\sigma_y^2} - \frac{z^2}{2\sigma_z^2}\right] \quad (7.33)$$

one can show that the advective transport of material by mean flow is given by

$$\bar{u}\frac{\partial \bar{c}}{\partial x} = \frac{\bar{u}}{2}\frac{d\sigma_y^2}{dx}\frac{\partial^2 \bar{c}}{\partial y^2} + \frac{\bar{u}}{2}\frac{d\sigma_z^2}{dx}\frac{\partial^2 \bar{c}}{\partial z^2} \quad (7.34)$$

Equation (7.34) can be recognized as being similar to the approximate diffusion equation

$$\bar{u}\frac{\partial \bar{c}}{\partial x} = K_y \frac{\partial^2 \bar{c}}{\partial y^2} + K_z \frac{\partial^2 \bar{c}}{\partial z^2} \quad (7.35)$$

in which diffusion in the direction of mean flow has been ignored. In fact, the two equations would be identical if eddy diffusivities were expressed in terms of dispersion parameters as

$$K_y = \frac{\bar{u}}{2}\frac{d\sigma_y^2}{dx} = \frac{1}{2}\frac{d\sigma_y^2}{dt}$$

$$K_z = \frac{\bar{u}}{2}\frac{d\sigma_z^2}{dx} = \frac{1}{2}\frac{d\sigma_z^2}{dt} \quad (7.36)$$

in which we have used the usual transformation between travel time and distance from the source. From Taylor's theorem (7.3), these "apparent eddy diffusivities" can also be expressed as

$$K_y = \overline{v'^2} \int_0^t R_{Lv}(\xi)d\xi$$

$$K_z = \overline{w'^2} \int_0^t R_{Lw}(\xi)d\xi \quad (7.37)$$

which clearly imply that apparent eddy diffusivities not only depend on the turbulence structure, but also on travel time or distance from the source. Because of the unexpected dependence of K_y and K_z on diffusion time or distance from the source, and possibly also on the source type (instantaneous or continuous), Taylor (1959) called the notion of apparent eddy diffusivity an "illogical conception."

Similar expressions for eddy diffusivities as given in Eq. (7.36), with the addition of $K_x = 0.5\ d\sigma_x^2/dt$, can be derived for puff diffusion (see e.g., Csanady, 1973), if one requires that the Gaussian puff formula (6.63) for an instantaneous point source in an infinite medium satisfy the mean diffusion equation (6.62). Earlier, Batchelor (1949) obtained a slightly more general expression for the eddy diffusivity tensor, without assuming the x, y, and z axes to be principal axes. It is obvious from Eq. (7.36) that only in the limit of large diffusion time, when $d\sigma_x^2/dt$, $d\sigma_y^2/dt$, and so on, become constants, that apparent eddy diffusivities may be considered as unique properties of turbulence, independent of diffusion time and source characteristics. In that limit, eddy diffusivity may be regarded as a genuine property of the turbulent flow, analogous to molecular diffusivity. Taylor's theorem gives the following useful relations between eddy diffusivities and other turbulence statistics:

$$K_x = \overline{u'^2}T_{Lu}; \quad K_y = \overline{v'^2}T_{Lv}; \quad K_z = \overline{w'^2}T_{Lw} \quad (7.38)$$

which are applicable only in the limit of large diffusion time $t \gg T_{Lu} = T_{iL}$.

7.4 STATISTICAL THEORY OF RELATIVE DIFFUSION

It has been argued in the previous section that the statistical theory of one-particle or absolute diffusion provides a suitable theoretical and conceptual framework for describing dispersion from maintained or continuous sources. Instantaneous releases of two or more particles in turbulent media and their subsequent dispersion relative to each other is more appropriately described by the statistical theory of relative diffusion. Although the basic conceptual framework for this theory was laid out by Richardson (1926), further substantive developments and generalizations were made much later, primarily by Batchelor (1950, 1952). The most general theoretical treatment deals with the three-dimensional relative dispersion of n particles. But, the basic results of the theory, as they are applied to practical diffusion problems, can more simply be derived from the consideration of one-dimensional dispersion of two particles in a homogeneous field of turbulence.

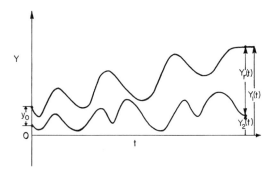

Figure 7.5 Schematic of the relative displacement of two marked fluid particles as a function of the time of travel in a stationary and homogeneous field of turbulence.

7.4.1 Kinematics of Relative Dispersion

Referring to the schematic of Figure 7.5, in which a particular realization of the lateral displacements of the two particles as well as their relative displacement (separation distance) in the Y-direction are shown as functions of travel time, we can write

$$Y_r = Y_1(t) - Y_2(t)$$
$$= y_0 + \int_0^t V_1'(t')dt' - \int_0^t V_2'(t')dt'$$
$$= y_0 + \int_0^t V_r'(t')dt' \quad (7.39)$$

where $V_r' = V_1' - V_2'$ is the relative Lagrangian velocity and y_0 is the initial value of the relative separation between the two particles in the Y direction. The analog of Taylor's (1921) theorem using relative Lagrangian displacements and velocities can easily be derived as

$$\frac{d\overline{Y_r^2}}{dt} = 2\frac{\overline{Y_r dY_r}}{dt} = 2\int_0^t \overline{V_r'(t)V_r'(t')}dt' \quad (7.40)$$

$$\overline{Y_r^2} = y_0^2 + 2\int_0^t \int_0^{t'} \overline{V_r'(t)V_r'(t+\xi)}d\xi\, dt' \quad (7.41)$$

Although these results look very similar to Taylor's results for one-particle (absolute) diffusion, there are some fundamental differences between the two cases. The first is the addition of y_0^2 term in Eq. (7.41) corresponding to the initial separation between the two particles at the time of release. If the two particles are released from the same initial position ($y_0 = 0$), they will be acted on by the same eddy motions and their relative velocity will remain zero at all times. Hence, the particles released from the same point, at the same time (instantaneously), will not disperse relative to each other, although they will disperse together relative to a fixed axis. Following this reasoning, Richardson (1926) concluded that an idealized instantaneous point source cannot be a suitable model for puff diffusion. For the puff to disperse about its center of gravity, it must have a finite size at the time of release. All practical release scenarios satisfy this requirement, in spite of an apparent contradiction presented by the instantaneous point source.

The other basic difference between the results of the absolute diffusion theory and Eqs. (7.40) and (7.41) is that, unlike the absolute particle velocity in a stationary and homogeneous field of turbulence, the relative velocity of the two tagged particles is not necessarily a stationary and random function of time. If the two particles are released not too far apart, they will be influenced by very similar eddy motions at first, but as they disperse farther and farther apart the range of eddy sizes contributing to their relative velocity also increases. Consequently, the autocorrelation function $\overline{V_r'(t)V_r'(t+\xi)}$ is likely to be a function of the diffusion time, as well as the lag time ξ.

Despite the above mentioned complications in dealing with relative dispersion, several specific predictions can be made from Eqs. (7.40) and (7.41) for the limiting cases of small and large diffusion times:

1. For small diffusion times, the relative velocity of the two particles does not change appreciably from its initial value v_{r0}', so that

$$\frac{d\overline{Y_r^2}}{dt} = 2\overline{v_{r0}'^2}t \quad (7.42)$$

$$\overline{Y_r^2} = y_0^2 + \overline{v_{r0}'^2}t^2 \quad (7.43)$$

Thus, the mean-square separation between the two particles initially increases in proportion to t^2, which appears to be similar to the earlier predicted increase in mean-square displacement of a particle relative to the fixed axis passing through its initial position. An important difference between the predictions of absolute and relative diffusion theories for small diffusion time arises from the dependence of relative dispersion rate on $\overline{v_{r0}'^2}$, which in a given field of turbulence is expected to depend on the velocity variance $\overline{v'^2}$, as well as on the initial separation vector $\mathbf{r_0}$ (with components x_0, y_0, z_0, and magnitude r_0) between the two particles at the time of their release. Assuming that particles take on the instantaneous fluid velocities at the locations of their release, so that $v_{r0}' =$

$V'_1 - V'_2 = v'_1 - v'_2$ (v' and V' denote the Eulerian and Lagrangian velocity components), all referred to the same time of release ($t = 0$),

$$\overline{v'^2_{r0}} = \overline{(v'_1 - v'_2)^2} = 2\overline{v'^2}[1 - R_{vv}(\underline{r}_0)] \quad (7.44)$$

Here, $R_{vv}(\underline{r}_0) = \overline{v'_1 v'_2}/\overline{v'^2}$ is the Eulerian spatial correlation coefficient between the velocity components at two points, which in a homogeneous field of turbulence is only a function of the separation vector $\underline{r}_0$ (see e.g., Batchelor, 1953). Knowing the general behavior of $R_{vv}(\underline{r}_0)$, we can expect $\overline{v'^2_{r0}}$ to have a magnitude between zero (when $\underline{r}_0 \to 0$) and $2\overline{v'^2}$ (when $\underline{r}_0 \to \infty$), depending on the initial separation ($\underline{r}_0$) between the two particles. The asymptotic value of $\overline{v'^2_{r0}} = 2\overline{v'^2}$ is likely to be reached only if $\underline{r}_0$ becomes larger than the largest eddies in the flow. Then, one would not expect any correlation at all between the initial velocities of two particles, and from Eq. (7.43)

$$\overline{Y^2_r}(t) = y^2_0 + 2\overline{v'^2}t^2 \quad (7.45)$$

for large r_0, but small diffusion time.

2. At large diffusion time, when the two particles drift so far apart that their Lagrangian velocities are no longer correlated with each other, it is easy to show that

$$\overline{V'_r(t)V'_r(t+\xi)} =$$
$$\overline{[V'_1(t) - V'_2(t)][V'_1(t+\xi) - V'_2(t+\xi)]}$$
$$= \overline{V'_1(t)V'_1(t+\xi)} + \overline{V'_2(t)V'_2(t+\xi)}$$
$$= 2\overline{v'^2}R_L(\xi) \quad (7.46)$$

We still do not know the behavior of $\overline{V'_r(t)V'_r(t+\xi)}$ for small and intermediate travel times, so that the integral in Eq. (7.40) or (7.41) cannot strictly be determined. The relevant dispersion relation can, however, be derived from the definition of $\overline{Y^2_r}$ as

$$\overline{Y^2_r}(t) = \overline{[Y_1(t) - Y_2(t)]^2}$$
$$= \overline{[\{Y_1(t) - Y_1(0)\} -}$$
$$\overline{\{Y_2(t) - Y_2(0)\} + \{Y_1(0) - Y_2(0)\}]^2}$$
$$= \overline{[Y_1(t) - Y_1(0)]^2} + \overline{[Y_2(t) - Y_2(0)]^2} + y^2_0$$
$$= 2\overline{Y^2}(t) + y^2_0 \quad (7.47)$$

where $Y_1(t) - Y_1(0)$ and $Y_2(t) - Y_2(0)$ are the particle displacements from their initial positions, which are uncorrelated with each other, each having the same mean-square value $\overline{Y^2}(t)$ given by the one-particle (absolute) diffusion theory.

For small initial separation between the particles and large diffusion time, the y^2_0-term in Eq. (7.47) can be neglected and one can write

$$\overline{Y^2_r}(t) \simeq 2\overline{Y^2}(t) \quad (7.48)$$

This result implies that the mean-square separation between any two particles eventually becomes twice the value of the mean-square displacement of one particle from a fixed axis in the limit of very large diffusion time ($t \to \infty$). This limit is effectively reached when the relative separation between the particles becomes larger than the largest eddies or coherent structures in the flow. The factor of two in Eq. (7.48) is essentially due to the difference between the definitions of Y and Y_r. The former is measured from a fixed axis, while the latter is twice the distance of any one particle from the center of gravity of the pair. In fact, the distinction between absolute diffusion and relative diffusion disappears in the limit of large diffusion time.

7.4.2 Similarity Relations for Relative Dispersion

The above kinematic relations for relative dispersion do not tell much about the behavior of $\overline{Y^2_r}(t)$ during a rather wide range of intermediate diffusion times between the extreme limits of small and large t. Further predictions of relative dispersion, including its unique behavior during the intermediate diffusion time and small initial separation between particles, have been made on the basis of Kolmogorov's local similarity theory (Batchelor, 1950, 1952; Batchelor and Townsend, 1956). So long as the scalar separation between the two particles is small compared with the size of large energy-containing eddies, but much larger than the Kolmogorov's microscale (i.e., Y_r lies within the inertial subrange of eddy sizes), one can assume that

$$\frac{d\overline{Y^2_r}}{dt} = f(\underline{r}_0, \varepsilon, t) \quad (7.49)$$

Then, from dimensional considerations, Eq. (7.49) can also be written in the dimensionless form as

$$\frac{t_*}{r^2_0} \frac{d\overline{Y^2_r}}{dt} = f_*\left(\frac{t}{t_*}\right) \quad (7.50)$$

where $r_0 = |\underline{r}_0|$ and

$$t_* = r^{2/3}_0 \varepsilon^{-1/3} \quad (7.51)$$

is the time scale based on the initial separation and the rate of energy transfer or dissipation ε. According to Eq. (7.50), the dimensionless rate of relative dispersion is some unique function of dimensionless time t/t_*. Several, more definite, predictions can be made, again if we consider the limits of small, intermediate, and large diffusion times and also require that our similarity relations be consistent or matched with those based on kinematics of dispersion:

1. For small diffusion or travel times, such that $t \ll t_*$, matching of Eq. (7.50) with Eq. (7.42) requires that the function $f_*(t/t_*)$ be linear, that is,

$$\frac{t_*}{r^2_0} \frac{d\overline{Y^2_r}}{dt} \propto \frac{t}{t_*}$$

or,

$$\frac{d\overline{Y^2_r}}{dt} \propto r^{2/3}_0 \varepsilon^{2/3} t$$

which, after integration, yields

$$\overline{Y^2_r} = c r^{2/3}_0 \varepsilon^{2/3} t^2 + y^2_0 \quad (7.52)$$

Here, c is a constant that may depend on the orientation of the initial separation vector r_0. Equation (7.52) also follows directly from Eq. (7.43) if one assumes $\overline{v_{r0}'^2}$ to be a function of r_0 and ε only. From local isotropy and dimensional considerations (Batchelor, 1950),

$$\overline{v_{r0}'^2} = c_1\left(1 + \frac{1}{3}\frac{y_0^2}{r_0^2}\right)r_0^{2/3}\varepsilon^{2/3} \tag{7.53}$$

where c_1 is a universal constant. Substituting from Eq. (7.53) into Eq. (7.43) one obtains

$$\overline{Y_r^2} = c_1\left(1 + \frac{1}{3}\frac{y_0^2}{r_0^2}\right)r_0^{2/3}\varepsilon^{2/3}t^2 + y_0^2 \tag{7.54}$$

A comparison of Eq. (7.54) with Eq. (7.52) indicates that

$$c = c_1\left(1 + \frac{1}{3}\frac{y_0^2}{r_0^2}\right) \tag{7.55}$$

Note that c_1, rather than c, is a universal constant of Kolmogorov's local similarity theory, which is empirically determined to be about 1.6.

2. For intermediate diffusion times when Y_r becomes large enough for it to become independent of r_0, but it is still small enough to lie within the inertial subrange of eddy sizes, a more appropriate similarity hypothesis might be (Batchelor, 1950)

$$\frac{d\overline{Y_r^2}}{dt} = f(\varepsilon, t) \tag{7.56}$$

instead of Eq. (7.49). Then, dimensional analysis yields

$$\frac{1}{\varepsilon t^2}\frac{d\overline{Y_r^2}}{dt} = \text{const.} = 3c_2 \tag{7.57}$$

which, after integration, gives

$$\overline{Y_r^2}(t) = c_2\varepsilon t^3 + y_0^2 \tag{7.58a}$$

$$\overline{Y_r^2}(t) \simeq c_2\varepsilon t^3, \text{ for } \overline{Y_r^2}(t) \gg y_0^2 \tag{7.58b}$$

In this dispersion relation, the y_0^2 term representing the constant of integration is usually neglected because of the assumed independence of $\overline{Y_r^2}(t)$ on initial separation. However, since Eq. (7.57) or (7.58a) is not expected to be valid for small diffusion time ($t \to 0$), the constant of integration should not be based on the initial condition $\overline{Y_r^2}(t) \to y_0^2$ as $t \to 0$. Another condition corresponding to the beginning of the more rapid (intermediate) phase of diffusion, $\overline{Y_r^2} = \overline{Y_{r1}^2}$ at $t = t_1$, might be more appropriate. Using the latter condition, a more accurate dispersion relation can be obtained from Eq. (7.57):

$$\overline{Y_r^2}(t) = c_2\varepsilon(t^3 - t_1^3) + \overline{Y_{r1}^2} \tag{7.59}$$

As a first approximation, $\overline{Y_{r1}^2}$ can be estimated from the matching of Eq. (7.59) with Eq. (7.43) or Eq. (7.52) at $t = t_1$, so that

$$\overline{Y_{r1}^2} = y_0^2 + \overline{v_{r0}'^2}t_1^2$$

$$= y_0^2 + cr_0^{2/3}\varepsilon^{2/3}t_1^2 \tag{7.60}$$

The difference between the dispersion relations (7.58a) and (7.59) is likely to be significant only in the early phase of intermediate diffusion. When travel time becomes much larger than t_1, the use of the simpler formula (7.58b) may be justified.

It can be argued, from similarity considerations, that the time t_1, which marks the division between small and intermediate diffusion times, must be proportional to the time scale t_* defined in Eq. (7.51), that is,

$$t_1 = c_3 r_0^{2/3}\varepsilon^{-1/3} \tag{7.61}$$

Substituting from Eq. (7.61) into Eq. (7.60) yields

$$\overline{Y_{r1}^2} = y_0^2 + cc_3^2 r_0^2$$

$$= y_0^2 + c_4 r_0^2 \tag{7.62}$$

where $c_4 = cc_3^2$ is another constant. Substituting from Eq. (7.62) into Eq. (7.59), we get

$$\overline{Y_r^2}(t) = c_2\varepsilon t^3 + y_0^2 + c_5 r_0^2 \tag{7.63}$$

in which $c_5 = c_3^2(c - c_2c_3) = c_4 - c_2c_3^3$ is a constant, which is simply related to c, c_2, and c_3. It can be argued that all of these constants are of the order of unity, but more definite values of the same have not been determined from relative diffusion observations or experiments. Note that the last term in Eq. (7.63) is a correction to the approximate dispersion relation (7.58a), which could be particularly significant for large initial separations between dispersing particles.

The above results for intermediate diffusion times bring out the unique accelerative nature of relative diffusion, which is expected to occur as long as particle separations are small compared with the length scale of large energy-containing eddies. This is quite in contrast to the case of absolute (one-particle) diffusion relative to a fixed axis for which Taylor's theory predicts a smooth and monotonic transition between the dispersion relations (7.7) and (7.9) for small and large diffusion times. This fundamental difference between the growth rates of absolute and relative dispersion parameters as functions of travel time was originally pointed out by Richardson (1926), who examined the available data at that time on diffusion of particle pairs and clusters over a wide range of scales. Richardson actually deduced horizontal diffusivities and proposed an empirical expression $K = 0.2\, l^{4/3}$, where l is the characteristic horizontal dimension of particle clusters in centimeters and K is expressed in cm^2/s. Rather unexpectedly, the diffusion of particle clusters was found to depend on the cluster size or the separation between representative pairs of particles. A formal explanation of Richardson's empirical law of relative diffusion was given by Obukhov (1941), who pointed out that it naturally follows from Kolmogorov's local similarity hypothesis for inertial subrange, which predicts

$$K = c'\varepsilon^{1/3}l^{4/3} \tag{7.64}$$

in which c' is a dimensionless, universal constant, unlike the dimensional, empirical constant of Richardson. Still, Richardson is credited to have foreseen or intuitively discovered some of the basic concepts of energy transfer and dispersion through a hierarchy of turbulent eddies, which were later formalized by Kolmogorov (1941) and Obukhov (1941) under their local similarity theory (see e.g., Chapter 4; Monin and Yaglom, 1975). Note that the dispersion relation (7.57) also implies an apparent eddy diffusivity

$$K_y = \frac{1}{2} \frac{d\overline{Y_r^2}}{dt} = \frac{3}{2} c_2^{1/3} \epsilon^{1/3} (\overline{Y_r^2})^{2/3} \qquad (7.65)$$

which is consistent with Eq. (7.64) if we recognize that the cluster size l is proportional to the relative dispersion parameter $(\overline{Y_r^2})^{1/2}$.

The relative dispersion relations based on the local similarity hypothesis for inertial subrange may not remain valid at large diffusion times when $(\overline{Y_r^2})^{1/2}$ attains the same order of magnitude as the characteristic scale of energy-containing (large) eddies. More recent analyses of extensive laboratory and atmospheric data on puff diffusion, on the other hand, have indicated the existence of a much wider range of diffusion times or cluster sizes over which the accelerative phase of diffusion is observed (see e.g., Gifford, 1977). This has raised some questions about the necessity of restricting the puff size to inertial subrange eddies for the validity of the t^3 law ($\overline{Y_r^2} \propto t^3$). Another derivation by Lin (1960) of a similar dispersion relationship

$$\overline{Y_r^2} = \frac{2}{3} D t^3 \qquad (7.66)$$

in which D is a Lagrangian quantity having the dimension of ϵ, does not explicitly use the assumption of Y_r to be within the inertial subrange. Lin's derivation is based on the consideration of relative accelerations, rather than relative velocities, of particle pairs. It shows that the relative dispersion parameter does not have to lie within the inertial subrange for the t^3 law to be valid. It is not easy, however, to measure or parameterize D in Eq. (7.66). A comparison of Eqs. (7.58b) and (7.66) indicates that $D \propto \epsilon$, so long as Y_r lies within the inertial subrange.

3. For large diffusion times, when the separation between the two particles becomes greater than the largest eddy size and is essentially determined by continuous but random movements of each particle, which become uncorrelated with those of the other particle, the dispersion relations (7.46) through (7.48) have been derived earlier. In particular, Eq. (7.47) with Taylor's theorem (7.9) for large diffusion time gives

$$\overline{Y_r^2}(t) = y_0^2 + 4\overline{v'^2} T_{iL} t \qquad (7.67)$$

The apparent relative diffusivity implied by the above dispersion relation is given by

$$K_{yr} = \frac{1}{2} \frac{d\overline{Y_r^2}}{dt} = 2\overline{v'^2} T_{iL} \qquad (7.68)$$

which is exactly twice the absolute diffusivity and becomes independent of diffusion time and, hence, independent of σ_{yr}. Thus, the constant K-model of puff diffusion might be valid only in this final stage when the distinction between relative and absolute diffusion apparently disappears.

The Fickian diffusion model utterly fails in predicting the earlier stages of relative diffusion corresponding to more rapid growth rates of puff or cluster of particles. The statistical theory of relative diffusion, on the other hand, is found to be more satisfactory in explaining the available data on puff diffusion. The theory leads to different dispersion relations for the three stages of diffusion, which can be summarized as:

$$\overline{Y_r^2}(t) = \begin{cases} y_0^2 + \overline{v'^2_{r0}} t^2, & \text{for } t < t_1 \\ y_0^2 + c_5 r_0^2 + c_2 \epsilon t^3, & \text{for } t_1 \leq t \leq t_2 \\ y_0^2 + 4\overline{v'^2} T_{iL} t, & \text{for } t > t_2 \end{cases} \qquad (7.69)$$

Here, t_1 and t_2 may simply be considered as the lower and upper limits of diffusion or travel time for the applicability of the intermediate, accelerative stage of relative diffusion. The actual boundaries between different stages (initial, intermediate, and final) may not be as sharp and well defined as implied in the above dispersion relations. It is more likely that smooth transitions will occur in dispersion relations in going from one stage to the next in the neighborhood of t_1 and t_2. Such transitions are ignored in the above theoretical predictions of $\overline{Y_r^2}(t)$ according to Eqs. (7.69). In terms of mean-square relative dispersions $\overline{Y_{r1}^2}$ and $\overline{Y_{r2}^2}$ at travel times t_1 and t_2, respectively, the above dispersion relations can also be written in the forms

$$\overline{Y_r^2}(t) = \begin{cases} y_0^2 + \overline{v'^2_{r0}} t^2, & \text{for } t < t_1 \\ \overline{Y_{r1}^2} + c_2 \epsilon (t^3 - t_1^3), & \text{for } t_1 \leq t \leq t_2 \\ \overline{Y_{r2}^2} + 4\overline{v'^2} T_{iL}(t - t_2) & \text{for } t > t_2 \end{cases} \qquad (7.70)$$

Note that in the lower atmosphere (troposphere), t_1 is expected to be of the order of 10 s, and t_2 of the order of an hour, implying an extensive intermediate range of accelerated diffusion (Gifford, 1977).

7.5 PUFF DIFFUSION FROM INSTANTANEOUS RELEASES

Just as the continuous point source is often used as a fundamental source geometry for describing diffusion from other continuous sources, the instantaneous point source can be used as a basis for describing diffusion from other instantaneous sources. In the context of solving the linear diffusion equation, the instantaneous point source is considered to be a more fundamental source geometry, because even a continuous point source can be thought of as a series of instantaneous puffs released in quick succession. Thus, concentration fields due to continuous sources can be derived from the appropriate integrations of concentrations in expanding puffs with time (see e.g., Sutton, 1953; Csanady, 1973; Seinfeld, 1986). One can also handle, in this way, any real sources with variable or intermittent emissions, as well as those with constant emission rates for limited (finite) durations.

Unlike the ideal continuous point source, which can be realized in practice by continuously releasing marked fluid particles, one after another, in a sequence, an idealized instantaneous point source is merely a theoretical concept that cannot strictly be realized in practice. The practical difficulty arises from the fact that only one fluid particle can occupy a point at any given time. Therefore, one cannot release more than one particle instantaneously from the same point. If two or more marked (tracer) particles are released instantaneously as a puff, which will subsequently expand and diffuse in a turbulent flow, those particles must have finite initial separations between them at the time of release. Thus, an instantaneous release must have finite dimensions and should not be characterized as an ideal point, line, or area source. Still, such idealized source geometries are commonly used or implied in the Gaussian puff diffusion model, as well as in other diffusion models based on the solution of mean diffusion equation, mainly because idealized source geometries yield simpler models and solutions. The solutions of the diffusion equation with constant molecular or eddy diffusivities have been given in Chapter 6. Their applicability to puff diffusion problems of the atmospheric environment appears to be very limited.

The results of the statistical theory of relative diffusion suggest that the gradient transport (K) theory is not likely to be valid during a wide range of short and intermediate diffusion times when the puff goes through an accelerative diffusion phase, unless eddy diffusivities are made dependent on growing puff dimensions, in consistency with the results of the statistical theory. Only in the final stage of the reduced puff growth does the Fickian (constant-K) theory become consistent with the statistical theory in homogeneous turbulence. At this stage, the distinction between absolute and relative diffusion theories also disappears. However, the available data on puff diffusion in the troposphere (Gifford, 1977) indicates that it takes several hours after release before puff diffusion may be considered to have reached the final stage of large diffusion times. This is in marked contrast to the case of plume diffusion in which the final stage of plume growth is reached much earlier (a few minutes after release), because there is no intermediate stage of accelerated growth in a continuous source plume. For this reason, empirically estimated plume-diffusion parameters are not likely to be valid and applicable for modeling puff diffusion from instantaneous releases, even under the same environmental conditions. The statistical theory of relative diffusion provides a much better alternative for specifying the puff diffusion parameters for estimating mean concentrations in ensemble-averaged puffs.

No theory can predict, however, the time-dependent instantaneous concentrations in a single puff or cloud. Even the trajectory of the center of an instantaneous puff is theoretically unpredictable due to the random or stochastic nature of turbulence that is superimposed on the average flow. The mean transport velocity can only predict the trajectory of an ensemble-averaged puff. This is another important point of contrast between puff and plume diffusion. In the latter case, the time-averaged concentrations in a fixed (Eulerian) frame of reference are approximately equivalent to the ensemble-averaged concentrations, and the statistical theory provides a sound basis for predicting diffusion parameters and, hence, mean concentrations in continuous source plumes, at least in a quasi-stationary and homogeneous flow. Theoretically predicted relative diffusion parameters, on the other hand, are not directly applicable to any particular instantaneous source or release scenario, but only to an ensemble average of many such releases under the same environmental conditions. Consequently, the predicted concentrations may at best represent only the expected values in a probability sense, which may deviate considerably from the actual instantaneous concentrations in a particular realization.

Practical puff diffusion problems involve accidental or deliberate releases of material to the atmosphere instantaneously, or for short durations. They are single (one-time) releases, rather than ensembles of similar releases under the same environmental conditions. Consequently, the outline of the material puff or cloud, in reality, is quite irregular as it travels along a jagged trajectory under the influence of turbulent eddies. On the other hand, the theory can describe only the ensemble-averaged puff or cloud with a smooth and regular outline and a predictable trajectory (e.g., straight line for short-range diffusion).

In applying the results of the statistical theory of relative diffusion to puff diffusion problems, the following assumptions or approximations are usually made or implied:

1. The mean concentration distribution in the puff is Gaussian in all the directions, so long as the puff remains unaffected by any physical barrier or boundary (e.g., the ground surface, the capping inversion). The Gaussian distribution can be specified in terms of its standard deviations or diffusion parameters σ_x, σ_y, and σ_z.

2. The puff diffusion parameters are directly related (proportional) to the two-particle relative dispersion parameters of the statistical theory, that is,

$$\sigma_x = \left(\frac{1}{2}\overline{X_r^2}\right)^{1/2}; \ \sigma_y = \left(\frac{1}{2}\overline{Y_r^2}\right)^{1/2};$$

$$\sigma_z = \left(\frac{1}{2}Z_r^2\right)^{1/2} \quad (7.71)$$

in which the factor 1/2 is essentially due to the difference in the definitions of absolute and relative displacements of dispersing particles (see also the discussion following Eq. (7.45). Equations (7.71) permit the specification of puff diffusion parameters on the basis of the statistical theory relations described in section 7.4.

The puff-diffusion parameters may differ considerably from the plume-diffusion parameters that are given by the statistical theory of one-particle (absolute) diffusion. This important distinction is often ignored in regulatory applications of Gaussian dispersion models when specific guidance on the specification of puff-diffusion parameters is lacking. The constant K or Fickian diffusion theory also fails to make any distinction between absolute and relative diffusion. However, the statistical theory clearly shows that the apparent eddy diffusivities, when related to the particle dispersion parameters as in Eq. (7.36), must be functions of diffusion time, particularly in the range of short and intermediate diffusion times.

7.6 FLUCTUATING PLUME MODELS

It has been pointed out that the statistical theory of relative diffusion is not only applicable to puff diffusion from instantaneous releases, but also to plume diffusion on instantaneous or a short-term basis (Gifford, 1959a; Csanady, 1973; Pasquill and Smith, 1983). Instantaneously, a continuous source plume has the appearance of a meandering or fluctuating plume in which plume meanders in the lateral and vertical directions are caused by the spatial distribution of large eddies in the flow. Instantaneous or short-term-averaged concentrations in meandering plumes far exceed the long-term-averaged concentrations in the steady plume at comparable distances from the source. This is because the instantaneous plume is relatively thin compared with the average plume and its dimensions increase as the averaging time increases. The center of gravity of the instantaneous plume meanders or fluctuates around the axis of the steady or long-term-averaged plume in an irregular manner. The Gaussian plume diffusion model with dispersion parameters given by the statistical theory of absolute diffusion can reasonably describe concentration distribution only in the steady plume. For the concentration distribution in an instantaneous or a short-term-averaged plume, meandering or fluctuating plume models have been proposed in the literature.

Gifford (1959a) proposed a fluctuating plume model based on the hypothesis that the absolute or total plume dispersion can be separated into two components, the spreading component that represents the relative dispersion about the center of gravity or local axis of the instantaneous plume, and the meandering component representing the variance of the displacement of the local (fluctuating) plume axis from the fixed axis of the steady plume. Mathematically, this can be expressed as

$$\sigma_y^2 = \sigma_{yi}^2 + \sigma_{ym}^2$$
$$\sigma_z^2 = \sigma_{zi}^2 + \sigma_{zm}^2 \quad (7.72)$$

where the subscript i denotes instantaneous or short-term-averaged plume and m denotes the meandering component. Our notation also implies the usual relations between the meandering component of dispersion and the mean-square displacement of the fluctuating plume axis, that is,

$$\sigma_{ym}^2 = \overline{y_m^2}; \quad \sigma_{zm}^2 = \overline{z_m^2}$$

An elegant proof of Gifford's hypothesis, and derivation of Eq. (7.72), is given by Csanady (1973). It implies that absolute dispersion is the sum of relative dispersion and plume meandering. The total (absolute) dispersion in a steady plume from a continuous source in a homogeneous flow is well described by Taylor's one-particle dispersion theory, while the concentration distribution in the plume is given by the Gaussian formula

$$\bar{c} = \frac{Q}{2\pi \bar{u} \sigma_y \sigma_z} \exp\left[-\frac{y^2}{2\sigma_y^2} - \frac{z^2}{2\sigma_z^2}\right] \quad (7.73)$$

For the fluctuating plume, Gifford (1959) argued that it is reasonable to assume that the concentration distribution in the instantaneous plume is also Gaussian, at least in the average sense, that is,

$$c = \frac{Q}{2\pi \bar{u} \sigma_{yi} \sigma_{zi}} \exp\left[-\frac{(y - y_m)^2}{2\sigma_{yi}^2} - \frac{(z - z_m)^2}{2\sigma_{zi}^2}\right] \quad (7.74)$$

This implies that instantaneous concentration in the fluctuating plume is a random function of time, as a consequence of the random variability of y_m and z_m. The possible effects of the variability of concentration distribution have been ignored. For the sake of simplicity, y_m and z_m are also assumed to have Gaussian probability density functions, that is,

$$p(y_m) = \frac{1}{(2\pi)^{1/2} \sigma_{ym}} \exp\left(-\frac{y_m^2}{2\sigma_{ym}^2}\right)$$

$$p(z_m) = \frac{1}{(2\pi)^{1/2} \sigma_{zm}} \exp\left(-\frac{z_m^2}{2\sigma_{zm}^2}\right) \quad (7.75)$$

Note that the average concentration in the steady plume is related to the instantaneous concentration in the fluctuating plume as

$$\bar{c} = \int_{-\infty}^{\infty} \int_{-\infty}^{\infty} c\, p(y_m) p(z_m) dy_m\, dz_m \quad (7.76)$$

which is indeed satisfied by Eqs. (7.72) through (7.75).

An obvious practical application of the above fluctuating plume model is to estimate the ratio of the maximum centerline concentrations in fluctuating and steady plumes. From Eqs. (7.73) and (7.74) we have

$$\frac{c}{\bar{c}} = \frac{\sigma_y \sigma_z}{\sigma_{yi} \sigma_{zi}}$$

$$= \left(1 + \frac{\sigma_{ym}^2}{\sigma_{yi}^2}\right)^{1/2} \left(1 + \frac{\sigma_{zm}^2}{\sigma_{zi}^2}\right)^{1/2} \quad (7.77)$$

which is often referred to as the ratio of peak to average concentrations. This ratio depends on the ratios of long-term-averaged and instantaneous plume dispersion parameters. Recognizing that σ_y and σ_z are given by the statistical theory of absolute diffusion, while σ_{yi} and σ_{zi} are given by the relative diffusion theory, the ratio $c/\bar{c}$ is expected to vary in inverse proportion to the travel time or distance from the source, particularly during the intermediate stage of relative diffusion when $\sigma_y\sigma_z \propto t^2$ and $\sigma_{yi}\sigma_{zi} \propto t^3$. At large travel times or distances, however, the ratio of peak to average concentration should become a constant, because absolute and relative diffusion parameters have similar variations with diffusion time. Gifford (1959a) presented some plume diffusion data in support of his model predictions. The ratio of peak to average concentrations may vary from several hundred for very small travel times to unity for very large travel times (Pasquill and Smith, 1983).

7.7 EXPERIMENTAL VERIFICATION OF STATISTICAL THEORIES

7.7.1 Lagrangian Particle Trajectory and Diffusion Measurements

For a direct experimental verification of the basic results of the statistical theories of absolute and relative diffusion, one needs to make Lagrangian measurements of displacements or trajectories of neutrally buoyant tagged particles in stationary and homogeneous flows. Such measurements, involving hundreds or, preferably, thousands of tagged particles or particle pairs to get ensemble-averaged statistics of turbulence and diffusion that appear in theoretical relations, have not been made, even in controlled laboratory conditions.

Some Lagrangian measurements have been made in the atmosphere using balloon and tetroons (Gifford, 1953, 1955, 1977; Angell, 1964; Islitzer and Slade, 1968; Hanna, 1981). Before 1960, no-lift balloons having the same density as that of air at the equilibrium or release height were used. Later, tetrahedron-shaped constant-level or constant-volume balloons, called tetroons, have more frequently been used for estimating horizontal trajectories and diffusion, particularly at long range. Tetroons are best followed or positioned by radars, either through the use of passive reflectors or radio transponders attached to the tetroons. With one or two doppler radars, tetroon positions can be obtained at intervals as small as 0.1 s over distances of well over 100 km. The more conventional and cheaper method of tracking balloons with theodolites is limited to larger sampling intervals of the order of 10 s and shorter distances of a few kilometers. With radars using fast sampling rates, Lagrangian velocity measurements can be made that include contributions from frequencies extending well into the inertial subrange. Such measurements, in conjunction with fixed point (Eulerian) turbulence measurements, have been used to obtain useful empirical relationships between Lagrangian and Eulerian statistics, such as autocorrelations, spectra, and integral scales (Gifford, 1955; Angell, 1964; Hanna, 1981).

The most frequent use of constant-density balloons and tetroons in the atmosphere has been for estimating the mean transport winds along the observed trajectories of balloons or tetroons in variable or complex terrains (Islitzer and Slade, 1968). For this purpose, measurements of positions at a few minutes to hourly intervals along each trajectory should be adequate. Also, balloons can be released at small time intervals when mean transport velocities are desired.

Only in a few cases, balloon or tetroon measurements have been made for estimating atmospheric diffusion (see e.g., Fig. 7.6). The most serious limitation of these estimates is the small number (3–10) of tetroons released over a long duration (8–48 hr). Such a small number of realizations extending over the course of a day or longer, during which time atmospheric conditions may change substantially due to diurnal variations, is not sufficient to give reliable estimates of the mean-square displacement or the lateral dispersion parameter σ_y. In Figure 7.6 the estimated values of σ_y reflect not only turbulent diffusion, but also the effects of changes in the mean wind speed and direction with time. Such measurements are, thus, not very suitable for verifying the statistical theory predictions. Nevertheless, the estimated values of the lateral dispersion parameter for most of these experiments are consistent with the statistical theory prediction, $\sigma_y \propto x$, up to distances of several kilometers. At larger distances, the rate of spread is reduced, but not to the extent predicted by $\sigma_y \propto x^{1/2}$.

Estimates of relative diffusion have also been made from observations of tetroon pairs, smoke puffs, tracer clouds, balloon clusters, and radioactive clouds. Observations of relative diffusion have mostly relied on visual and photographic methods. Earliest diffusion experiments and observations have been discussed by Roberts (1923), Richardson (1926), and Sutton (1932, 1953). Richardson derived his famous law of relative diffusion empirically from observations of puff spreading in the atmosphere over a very large range of scales. The implied rate of spread, $\sigma_y \propto t^{3/2}$, is found to be consistent with that predicted by the relative diffusion theory (Batchelor, 1949, 1952). Later diffusion data have been critically examined and reviewed by Gifford (1957a, 1957b, 1977), among others. In particular, the tropospheric data on cloud spread as a function of travel time is shown in Figure 7.7. Further details of the original sources of these and other diffusion data are given by Gifford (1977), Pasquill and Smith (1983), and Draxler (1984).

It has been observed that, for puffs and instantaneous plumes (see Fig. 7.8) diffusing in the PBL and the troposphere, there is an initial linear growth regime ($\sigma \propto t$) for travel times up to 10 to 50 s, which is followed by the more extensive accelerated growth regime ($\sigma \propto t^{3/2}$) for travel times extending up to 3000 s. This is followed by a slower growth regime, characteristic of mesoscale diffusion, which might also be indicative of the transition to the still slower $t^{1/2}$-spreading regime predicted by the statistical theory. Other experimental data confirming the asymptotic puff-growth regime ($\sigma \propto t^{1/2}$) for large travel times ($t > 1$ hr), typical of long-range transport and diffusion, have been reviewed by Hage (1964) and Heffter (1965).

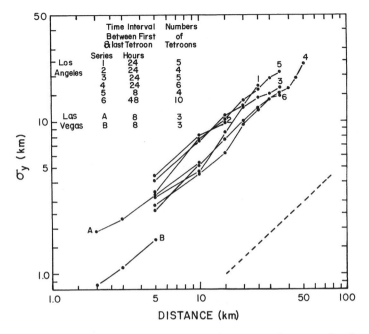

Figure 7.6 Estimates of lateral dispersion parameter from successive tetroon releases as a function of downwind distance from the release point. Dashed line gives typical variation of σ_y with x obtained from 1-hr tracer diffusion experiments. After Islitzer and Slade, 1968.

Figure 7.8, based on an analysis of the photographs of instantaneous and time-averaged smoke plumes, shows a clear distinction between the observed dispersions of these different types of plumes. The strong evidence for the accelerated diffusion regime in the instantaneous-plume dispersion further confirms the relative diffusion theory.

One may think that more extensive and systematic Lagrangian measurements of lateral and vertical diffusion should be possible in well-controlled laboratory experiments in wind tunnels and water channels. Two major difficulties have been encountered, however, in attempts to make such measurements. First, there is a practical problem of producing tracer particles or droplets of appropriate size, shape, and density that would faithfully follow the turbulent fluid motion with negligible gravitational settling and fluid drag (Merzkirch, 1974, 1987). Another, even more serious, problem is to track

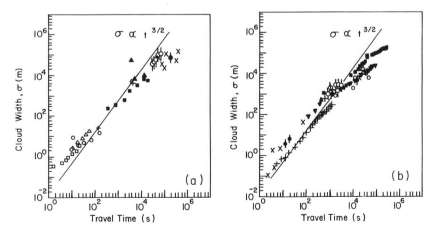

Figure 7.7 Tropospheric relative diffusion data showing cloud width as a function of travel time: (a) Crawford's (1966) data. (b) Other data. After Gifford, 1977.

172 STATISTICAL THEORIES OF DIFFUSION

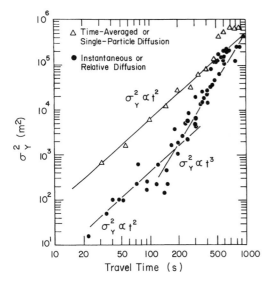

Figure 7.8 Mean-square dispersion as a function of travel time for instantaneous and time-averaged plumes. After Nappo, 1981.

the particles as they follow the highly irregular and tortuous paths in turbulent flows. With the rapid development of efficient particle imaging techniques in recent years (see e.g., Adrian, 1991), Lagrangian measurements of turbulence and diffusion in laboratory facilities have become practically more feasible.

Using neutrally buoyant 0.5-mm-diameter spheres in an open channel flow, Sullivan (1971) used a photographic technique to determine the particle motions at small (0.07 s) time intervals and obtained the Lagrangian autocorrelation functions of three velocity components. Snyder and Lumley (1971) have reported similar measurements in grid turbulence. Extensive measurements of particle motion and diffusion in a simulated convective mixed layer have been reported by Deardorff and Willis (1974, 1975, 1982) and Willis and Deardorff (1976, 1978, 1981). These have been used to estimate the dimensionless mean particle height ($\bar{Z}/h$), the standard deviations of lateral and vertical spread (σ_y/h, σ_z/h), mean concentration distributions, and probability distributions of particle displacements as functions of the dimensionless travel time or distance from the source (X_*) and the release height (H/h).

The pioneering experiments by Deardorff and Willis were performed in a water tank of horizontal dimensions 122 cm by 114 cm in which mixed-layer depths of up to 60 cm were produced in a stably stratified environment (water) by heating from below at a constant heating rate. Diffusion in the simulated convectively mixed layer was studied by releasing a large number of neutrally buoyant oil droplets at different heights along a horizontal line extending the length of the tank. Droplet positions were photographed at 2 to 10-s intervals until turbulent diffusion spread a significant

number of particles to the vicinity of side walls. Although particles were released instantaneously in a line parallel to the x direction, an instantaneous view (photograph) of particle distribution at some time t after release, in the y–z plane integrated along the line of sight x axis, can be considered analogous to that of the y–z cross section of a continuous source plume at a distance $x = \bar{u}t$ from the source. This transformation from an instantaneous line source to a simulated continuous point source (CPS) assumes that Taylor's frozen-turbulence hypothesis is valid. This assumption implies that diffusion in the longitudinal direction from the CPS is negligible in comparison with mean advection. The limitations on the mean velocity for the applicability of this approach has been estimated to be (Deardorff and Willis, 1974)

$$1.5 W_* \leq \bar{u} \leq 6 W_* \qquad (7.78)$$

in which the lower limit is due to the breakdown of Taylor's hypothesis in too weak a mean wind and the upper limit is to ensure that the convective turbulence dominates over the mechanical or shear-generated turbulence.

Figure 7.9, based on Willis and Deardorff's (1981) experiments, confirms the existence of the linear growth regime for both the lateral and vertical dispersion parameters. The asymptotic growth in proportion to $t^{1/2}$ or $x^{1/2}$ for large diffusion times is more clearly indicated by the lateral dispersion parameter, while the vertical dispersion parameter attains an equilibrium value, depending on the release height, due to the upper limit of spread imposed by the inversion at the top of the mixed layer.

7.7.2 Eulerian Measurements

Most of the experimental diffusion data is based on the mean concentration measurements in an Eulerian framework, either with instantaneous releases of material or with continuous releases for limited durations. Most often, such measurements in the field are made using only ground-level samplers. A few vertical profiles of mean concentration have also been measured from instruments mounted on towers at relatively short distances from sources. Thus, while direct estimates of σ_y at or close to the ground surface are quite abundant, there are only limited data available on concentrations and σ_y at different heights above the surface. Comprehensive reviews of atmospheric diffusion experiments are given by Islitzer and Slade (1968), Pasquill and Smith (1983), and Draxler (1984). Only a few of the experimental data or studies that are particularly suitable for the validation of the statistical theories of diffusion will be discussed here.

7.7.3 Cross-wind Spread from Continuous Point Sources

Diffusion of passive tracers from continuous point sources over flat and homogeneous terrains under sta-

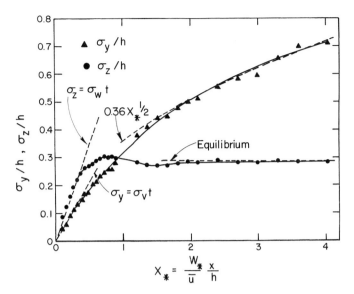

Figure 7.9 Normalized standard deviations of lateral and vertical plume spread as functions of the dimensionless travel time or distance in the simulated convective boundary layer in a water tank heated from below. After Willis and Deardorff, 1981.

tionary flow conditions prevailing during the time of release and sampling can be considered to be essentially homogeneous and stationary, provided the sampling and release times are much larger than the integral time scale of large eddies in the flow. For the verification of statistical theory predictions, diffusion experiments should conform to the above theoretical assumptions. At the same time, accurate measurements of turbulence should be made, so that turbulence intensities and scales appearing in theoretical relations can be determined. Very few atmospheric diffusion experiments can satisfy all the assumptions implied in theory, as well as the requirements for experimental verification of theory.

The direct correspondence or proportionality between the lateral plume spread (e.g., the plume width based on 10 percent of peak concentration) and turbulence intensity was first demonstrated by early (1923–25) smoke diffusion experiments at Porton, England. More extensive turbulence and diffusion data from experiments carried out during 1954–57 at Round Hill, Massachusetts, and at O'Neil, Nebraska, during Project Prairie Grass (PPG) further confirmed that relationship, at least at short distances of up to 800 m from near surface sources. Figure 7.10 shows the linear dependence of plume width on standard deviation of wind direction (σ_θ) at 100 m downwind of a point source. Both the turbulence and diffusion data refer to a sampling time of about 600 s. Note that this short-range diffusion relationship does not depend on the surface roughness, which varied by more than an order of magnitude between the two sites, or on atmospheric stability.

Next, we consider the lateral diffusion data from PPG and several other classical diffusion experiments, in which both the surface and elevated sources were used and surface concentration measurements extended to distances of several tens of kilometers. Islitzer and Slade (1968) plotted the data in the form of σ_y/σ_θ versus x, which is reproduced here in Figure 7.11. The data confirm again the statistical theory prediction, $\sigma_y = \sigma_\theta x$, for the short-range diffusion, extending to several hundred meters. There is slight flattening of the slope of σ_y/σ_θ versus x at larger distances,

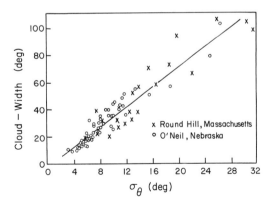

Figure 7.10 Relationship between angular plume width and the standard deviation of wind direction at 100 m downwind of a continuous point source at two different sites. After Cramer, 1957.

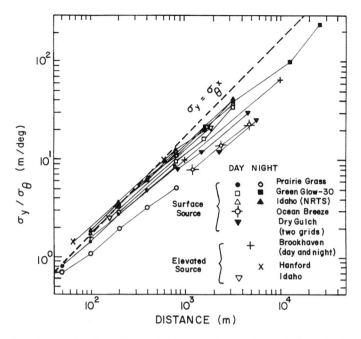

Figure 7.11 The ratio σ_y/σ_θ as a function of downwind distance from surface and elevated releases at different sites. After Islitzer and Slade, 1968.

but it does not quite lead to the theoretically predicted asymptotic behavior for the long-range diffusion, at least within the limited range of these diffusion experiments.

It has been argued that the effect of wind shear, in both speed and direction, in the lower part of the atmospheric boundary layer is to enhance the lateral diffusion as the distance from the source increases and plume spreads in the vertical. This shear-enhanced diffusion apparently delays the much slower growth of σ_y with x, as predicted by the statistical theory relation (7.25) for long-range diffusion. For the same reason, the simpler relation (7.24) for short-range diffusion is found to be approximately valid for distances extending to several kilometers.

7.7.4 Vertical Spread from Continuous Sources

The vertical diffusion from near-surface and elevated sources in the PBL cannot strictly be described in terms of the classical statistical theory, because the basic assumption of homogeneity of mean flow and turbulence implied in theory is not generally valid in the PBL. On the contrary, observations show that mean wind speed, vertical turbulence intensity, and scales of turbulence vary with height in the PBL, especially in the surface layer. Still, measurements of the vertical plume spread (σ_z) in field experiments as well as in laboratory experiments are found to be mostly consistent with the statistical theory relations (predictions) for small travel times or distances from the source (see e.g., Fig. 7.9).

The best agreement between the theory and experiments is found for the elevated plumes diffusing in approximately homogeneous layers, such as the convective mixed layer, the middle and upper parts of the near-neutral boundary layer, and possibly shallow turbulent layers within stably stratified flows to which plumes might be confined. The initial phase of the lin-

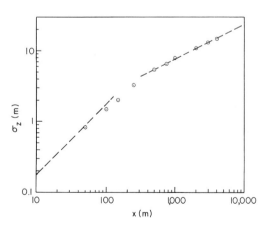

Figure 7.12 Vertical spread (σ_z) as a function of distance (x) from an elevated source ($H = 50$ m) in a very stable atmospheric boundary layer at Agesta, Sweden. The dashed lines have slopes of unity and one-half corresponding to the short-time and long-time limits of the statistical theory. After Hogstrom, 1964).

ear plume spread in the vertical is observed in all turbulent flows (see Figs. 7.9–7.12). An extensive asymptotic growth regime with $\sigma_z \propto x^{1/2}$ is observed only under certain conditions when a plume can travel long distances (far beyond the large-eddy length scale, L_x) without being influenced by the surface or an elevated inversion. This is more likely to occur in a stably stratified boundary layer in which turbulence is weak and the vertical plume spread is slow, as shown in Figure 7.12. The vertical diffusion under convective conditions, on the other hand, is so rapid that the plume reaches the ground surface as well as the top of the CBL and, hence, becomes trapped between the surface below and the inversion above before the predicted asymptotic slow-growth regime ($\sigma_z \propto x^{1/2}$) is reached.

7.8 APPLICATIONS TO ATMOSPHERIC DISPERSION AND LIMITATIONS

7.8.1 Plume Diffusion

The applicability of the statistical theory of one-particle (absolute) dispersion to plume diffusion from a continuous point source in homogeneous flow has already been discussed in section 7.3. Its application to modeling dispersion from other continuous sources (e.g., line, area, and volume) utilizes the principle of superposition. It has also been pointed out that the particle dispersion parameters σ_y and σ_z are identical to the standard deviations of mean concentration distributions across the plume in y and z directions, respectively. Thus, the statistical theory relations for σ_y and σ_z provide a sound theoretical basis for specifying the plume-dispersion parameters, particularly in simpler Gaussian models.

The original theoretical relations contain Lagrangian integral time scales or the corresponding large-eddy length scales, which must be parameterized in terms of more easily measured PBL height and velocity scales (see e.g., section 5.10). Alternatively, modified relations containing only the Eulerian statistics can be used.

Most of the proposed empirical formulas (Briggs, 1973; Cramer, 1976; Draxler, 1976) or graphical representations of σ_y and σ_z (e.g., the Pasquill–Gifford sigma scheme) for use in the Gaussian plume-diffusion model incorporate the statistical theory results to varying degrees. For example, the widely recommended formulas for rural and urban areas by Briggs (1973) were based on not only the empirical data from several diffusion experiments but also the statistical theory predictions for small and large diffusion times. Some allowance was made for the enhanced vertical dispersion under unstable conditions due to the effects of wind shear and positive buoyancy, and suppressed vertical diffusion by negative buoyancy under stable conditions. Different lateral and vertical turbulence intensities are implied for each of the Pasquill's stability/turbulence classes. But, in the atmosphere, turbulence intensities are also found to strongly depend on the terrain roughness, and specifying one value for the rural terrain and another for the urban area may not be entirely satisfactory.

Dispersion formulas explicitly containing turbulence intensities or standard deviations of velocity or direction fluctuations are more directly based on the statistical theory with some input from the experimental diffusion data (Cramer, 1976; Draxler, 1976, 1984). These are not dependent on terrain roughness and atmospheric stability and, therefore, have more general validity, provided the terrain is homogeneous. However, their use in Gaussian dispersion models requires measurements or estimates of turbulence quantities involved in dispersion formulas.

7.8.2 Puff Diffusion

Applications of the statistical theory of relative diffusion to atmospheric dispersion problems have been hampered by the lack of information on turbulence quantities involved in relative dispersion relations (7.69) or (7.70). There are also large uncertainties in the estimated values of empirical constants and the lower and upper limits of travel time for the applicability of the intermediate, accelerative regime of relative diffusion. Systematic puff diffusion experiments are needed to estimate more accurately the relative diffusion parameters during different regimes of small, intermediate, and large travel times. More accurate measurements of relative particle velocities are also needed for determining their variance and the autocorrelation function and for verifying the relative diffusion relations involving the same.

The most practical application of the relative diffusion theory has been in the development of fluctuating plume models. But their use in practical problems of estimating instantaneous or short-term concentration fluctuations is again hampered by the lack of an accurate parameterization for the relative diffusion parameters.

7.8.3 Limitations

Limitations of the statistical theories largely arise from the simplifying assumption made to derive the various kinematic and similarity relations. For example, the basic assumptions of uniform mean winds, homogeneous turbulence, and stationarity cannot be strictly satisfied in the atmospheric boundary layer. The mixed layer of the CBL probably comes closest to meeting these requirements, but the observed asymmetry in convective updrafts and downdrafts (skewed distribution of vertical velocity) causes significant deviations in the vertical concentration profiles from the expected Gaussian distribution in homogeneous flows. The vertical inhomogeneity of turbulence and mean flow shear are characteristic features of the atmospheric surface layer, in which the vertical dispersion may not be well described by the classical statistical theory. Later extensions of the theory to include the effects of wind shear and inhomogeneity of turbulence have been proposed using the Lagrangian similarity theory relations, which will be discussed in the next chapter.

The effects of moderate wind shear and vertical inhomogeneities associated with the near-neutral PBL are often ignored and the statistical theory can be applied to modeling vertical dispersion from elevated sources. Under moderately and strongly stratified conditions generally prevailing at night, however, both the lateral and vertical dispersions are enhanced by the stronger wind shears present in the SBL. The shear-induced dispersion can be added to the turbulent diffusion, which is adequately described by the statistical theory.

A limitation of the statistical theory arising from the assumption of homogeneity is that the effects of terrain inhomogeneities, such as a sudden change in surface roughness, temperature, and wetness; isolated hills and buildings; and more complex terrain features cannot be considered in theoretical dispersion relations. The theory is only applicable to relatively homogeneous terrains.

Other limitations arise from the turbulence parameters contained in theoretical dispersion relations. As a minimum, the standard deviations of velocity fluctuations must be measured or otherwise estimated. The Lagrangian time scales or the corresponding large-eddy length scales, appropriate for describing lateral and vertical dispersion, must also be estimated. The relative dispersion relations contain ever more complex Lagrangian statistics that are not generally available on a routine basis.

PROBLEMS AND EXERCISES

1. a. Derive Taylor's relation (theorem) for the mean-square displacement of particles as a function of travel time in homogeneous turbulence.
   b. Considering the properties of the Lagrangian autocorrelation function, obtain simplified expressions for the mean-square displacement for small and large travel times.
2. a. Starting from Taylor's theorem, derive the alternative forms of the particle dispersion relations (7.17) and (7.18). Use the latter relation to compute and plot the weighted spectra as functions of the dimensionless frequency nT_{iL} at different dimensionless travel times (say, $t/T_{iL} = 0, 1, 2, 4,$ and 8). You may use the usual form of the Lagrangian velocity spectrum, which corresponds to an exponential correlation function.
   b. Discuss the relative roles of different frequency components (eddies) of turbulence on particle dispersion as a function of travel time.
3. Using Taylor's theorem or an equivalent dispersion relation, obtain the expressions for the mean-square dispersion $\overline{Y^2}(t)$ that correspond to the following forms of Lagrangian autocorrelation function:

   a. $R_L(\xi) = \exp\left(-\dfrac{\xi}{T_{iL}}\right),$ for $\xi \geq 0$

   b. $R_L(\xi) = 1 - \dfrac{\xi}{2T_{iL}},$ for $\xi \leq 2T_{iL}$
   $\quad\quad\quad\quad = 0,$ for $\xi > 2T_{iL}$

   c. $R_L(\xi) = 1,$ for $\xi \leq T_{iL}$
   $\quad\quad\quad\quad = 0,$ for $\xi > T_{iL}$

   Compare the resulting dispersion relations by plotting $\sigma_Y/\sigma_v T_{iL}$ versus t/T_{iL} on a log-log graph and verify the relative insensitivity of dispersion to the shape of $R_L(\xi)$.
4. a. State the basic assumptions implied in the Hay–Pasquill approach to modeling diffusion from continuous sources.
   b. Discuss the merits and limitations of the Hay–Pasquill model for atmospheric dispersion.
5. A simple interpolation form of the dispersion relation, which is consistent with Taylor's expressions for small and large travel times, is

   $$\sigma_y = \sigma_v t [1 + 0.5(t/T_{iL})]^{-1/2}$$

   a. Compare it with that resulting from the exponential form of the autocorrelation function.
   b. Derive an expression for the apparent eddy diffusivity that is consistent with the above interpolation formula.
   c. Rewrite the above dispersion relation in the dimensionless form, using the mixed-layer similarity scaling for the CBL, and suggest further simplification of the same.
6. For modeling the near-source dispersion from an elevated continuous point source in the CBL, the plume dispersion parameters can be specified as $\sigma_y = \sigma_z = axW_*/\overline{u}$, where $a \simeq 0.6$.
   a. Using the Gaussian plume-dispersion model, obtain an expression for the dimensionless g.l.c. $\overline{c}_{*0} = \overline{c}_0 h^2 W_*/Q$, using the mixed-layer similarity scaling.
   b. Show that the maximum g.l.c. and its downwind distance from the source are given by

   $$\overline{c}_{*0\max} = 0.234 \, \frac{W_*}{\overline{u}} \left(\frac{h}{H}\right)^2; \quad x_{*\max} = 1.18 \, \frac{\overline{u}\,H}{W_* h}$$

   c. Discuss the practical implications of the above results, as well as the limitations of the model on which they are based.
7. At large enough distances from the source, the vertical dispersion is limited by the presence of the inversion at the top of the mixed layer, so that σ_z/h approaches a constant equilibrium value (see Fig. 7.9). The lateral dispersion also slows down to the theoretical prediction for large diffusion times. The following empirical relations can be used to specify the plume dispersion parameters in the CBL:

   $$\sigma_y/h = 0.6 X_*(1 + 2X_*)^{-1/2}$$
   $$\sigma_z/h = 0.6 X_*(1 + 5X_*^2)^{-1/2}$$

   a. Using the Gaussian plume-dispersion model, calculate and plot the dimensionless g.l.c. $\overline{c}_{*0} = \overline{c}_0 h^2 W_*/Q$ as a function of the dimensionless distance $X_* = xW_*/h\overline{u}$, from the source for different dimensionless source heights (say, $H/h = 0.1$, 0.3, and 0.5)

b. What conclusions can you draw from the results of your computations?

c. Compare the computed values of $\bar{c}_{*0max}$ and x_{*max} with those obtained from the simpler expressions in Prob. 6(b).

8. a. Consider the two-particle relative diffusion in the direction of mean flow and derive an expression for the mean-square relative dispersion $\overline{X_r^2}$ in homogeneous turbulence.

b. Point out the basic differences between the one-particle and two-particle dispersion relations.

9. a. The relative dispersion relation for small travel times contains the variance of relative particle velocity at the time of release. How would you estimate it in the absence of Lagrangian particle velocity measurements?

b. Discuss the importance of initial separation between the two particles in their subsequent relative dispersion for short times.

10. a. Discuss the similarities and differences between the one-particle and two-particle dispersion relations for large travel times.

b. Under what conditions can one ignore the initial separation between the particles, while considering their relative dispersion at large travel times?

11. a. Discuss Batchelor's (1950) similarity hypothesis and derive the implied relative dispersion relation for intermediate travel times.

b. Out of the several alternative expressions of the relative dispersion for intermediate travel times, which one is the most suitable for practical applications? Give reasons for your answer.

c. Discuss the implications of the intermediate stage of accelerated diffusion to puff diffusion in the atmosphere and show why the gradient transport theory may not be applicable to the same.

12. a. Discuss the limitations of the relative dispersion relations (7.69) or (7.70) in modeling puff diffusion from instantaneous releases in the atmosphere.

b. What are the basic assumptions implied in the statistical theory of relative dispersion?

13. a. How would you distinguish between an instantaneous, a short-time-averaged, and a long-time-averaged plume?

b. Discuss the applicability of the statistical theories of absolute and relative diffusion to plume diffusion from a continuous point source.

14. a. What are the limitations of the fluctuating plume model proposed by Gifford (1959)?

b. Discuss the application of the fluctuating plume model to estimate the ratio of the maximum concentration in the instantaneous plume to that in the long-time-averaged plume.

15. a. Discuss some of the difficulties encountered in making Lagrangian measurements of turbulence and dispersion for direct verification of the statistical theories of diffusion.

b. What are the major limitations of the statistical theories for describing atmospheric dispersion?

8

Similarity Theories of Dispersion

8.1 DISPERSION IN STRATIFIED SHEAR FLOWS

It has been shown in Chapter 7 that turbulent diffusion in uniform flows with homogeneous turbulence is well described by statistical theories of diffusion. Theoretical limitations arise, however, in the presence of significant wind shears and stratification, which give rise to vertical inhomogeneities in the flow. Mean wind shears and thermal stratification commonly occur in the PBL, particularly in the surface layer, due to the presence of the dynamically and thermodynamically active surface. Surface friction and heating or cooling have strong influences on the mean flow and turbulence structure of the PBL. Consequently, the dispersion of contaminants from near-surface sources is also strongly influenced by the above factors.

Both the magnitude and the direction of mean wind change with height. We first discuss, in qualitative terms, the possible effects of both the wind speed shear and the wind direction shear on vertical and horizontal dispersion in the PBL. More quantitative description will be given later, following the development of Lagrangian similarity theories.

8.1.1 Effects of Wind-speed Shear

Csanady (1973) has given a particularly simple schematic illustration (Fig. 8.1) and qualitative description of the effects of mean flow shear and the typical linear variation of the large-eddy length scale on plume dispersion in a neutral surface layer. In this layer, both mean wind speed and large-eddy length increase with height above the surface. How do these affect the mean concentration distribution in a continuous point or line source plume in comparison with that in a homogeneous flow?

Referring to the schematic of Figure 8.1 showing an elevated plume in the surface layer, the mean vertical concentration distribution at a very short distance from the source may be expected to remain nearly Gaussian, because the inhomogeneities of wind and eddy size may not have any significant effect over the relatively thin plume developing over a short distance from the surface. Farther downstream, however, particles above the release height are dispersed by larger eddies than those below the release height. Consequently, the particles aloft are stretched out much farther from the initial plume axis than the particles diffusing downward below the plume axis. Also, particles at greater heights are transported at higher velocities and therefore are stretched out farther in the mean flow direction than those at lesser heights. The combined influence of both the factors is to reduce mean concentrations just above the source height relative to those at or just below the release height. At the same time, the plume is more spread out above the release height H than below H (see Fig. 8.1) and becomes more and more asymmetrical with increasing distance from the source. This upward stretching of the plume moves the center of gravity or mass of the plume upward with increasing distance from the source. This explains why the mean vertical velocity of particles is finite and positive in the near-surface shear layer, compared with the zero mean particle velocity in a homogeneous flow with no shear.

Another effect of increasing eddy size with height is that a puff or a plume released from a near-surface source comes under the influence of larger and larger eddies as it grows in size. This is similar to the process described by the statistical theories of absolute and relative diffusion in homogeneous flows. We have seen that such a process always leads to a regime of accelerated diffusion in the sense that the effective appar-

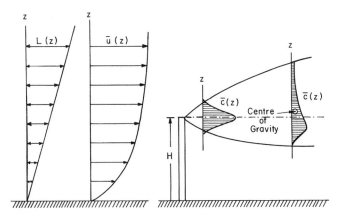

Figure 8.1 Schematic illustration of the effects of increase in wind speed and the large-eddy length scale with height on plume dispersion. From Csanady, 1973.

ent diffusivity increases with travel time or distance from the source. Thus, one might expect the short-term linear growth regime of a continuous source plume in homogeneous flow to be further extended due to the inhomogeneities of mean flow and turbulence in the lower part of the PBL. There is considerable observational evidence for the extended regime of linear plume growth in the atmospheric boundary layer.

8.1.2 Effect of Wind-direction Shear

Above the surface layer, wind direction often changes with height due to the combined effects of frictional veering and backing or veering of geostrophic winds with height. Wind-direction changes of 10° to 50° commonly occur across the PBL depth, with larger values occurring under very stable conditions. Wind-direction shears are not considered to be very significant in the daytime unstable and convective boundary layers, except, perhaps, within the shallow transition layer capping the convectively mixed layer. Large wind-direction shears are commonly observed, however, in the stable atmospheric boundary layer, due to strong wind veering over the comparatively shallow PBL depth.

It is a well-known phenomenon that variation of wind direction with height results in the enhanced lateral dispersion over and above that due to turbulent diffusion. A simple physical explanation for this can be given in terms of the diffusing material at different heights being transported by mean winds in different directions. The lateral plume spread due to wind-direction shear is expected to be directly proportional to the change in mean wind direction across the vertical depth of the plume. This is indeed the basis for the empirical correction to the lateral plume spread proposed by Pasquill (1978) and used in simple Gaussian models for regulatory applications. The correction amounts to adding a shear contribution $0.03(\Delta\bar{\theta})^2 x^2$, where $\Delta\bar{\theta}$ (radians) is the total change of mean wind direction across the whole depth of the plume, to the estimated or parameterized value of σ_y^2 without the shear effect.

8.1.3 Effects of Stratification

Stratification of the PBL due to the diurnal heating and cooling cycle and the energy exchange processes occurring near the earth's surface has profound effects on its mean wind and turbulence structure. Consequently, the diffusion and dispersion processes in the PBL are also strongly influenced by stratification. It is well known that an unstable stratification generally leads to decreased wind shears and enhanced turbulence and diffusion. Stable stratification, on the other hand, leads to strong wind shears and much reduced turbulence and diffusion. It is no wonder that atmospheric stability is frequently the most important characteristic utilized for modeling and parameterization of dispersion in the PBL. For example, in the Gaussian plume-diffusion model with an empirical dispersion scheme, dispersion parameters are specified on the basis of stability or turbulence characterization.

The above mentioned effects of mean wind shear and stratification on dispersion in the PBL can be more quantitatively described with the aid of statistical theories, Lagrangian similarity theories, gradient-transport theories, higher order closure models, or large-eddy simulations. Here we discuss mainly the formulation and development of Lagrangian similarity theories of dispersion and how these theories can be used in conjunction with other theories and models to express mean concentration fields in dispersing plumes. By their very nature, similarity theories need considerable input from experiments or numerical models for determining the various unspecified similarity functions.

The general Lagrangian similarity hypothesis states that the physical parameters on which Eulerian characteristics of turbulence depend also determine the Lagrangian characteristics. Batchelor (1950) was the first to demonstrate the power of similarity arguments in describing relative dispersion of particles in homogeneous flows. Later, he used the same approach to describe dispersion of particles in steady, self-preserving free shear flows (Batchelor and Townsend, 1956; Batchelor, 1957). Applications to dispersion in the turbulent surface layer and boundary layer followed shortly thereafter.

8.2 LAGRANGIAN SIMILARITY THEORY FOR THE NEUTRAL SURFACE LAYER

The foundation of the Lagrangian similarity theory of dispersion in the surface layer was laid by Kazanski and Monin (1957), Monin (1959a&b), Ellison (1959), and Batchelor (1959, 1964). Further developments and extensions came from the contributions of Gifford (1962), Cermak (1963), Chatwin (1968), Hunt and Weber (1979) and others (see e.g., Monin and Yaglom, 1971; Pasquill and Smith, 1983).

For the simpler case of a horizontally homogeneous neutral surface layer over an aerodynamically rough surface, the mean velocity profile is given by the log-law, with friction velocity as the velocity scale for both the mean and fluctuating velocities. According to the Lagrangian similarity hypothesis, u_* is also the appropriate scale for the Lagrangian motion.

8.2.1 Mean Lagrangian Velocities and Particle Displacements

To consider the dispersion from an elevated line or point source in the surface layer, one can assume, without loss of generality, that a fluid particle initially (at $t = 0$) at $\underline{X}(0) = (0,0,H)$ will be displaced to a new position $(\underline{X}(t) = [X(t), Y(t), Z(t)]$ at a time t after release and its velocity will be $\underline{V}(t) = [U(t), V(t), W(t)]$. The Lagrangian velocity of the particle $\underline{V}(t)$ obviously is not stationary; after sufficiently long time the particle would rise to a significant height $Z(t)$, thereby increasing its longitudinal velocity $U(t)$, as well as its vertical velocity $W(t)$. However, it is natural to assume that the Lagrangian characteristics of mean flow and turbulence in the neutral surface layer will depend only on the travel time t, friction velocity u_*, and initial release height H. The role of the surface roughness z_0 is considered important only in determining the longitudinal mean velocity. In addition, after a sufficiently large time t the particle must forget its initial height H, so that H also ceases to be relevant for $t \gg H/u_*$ (this condition can be quite restrictive, however, for releases above a few tens of meters in the presence of weak winds).

First, let us consider the ensemble-averaged Lagrangian velocity $\underline{\overline{V}}(t) = [\overline{U}(t), \overline{V}(t), \overline{W}(t)]$ at time t. The symmetry of the unidirectional surface-layer flow with respect to the x–z plane implies that $\overline{Y}(t) = 0$ and, consequently, the mean lateral velocity $\overline{V}(t) = d\overline{Y}/dt = 0$ for all t. Using the above stated Lagrangian similarity hypothesis, the Lagrangian mean vertical velocity can be expressed as

$$\overline{W}(t) = \frac{d\overline{Z}(t)}{dt} = b\, u_* \tag{8.1}$$

in which b is a universal constant. Integrating Eq. (8.1) with respect to t and using the initial condition $\overline{Z}(0) = H$, one gets

$$\overline{Z}(t) = bu_* t + H \tag{8.2}$$

Equations (8.1) and (8.2) are remarkable results (predictions) of the Lagrangian similarity theory. According to the former, the particle experiences on the average a constant vertical velocity $\overline{W}$ in spite of the zero average Eulerian velocity ($\overline{w} = 0$) at all points of the horizontally homogeneous flow. This may be explained by the inability of the particle to drop below the release height due to the presence of the ground surface below, while nothing restricts its ascent above the release height. This argument is quite obvious for the particles released at the ground level whose average height increases linearly with the time after release, as long as the particles remain in the surface layer.

The mean particle velocity in the longitudinal (mean flow) direction is generally assumed to be equal to the Eulerian mean flow velocity at some height proportional to $\overline{Z}$, that is,

$$\frac{d\overline{X}}{dt} = \overline{U} = \overline{u}(c\overline{Z}) \tag{8.3}$$

where c is another empirical similarity constant that is expected to be less than 1 (Monin and Yaglom, 1971; Pasquill and Smith, 1983). Using the logarithmic velocity-profile law for the neutral surface layer, Eq. (8.3) can also be written as

$$\frac{d\overline{X}}{dt} = \frac{u_*}{k} \ln \frac{c\overline{Z}}{z_0} \tag{8.4}$$

which can be integrated with respect to t after substituting for $\overline{Z}$ from Eq. (8.2). Of greater interest is the mean particle trajectory, which can be obtained from Eqs. (8.1) and (8.4)

$$\frac{d\overline{X}}{d\overline{Z}} = \frac{1}{kb} \ln \frac{c\overline{Z}}{z_0} \tag{8.5}$$

Integration of Eq. (8.5) with respect to Z from H to $Z(t)$ yields the following similarity expression for the mean particle trajectory:

$$kb\overline{X}_* = \overline{Z}_* \ln(c\overline{Z}_*) \\ - (\overline{Z}_* - H_*) - H_* \ln(cH_*) \tag{8.6}$$

where $\overline{X}_* = \overline{X}/z_0$ and $\overline{Z}_* = \overline{Z}/z_0$ are the normalized coordinates of the mean particle trajectory, and $H_* = H/z_0$ is the dimensionless release height.

Cermak (1963) derived a slightly different expression by assuming $c = 1$ and incorporating an approximate particle relaxation time t_r of the order of H/u_*, which is elapsed before particles released at H when $t = 0$ acquire a motion identical with the fluid motion. He estimated the corresponding relaxation distance $\overline{X}_r = \overline{u}(H)H/u_*$, assuming the proportionality coefficient in $t_r \propto H/u_*$ to be unity. Actual relaxation time for tiny fluid particles is likely to be much smaller and Cermak's uncertain correction can be avoided altogether by assuming an isokinetic release of particles at $t = 0$. In any case, Eq. (8.4) is probably valid only for $t > H/u_*$, because the dependence of $d\overline{X}/dt$ on H is not explicitly considered. Consequently, Eqs. (8.5) and (8.6) also have the same restriction ($t > H/u_*$) for their validity.

For the ground-level source, Eq. (8.5) should be integrated with respect to $\overline{Z}$ from z_0/c to $\overline{Z}(t)$ to obtain the simpler expression

$$kb\overline{X}_* \simeq \overline{Z}_*[\ln(c\overline{Z}_*) - 1], \text{ for } \overline{Z}_* \gg 1 \quad (8.7)$$

for the mean particle trajectory. The constants b and c have been evaluated empirically as well as theoretically (see e.g., Chatwin, 1968; Yaglom, 1972; Pasquill and Smith, 1983); the most widely accepted values for the neutral surface layer are $b \simeq 0.40$ and $c \simeq 0.6$. Monin and Yaglom (1971) have given good physical reasoning for the values of b and c being less than unity.

Figure 8.2 compares the mean particle trajectories ($\overline{Z}_*$ vs. $\overline{X}_*$) for different dimensionless release heights. As expected, mean particle trajectory is quite sensitive to H_*, particularly for small travel times or distances. Batchelor (1964) pointed out that a particle forgets its initial position while traveling downwind for a sufficiently long time compared with H/u_*. Then, Eq. (8.7) should be adequate for describing the mean particle trajectory. For short travel times, Eq. (8.6) may also be expressed in a similar form, that is,

$$kb(\overline{X}_* + x_{0*}/x_{0*}) = \overline{Z}_*[\ln(c\overline{Z}_*) - 1] \quad (8.8)$$

where $x_{0*} = H_*[\ln(cH_*) - 1]/kb$ represents the dimensionless upwind distance where an equivalent surface source may be assumed to be located, such that it would yield an identical mean plume trajectory to that of the elevated source at $x = 0$. This concept of virtual source location upwind of the actual source is commonly used in dispersion modeling (see e.g., Cermak, 1963; van Ulden, 1978). The mean particle trajectory in terms of $\overline{X}_* + x_{0*}$ versus $\overline{Z}_*$ given by Eq. (8.8) is represented by the curve in Figure 8.2 corresponding to $H_* = 0$; for large travel times the plume trajectory may also be approximated by a simple power law of the form

$$\overline{Z}_* \simeq 0.078(\overline{X}_* + x_{0*})^{0.9} \quad (8.9)$$

8.2.2 Probability Density of Particle Displacements

Using dimensional considerations, one can also derive similarity relations for the probability density of particle displacements, $\underline{X}'(t) = \underline{X}(t) - \overline{X} = [X'(t), Y'(tr), Z'(t)]$, from their mean position at any travel time t, or the probability density of the Lagrangian velocity

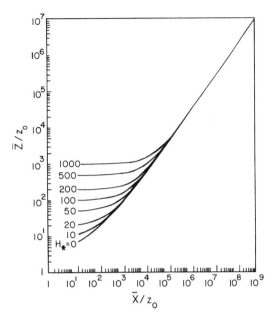

Figure 8.2 Normalized mean particle trajectories for different dimensionless release heights.

fluctuations $\underline{V}'(t) = [U'(t), V'(t), W'(t)]$. When $t \gg z_0/u_*$ for the near-surface release and $t \gg H/u_*$ for an elevated release in the neutral-surface layer, the joint probability density functions may depend only on u_* and t. Consequently, they must have the forms (Monin and Yaglom, 1971)

$$p(U',V',W'; t) = \frac{1}{u_*^3} f\left(\frac{U'}{u_*}, \frac{V'}{u_*}, \frac{W'}{u_*}\right) \quad (8.10)$$

$$p(X',Y',Z'; t) = \frac{1}{u_*^3 t^3} F\left(\frac{X'}{u_*t}, \frac{Y'}{u_*t}, \frac{Z'}{u_*t}\right) \quad (8.11)$$

where f and F are some universal functions. According to Eq. (8.2), u_*t may be replaced by $\overline{Z}$ without any loss of generality, so that Eq. (8.11) can also be expressed as

$$p(X',Y',Z'; t) = \frac{1}{\overline{Z}^3} F_1\left(\frac{X'}{\overline{Z}}, \frac{Y'}{\overline{Z}}, \frac{Z'}{\overline{Z}}\right) \quad (8.12)$$

in which F_1 is totally independent of the particle travel time.

8.2.3 Mean Concentrations

Inquiring about the mean concentration at a fixed point (x,y,z) in a continuous source plume is equivalent to determining the probability of a particle reaching the point (x,y,z) after a travel time t when the mean particle position is $(\overline{X},0,\overline{Z})$. From Eq. (8.12), one can express this probability as

$$p(x,y,z; t) = \frac{1}{\overline{Z}^3} F_2\left(\frac{x - \overline{X}}{\overline{Z}}, \frac{y}{\overline{Z}}, \frac{z - \overline{Z}}{\overline{Z}}\right) \quad (8.13)$$

in which F_2 is, presumably, a universal function that is directly related to F_1 in Eq. (8.12).

The mean concentration in a continuous point source plume is directly related to the above probability density as

$$\bar{c}(x,y,z) \propto \int_0^\infty p(x,y,z; t)dt \qquad (8.14)$$

Again, changing the variable from t to $\bar{Z}$, and incorporating the expected proportionality between concentration and source strength, one can write from Eqs. (8.13) and (8.14)

$$\bar{c}(x,y,z)$$
$$= \frac{Q}{bu_*} \int_0^\infty F_3\left(\frac{x-\bar{X}}{\bar{Z}}, \frac{y}{\bar{Z}}, \frac{z-\bar{Z}}{\bar{Z}}\right)\frac{d\bar{Z}}{\bar{Z}^3} \qquad (8.15)$$

The similarity functions F_1 and F_3 have not been determined, so that the above integral cannot be evaluated. Some useful inferences can be made, however, about the ground-level concentration along the plume centerline, which can be expressed in the similarity form as

$$\bar{c}_0(x,0,0) = \frac{Q}{bu_*} \int_0^\infty F_4\left(\frac{x-\bar{X}}{\bar{Z}}\right)\frac{d\bar{Z}}{\bar{Z}^3} \qquad (8.16)$$

Gifford (1962) simplified and evaluated the above integral after transforming the integration variable from $\bar{Z}$ to $\xi = (x - \bar{X})/\bar{Z}$, so that

$$\bar{c}_0 = \frac{Q}{bu_*} \int_0^\infty \frac{F(\xi)}{\bar{Z}^2(\xi + d\bar{X}/d\bar{Z})}d\xi \qquad (8.17)$$

Gifford argued that at a sufficient distance downwind, $x - \bar{X} \ll \bar{X}$, so that the function $F(\xi)$ may be expected to have a fairly sharp maximum near $\xi = 0$, or $x = \bar{X}$ (note that $F(\xi)$ is directly proportional to the probability of a particle displaced from its mean position by $x - \bar{X}$ after a travel time t). Therefore, the integral in Eq. (8.17) is primarily determined by values of $\bar{Z}$ near $\bar{Z}_x$, at which $\bar{X} = x$. Then the above integral can be simplified to

$$\bar{c}_0 \simeq dQ/bu_*\bar{Z}_x^2 \left(\frac{d\bar{X}}{d\bar{Z}}\right)_x \qquad (8.18)$$

in which d is another constant and $(d\bar{X}/d\bar{Z})_x$ represents the inverse of the slope of the mean particle trajectory at $\bar{X} = x$. Evaluating $(d\bar{X}/d\bar{Z})_x$ from Eq. (8.5) and substituting in Eq. (8.18), one obtains

$$\bar{c}_0 = \frac{dkQ}{u_*\bar{Z}_x^2 \ln (c\bar{Z}_x/z_0)} \qquad (8.19)$$

Further noting that, according to Eq. (8.7), $\bar{Z}_x \ln (c\bar{Z}_x/z_0) \simeq kbx$, Eq. (8.19) can also be written as

$$\bar{c}_0 = \frac{dQ}{bu_*x\bar{Z}_x} \propto \frac{Q \ln (c\bar{Z}_x/z_0)}{u_*x^2} \qquad (8.20)$$

Similarly, for a cross-wind continuous line source one can obtain (see e.g., Pasquill and Smith, 1983)

$$\bar{c}_0 \propto \frac{Q_{cl}}{u_*\bar{Z}_x \ln (c\bar{Z}_x/z_0)} \propto \frac{Q_{cl}}{u_*x} \qquad (8.21)$$

This result is consistent with the solution (6.92) of the diffusion equation for a linear variation of eddy diffusivity ($K_z = ku_*z$).

8.2.4 Analysis Based on the Statistical Theory

Statistical dispersion relations analogous to Taylor's theorem can also be derived for the vertical dispersion in shear flow (Csanady, 1973; Hunt and Weber, 1979):

$$\frac{d\overline{Z^2}}{dt} = 2\int_0^t \overline{W(t)W(t')}dt' \qquad (8.22)$$

$$\frac{d\overline{Z'^2}}{dt} = 2\int_0^t \overline{W'(t)W'(t')}dt' \qquad (8.23)$$

in which overbar implies an ensemble average and the particle displacement is measured from the ground surface where particles are released. The main limitation of the above relations is that there is not much quantitative information on the Lagrangian velocity covariances in shear flows, particularly the atmospheric boundary layer. Since the Lagrangian vertical velocity is not a stationary function of travel time, the autocovariances $\overline{W(t)W(t')}$ and $\overline{W'(t)W'(t')}$ are expected to be functions of not only the time lag $\xi = t' - t$, but also the diffusion time t. One can still formally define the dimensionless autocorrelation functions

$$R_{WW}(\xi,t) = \frac{\overline{W(t)W(t+\xi)}}{\overline{W^2}(t)} \qquad (8.24)$$

$$R_{W'W'}(\xi,t) = \frac{\overline{W'(t)W'(t+\xi)}}{\overline{W'^2}(t)} \qquad (8.25)$$

The differences between the Lagrangian autocorrelation functions of the total velocity and the fluctuating velocity are schematically shown in Figure 8.3. Note that $R_{WW}(\xi,t)$ has a finite value at $\xi = -t$ when the particle is released, because $W(t)$ contains a mean component. This function is highly asymmetrical about $\xi = 0$ and is expected to be nonstationary, that is, to be a function of travel time, because both the autocovariance $\overline{W(t)W(t+\xi)}$ and the variance $\overline{W^2}(t)$ evolve with the particle travel time. The dependence of $R_{W'W'}(\xi,t)$ on travel time is expected to be much weaker, because the fluctuating Lagrangian velocity is more likely to be closer to stationarity than the total velocity, particularly in the neutral-surface layer in which σ_w is more or less independent of height. For the same reason, $R_{W'W'}(\xi,t)$ is also expected to be only weakly asymmetrical. One can also define an integral time scale

$$T_{iL}(t) = \int_0^\infty R_{W'W'}(\xi,t)d\xi \qquad (8.26)$$

that may be a weak function of travel time t.

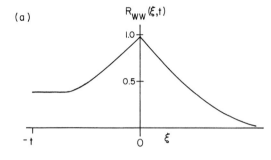

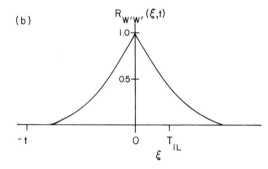

Figure 8.3 Schematics of Lagrangian autocorrelation functions for the total and fluctuating vertical velocities: (a) $R_{WW}(\xi,t)$. (b) $R_{W'W'}(\xi,t)$.

Expressing $W(t) = \overline{W}(t) + W'(t)$ and substituting in Eq. (8.22), the latter can also be expressed in terms of the Lagrangian autocorrelation of fluctuating vertical velocity (see Hunt and Weber, 1979):

$$\frac{d\overline{Z^2}}{dt} = 2\overline{W}(t)\int_0^t \overline{W}(t')dt'$$
$$+ 2\overline{W'^2}(t)\int_{-t}^0 R_{W'W'}(\xi,t)d\xi \qquad (8.27)$$

Equation (8.27) is a general dispersion relation for shear flows with or without stratification. It can be further simplified for practical dispersion estimates, using the Lagrangian similarity relations. In particular, for the neutral-surface layer, we have shown that the mean Lagrangian vertical velocity is independent of travel time ($\overline{W} = bu_*$), so that the first term in Eq. (8.27) can be simplified to $2b^2u_*^2 t$. Furthermore, in the limit of large diffusion time, the integral in the second term can be approximated by T_{iL}. Thus, Eq. (8.27) can be simplified to

$$\frac{d\overline{Z^2}}{dt} = 2b^2u_*^2 t + 2\overline{W'^2}(t)T_{iL}(t) \qquad (8.28)$$

Hunt and Weber (1979) made several plausible assumptions in order to evaluate the second term in the above dispersion relation. Arguing that, in the neutral surface layer, turbulence is only weakly inhomogeneous in the sense that only the integral scales of turbulence change with height whereas the Eulerian velocity variances do not, they assumed that the variances of Lagrangian and Eulerian vertical velocities are approximately equal, that is, $\overline{W^2} = \sigma_w^2$. This implies that $\overline{W'^2}$ is also independent of t and can be evaluated from similarity theory relations

$$\overline{W'^2} = \sigma_w^2 - \overline{W}^2 = b'^2 u_*^2$$

in which $b' = [(1.3)^2 - (0.4)^2]^{1/2} \simeq 1.24$. Thus, the variance of Lagrangian velocity fluctuations is slightly smaller than that of Eulerian velocity fluctuations. The difference between the two may not be detectable from turbulence measurements in the surface layer.

Another assumption made by Hunt and Weber (1979) for estimating the Lagrangian time scale is that it is proportional to the integral length scale. Since the latter increases linearly with height in the neutral surface layer, it can be shown that

$$T_{iL} \propto \frac{\overline{Z}}{\sigma_w} \propto t$$

or $\qquad T_{iL} = \gamma t$

in which $\gamma \simeq 0.10$ is an empirical similarity constant.

Substituting the above similarity relations into Eq. (8.27) and integrating with respect to t, the total mean-square dispersion can be expressed as

$$\overline{Z^2} = b^2 u_*^2 t^2 + b'^2 \gamma u_*^2 t^2 \simeq 0.31 u_*^2 t^2 \qquad (8.29)$$

Note that $\overline{Z^2}$ consists of two parts; the first part, $\overline{Z}^2 = b^2 u_*^2 t^2 \simeq 0.16\, u_*^2 t^2$, represents the contribution due to the Lagrangian mean vertical velocity, whereas the second part represents the contribution due to turbulence.

$$\overline{Z'^2} = b'^2 \gamma u_*^2 t^2 \simeq 0.15\, u_*^2 t^2 \qquad (8.30)$$

Note that the mean Lagrangian motion and turbulence contribute equally to total dispersion, even though $\overline{W}^2 \simeq 0.1\, \overline{W'^2}$.

8.3 LAGRANGIAN SIMILARITY THEORY FOR THE STRATIFIED SURFACE LAYER

The extension and generalization of the above results for the neutral-surface layer to more general stratified conditions has been considered very important, because thermal stratification occurs most of the time in the atmospheric boundary layer, including the surface layer. The Monin–Obukhov similarity theory has been the cornerstone of micrometeorological studies in the latter half of this century. The basic assumptions and implications of the M–O similarity theory have already been discussed in Chapter 4. Its application to describe turbulent diffusion in the stratified surface layer was originally proposed by Kazanski and Monin (1957) and further refined later by Monin (1959), Gifford (1962), Cermak (1963), and others.

8.3.1 Mean Particle Velocities and Displacements

Using the original M–O similarity hypothesis and considering the vertical dispersion of passive particles released from a point at the surface ($H = 0$), the mean and the standard deviation of the particle displacement in the vertical as functions of travel time can be expressed by the similarity relations

$$\overline{Z} = bu_* t \phi_1\left(\frac{\overline{Z}}{L}\right) \tag{8.31}$$

$$\sigma_z = b_2 u_* t \phi_2\left(\frac{\overline{Z}}{L}\right) \tag{8.32}$$

in which b and b_2 are the dimensionless similarity constants ($b \simeq b_2 \simeq 0.4$, according to our discussion in section 8.2), and $\phi_1(\overline{Z}/L)$ and $\phi_2(\overline{Z}/L)$ are universal similarity functions of the dimensionless parameter $\overline{Z}/L$, such that under the neutral condition $\phi_1(0) = 1$ and $\phi_2(0) = 1$.

The mean vertical velocity of the particles is also given by the generalized Lagrangian similarity hypothesis as

$$\frac{d\overline{Z}}{dt} = \overline{W} = bu_* \phi\left(\frac{\overline{Z}}{L}\right) \tag{8.33}$$

in which $\phi(\overline{Z}/L)$ is also a universal similarity function related to $\phi_1(\overline{Z}/L)$. The mean particle velocity in the longitudinal (mean flow) direction is still assumed to be equal to the Eulerian mean flow velocity at some height lower than $\overline{Z}$, so that the mean particle trajectory can be obtained from Eqs. (8.3) and (8.33) as

$$\frac{d\overline{X}}{d\overline{Z}} = \frac{\overline{u}(c\overline{Z})}{bu_* \phi(\overline{Z}/L)} \tag{8.34}$$

Here, c could, in general, be a weakly dependent function of stability, but deviations from its neutral value of about 0.6 may not be considered significant (Pasquill and Smith, 1983).

Equation (8.34) can be integrated if the similarity function $\phi(\overline{Z}/L)$ is specified; the velocity $\overline{u}(c\overline{Z})$ at the height $c\overline{Z}$ can be expressed in terms of the Monin–Obukhov similarity theory. The similarity form of the mean velocity profile is well known and has been discussed in Chapter 4. Substituting from Eq. (4.35) into Eq. (8.34) and integrating the latter with respect to $\overline{Z}$, one obtains

$$\overline{X} = \frac{1}{kb} \int_{\overline{Z}_{0/c}}^{\overline{Z}} \left[\ln\left(\frac{c\overline{Z}'}{z_0}\right) - \psi_m\left(\frac{c\overline{Z}'}{L}\right)\right] \phi^{-1}\left(\frac{\overline{Z}'}{L}\right) d\overline{Z}' \tag{8.35}$$

in which the form of the ψ_m function has been given in Chapter 4, Eq. (4.46).

The form of $\phi(\overline{Z}/L)$ is still unknown. Ideally, it should be evaluated from the Lagrangian observations of particle dispersion in stratified surface layers under different stability conditions. However, most of the experimental diffusion data are Eulerian, for example, mean concentrations at fixed points in continuous source plumes. In the absence of Lagrangian dispersion data, theoretical and semiempirical methods have been used to determine $\phi(\overline{Z}/L)$. Monin (1959) used the simplified turbulent kinetic energy equation, without the turbulent-transport term, to obtain a simple expression for $\phi(\overline{Z}/L)$. But his derivation was based on the questionable assumption that the Lagrangian velocity $\overline{W}$ is the appropriate velocity scale for expressing the energy dissipation as $\varepsilon \propto \overline{W}^3/\ell$, and $K_m \propto \overline{W}\ell$. If one uses the right scaling velocity u_* or σ_w, instead of $\overline{W}$, one cannot derive an expression for $\overline{W}/u_*$ or $\phi(\overline{Z}/L)$ from the TKE equation.

Other estimates of $\phi(\overline{Z}/L)$ and the constant b are based on the use of gradient transport theory for describing vertical diffusion in the surface layer. Using the solution (6.91) of the diffusion equation for an instantaneous area source at the ground in the neutral surface layer, one can obtain the mean height of the diffusing material

$$\overline{z} = \frac{\int_0^\infty \overline{c} z \, dz}{\int_0^\infty \overline{c} \, dz} = ku_* t \tag{8.36}$$

which implies that $b = k \simeq 0.4$, if $\overline{z}$ can be identified with the mean particle displacement $\overline{Z}$ after the same travel or diffusion time t. The above theoretical determination of b was given by Ellison (1959). Extending the same approach and that of Chatwin (1968) to the stratified surface layer in which the vertical-eddy diffusivity is assumed to be given by the M–O similarity relation $K_z = ku_* z/\phi_h(z/L)$, Chaudhry and Meroney (1973) have shown that

$$\frac{d\overline{Z}}{dt} \simeq \frac{ku_*}{\phi_h(\overline{Z}/L)} \tag{8.37}$$

which implies

$$\phi\left(\frac{\overline{Z}}{L}\right) \simeq \frac{1}{\phi_h(\overline{Z}/L)} \tag{8.38}$$

Substituting from Eq. (8.38) into Eq. (8.35), the mean particle trajectory from a ground-level source in the atmospheric surface layer is given by

$$\overline{X}_* = \frac{1}{kb\zeta_0} \int_{\zeta_0}^{\overline{Z}/L} \left[\ln\left(\frac{c\zeta}{\zeta_0}\right) - \psi_m(c\zeta)\right] \phi_h(\zeta) d\zeta \tag{8.39}$$

in which $\zeta_0 = z_0/L$ is the stability parameter and $\zeta = \overline{Z}/L = \overline{Z}_* \zeta_0$. After specifying the M–O similarity functions ψ_m and ϕ_h (e.g., Eqs. (4.46) and (4.38) of Chapter 4), Eq. (8.39) can be integrated numerically for different values of z_0/L. The results of computation for different values of ζ_0 are plotted in Figure 8.4 in the dimensionless form $\overline{Z}/z_0$ versus $\overline{X}/z_0$, using the value of $c = 1$ and the empirical forms of the similarity functions reported by Businger et al. (1971).

Note that, as expected, the mean particle trajectory becomes flatter with increasing stability (z_0/L) and

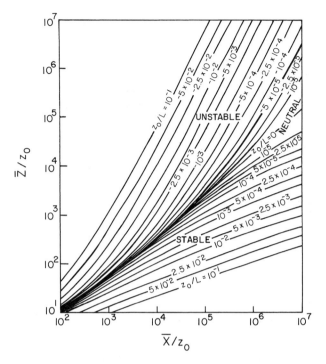

Figure 8.4 Normalized mean particle trajectory as a function of the stability parameter z_0/L. From Chaudhry and Meroney, 1973.

steeper with increasing instability ($-z_0/L$). The validity of these trajectory calculations is limited to rather short distances from the source, at which dispersing particles are likely to stay within the surface layer. Assuming a surface layer thickness of about $0.1\ h$, and realizing that particles may disperse up to $2\overline{Z}$ in the vertical, an approximate condition of the validity of the above similarity theory results might be expressed as $\overline{Z} < 0.05\ h$, where h is the PBL depth.

8.3.2 Mean Concentration in Continuous Source Plumes

The similarity relations (8.10) through (8.16) can also be generalized in a formal but straightforward manner by including an additional stability-dependent parameter $\overline{Z}/L$ (Gifford, 1962; Cermak, 1963; Monin and Yaglom, 1971). But not much progress has been made in the determination of the various similarity functions. Gifford (1962) implicitly ignored the dependence of these functions on $\overline{Z}/L$ and suggested that his derivations for g.l.c.,

$$\overline{c}_0 \propto \frac{Q}{u_* \overline{Z}_x^2 (d\overline{X}/d\overline{Z})_x} \tag{8.40}$$

for the continuous point source and

$$\overline{c}_0 \propto \frac{Q_{cl}}{u_* \overline{Z}_x (d\overline{X}/d\overline{Z})_x} \tag{8.41}$$

for the continuous cross-wind line source in the neutral surface layer should be applicable to stratified surface layers as well. The dependence of $\overline{Z}/L$ is implicit in $\overline{Z}_x$ and $(d\overline{X}/d\overline{Z})_x$ and, presumably, also in the coefficients of proportionality in the above expressions. Gifford's numerical evaluations of the g.l.c. showed that for a point source at the ground, $\overline{c}_0 u_*/Q \propto x^{-\beta}$, in which β is substantially less than 2 in stable conditions and greater than 2 in unstable conditions.

8.3.3 Analyses Based on the K-Theory

Further applications of the similarity theory to the computation of the mean plume height and ground-level concentrations due to near-surface sources have been reported by Nieuwstadt and van Ulden (1978), van Ulden (1978), and Horst (1979), among others. In these studies the vertical concentration profile is assumed to have the modified exponential form

$$\frac{\overline{c}}{\overline{c}_0} = \exp\left[-B_\alpha \left(\frac{z}{\overline{z}}\right)^\alpha\right] \tag{8.42}$$

in which $\overline{z}$ is the mean plume height, defined as

$$\overline{z} = \frac{\int_0^\infty z\overline{c}\,dz}{\int_0^\infty \overline{c}\,dz} \tag{8.43}$$

and B_α is a dimensionless constant that is related to the profile exponent α as

$$B_\alpha = \left[\Gamma\left(\frac{2}{\alpha}\right)/\Gamma\left(\frac{1}{\alpha}\right)\right]^\alpha \quad (8.44)$$

where Γ is the gamma function. Note that Eq. (8.42) is consistent with the solutions of the diffusion equation for the power-law profiles of wind speed and eddy diffusivity, as well as with the available concentration profile data (see e.g., Nieuwstadt and van Ulden, 1978; Brown et al., 1993). As discussed earlier in Chapter 6, the profile exponent α may vary from 1 under very unstable conditions to more than 2 under very stable conditions. The Gaussian distribution, corresponding to $\alpha = 2$, is observed only under a rather narrow range of moderately stable conditions.

Nieuwstadt and van Ulden (1978) solved the diffusion equation numerically for a continuous line source, using the Monin–Obukhov similarity relations for mean velocity and eddy diffusivities of heat and momentum. In particular, they used the expressions proposed by Businger (1973). Two alternative specifications of the vertical eddy diffusivity were tested: (1) $K_z = K_h$ and (2) $K_z = 1.35\ K_m$. The numerical model results were compared with the vertical concentration profiles measured at a distance of 100 m from the source during the Project Prairie Grass (PPG) and Porton experiments. A comparison of the computed and measured $\bar{z}$ as a function of stability $(1/L)$ showed good agreement, with a preference for $K_z = 1.35\ K_m$ over $K_z = K_h$ under unstable conditions (the two parameterizations give comparable results for near-neutral and stable conditions). More extensive comparisons with the PPG data at 50, 200, 400, and 800 m, however, showed that the specification $K_z = K_h$ gives better agreement than the alternative assumptions of $K_z = K_m$ or $K_z = \alpha K_m$ (Horst, 1979, 1980a).

van Ulden (1978) extended the above comparison study to include the PPG experimental data at $x = 50$, 200, and 800 m for many runs covering a wide range of stability conditions. He used the analytical solution to the diffusion equation in the form

$$\bar{c}(x,z) = \frac{A_\alpha Q_{cl}}{\bar{u}_p \bar{z}} \exp\left[-B_\alpha\left(\frac{z}{\bar{z}}\right)^\alpha\right] \quad (8.45)$$

where

$$A_\alpha = \frac{\alpha\ \Gamma(2/\alpha)}{[\Gamma(1/\alpha)]^2} \quad (8.46)$$

is another constant related to the profile exponent α and

$$\bar{u}_p = \int_0^\infty \bar{u}\bar{c}\ dz / \int_0^\infty \bar{c}\ dz \quad (8.47)$$

is the average horizontal transport velocity across the plume. The dependence of $\bar{c}$ on x is implicit through the mean plume height $\bar{z}$. From the definition of $\bar{z}$ in Eq. (8.43), one can write

$$\frac{d\bar{z}}{dx} = \int_0^\infty (z - \bar{z})\frac{\partial \bar{c}}{\partial x}dz / \int_0^\infty \bar{c}\ dz \quad (8.48)$$

After evaluating the integrals in Eq. (8.48) using Eq. (8.45), van Ulden (1978) obtained

$$\frac{d\bar{z}}{dx} = \frac{k^2}{[\ln(c_\alpha \bar{z}/z_0) - \psi_m(c_\alpha \bar{z}/L)]\phi_h(c_\alpha \bar{z}/L)} \quad (8.49)$$

in which c_α is another constant related to α; it is rather insensitive to the actual value of $\alpha > 1$, and a constant value of $c_\alpha \simeq 1.55$ has been suggested for the expected range of α values ($1 < \alpha < 3$). Equation (8.49) can be integrated to yield

$$x_* = \frac{1}{k^2 \zeta_0}\int_{\zeta_0}^{\bar{z}/L}\left[\ln\left(\frac{c_\alpha \zeta}{\zeta_0}\right) - \psi_m(c_\alpha \zeta)\right]\phi_h(c_\alpha \zeta)d\zeta \quad (8.50)$$

in which $x_* = x/z_0$, $\zeta = \bar{z}/L$, and $\zeta_0 = z_0/L$.

Note that Eq. (8.50) is quite similar to Eq. (8.39), except for the difference in the constants c and c_α. The integral in Eq. (8.50) can be evaluated numerically for the specified forms of the Monin–Obukhov similarity functions ψ_m and ϕ_h. Van Ulden (1978) has given the approximate analytical expressions for the mean plume height as a function of distance from the source and plotted them in the similarity theory framework (see Fig. 8.5).

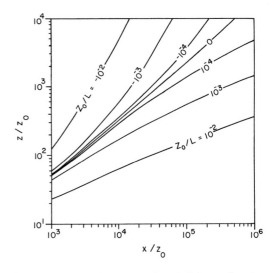

Figure 8.5 Normalized mean plume height as a function of the dimensionless distance from the continuous source and the stability parameter z_0/L. From van Ulden, 1978.

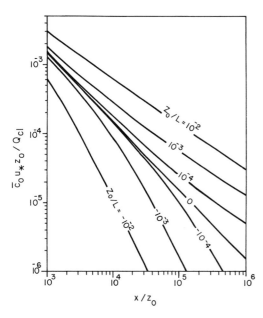

Figure 8.6 Normalized ground-level concentration as a function of the dimensionless distance from the source and stability. From van Ulden, 1978.

Approximate analytical expressions for $\bar{z}/z_0$ and $\bar{u}_p/u_*$ are given by van Ulden (1978), and his plot of the dimensionless g.l.c. as a function of x/z_0 and z_0/L is shown in Figure 8.6. As expected, the g.l.c. decreases with distance more and more slowly as stability increases. Similar results have been reported by Horst (1979) using a slightly different expression for the mean plume height, which implies that $\bar{z} = \bar{Z}$ at $x = \bar{X}$. Theoretically predicted ground-level concentrations are shown to be in good agreement with the PPG experimental data.

The above mentioned applications of the K-theory in conjunction with the Lagrangian similarity theory may be justified for neutral, stable, and slightly unstable conditions. The use of the K-theory becomes particularly questionable under free convective conditions for which the local free convection similarity scaling would be more appropriate.

8.3.4 Local Free Convection Similarity Theory

Monin and Yaglom (1971) have extended the local free convection similarity theory for the surface layer under free convection conditions ($-L \ll z \geq 0.1\,h$) to diffusion from surface sources. They argued that, in this regime, buoyant production of turbulence dominates over the shear production and u_* ceases to be important. Using the simple Lagrangian similarity hypothesis that diffusion is essentially governed by the buoyancy variable g/T_0 and surface heat flux $H_0/\rho c_p$, one obtains from dimensional considerations

$$\bar{Z}(t) = a\left(\frac{g}{T_0}\frac{H_0}{\rho c_p}\right)^{1/2} t^{3/2} \tag{8.52}$$

in which a is a universal similarity constant. From Eq. (8.52) one can also write

$$\bar{W}(t) = \frac{d\bar{Z}}{dt} = \frac{3}{2}a\left(\frac{g}{T_0}\frac{H_0}{\rho c_p}\right)^{1/2} t^{1/2} \tag{8.53}$$

or, alternatively,

$$\bar{W}(t) = \frac{3}{2}a^{2/3}\left(\frac{g}{T_0}\frac{H_0}{\rho c_p}\bar{Z}\right)^{1/3}$$

$$= \frac{3}{2}a^{2/3}u_f(\bar{Z}) \tag{8.54}$$

in which $u_f(\bar{Z})$ is the local free convection velocity at the mean particle height. Thus, the mean Lagrangian vertical velocity in the free convective surface layer is proportional to $u_f(\bar{Z})$.

Similar dimensional reasoning can be applied to the standard deviation of the particle spread to obtain

$$\sigma_z = (\overline{Z'^2})^{1/2} = a'\left(\frac{g}{T_0}\frac{H_0}{\rho c_p}\right)^{1/2} t^{3/2} \tag{8.55}$$

in which a' is another universal similarity constant. The above theoretically predicted accelerated growth

Despite the similarities between Eqs. (8.39) and (8.50) and their corresponding plots in figures 8.4 and 8.5, respectively, the differences between the two should also be recognized (van Ulden and Nieuwstadt, 1980). While $\bar{Z}$ represents the ensemble mean height of particles that have traveled a certain time t after release and an average horizontal distance $\bar{X}$, $\bar{z}$ represents the mean height of particles at a certain distance x downwind from a maintained source (point or line). Different particles pass by a fixed location downwind of the source at different times, because particles dispersing to greater heights are subjected to larger velocities than those traveling at lower heights. The former will take less time than the latter to reach the same distance x. This difference is essentially due to mean wind shear and can become quite significant in the stably stratified surface layer with almost linear variation of wind speed with height. If one's interest is primarily in the mean concentration distribution in the plume due to a maintained source, then the representation of mean plume height $\bar{z}$ as a function of distance x from the source is more relevant than that of the mean particle height $\bar{Z}$ as a function of t or $\bar{X}(t)$. Using the former, the concentration field can be easily obtained as an appropriate solution of the diffusion equation, such as Eq. (8.45) for the continuous line source at the ground. In particular, the g.l.c. is given by

$$\frac{\bar{c}_0 u_* z_0}{Q_{c\ell}} = A_\alpha \left(\frac{\bar{z}}{z_0}\right)^{-1}\left(\frac{\bar{u}_p}{u_*}\right)^{-1} \tag{8.51}$$

of plume spread ($\sigma_z \propto t^{3/2}$) is associated with the increase of σ_w with height in the free-convective surface layer (according to local free convection similarity theory, $\sigma_w \sim u_f \propto z^{1/3}$). It is not to be confused with the $t^{3/2}$ law for the intermediate range of travel time according to the relative diffusion theory of homogeneous turbulence. The accelerated plume growth is expected to occur only during the relatively short range of travel times in which particles may remain in the surface layer. Particles will eventually disperse out of the surface layer into the mixed layer as $Z \geq 0.1\,h$, or $Z \geq 0.05\,h$.

Yaglom (1972) has estimated the similarity constants a and a' by utilizing the solution of the vertical diffusion equation with the eddy diffusivity given by the local free convection similarity relation, that is,

$$K_z = a_1 \left(\frac{g}{T_0} \frac{H_0}{\rho c_p} \right)^{1/3} z^{4/3} \qquad (8.56)$$

in which a_1 is another empirical similarity constant. His estimates of the above constants are: $a \simeq 0.67\, a_1^{3/2}$ and $a' \simeq 0.84\, a_1^{3/2}$. The constant a_1 in Eq. (8.56) can be determined from the micrometeorological measurements of fluxes and gradients in the unstable surface layer. If one assumes $K_z = K_w$ or $K_z = K_h$, experimental data on diffusivities of heat and water vapor give $a_1 \simeq 1.1$ in the local free convection regime (Dyer, 1965, 1967). Eddy diffusivities of heat and water vapor may differ considerably under conditions of sensible heat advection (Verma et al., 1978), in which case $K_z = K_w$ would be a better assumption.

There is also some conflicting experimental evidence about the eddy diffusivity of heat under very unstable conditions approaching free convection. The empirical forms of the Monin–Obukhov similarity function $\phi_h(z/L)$ based on the micrometeorological data over homogeneous sites (Dyer, 1965, 1967; Businger et al., 1971) imply

$$K_h \simeq 4k^{3/2} u_*^{-1/2} \left(\frac{g}{T_0} \frac{H_0}{\rho c_p} \right)^{1/2} z^{3/2} \qquad (8.57)$$

for $-z/L \gg 1$. However, Eq. (8.57) cannot be expected to remain valid in the free convection limit of $u_* \to 0$; Eq. (8.56) would be more appropriate in that limit. Yet, using an expression similar to Eq. (8.57), Venkatram (1992) derived different expressions for the asymptotic behavior of the vertical plume spread and the cross-wind integrated ground-level concentration

$$\sigma_z \sim \frac{x^2}{|L|} \qquad (8.58)$$

$$\bar{c}_y \sim \frac{Q|L|}{u_* x^2} \qquad (8.59)$$

He also presented limited observational data from the Project Prairie Grass experiment in support of Eq. (8.59). More extensive and broader experimental support has been shown for the local free convection similarity expression (8.55), which implies $\bar{c}_y \propto x^{-3/2}$ and is consistent with observed data (see e.g., Nieuwstadt, 1980b; Briggs, 1988).

8.4 THE MIXED-LAYER SIMILARITY THEORY

8.4.1 Convective Similarity Scaling

Following the introduction of the mixed-layer similarity theory in Chapter 4, section 4.7, it is clear that Deardorff (1970b, 1972) originally proposed the use of the mixed-layer depth h and the convective velocity W_* as the appropriate length and velocity scales for representing turbulence in the convective boundary layer (CBL). The corresponding time scale h/W_* was shown to be the characteristic time for large-eddy motions (convective updrafts and downdrafts) in the CBL. Later, Deardorff and Willis (1974, 1975) extended and applied the same mixed-layer scaling ideas to modeling diffusion in the CBL using a convection tank or chamber. They defined the appropriate dimensionless travel (diffusion) time and distance as

$$t^* = \frac{t\,W_*}{h} \qquad (8.60a)$$

$$X^* = \frac{W_*}{\bar{u}} \frac{x}{h} \qquad (8.60b)$$

Note that t^* and X^* are equivalent only if one can invoke Taylor's hypothesis that $t = x/\bar{u}$, which implies that the turbulent diffusion in the direction of mean flow is negligible in comparison with the mean transport. This puts some limit on the value of $W_*/\bar{u} \leq 1.5$ for the validity of Taylor's hypothesis and, hence, on the use of X^*, in place of t^*, for representing experimental diffusion data for the convective mixed layer.

8.4.2 Similarity Relations

The mixed-layer similarity theory utilizes the convective scaling for representing turbulence and diffusion data in the convective mixed layer and empirically determining the forms of the various similarity functions. For example, dimensionless mean concentration and cross-wind integrated concentration in a continuous point source plume can be expressed as

$$\frac{\bar{c}\,\bar{u}\,h^2}{Q} = C^*(X^*, Y^*, Z^*; H^*)$$

$$\frac{\bar{c}_y\,\bar{u}\,h}{Q} = C_y^*(X^*, Z^*, H^*) \qquad (8.61)$$

where X^* is defined in Eq. (8.60b), $Y^* = y/h$, $Z^* = z/h$, and $H^* = H/h$. In particular, the normalized ground-level concentration along the plume centerline C_0^* is expected to be a function of X^* and H^* only. One can also represent the dimensionless plume dispersion parameters as functions of the same parameters, that is,

$$\frac{\sigma_y}{h} = F_y(X^*, H^*)$$

$$\frac{\sigma_z}{h} = F_z(X^*, H^*) \qquad (8.62)$$

These are generally based on the assumption of Gaussian distribution of concentration in both the lateral and vertical directions (for an elevated source above a reflecting ground surface, the reflected Gaussian distribution accounting for the reflection from the surface is implied). The lateral distribution of C^* is found to be closely Gaussian, particularly for neutrally and positively buoyant releases. Vertical concentration distributions in surface and elevated plumes in the CBL often deviate from the Gaussian or the reflected Gaussian form (see e.g., Willis and Deardorff, 1976, 1978; Eberhard et al., 1988; Briggs, 1993). For these, one can define the mean particle height

$$\overline{Z} = \int_0^h Z p(Z) dZ \qquad (8.63)$$

and the standard deviation of particle displacement

$$\sigma_{z'} = (\overline{Z'^2})^{1/2} = \left[\int_0^h (Z - H)^2 p(Z) dZ \right]^{1/2} \qquad (8.64)$$

where $p(Z)$ is the probability density function of vertical particle displacement Z. The standard deviation of lateral particle displacement Y' from the mean lateral position (also release position) is similarly defined. These particle dispersion parameters, when normalized by the mixing height h, are expected to be similarity functions of X^* and H^*, that is,

$$\frac{\sigma_{Y'}}{h} = F_Y(X^*, H^*)$$

$$\frac{\overline{Z}}{h} = F_Z(X^*, H^*)$$

$$\frac{\sigma_{Z'}}{h} = F_{Z'}(X^*, H^*) \qquad (8.65)$$

for passive, nonbuoyant releases in the CBL. For buoyant releases, an additional dimensionless source buoyancy flux parameter F^* also enters the above similarity relations.

8.4.3 Empirical Forms of Similarity Functions

The forms of the various similarity functions representing diffusion in the CBL have been empirically determined from diffusion experiments in laboratory convection tanks (Deardorff and Willis, 1974, 1975; Willis and Deardorff, 1976, 1978, 1981) and, more recently, from atmospheric diffusion experiments (Briggs, 1985, 1988, 1993; Eberhard et al., 1988). Concentration distributions as well as particle statistics were measured in the laboratory experiments, whereas at-

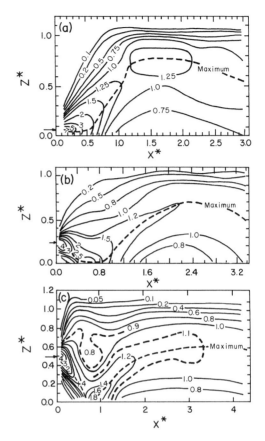

Figure 8.7 Contours of the normalized cross-wind integrated concentration c_y^* as functions of X^* and Z^* for non-buoyant releases at different heights indicated by horizontal arrows: (a) $H^* = 0.067$. (b) $H^* = 0.24$. (c) $H^* = 0.49$. From Willis and Deardorff, 1976, 1978, 1981.

mospheric diffusion data represent only the concentration fields. The latter also contain more variability due to inadequate sampling and the influence of large convective eddies (updrafts and downdrafts). The validity and usefulness of mixed-layer similarity scaling have been amply demonstrated by these diffusion experiments, as well as by sophisticated numerical simulations such as large-eddy simulation (LES) of diffusion in the CBL (Deardorff and Willis, 1974; Lamb, 1978, 1979, 1982; Nieuwstadt, 1992a&b; Henn and Sykes, 1992). Several interesting and novel aspects of dispersion in the CBL have been discovered.

The most surprising and somewhat unexpected result was the lifting off of the height of maximum concentration in plumes from near-surface releases ($H^* < 0.1$). This is clearly shown in the observed concentration fields in Willis and Deardorff's convection tank experiments represented in Figure 8.7(a). Here, contours of the dimensionless cross-wind integrated concentration

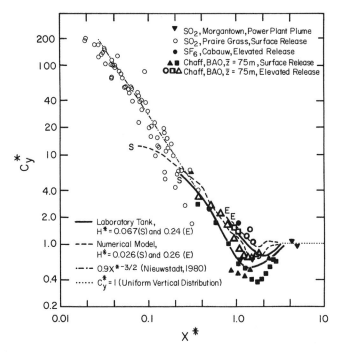

Figure 8.8 Normalized cross-wind integrated ground-level concentration c_{y0}^* as a function of X^* and H^*. Symbols E and S represent best-fit curves for elevated and surface sources, respectively. BAO, Boulder Atmospheric Observatory. From Briggs, 1988.

$$C_y^* = \int_{-\infty}^{\infty} C^*(X^*, Y^*, Z^*) dY^* \qquad (8.66)$$

are plotted in a vertical cross section through the plume centerline. The height of the maximum in C_y^*, or the plume centerline, coincides with the release height only up to a dimensionless distance $X^* \simeq 0.5$, after which it lifts off rapidly, attains a maximum around $Z^* = 0.8$, and then gradually comes down to the surface. This implies a highly non-Gaussian vertical distribution of concentration for $X^* > 0.5$.

In contrast to near-surface releases, mean concentration fields in plumes from elevated releases show the plume centerline (height of the maximum C_y^*) descending from the release height to the surface, with a speed of descent of about $0.5 \, W_*$, and then ascending, similar to the case of near-surface release (see Fig. 8.7(b) and (c)). The resulting ground-level concentrations where plume descends to the surface are much higher than the maximum g.l.c. predicted by standard Gaussian plume diffusion models. At large X^*, material released from any height becomes vertically well mixed, with nearly uniform concentration in the CBL.

The above observed plume behaviors in convection tank experiments have also been simulated or discovered in sophisticated numerical models of dispersion in the CBL (see e.g., Lamb, 1978, 1979, 1982) and later confirmed by atmospheric diffusion measurements (Briggs, 1985, 1988, 1993). These have been explained largely in terms of the large-eddy circulations (convective updrafts and downdrafts) and their role in transport and diffusion of material in the CBL. It is well known that updrafts typically occupy about 40 percent of the horizontal area and are characterized by strong vertical motions, convergence of horizontal flow near the surface, and divergence at the top of the CBL. In contrast, downdrafts occupy a much larger (~60%) area and are characterized by relatively weaker downward motion, divergence near the surface, and convergence near the top of the CBL.

For a near-surface release, the material emitted into the base of an updraft begins to ascend, whereas that released into a downdraft remains near the ground and spreads horizontally. Updrafts and downdrafts are alternately advected over the location of the continuous release. Since downdrafts occupy most of the horizontal area, more material remains near the ground initially, as does the maximum concentration. After a substantial amount of material is swept out of downdrafts into the neighboring updrafts, however, the plume centerline begins to rise or "lift off" the ground. The subsequent descent of the plume centerline or the height of maximum concentration is due to the large area occupied by downdrafts and the higher probability of material getting into them. This also explains the observed initial descent of plume centerline for elevated releases ($H^* > 0.1$) in the CBL. Once the bulk of the material reaches the surface, it behaves like that from a near-surface release and subsequently lifts off after being caught in updrafts.

Similar features of cross-wind integrated concentration fields in the CBL have been observed in the CONDORS (convective diffusion observed with remote sensors) field experiment conducted at Boulder Atmospheric Observatory (Briggs, 1988, 1993). It is interesting to examine the variation of dimensionless cross-wind integrated ground-level concentration with dimensionless distance from the source. Figure 8.8 compares the results of several field experiments, laboratory convection tank experiments, and a numerical (LES) model. Note that for surface releases, C_{y0}^* decreases initially as $X^{*-3/2}$ (which is consistent with the prediction of the local free convection similarity theory), attains a minimum value less than 1, and then approaches to the constant value (1.0) corresponding to the uniform concentration distribution in the far field. The data for elevated releases show similar behavior, except for their different behavior near the source where C_{y0}^* is expected to increase and attain a maximum before dropping off with further increase in distance. All the experimental and numerical model results show a dip in C_{y0}^* before it levels off to its constant value of unity. This dip is caused by the lift-off of the maximum in C_y^* from the surface by convective updrafts.

The strong influence of convective updrafts and compensating downdrafts on vertical transport and diffusion of material in the CBL results in non-Gaussian distributions of C_y^* in the vertical, particularly at $X^* > 0.5$, as shown in Figure 8.9. Here the measured distributions in a convection tank are compared with those based on the reflected Gaussian plume model for an elevated release with $H/h = 0.24$. Similar deviations from the Gaussian distribution are also observed for the near-surface releases, especially downwind of the location where the plume centerline lifts off ($X^* > 0.5$) (Willis and Deardorff, 1976).

The ensemble-averaged lateral concentration profiles in surface and elevated plumes from nonbuoyant releases are observed to be closely Gaussian (Willis and Deardorff, 1976, 1978). Measured concentration distributions in field experiments (see e.g., Briggs, 1993) usually suffer from inadequate sampling and averaging and rarely resemble the smooth Gaussian form. However, the standard deviations of lateral and vertical particle or plume spread are considered to be more robust statistics. Figure 8.10 shows the particle dispersion parameters measured by Willis and Deardorff in their convection tank experiments. These are consistent with the statistical theory prediction (Taylor's theorem) for small travel times. Following the initial linear growth, the normalized standard deviations $\sigma_{Y'}/h$ and $\sigma_{Z'}/h$ follow different trends with increasing dimensionless time or distance X^*. The lateral dispersion parameter grows monotonically with X^*, but the approach toward the theoretically predicted behavior for large travel times or distances seems to depend on the dimensionless release height; it is more clearly seen for

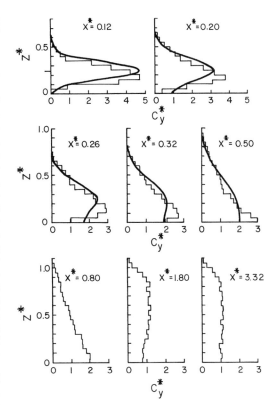

Figure 8.9 Observed vertical distributions of normalized cross-wind integrated concentration c_y^* histogram compared with the best-fit Gaussian distribution for different values of X^*. From Willis and Deardorff, 1978.

the release in the middle of the convective mixed layer. The vertical dispersion parameter, on the contrary, rapidly flattens out to its equilibrium value, which depends on the dimensionless release height. This is obviously due to the reflection of the material from the surface as well as from the inversion base.

A mixed-layer similarity plot of σ_y/h versus X^* for surface and elevated releases in the CBL is shown in Figure 8.11. It includes some field experimental data and compares them with laboratory tank and numerical (LES) model results. On the basis of this, Briggs (1985, 1988) suggested the following empirical relations:

$$\frac{\sigma_y}{h} = 0.6X^*, \qquad \text{for } H^* < 0.1 \qquad (8.67)$$

$$\frac{\sigma_y}{h} = \frac{0.6 X^*}{(1 + 2X^*)^{1/2}}, \qquad \text{for } H^* \geq 0.1 \qquad (8.68)$$

which are also consistent with the predictions of the statistical diffusion theory. The coefficient in the linear relationship for small X^* represents the ratio σ_v/W_*, which is observed to be around 0.6 in the convective mixed layer and between 0.6 and 1.0 in the

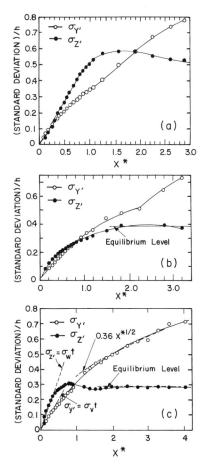

Figure 8.10 Normalized standard deviations of particle displacements in the lateral and vertical directions for different dimensionless release heights: (a) $H^* = 0.067$. (b) $H^* = 0.24$. $H^* = 0.49$. From Willis and Deardorff, 1976, 1978, 1981.

surface layer, depending on the contribution of the shear-generated turbulence. Using some additional power-plant plume data, extending the range of X^* up to 20, Briggs (1993) proposed a different empirical relationship for elevated releases

$$\frac{\sigma_y}{h} = \frac{0.6X^*}{(1 + 2X^{*2})^{1/6}} \quad (8.69)$$

and a more complicated interpolation formula for the surface release, which asymptotically tends to $\sigma_y/h = X^*$ for very small X^*, and to Eq. (8.69) for large values of X^*.

Empirical estimates of σ_z based on vertical concentration distributions in the CBL are very limited. Figure 8.12 shows a convective similarity plot of σ_z/h versus X^* for the CONDORS field experiment, which was conducted primarily for the purpose of verifying the mixed-layer similarity theory of diffusion and determining σ_y/h, σ_z/h, and so on, as functions of X^* and H^*. In spite of the large scatter, these data confirm the linear increase of σ_z/h with X^* for small values of the latter. The convergence of σ_z/h to a constant value for large X^* is a consequence of the lid on vertical mixing (a uniform concentration distribution due to complete mixing at $X^* \gg 1$ implies a constant value of $\sigma_z/h = 1/\sqrt{12} \simeq 0.29$). Note that σ_z is defined here as

$$\sigma_z = \left[\int_0^h (z - \bar{z})^2 \bar{c}(z) dz / \int_0^h \bar{c}(z) dz \right]^{1/2} \quad (8.70)$$

which includes the influence of reflections from the surface and the top of the CBL and is not the same as the unbounded σ_z used in Gaussian plume models.

Considering all the laboratory and field diffusion data, the simplest parameterization that can be suggested for σ_z is for small $X^* < 1$, that is,

$$\frac{\sigma_z}{h} \simeq 0.9 X^{*3/2}, \quad \text{for } H^* < 0.1 \quad (8.71)$$

$$\frac{\sigma_z}{h} \simeq 0.6 X^*, \quad \text{for } H^* \geq 0.1 \quad (8.72)$$

Note that Eq. (8.71) is also consistent with the prediction (8.55) of the local free convection similarity theory discussed earlier in this chapter. This may be strictly valid only for the surface release ($H^* = 0$), rather than for the significantly elevated releases in the surface layer.

Convective similarity plots of $\bar{z}/h$ versus X^* for different release heights are shown in Figure 8.13. There is considerable scatter in the field data for small X^*, but these are in fair agreement with Willis and Deardorff's (1976, 1978, 1981) convection tank results. As expected, all data approach the asymptotic value of $\bar{z}/h = 0.5$ for uniform concentration distribution at large X^*. Most data also indicate an overshoot in $\bar{z}/h$ over the above value in the range $1 < X^* < 3$, with the exception of the CONDORS oil-fog data for the periods 32–33, which had large mean downward motion (Fig. 8.13(a)). For elevated sources, the mean plume height trajectory is expected to be strongly dependent on the dimensionless release height H^*. The field experimental data shown in Figure 8.13(b) correspond to values of H^* between 0.32 and 0.34, while the two curves from Willis and Deardorff (1978, 1981) are for $H^* = 0.24$ and 0.49. The chaff data also contain some bias due to the chaff settling velocity of nearly -0.3 m s^{-1}. The hypothetical mean particle trajectory due to settling alone is also shown in Figure 8.13(b).

8.5 EXPERIMENTAL VERIFICATION OF SIMILARITY THEORIES

Usefulness of similarity scaling in systematic ordering and comparison of diffusion data from the various laboratory and field experiments has been amply demonstrated. Most of these data have also been used

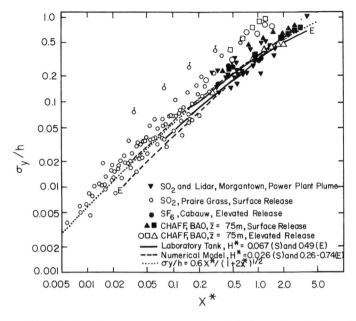

Figure 8.11 Normalized lateral dispersion parameter as a function of X^* for surface and elevated releases in the convective boundary layer. BAO, Boulder Atmospheric Observatory. From Briggs, 1985.

to verify the predictions of the similarity theories of dispersion and for empirically determining the forms of the various similarity functions. For example, diffusion experiments in wind tunnels provided an early verification of the Lagrangian similarity hypothesis for the near-neutral surface layer (Cermak, 1963). Field experimental data on diffusion in the surface layer provided more general and broader confirmation of the surface-layer similarity theory (Pasquill, 1966; Chaudhry and Meroney, 1973; Nieuwstadt and van Ulden, 1978; Horst, 1979; Briggs, 1993). These data were limited, however, to mean concentration measurements in a fixed frame of reference. Lagrangian measurements to determine the statistics of particle velocities and displacements and, hence, the various similarity functions describing particle dispersion in the stratified surface layer are still not available. Mean concentration measurements in tracer plumes from near-surface releases have generally confirmed the predictions of the similarity theory of diffusion in the surface layer, especially

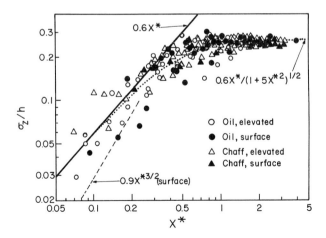

Figure 8.12 Normalized vertical dispersion parameter as a function of X^* for surface and elevated releases in the convective boundary layer during the CONDORS (convective diffusion observed with remote sensors) field experiment. From Briggs, 1993.

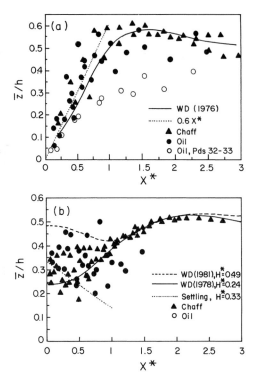

Figure 8.13 Dimensionless mean particle or plume height as a function of X^* for different dimensionless release heights in the convection tank and during CONDORS field experiment. WD, Willis and Deardorff. From Briggs, 1993.

those obtained from the joint application of similarity and gradient transport theories (van Ulden, 1978; Nieuwstadt and van Ulden, 1978; Horst, 1979).

Diffusion data taken during unstable and convective conditions have been used to verify the predictions of the local free convection similarity theory (Nieuwstadt, 1980b), although there is also some evidence of deviation from the same due to the effects of finite shear (Venkatram, 1992). The local free convection similarity relations can also be recast in terms of more widely used mixed-layer similarity theory.

The very basis and early support for the mixed-layer similarity theory came from three-dimensional numerical modeling (LES) of turbulence and diffusion in the CBL by Deardorff (1972) and physical modeling of the same in a convection tank by Deardorff and Willis (1974, 1975). These pioneering studies were followed by other experimental studies (Willis and Deardorff, 1976, 1978, 1981), as well as by numerical modeling studies (Lamb, 1978, 1979, 1982; Sun, 1989; van Haren and Nieuwstadt, 1989; Henn and Sykes, 1992), all confirming the usefulness and validity of the mixed-layer similarity theory. Further support for the theory has been provided by the extensive diffusion data collected during the CONDORS field experiment (Briggs, 1985, 1993). The empirical forms of the various similarity functions have been given in the previous section.

8.6 APPLICATIONS TO DISPERSION IN THE PBL

In spite of the fairly long and rich history of the development of the Lagrangian similarity theory of diffusion in the surface layer, applications of this theory to atmospheric dispersion modeling have been limited. This is largely due to the lack of diffusion parameterization in terms of the basic similarity parameter z/L and also to the restriction of the theory to small travel times and near-surface sources, to ensure that plume remains within the surface layer.

The usefulness and validity of the surface-layer similarity analysis has been amply demonstrated (Pasquill, 1966; Horst, 1979; Briggs, 1988). Also, simple ways of estimating the Obukhov length L using on-site measurements of wind speed at one level and temperature difference between two levels have been proposed (Irwin and Binkowski, 1981; Arya, 1995). One can also estimate L from the simpler determinations of Pasquill's stability class and the surface roughness parameter, using some empirical nomograms (Golder, 1972). A generally valid parameterization of σ_y and σ_z as functions of the appropriate similarity parameters is still not available. The development of such a parameterization scheme for use in Gaussian plume models would require new diffusion experiments to be conducted at sites of different roughness where suitable micrometeorological measurements are also made. Analytical non-Gaussian models might be more appropriate for representing dispersion in the surface layer (van Ulden, 1978).

The use of the mixed-layer scaling and similarity theory has found more real and potential applications in dispersion modeling, because it has become more widely accepted and covers the whole boundary layer under unstable and convective conditions. It has been proposed for Gaussian models with new and generalized dispersion parameterization schemes (Irwin, 1979a&b; Weil and Brower, 1984; Briggs, 1985, 1988, 1993). The basic stability parameter for dispersion parameterization is the ratio $W_*/\overline{u}$ or W_*/u_*; the latter is also uniquely related to the PBL stability parameter h/L. This parameter is considered to be important only for parameterizing diffusion in the unstable surface layer in which it can also be related to Pasquill's stability classes A, B, C, and D (see e.g., Weil and Brower, 1984). Turbulence and diffusion in the convective mixed layer become more or less independent of $W_*/\overline{u}$ or h/L. According to the similarity scaling, dimensionless dispersion parameters σ_y/h, σ_z/h, and $\overline{z}/h$ are unique functions of only the dimensionless diffusion time or distance X^*, the dimensionless release height H^*, and possibly also the dimensionless plume buoyancy flux F^*. For the passive nonbuoyant plumes in the CBL, the approximate empirical forms of these functions have already been discussed in the previous section. These can be easily used in Gaussian or non-Gaussian plume dispersion models during most of the daytime unstable and convective conditions.

8.7 LIMITATIONS OF SIMILARITY THEORIES

Similarity theories of diffusion have several limitations that arise from their underlying similarity hypotheses and other simplifying assumptions. The most serious limitation of all the similarity theories discussed in this chapter is that these theories are applicable only to diffusion in horizontally homogeneous turbulent flows. For diffusion in the atmospheric boundary layer or the surface layer, this also implies flat and homogeneous terrain with a uniform surface roughness. The assumption of stationarity of flow is also implied in all the similarity hypotheses. This may not constitute a severe limitation for applying the similarity theories to commonly observed diurnal variations of the boundary layer in response to the diurnal cycle of surface heating and cooling. The flow can be assumed to be quasi-stationary, at least for short periods of the order of an hour or so. This assumption may not be justified, however, during the morning and evening transition periods and also during the rapid passage of fronts and storms, when rapid changes might occur in the boundary layer height, mean winds, and turbulence.

Other limitations of similarity theories arise from the implied restrictions of stability and flow regime in their formulations. For example, the surface-layer similarity theories can be applied only to the near-surface releases and relatively short travel times or distances, so that plume remains within the surface layer. The mixed-layer similarity theory, on the other hand, is applicable to all elevated sources in the convective boundary layer and for much longer travel times or distances. A generalized similarity theory of diffusion in the PBL under all stability conditions is still lacking. In extremely stable conditions, in particular, turbulence becomes very sporadic in space and intermittent in time. It may not be possible to describe diffusion in such an environment, using a simple similarity approach.

PROBLEMS AND EXERCISES

1. a. Explain how veering of wind with height might contribute to the lateral spreading of a continuous point source plume in the PBL.
   b. Consider the lateral plume dispersion from an elevated release in the middle of a nocturnal stable boundary layer of 300 m depth over an urban area. If σ_y and σ_z due to turbulence diffusion alone can be estimated from Briggs' (1973) formulas $\sigma_y = 0.11\ x(1 + 0.0004\ x)^{-1/2}$, $\sigma_z = 0.08\ x(1 + 0.0015\ x)^{-1}$, in which σ_y, σ_z, and x are expressed in meters, estimate the lateral plume spread at distances of 5, 10, 20, and 40 km from the source with and without the effect of wind-direction shear when the total wind veering across the PBL is 30°.
2. a. Using the Lagrangian similarity theory of diffusion in the neutral surface layer, derive Eq. (8.6) for the mean particle trajectory and reduce it to Eq. (8.8) using the concept of virtual source at the surface.
   b. Calculate and plot mean particle trajectories in dimensionless coordinates for dimensionless release heights $H/z_0 = 0$, 50, and 500 in the neutral surface layer of height $h_s/z_0 = 10^5$.
3. a. Following the steps in the derivation of Eq. (8.20) for the g.l.c. due to a continuous point source, derive a similar expression for the g.l.c. due to a continuous line source.
   b. Compare the above expressions for the g.l.c. based on the similarity theory with those based on the gradient transport theory.
4. a. Derive the statistical theory expression (8.22) for the mean-square displacement of particles released at the surface and simplify it further, considering the observed properties of turbulence in the neutral surface layer.
   b. Explain why mean and fluctuating Lagrangian vertical motions contribute equally to the vertical dispersion, even though $\overline{W} \ll \sigma_w$.
5. a. Discuss the effect of stability on the mean particle trajectory in a stratified surface layer.
   b. Determine the slopes of mean particle trajectories shown in Figure 8.4 at large distances from the release point for different values of z_0/L and plot the same as a function of z_0/L. What conclusions can you draw from this plot?
6. a. Using van Ulden's (1978) formula (8.42), plot and compare the dimensionless concentration profiles ($\overline{c}/\overline{c}_0$ vs. $z/\overline{z}$) for different values of the profile exponent $\alpha = 1$, 1.5, 2.0, and 2.5.
   b. How is the 10-percent plume thickness z_p related to $\overline{z}$ for the above profile formula?
7. a. Using the appropriate empirical forms of the Monin–Obukhov similarity functions, given in Chapter 4, in conjunction with van Ulden's formula (8.49) or (8.50), derive an analytical expression for x/z_0 as a function of $\overline{z}/z_0$ and z_0/L for the neutral and stably stratified surface layers.
   b. Derive an analytical expression for the mean transport velocity $\overline{u}_p$ in the plume from surface releases under neutral and stable conditions.
8. a. Using the expressions for $\overline{z}$ and $\overline{u}_p$ derived in problem 7, express the cross-wind integrated g.l.c. as a function of x/z_0 and z_0/L.
   b. Plot the cross-wind integrated g.l.c. due to a continuous point source at the surface, using the surface-layer similarity framework for neutral and stable conditions.
9. a. A passive tracer is released continuously from a smooth ground surface ($z_0 = 0.01$ m) at the rate of 10 g s^{-1} when the wind speed at the 10-m height is 10 m s^{-1} and sky is overcast with clouds. Calculate the cross-wind integrated g.l.c. at a distance of 500 m downwind of the source, using the van Ulden model.

b. What would be the cross-wind integrated g.l.c. at 500 m under moderately stable conditions (say, $L = 20$ m) when wind speed decreases to 2 m s^{-1}.

10. a. Using the Lagrangian similarity hypothesis for the free convective surface layer, show that the Lagrangian mean vertical velocity is proportional to the local free convection velocity u_f.

b. Using the above similarity hypothesis, derive an expression from the cross-wind integrated g.l.c. due to a continuous point release at the surface in the convective boundary layer. Express the same in terms of the mixed-layer similarity scaling, pointing out its obvious limitations.

11. a. What are the alternative definitions of the vertical dispersion parameter used in the literature? Discuss the usefulness and limitations of each in modeling dispersion in the CBL.

b. Discuss the possible causes for the observed non-Gaussian vertical concentration distributions in the CBL.

12. A chimney emits sulfur dioxide at a rate of 0.5 kg s^{-1} at an effective height of 360 m in the CBL of depth 1500 m and mean wind speed 5 m s^{-1}. The surface heat flux is estimated to be 400 W m^{-2} and the near-surface air temperature is 27°C. Estimate the maximum g.l.c. and its distance from the source using the following models/simulations: (a) Willis and Deardorff's (1978) convection tank simulation of diffusion from an elevated source; (b) the Gaussian plume model with the simple dispersion parameterization $\sigma_y/h = \sigma_z/h \simeq 0.6\ X^*$, for short distances in the CBL.

13. a. Discuss the usefulness and practical applications of Lagrangian similarity theories in modeling dispersion from near-surface sources.

b. What are the limitations of the surface-layer similarity theories?

14. a. Contrast between the local free convection and mixed-layer similarity theories of diffusion.

b. Discuss the usefulness and applications of the mixed-layer similarity theory to describing dispersion in the CBL.

9

Gaussian Diffusion Models

9.1 BASIS AND JUSTIFICATION FOR GAUSSIAN MODELS

9.1.1 Theoretical Basis

Gaussian diffusion models are extensively used in assessing the impacts of existing and proposed sources of air pollution on local and urban air quality, particularly for regulatory applications. The history of Gaussian diffusion modeling goes back to the early 1920s when foundations of gradient transport and statistical theories of diffusion were laid (Taylor, 1921; Roberts, 1923; Richardson, 1926). Earlier, temperature distributions in certain heat conduction problems were shown to be Gaussian, as were the concentration distributions in problems of molecular (Brownian) diffusion. In his classical paper on scattering of smoke in a turbulent atmosphere, Roberts (1923) obtained the solutions to the mean diffusion equation with constant-eddy diffusivities for different source configurations. His solutions showed Gaussian distributions in ensemble-averaged smoke puffs from an instantaneous point source. Gaussian concentration distributions were also derived in plumes from continuous point and line sources under certain simplifying assumptions (equivalent to the slender-plume approximation). Although Roberts (1923) did not put his solutions in the standard Gaussian forms, using σ_x, σ_y, and so on, he provided the original theoretical basis for what came to be known as Gaussian diffusion models. A stronger theoretical basis, without recourse to the questionable gradient transport hypothesis, was later provided by the statistical theories and random-walk models of particle dispersion in homogeneous turbulent flows (Csanady, 1973; Pasquill and Smith, 1983).

Theoretical basis for Gaussian diffusion models is limited to idealized uniform flows with homogeneous turbulence. For continuous point and line sources, mean wind speed is also required to be larger than the standard deviations of turbulent velocity fluctuations, so that the upstream or longitudinal diffusion can be neglected. Mean winds and turbulence encountered in the atmosphere, particularly in the PBL, rarely satisfy the above simplifying assumptions of the theory. Frequently, one encounters significant wind shears, inhomogeneities of turbulence, and weak winds to make the theoretical basis of Gaussian diffusion modeling somewhat tenuous, if not totally invalid.

9.1.2 Other Justifications

The primary justification for the use of Gaussian diffusion models in regulatory applications comes from their evaluation and validation against experimental diffusion data. Much of the data used for this purpose are, however, limited to near-field or maximum ground level concentrations (g.l.c.) and the meteorological and source parameters that are used as input to Gaussian models. Since the regulatory emphasis is also on near-source maximum g.l.c., the limited evaluation and validation of models against observed g.l.c. may be well justified. Other reasons for using Gaussian diffusion models in regulatory applications are (Hanna, 1982):

1. They are analytical and conceptually appealing.
2. They are consistent with the random nature of turbulence.
3. They are computationally cheaper to use.
4. They have acquired an official "blessed" status in regulatory guidelines (U.S. Environmental Protection Agency, 1986, 1996).

9.2 GAUSSIAN PLUME AND PUFF DIFFUSION MODELS

9.2.1 The Gaussian Plume Model

The cornerstone of most dispersion calculations in regulatory applications is the Gaussian plume formula for a continuous point source in a uniform flow with homogeneous turbulence:

$$\bar{c}(x,y,z) = \frac{Q}{2\pi\bar{u}\sigma_y\sigma_z}$$
$$\exp\left[-\frac{y^2}{2\sigma_y^2} - \frac{z^2}{2\sigma_z^2}\right] \quad (9.1)$$

Here Q is the source strength or emission rate, $\bar{u}$ is the mean transport velocity across the plume, and σ_y and σ_z are the Gaussian plume dispersion parameters. Equation (9.1) can be derived simply from the assumption of Gaussian concentration distributions in y and z directions at any cross section in the plume downwind of the source, and the integral mass-conservation condition

$$\int_{-\infty}^{\infty}\int_{-\infty}^{\infty} \bar{c}\bar{u}\, dy\, dz = Q \quad (9.2)$$

Since most point sources of pollution in the atmosphere are located at or near the earth's surface (e.g., vents, valves, stacks, and chimneys), it is necessary to account for the presence of the ground surface. Assuming perfectly reflecting surface as the most conservative boundary condition, addition of an "image" source (see e.g., the method of images described in Chapter 6) yields the following reflected-Gaussian formula for a surface or an elevated release:

$$\bar{c}(x,y,z; H) = \frac{Q}{2\pi\bar{u}\sigma_y\sigma_z}\exp\left[-\frac{y^2}{2\sigma_y^2}\right]$$
$$\left\{\exp\left[-\frac{(z-H)^2}{2\sigma_z^2}\right] + \exp\left[-\frac{(z+H)^2}{2\sigma_z^2}\right]\right\} \quad (9.3)$$

where H is the effective height of release above the ground. Note that Eq. (9.3) satisfies the mass-conservation condition (9.2) with the lower limit on z changed from $-\infty$ to 0. Following D. B. Turner (1994), Eq. (9.3) can be thought of as a product of four factors:

1. The emission factor, Q, which indicates that concentration at any point is directly proportional to the emission rate or the source strength.
2. The inverse mean wind factor, $1/\bar{u}$, which indicates that concentrations are inversely proportional to mean wind speed.
3. The cross-wind dispersion factor

$$F_y = \frac{1}{\sqrt{2\pi}\sigma_y}\exp\left[-\frac{y^2}{2\sigma_y^2}\right] \quad (9.4)$$

which indicates that along the y axis, concentration is normally distributed, with the maximum value at the plume centerline which is inversely proportional to the lateral dispersion parameter, σ_y.

4. The vertical dispersion factor

$$F_z = \frac{1}{\sqrt{2\pi}\sigma_z}$$
$$\left\{\exp\left[-\frac{(z-H)^2}{2\sigma_z^2}\right] + \exp\left[-\frac{(z+H)^2}{2\sigma_z^2}\right]\right\} \quad (9.5)$$

which indicates that, in the vertical direction (z axis), concentration distribution is given by the sum of the Gaussian distributions due to real and image sources, respectively, and concentration is inversely proportional to the vertical dispersion parameter, σ_z.

In terms of the above factors, Eq. (9.3) can simply be expressed as

$$\bar{c}(x,y,z; H) = \frac{Q}{\bar{u}}F_y F_z \quad (9.6)$$

In particular, the ground-level concentration (g.l.c.) along the plume centerline is given by

$$\bar{c}_o(x; H) = \frac{Q}{\pi\bar{u}\sigma_y\sigma_z}\exp\left(-\frac{H^2}{2\sigma_z^2}\right) \quad (9.7)$$

which is the basic Gaussian plume formula for computing ground-level concentrations due to surface ($H = 0$) and elevated sources and, for the latter, determining the maximum g.l.c. ($\bar{c}_{o\max}$) and its location ($x_{\max}$) downwind of the source.

One can also determine the cross-wind integrated g.l.c. from Eq. (9.3) as

$$\bar{c}_{yo} = \left(\frac{2}{\pi}\right)^{1/2}\frac{Q}{\bar{u}\sigma_z}\exp\left(-\frac{H^2}{2\sigma_z^2}\right) \quad (9.8)$$

which is obviously related to the plume-centerline concentration. A similar expression is obtained for the g.l.c. due to an infinite continuous line source, that is,

$$\bar{c}_o(x; H) = \left(\frac{2}{\pi}\right)^{1/2}\frac{Q_{cl}}{\bar{u}\sigma_z}\exp\left(-\frac{H^2}{2\sigma_z^2}\right) \quad (9.9)$$

where $\bar{u}$ is the mean wind component perpendicular to the line source. Concentration fields resulting from more realistic finite line and area sources can be obtained through numerical integration of the point source formula along the line segment or the source area of interest, using the principle of superposition, which has been discussed earlier in Chapter 6.

Many assumptions and approximations are implied in the Gaussian plume model. Some of the important assumptions are (Lyons and Scott, 1990):

1. Continuous emission from the source at a constant rate, at least for a time equal to or greater than the time of travel to the location (receptor) of interest. The plume diffusion formulae assume that release and sampling times are long compared with the travel time

to receptor, so that the material is spread out in the form of a steady plume between the source and the farthest receptor. A shorter release will result in an elongated puff with a time-dependent concentration field.

2. Steady-state flow and constant meteorological conditions, at least over the time of transport (travel) from the source to the farthest receptor. This assumption may not be valid during rapidly changing meteorological conditions, such as during the passage of a front or a storm and also during the morning and evening transition periods.

3. Conservation of mass in the plume. The continuity equation satisfied by the Gaussian plume formula (9.3) is a mathematical expression of the condition that the mass flow rate through any plume cross section is equal to the source emission rate. This implies that none of the material is removed through chemical reaction, gravitational settling, or deposition at the surface. All the material reaching the surface through turbulent diffusion is reflected back and none is absorbed there.

4. Gaussian or reflected Gaussian distribution of mean concentration in the lateral (cross-wind) and vertical directions at any downwind location in the plume. The assumption of Gaussian distribution in the vertical direction is somewhat questionable, but does not appear to affect adversely the model predicted ground-level concentrations.

5. A constant mean transport wind in the horizontal (x–y) plane. This implies horizontal homogeneity of flow and the underlying surface and becomes invalid over a complex terrain.

6. No wind shear in the vertical. This assumption is implicit in the constant mean transport velocity $\bar{u}$ in the Gaussian plume formulae. In practice, $\bar{u}$ is often taken as the wind speed at 10 m height for near-surface sources ($H < 10$ m) and the wind speed at the effective release height for elevated sources. The variation of wind speed with height can also be considered in more accurately estimating the effective transport velocity, but this requires the knowledge of vertical concentration distribution in the plume at each receptor location. The variation of wind direction with height is ignored, although its effect on the lateral plume spread and concentration field can be considered superficially through an appropriate parameterization of σ_y.

7. Strong enough winds to make turbulent diffusion in the direction of flow negligible in comparison with mean transport. This assumption, also known as the slender-plume approximation, which is implicit in the Gaussian plume model, generally becomes invalid very close to the source where material diffuses upwind of the source due to longitudinal velocity fluctuations. The assumption becomes invalid farther and farther away from the source as mean wind becomes weaker and vanishes entirely (e.g., under extremely stable and free convection conditions).

Additional assumptions are introduced in the specification of meteorological and dispersion parameters in the Gaussian plume model. These will be discussed in the next section.

9.2.2 The Gaussian Puff Model

The assumption of Gaussian concentration distributions in all directions in the ensemble-averaged puff and the integral mass-conservation condition

$$\int_{-\infty}^{\infty}\int_{-\infty}^{\infty}\int_{-\infty}^{\infty} \bar{c}\, dx\, dy\, dz = Q_{ip} \quad (9.10)$$

lead to the Gaussian puff formula for an instantaneous point source

$$\bar{c}(x,y,z,t) = \frac{Q_{ip}}{(2\pi)^{3/2}\sigma_x\sigma_y\sigma_z}$$
$$\exp\left[-\frac{x^2}{2\sigma_x^2} - \frac{y^2}{2\sigma_y^2} - \frac{z^2}{2\sigma_z^2}\right] \quad (9.11)$$

in which σ_x, σ_y, and σ_z are the puff-diffusion parameters, which are, in general, different from the plume-dispersion parameters.

From an instantaneous point source near a reflecting surface, the above formula is modified to

$$\bar{c}(x,y,z,t;H) = \frac{Q_{ip}}{(2\pi)^{3/2}\sigma_x\sigma_y\sigma_z}$$
$$\exp\left[-\frac{x^2}{2\sigma_x^2} - \frac{y^2}{2\sigma_y^2}\right]\left\{\exp\left[-\frac{(z-H)^2}{2\sigma_z^2}\right]\right.$$
$$\left.+ \exp\left[-\frac{(z+H)^2}{2\sigma_z^2}\right]\right\} \quad (9.12)$$

which, again, can be considered as a product of four factors—the source factor and the longitudinal, lateral, and vertical diffusion factors. In particular, the g.l.c. directly below the puff center is given by

$$\bar{c}_0(t;H) = \frac{2Q_{ip}}{(2\pi)^{3/2}\sigma_x\sigma_y\sigma_z}\exp\left(-\frac{H^2}{2\sigma_z^2}\right) \quad (9.13)$$

The above puff-diffusion equations are in the moving frame of reference with its origin located at the puff center.

Using the principle of superposition, it is easy to derive similar equations for instantaneous line, area, and volume sources. Most of the instantaneous and short-term accidental releases involve finite area or volume configurations, which can also be handled in the same manner. A more difficult problem in predicting the likely ground-level impact of such releases is estimating the puff trajectory and diffusion parameters as functions of time.

Most of the assumptions implied in the Gaussian plume model are also implied in the puff model. The only exceptions are the requirements of steady emission and strong winds in the former. The Gaussian puff model can only predict the ensemble-averaged puff outline and concentration field as functions of time, but not for any particular realization of the puff or cloud. The latter scenario is of greater and immediate interest in any accidental release of hazardous material.

9.3 DIFFUSION EXPERIMENTS

A number of early diffusion experiments were conducted for the expressed purpose of verifying certain diffusion theories and models and empirically estimating the dispersion or diffusion parameters. More comprehensive reviews of these experiments are given elsewhere (see e.g., Islitzer and Slade, 1968; Pasquill and Smith, 1983; Draxler, 1984). Here we give only the brief descriptions of some of the classical diffusion experiments, as well as more recent special-case diffusion experiments, involving continuous and instantaneous or quasi-instantaneous sources. For further details of the experiments and measurement techniques, the reader should refer to the original references or the above mentioned review articles.

9.3.1 Classical Diffusion Experiments

Continuous source plume-dispersion experiments conducted in the fifties and sixties are described in detail by Islitzer and Slade (1968). They have also discussed the early instantaneous or quasi-instantaneous puff diffusion experiments. Of these, only some of the better known experiments are briefly discussed here.

Continuous Near-surface Releases Probably the best known plume dispersion experiments involving near-surface releases of neutrally buoyant passive tracers were those conducted during the 1956 Project Prairie Grass at O'Neill, Nebraska. This was a comprehensive field program on turbulence and diffusion in the atmospheric surface layer over a large flat field in the Great Plains. A detailed description of the experiment and tabulations of meteorological and diffusion data are given by Barad (1958) and Haugen (1959). About seventy continuous releases of 10-min duration were made of SO_2 at a height of 0.5 m above the grass surface. Concentrations at 1.5 m above ground were measured along semicircular arcs at various distances between 50 and 800 m from the source. In addition, the vertical concentration distributions were measured from six 20-m towers mounted along the 100-m arc. Similar experiments were conducted at a much rougher and uneven site at Round Hill, Massachusetts, and reported by Cramer et al. (1958).

Another series of sixty-six diffusion experiments were conducted under the Project Green Glow at Hanford, Washington. In these experiments, the fluorescent tracer zinc sulfide was released at 1.5 m above the ground and concentrations were measured on arcs between 200 m and 25.6 km from the source. Vertical concentration profiles up to heights of 62 m were also measured at a number of 62-m towers. The test site was quite flat to gently rolling with sagebrush cover 1 to 2 m high. The zinc sulfide tracer, with water as a carrier, was released continuously for a 30-min period through two fog generators. The fluorescent particles were collected on filters. The experimental details and results have been reported by Fuquay et al. (1964). Islitzer and Slade (1968) have given a more comprehensive review of these and other continuous surface-release experiments, such as those conducted under Project Ocean Breeze at Cape Canaveral, Florida, and Project Dry Gulch at Vandenberg Air Force Base, California, and at the National Reactor Testing Station, Idaho Falls, Idaho.

Continuous Elevated Releases Plume diffusion from continuous elevated releases is of considerable interest in practical applications because many real emissions from industrial sources are elevated, not only because of the physical stack or chimney height but also due to the momentum and buoyancy effects of release. Effective release heights vary from tens of meters to several hundred meters. Dispersion of material from such sources takes place largely in the middle and upper parts of the PBL, and the maximum impact at the ground level occurs some distance downwind of the source.

Some of the earliest atmospheric diffusion experiments were conducted by the British Chemical Defense Experimental Station at Porton, England. These were aimed at the determination of the lateral plume spread as a function of atmospheric stability and downwind distance from the source. Hay and Pasquill (1957) studied dispersion of *Lycopodium* spores, released at 150 m above ground, to verify the statistical theory-predicted relationships between turbulence and diffusion. A detailed discussion of these and other early diffusion experiments has been given by Pasquill (1962, 1974).

The National Reactor Testing Station, Idaho Falls, conducted a series of diffusion experiments during unstable atmospheric conditions. A fluorescent tracer, uranin dye in solution, was released at 46 m above ground and mean concentrations were measured along six arcs of ground samplers extending from 150 to 1800 m from the release point. Mean wind and turbulence measurements were also made at several levels of a 46-m tower. The lateral and vertical plume spreads, σ_y and σ_z, were found to be proportional to the standard deviations of wind direction fluctuations σ_θ and σ_ϕ, respectively. Diffusion measurements in elevated plumes of a fluorescent pigment during various conditions of inversion were made at Hanford, Washington. The results of both the above mentioned experiments have been summarized by Islitzer and Slade (1968).

An extensive series of elevated-source plume-dispersion experiments conducted over a period of 15 years at the Brookhaven National Laboratory (BNL) have been summarized by Singer and Smith (1966). The BNL site can be characterized as essentially flat and rough, with a mixed land use of developed areas, open fields, and trees. Diffusion measurements were made using several different tracers released from a height of 108 m above ground, while mean winds and temperatures were measured at different levels of the 130-m meteorological tower and wind direction fluctuations were recorded at the release height. Some of the early diffusion experiments at BNL using oil fog as the tracer were conducted during daytime unstable conditions when the plume could easily diffuse down to the ground for surface sampling. Aircraft sampling became necessary, however, during nighttime stable conditions. Aircraft sampling was also done in the elevated plume released from the Brookhaven reactor during stable conditions. The results of the BNL plume dispersion experiments have been summarized by Islitzer and Slade (1968) and Draxler (1984).

Carpenter et al. (1971) have summarized 20 years of Tennessee Valley Authority (TVA) experience with the measurement of SO_2 concentration distributions and related meteorological parameters for a number of their power-plant plumes. These were probably the first dispersion measurements involving large buoyant plumes from industrial facilities. The sampling was done by instrumented helicopters that were flown to take lateral and vertical cross sections through the plumes at various distances from the stacks. The stack heights ranged from 75 to 250 m, sampling distances from 0.8 to 32 km, and the sampling time of 2 to 5 min was limited by the time it took the helicopter to pass through the plume. The temperature gradient at the plume height was used as a measure of atmospheric stability.

Quasi-instantaneous Releases There have been only a few puff-diffusion experiments involving instantaneous or quasi-instantaneous releases. Because of the difficulty of measuring instantaneous concentrations simultaneously at a sufficient number of points in an expanding puff, the observations of puff diffusion rely mainly on visual or photographic methods. Only some of the more extensive series of observations of puff diffusion are discussed here; for more comprehensive reviews the reader should refer to Islitzer and Slade (1968), Pasquill and Smith (1983), and Draxler (1984).

Kellog (1956) reported observations of smoke puffs in the stratosphere, using vials of titanium tetrachloride and water that were attached to trains of balloons and exploded at predetermined altitudes between 7 and 19 km. The resulting puffs of smoke were observed by phototheodolites, whose records gave the positions and visible sizes of the puffs at various diffusion times up to a maximum of 11 min. Similar, photographic observations of smoke puffs released within 100 m above a water surface have been reported by Frenkiel and Katz (1956). The puffs were generated by exploding small pillboxes of gunpowder carried aloft by a tethered balloon. Positions and sizes of smoke puffs at 1-s intervals were determined from cine-photographs.

Smith and Hay (1961) carried out two sets of puff diffusion experiments. In the first set, clusters of *Lycopodium* spores were released near the ground and concentration measurements were made by collecting samples of spores on adhesive cylinders set out on cross-wind arcs at different distances up to 300 m. Simultaneous measurements of wind speed and directions at a height of 2 m were also made at four different positions in the cross-wind direction. In the second set of medium-range diffusion experiments reported by Smith and Hay (1961), a cluster of zinc cadmium sulfide particles was released from an aircraft flying on a cross-wind track and an apparatus mounted on the cable of a captive balloon was used for sampling particle concentrations in the cloud. Measurements of wind speed and vertical wind direction fluctuations were made from instruments mounted on the cable. The estimated rate of growth of the cluster is found to be related to the turbulence intensity in accordance with a theoretical expression derived by Smith and Hay (1961).

MacCready et al. (1961) conducted a series of diffusion experiments during unstable conditions, using elevated line releases from a low-flying (100–320 m) aircraft in the vicinity of the 430-m television tower at Cedar Hill, Texas. Vertical sampling of zinc cadmium sulfide tracer was done by sequential filter samplers located at various levels on the tower and by rotorod samplers located on the ground 1.6 km apart to a distance of 48 km downwind of the line source. Source lines, ranging between 14.5 and 42 km in length, were released at heights between 110 and 320 m above ground and locations between 1.6 and 11 km upwind of the tower. The measured diffusion characteristics of the cloud were related to the observed wind speed, turbulence, stability, and diffusion time. The lateral spread of the cloud from the ends of the release line was also measured in some of the experiments.

An atmospheric diffusion program, called Project Sand Storm, conducted forty-three diffusion experiments in unstable atmospheric conditions at Edwards Air Force Base, California (Taylor, 1965). Solid-propellant rocket motors, which contained known amounts of finely divided metallic berylium in mixture with the propellant, were fired for short durations ranging from 2 to 8 s, and puffs were produced with initial visible diameters of 15 to 45 m. The tracer was collected on membrane filters located along circular arcs about the firing point, extending up to 2400 m. Phototheodolites were used to obtain information on the visual appearance and dimensions of the puffs. Meteorological data (mean winds, turbulence, and temperatures) were obtained from a 60-m tower located 60 m upwind of the release point.

An extensive set of observations of the diffusion of very short plume segments released from elevated sources during near-neutral and stable atmospheric conditions has been reported by Hogstrom (1964). These tests were conducted at a coastal and an inland site near Stockholm, Sweden. Oil-fog plumes were generated over a 30-s period and released at heights ranging between 24 and 87 m above the surface. Photographs of the plume segments were used to estimate horizontal and vertical spreads. A more detailed description of these experiments and the observed relative and total diffusion parameters as functions of distance from the source and stability is given by Hogstrom (1964). Islitzer and Slade (1968) have further summarized these and other puff diffusion experiments.

Special-case Diffusion Experiments Diffusion experiments carried out earlier for the specific purpose of empirically estimating the diffusion parameters for use in the Gaussian plume and puff diffusion models were generally restricted to simpler terrain (flat with uniform roughness and thermal characteristics) and meteorological conditions (steady and homogeneous flow).

In practice, the air pollution meteorologist has to deal with the dispersion of material in the atmosphere over a variety of complex terrain and flow conditions. The use of the Gaussian diffusion models can hardly be justified under such complex terrain and flow conditions. Special-case diffusion experiments have been carried out for a better understanding of the dispersion processes and for testing and validating more sophisticated dispersion models. An excellent review and summary of the special-case experiments dealing with shoreline and coastal diffusion, rough and complex terrain diffusion, urban diffusion, low-wind-speed diffusion, and long-range diffusion are given by Draxler (1984).

9.4 EMPIRICAL DISPERSION PARAMETERIZATION SCHEMES

A number of early diffusion experiments discussed in the previous section were conducted with the specific aim of evaluating the plume or puff diffusion parameters for use in the simpler Gaussian models. The most commonly used or recommended empirical dispersion schemes are reviewed here. Their experimental and theoretical bases as well as limitations are also pointed out. Most of these dispersion parameterization schemes are based on continuous-source diffusion experiments and should be used only in the plume dispersion models. Plume dispersion parameters were estimated either from photographic observations of time-averaged plumes, made visible by smoke or dye, or from mean concentration measurements of tracer in continuous-source plumes at various distances from the source.

From the measured concentration distribution in the lateral direction, one can directly determine σ_y, the plume centerline concentration, and the cross-wind integrated concentration. Measurements of mean concentration distribution in the vertical are rarely available for direct determination of σ_z. But σ_y and σ_z can be estimated indirectly from the Gaussian plume formulas (9.7) and (9.8), provided the source strength and the effective transport velocity across the plume are also known. In simpler dispersion schemes, σ_y and σ_z are expressed as functions of distance from the source and some stability or turbulence type or parameter. The so-called dispersion curves are plots of σ_y and σ_z versus x on a log-log scale. Alternatively, analytical power-law-type expressions are also available for use in the Gaussian dispersion models.

9.4.1 The Pasquill–Gifford (P–G) Scheme

This is the most widely used dispersion parameterization or sigma scheme. It was originally proposed as the British Meteorological Office 1958 system and published later by Pasquill (1961). Tentative estimates of the angular lateral plume spread θ_p and the height of vertical spread z_p were presented as functions of distance from the near-surface source for six broad stability categories (designated as Pasquill's stability classes A–F). The original rationale for Pasquill's discrete stability classes and their possible relationship to turbulence intensity and other, more quantitative, stability parameters have been discussed in Chapter 4, section 4.8. The applicable stability class can be chosen by reference to a table (see e.g., Table 4.2), using the routinely available information on wind speed at 10 m, cloud cover, and the strength of solar insolation. An objective procedure of classifying solar insolation as strong, moderate, and slight, as proposed by Turner (1961), is also used in current practice.

In proposing his plume spread curves and tables, Pasquill (1961) utilized theoretical guidance available at that time and also dispersion estimates from the Project Prairie Grass and British diffusion experiments. In these experiments, nonbuoyant, passive tracers were released near the surface and concentration measurements were limited to rather short distances (< 1 km) from the source. In addition, preliminary statistics of wind direction fluctuations over a 3-min sampling period and properties of wind profile over a natural rough surface were utilized in the estimates of lateral and vertical spreads, respectively. Diffusion parameters for long range (10–100 km) were estimated by extrapolation of short-range data in the light of limited special observations of tracer dispersion over a flat terrain of mixed roughness ($z_o \simeq 0.30$ m), with guidance from early turbulence data (Pasquill, 1976).

Gifford (1961) converted the plume spreading parameters θ_p and z_p of Pasquill into the more familiar standard deviations σ_y and σ_z of concentration distributions in the lateral and vertical directions, respectively, using the following relationships based on the Gaussian distribution:

$$\sigma_y \simeq \frac{x \tan(\theta_p/2)}{2.15} \qquad (9.14)$$

$$\sigma_z \simeq \frac{z_p}{2.15} \qquad (9.15)$$

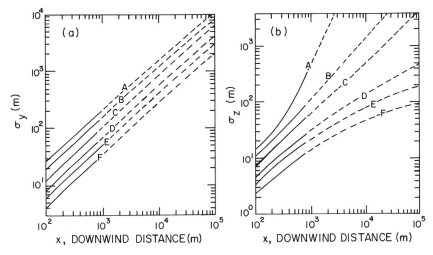

Figure 9.1 The Pasquill–Gifford dispersion curves for the various Pasquill stability classes. From Gifford, 1976.

The graphs of σ_y versus x and σ_z versus x on the log-log scale, presented by Gifford (1961), are called the Pasquill–Gifford (P–G) dispersion curves. These are reproduced here in Figure 9.1 with dotted parts emphasizing the larger uncertainty in the long-range extrapolations of the short-range experimental curves.

Limitations of the P–G dispersion scheme have been pointed out by Pasquill (1976), Gifford (1976), and others. These arise primarily from the limitations of the original diffusion data on which the curves were based, as well as from the limitations of Pasquill's stability or turbulence classes. This dispersion scheme is strictly valid for short-range dispersion from nonbuoyant near-surface sources over moderately rough and flat terrain with $z_o = 0.03$ to 0.3 m. A short sampling time of 3 to 10 min is also implied. For conditions deviating from the above, some adjustments in the estimated stability classes and/or dispersion parameters using the P–G scheme would be necessary (Gifford, 1976; Pasquill, 1976).

On the plus side, the P–G dispersion scheme has a sound theoretical and experimental basis, at least within the above stated restrictions of terrain roughness, sampling times and distances, and source type and height. The greatest advantage and the main reason for the continued use of the scheme in regulatory dispersion models is the ease with which stability or turbulence type can be determined from routine meteorological data. The use of discrete stability classes is also a drawback in the sense that the estimated dispersion parameters show rather abrupt and large changes in going from one stability class to the next.

9.4.2 The Brookhaven National Laboratory (BNL) Scheme

This scheme was based on more than 15 years of diffusion data collected at the BNL site in central Long Island, New York, using nonbuoyant, passive tracers released mostly at 108 m and some near the surface (Islitzer and Slade, 1968). The dispersion parameters σ_y and σ_z were determined from the analysis of extensive ground-level concentration measurements, some extending up to 60 km. These were plotted as functions of distance from the source for the four gustiness classes (B_2, B_1, C, and D) based on the range of fluctuations of horizontal wind direction (θ) trace, recorded over a 1-hr period by an aerovane located at the 108-m level of the BNL tower. These turbulence types were originally proposed and defined by Smith (1951) as follows:

B_2: Peak to peak fluctuations of θ ranging from 40° to 90°

B_1: Fluctuations of θ from 15° to 40°

C: Fluctuations of θ greater than 15° distinguished by the unbroken solid core of the trace

D: The θ trace approximating a line; short-term fluctuations not exceeding 15°

Later on, an A turbulence type was added for representing extremely convective conditions with peak to peak fluctuations of θ exceeding 90°, although no dispersion parameterization was suggested for this category. For the original four turbulence (gustiness) types, Singer and Smith (1966) proposed the power-law dispersion relations given in Table 9.1 below.

Here, x is the distance from the source in meters. The power-laws for σ_y and σ_z were assumed to have equal indices (exponents), which are found to decrease with decreasing turbulence or increasing stability. This contrasts with the P–G curves for σ_y, which can also be represented by power laws with a nearly constant exponent of 0.87 irrespective of the stability class.

A basic limitation of the BNL scheme is that both the turbulence and diffusion characterizations are site

Table 9.1 The BNL* Dispersion Parameterization Scheme

Turbulence Type	σ_y (m)	σ_z (m)
B_2	$0.40x^{0.91}$	$0.41x^{0.91}$†
B_1	$0.36x^{0.86}$	$0.33x^{0.86}$
C	$0.32x^{0.78}$	$0.22x^{0.78}$
D	$0.31x^{0.71}$	$0.06x^{0.71}$

*BNL, Brookhaven National Laboratory.
†x, distance from the source in meters.

specific and would apply strictly only to conditions equivalent to those found at BNL and to elevated nonbuoyant releases above the surface layer. The values of σ_y and σ_z correspond to a sampling time of the order of 1 hr. For easy comparison with the P–G dispersion curves, Figure 9.2 represents the BNL dispersion scheme in the graphical form.

9.4.3 The Tennessee Valley Authority (TVA) Scheme

In contrast to nonbuoyant tracer diffusion experiments used in formulating the P–G and the BNL dispersion schemes, Carpenter et al. (1971) summarized 20 years of TVA experience with the measurements of SO_2 concentrations in their power-plant plumes and the related meteorological parameters. These highly elevated plumes resulted from very buoyant emissions from stacks ranging in height from 75 to 250 m; the effective source heights were considerably larger (150–600 m). Measurements in plumes were made by helicopters traversing the plume in both the lateral and vertical directions at various distance from the source. The sampling time was basically the time it took the helicopter to pass through the plume; it varied between 2 and 5 minutes. The data from many different runs were averaged for six different stability classes (A–F) based on the potential temperature gradient $\partial\bar{\theta}/\partial z$ at the plume height. The average value of $\partial\bar{\theta}/\partial z$ for each stability class and the corresponding dispersion parameters for the TVA scheme are shown in Figure 9.3. The dashed portions of the dispersion curves are subjective extrapolations of the solid curves representing the best-fitted power laws to the experimental data.

The TVA dispersion scheme has several limitations. For shorter distances of the order of 5 to 10

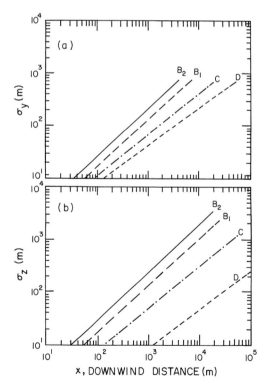

Figure 9.2 The Brookhaven National Laboratory dispersion curves for the various turbulence types. From Gifford, 1976.

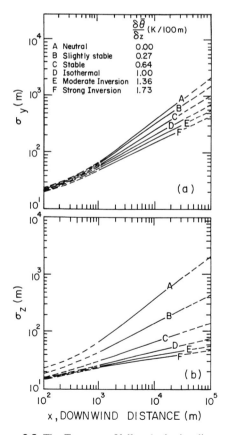

Figure 9.3 The Tennessee Valley Authority dispersion curves for the various stability classes based on the potential temperature gradient at the plume height. From Gifford, 1976.

source heights, σ_y and σ_z primarily reflect buoyancy and entrainment effects on plume spreading rather than the effect of ambient turbulence and, hence, may not be applicable to nonbuoyant or weakly buoyant releases. Another major limitation is the use of $\partial\theta/\partial z$ at the plume height as the stability index, which is not an adequate measure of the mixing capability of the atmosphere. It is well known that in the daytime convective mixed layer, $\partial\theta/\partial z$ is nearly zero or slightly positive, while it is negative in the surface layer. Thus, all the unstable and convective conditions appear to be in masked form amongst the neutral (A) and slightly stable (B) categories of the TVA scheme. These dispersion curves would not be applicable to near-surface releases. The lack of any input on wind speed, wind shear, or turbulence in the determination of stability precludes any broad generalization of this scheme to different terrain and source conditions. Short sampling times implied in the diffusion data also constitute a serious limitation. This scheme, however, reflects the observed behavior of highly buoyant industrial plumes.

9.4.4 The Urban Dispersion Schemes

It has been observed that diffusion over cities is considerably enhanced, compared with that over open country, due to increased surface roughness and increased temperature and heat capacity of cities. Thus, both mechanical (shear-generated) and thermal (buoyancy-generated) turbulence are increased, leading to enhanced dispersion, including the building wake effects. McElroy and Pooler (1968) analyzed diffusion of tracer plumes over St. Louis to obtain the Gaussian dispersion parameters for urban areas. The tracer material was released from the near-ground level and measured by ground samplers located on several circular arcs extending up to 16 km from the source. The estimated σ_y and σ_z for 1-hr sampling time were represented as functions of distance from the source and a modified gustiness class. The Pasquill stability scheme, Richardson number (Ri), and σ_θ were also used for characterizing stability and turbulence.

An excellent review and summary of experimental diffusion data from St. Louis, Missouri; Fort Wayne, Indiana; Johnstown, Pennsylvania; Columbus, Ohio; New York City; and other cities in the United States is given by Draxler (1984). The results of these urban diffusion experiments bring out several special features of plume dispersion over cities. The initial horizontal and vertical plume dimensions are closely approximated by those of nearby buildings, because eddies and wakes generated by buildings dominate the initial or short-range dispersion. Long-range dispersion is also considerably enhanced, compared with that over open country, due to the increased roughness as well as increased convective activity over the city. The latter is a consequence of the urban heat island, which is produced by the increased thermal capacity, anthropogenic heat sources, and decrease in evapotranspiration over the city.

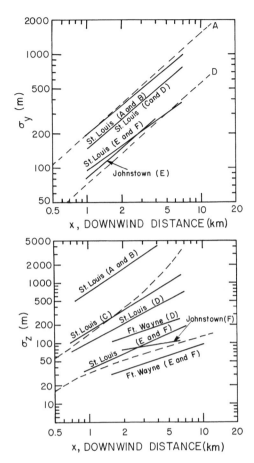

Figure 9.4 Comparison of plume-dispersion parameters for urban areas (solid curves) with the Pasquill–Gifford curves for open country. (dashed curves). From Draxler, 1984.

The effect of the urban heat island on dispersion is most pronounced at night when the stably stratified rural boundary layer approaching the city is modified by the relatively warmer surface. Consequently, an unstable urban boundary layer forms and grows with distance as an internal boundary layer, extending up to several hundred meters over the city center. Dispersion in the urban mixed layer is much faster than that in the rural stable boundary layer. This is clearly evident from Figure 9.4, which compares the horizontal and vertical dispersion parameters over urban areas with the Pasquill–Gifford curves for the open country. Note that horizontal dispersion during the day is similar to that for open country for the strong and moderately unstable classes A and B.

9.4.5 Briggs' Interpolation Formulas

Several empirical dispersion schemes for the open country have been compared and major differences

Table 9.2 Comparison of Different Empirical Dispersion Parameterization Schemes for Open Country

Basis for Comparison	Pasquill–Gifford (P–G) Scheme	Brookhaven National Laboratory (BNL) Scheme	Tennessee Valley Authority (TVA) Scheme
Experimental basis	Diffusion data from Project Prairie Grass at O'Neill, Neb., Porton Downs and Cardington, U.K.	Turbulence and diffusion data from Brookhaven National Lab., Long Island, N.Y.	Meteorological and diffusion data from several TVA power plants in northwestern Alabama
Theoretical basis	Gradient transport and statistical theories	Statistical theory	Statistical theory
Terrain type/ roughness	Rural, flat grassland with $z_0 = 0.03$ m for short range (<1 km) and $z_0 \simeq 0.3$ m for long range (10–100 km)	Combination of forested and developed areas with $z_0 \simeq 1$ m	Gently rolling terrain with variable roughness ($z_0 = 0.5$–2 m)
Source type and height	Nonbuoyant, ground level	Nonbuoyant, surface and elevated ($H = 108$ m)	Highly buoyant with effective release heights of 100–600 m
Sampling times and distances	3–10 min. for short range up to 800 m	30–90 min. for long range up to 50 km	2–5 min for distances of 10–30 km
Stability/ turbulence characterization	Pasquill's stability classes A–F based on $\bar{u}_{10}$, cloud cover, and solar insolation	Smith's gustiness classes A–D based on peak-to-peak fluctuations of wind direction or σ_θ	Static stability classes A–F based on $\partial \bar{\theta}/\partial z$ at plume centerline

among them have been pointed out by Briggs (1973), Gifford (1976), and others. These differences arise largely from differences in terrain roughness, release heights, sampling times/distances, and stability/turbulence characterizations for different diffusion experiments on which these schemes were based (see Table 9.2). The P–G curves were developed primarily from diffusion measurements at short distances (<1 km) from nonbuoyant releases near the surface. The BNL curves or formulas reflect the effects of elevated ($H = 108$ m) releases over a rougher surface and diffusion measurements extending to tens of kilometers. The TVA dispersion data reflected still higher effective source heights and highly buoyant power-plant emissions over gently rolling terrain. Briggs (1973) attempted to synthesize these different sets of diffusion data and parameterization schemes in proposing his interpolation formulas for open country. These are given in Table 9.3.

Briggs' interpolation formulas use the Pasquill stability classes and agree with the P–G curves for distances of up to several kilometers. At larger distances, they reflect more of the BNL and the TVA experience. The forms of his interpolation formulas for σ_y also conform to the statistical theory predictions for small and large diffusion times or distances. For σ_z, the differences in the exponents reflect the buoyancy enhancement in vertical dispersion under moderately unstable and convective conditions and suppression under stable conditions. For easy comparison with other dispersion curves, the dispersion curves corresponding to Briggs' interpolation formulas are shown in Figure 9.5. The flattening of σ_z curves at large distances for stable conditions is consistent with the observed lack of growth in σ_z under such conditions (see e.g., Islitzer and Slade, 1968).

Briggs (1973) also proposed similar interpolation formulas for parameterizing urban dispersion. These are largely based on the McElroy and Pooler's (1968) data from St. Louis and are given in Table 9.4. Briggs' recommendations for both the open country and urban areas apply up to 10 km and could perhaps be extended

Table 9.3 Briggs' (1973) Interpolation Formulas for Open Country

Pasquill Type	σ_y (m)	σ_z (m)
A	$0.22x(1 + 0.0001x)^{-1/2}$	$0.20x$*
B	$0.16x(1 + 0.0001x)^{-1/2}$	$0.12x$
C	$0.11x(1 + 0.0001x)^{-1/2}$	$0.08x(1 + 0.0002x)^{-1/2}$
D	$0.08x(1 + 0.0001x)^{-1/2}$	$0.06x(1 + 0.0015x)^{-1/2}$
E	$0.06x(1 + 0.0001x)^{-1/2}$	$0.03x(1 + 0.0003x)^{-1}$
F	$0.04x(1 + 0.0001x)^{-1/2}$	$0.016x(1 + 0.0003x)^{-1}$

*x is the distance downwind of the source in meters.

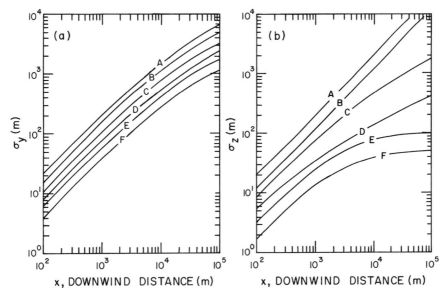

Figure 9.5 Plume-dispersion curves based on the interpolation formulas by Briggs for open country. From Gifford, 1976.

to 20 or 30 km (Gifford, 1976). In practice, these formulas, as well as the P–G curves, are often used well beyond their recommended range. Some commonly encountered errors in the coefficients and exponents of these formulas, as reported in the literature, have also been pointed out by Griffiths (1994).

9.4.6 Puff-diffusion Parameters

Experimental efforts directed toward estimation of puff-diffusion parameters from instantaneous or quasi-instantaneous releases have been far fewer than those for estimating plume-dispersion parameters. These have been reviewed by Islitzer and Slade (1968), Pasquill and Smith (1983), and Draxler (1984). Summary diagrams of the instantaneous dispersion parameters σ_{yI} and σ_{zI} as functions of downwind distance from the location of initial releases have been given by Islitzer and Slade (1968) who also suggested the approximate power laws given in Table 9.5 for the three broad stability categories. The lack of a uniform stability index or parameter and the paucity of diffusion data do not permit any finer categorization of diffusion parameters according to atmospheric stability or turbulence types.

These empirical diffusion parameterizations are based on limited diffusion data in puffs or plume segments observed over distances of less than 10 km. They do not include the effects of initial cloud dimensions, surface roughness, degree of stability or instability, and turbulence intensity on which relative diffusion is expected to depend. More accurate diffusion parameterizations should be obtained from systematic puff diffusion experiments under a wide range of atmospheric stability or turbulence conditions. The results of the relative diffusion theory should also be incorporated in any diffusion parameterization scheme. There is no justification for the current practice of using the plume dispersion curves or formulas for modeling puff diffusion.

9.5 FURTHER IMPROVEMENTS IN DISPERSION PARAMETERIZATION

Since several of the empirical dispersion schemes discussed in the previous section have been extensively used in operational dispersion models for regulatory

Table 9.4 Briggs' (1973) Interpolation Formulas for Urban Areas

Pasquill Type	σ_y (m)	σ_z (m)
A–B	$0.32x(1 + 0.0004x)^{-1/2}$*	$0.24x(1 + 0.001x)^{1/2}$
C	$0.22x(1 + 0.0004x)^{-1/2}$	$0.20\,x$
D	$0.16x(1 + 0.0004x)^{-1/2}$	$0.14x(1 + 0.0003x)^{-1/2}$
E–F	$0.11x(1 + 0.0004x)^{-1/2}$	$0.08x(1 + 0.0015x)^{-1/2}$

*x is the distance downwind of the source, in meters.

Table 9.5 Suggested Formulas for Instantaneous Puff Diffusion

Stability Condition	σ_{yI} (m)	σ_{zI} (m)
Unstable	$0.14x^{0.92}$*	$0.53x^{0.73}$
Neutral	$0.06x^{0.92}$	$0.15x^{0.70}$
Very stable	$0.02x^{0.89}$	$0.05x^{0.61}$

Source: Islitzer and Slade (1968).
*x is the distance downwind of release point in meters.

applications, further improvement or refinement of the same is considered crucial to the reduction of model uncertainties and improved model performance. Some of the original developers of dispersion parameterization schemes have pointed out the various shortcomings and limitations of these dispersion schemes and recommended further improvements in them (Pasquill, 1974, 1976; Gifford, 1976, 1981; Hanna et al., 1977). These recommendations are perhaps best summarized by Hanna et al. (1977) following a workshop on stability classification and dispersion schemes. Here we mention only some of the salient features of those recommendations.

9.5.1 Estimation of σ_y

For the best estimation of the lateral dispersion parameter, the statistical theory relation

$$\sigma_y = \sigma_v t f_y\left(\frac{t}{T_L}\right) \quad (9.16)$$

or an equivalent form after Taylor's transformation $t = x/\bar{u}$,

$$\sigma_y = \sigma_\theta x f_y\left(\frac{x}{\bar{u}T_L}\right) \quad (9.17)$$

is recommended. Here, T_L is the Lagrangian time scale, which may be specified as a function of stability, boundary layer height, and so on. An empirical form of the function $f_y(t/T_L)$, which is also consistent with the statistical theory prediction for small and large diffusion times, has been proposed by Draxler (1976). Due to the lack of routine information on T_L, Pasquill (1976) suggested an abbreviation of Eq. (9.17), that is,

$$\sigma_y = \sigma_\theta x f(x) \quad (9.18)$$

in which the empirical function $f(x)$ is specified as follows:

x (km)	0	0.1	0.2	0.4	1	2	4	10	>10
$f(x)$	1.0	0.8	0.7	0.65	0.6	0.4	0.4	0.33	$0.33(10/x)^{1/2}$

Note that the above recommendation would require the measurement of σ_v or σ_θ at the release height. Alternatively, the same can be estimated from the appropriate parametric relations for turbulence discussed in Chapter 4. The estimate of σ_θ should be consistent with the desired sampling time for concentration and dispersion estimates.

For unstable stratification, the dispersion relations based on the mixed-layer similarity theory may also be used. These are given in Chapter 8 and require an estimation of the convective velocity W_* for determining σ_y.

If direct measurements or estimates of σ_v or σ_θ are not available and it is necessary to use the routine meteorological data, the existing dispersion schemes with suitable adjustment can be used, for example, the Pasquill–Gifford curves for open country and the McElroy and Pooler (1968) dispersion curves for urban areas. Some adjustments in dispersion estimates from these existing schemes are recommended to account for the effects of different sampling time and surface roughness, compared with those in the original schemes. These adjustments may be based on the simple power-law relations for the dependence of σ_y on the sampling time T and the roughness length z_o, for example,

$$\sigma_y \propto T^p \quad (9.19)$$

$$\sigma_y \propto z_o^q \quad (9.20)$$

in which the empirically estimated values of the exponents are:

$p \simeq 0.2,$ for 3 min $< T <$ 1 h
$p = 0.2$–$0.3,$ for 1 h $< T <$ 100 h
$q \simeq 0.2$

The lateral dispersion is enhanced by the turning of wind direction with height, to an extent that is probably important only for $x >$ 20 km. A rough rule that has been suggested by Pasquill (1976) is to add to the σ_y^2 already estimated, beyond distances of 20 km, the contribution $0.03(\Delta\bar{\theta})x^2$ due to the change in wind direction $\Delta\bar{\theta}$ (in radians) over the whole depth of the plume. The plume depth at any distance from the source can be estimated from the knowledge of the vertical dispersion parameter σ_z and the effective release height.

9.5.2 Estimation of σ_z

To estimate the effect of stability on vertical diffusion in the PBL, it is necessary to consider the ratio, h/L, of the PBL height to the Obukhov length as the most appropriate stability parameter. The bulk Richardson number, Ri_h, based on the PBL height and the potential temperature difference across the whole PBL, is an equivalent alternative stability parameter. For unstable and convective conditions, the ratio W_*/u_* or $W_*/\bar{u}$ may also be used as a measure of convective turbulence and diffusion. There is little physical justification for the use of $\partial\bar{\theta}/\partial z$ or the static stability parameter s for parameterizing vertical diffusion.

The standard deviation of vertical wind direction fluctuations σ_ϕ is a good indicator of atmospheric stability and turbulence affecting vertical diffusion. In particular, the vertical diffusion from an elevated source, when σ_z is less than the source height, should be estimated from the statistical theory relationships

$$\sigma_z = \sigma_w t f_z\left(\frac{t}{T_L}\right) \quad (9.21)$$

$$\sigma_z = \sigma_\phi x f_z\left(\frac{x}{\bar{u}T_L}\right) \quad (9.22)$$

in which $f_z = 1$ for small x. Draxler (1976) has presented graphs of f_z over a larger range of x for elevated sources and unstable stratification. His data indicated too much scatter for stable conditions. Approximate empirical formulas for f_z for stable and unstable stratifications have also been suggested by Draxler (1976).

In the absence of turbulence measurements, σ_w or σ_ϕ may be estimated from the parametric relations suggested in Chapter 4. Alternatively, a suitable empirical dispersion scheme, such as the P–G curves for near-surface sources and rural areas with low roughness, BNL formulas for elevated sources and rougher open country, and the McElroy and Pooler (1968) curve for urban areas, may be used with adjustments for the surface roughness, effective release height, and buoyancy of release.

The roughness effect can be considered through the use of the power-law relation $\sigma_z \propto z_o^q$, in which q lies in the range of 0.10 to 0.25, the larger values being applicable to shorter distances and rougher surfaces (Pasquill and Smith, 1983). The height of the source and roughness of the site should be considered in relation to the data base for each of the available dispersion schemes. Regardless of which dispersion scheme is used, the plume thickness should not exceed the mixing depth, h, except when the initial plume buoyancy is large and there is only weak and shallow inversion capping the PBL. With reflections from both the surface and the inversion base, σ_z should not be allowed to exceed 0.8 h, a condition implied by the uniform concentration distribution through the PBL depth at large distances from the source. Finally, in the case of a buoyant plume, one should allow for the much enhanced initial plume dispersion due to plume buoyancy. A simple correction to account for plume buoyancy has been proposed by Pasquill (1976). He suggested adding to the original σ_z^2 estimate from the P–G or the BNL scheme the term $0.1(\Delta H)^2$, where ΔH is the buoyant plume rise above the stack. With this,

$$\sigma_z^2 = \sigma_{zo}^2 + 0.10(\Delta H)^2 \quad (9.23)$$

in which σ_{zo} is the dispersion parameter without the plume buoyancy effect. The above modification is not necessary if the TVA dispersion curves are used for estimating σ_z, because the effect of plume buoyancy is already included in those curves.

9.5.3 Generalized Dispersion Scheme

More observations are required for developing the most appropriate generalized dispersion scheme. The emphasis should be on more complete measurements of the vertical structure of the atmospheric boundary layer and on concentration and plume dispersion measurements at greater distances. Concentration measurements in the vertical are also very desirable under the whole wide range of stability conditions encountered in the atmosphere. New diffusion observations should be accompanied by measurements of velocity and temperature profiles throughout the PBL. Measurements of friction velocity, surface heat flux, mixing height, and one or more turbulence parameters should also be made. Simultaneous boundary layer, turbulence, and diffusion measurements would allow dispersion parameters to be expressed as functions of travel time or distance from the source, of the quantitative stability and turbulence parameters (e.g., h/L, $W_*/\bar{u}$, or $u_*/\bar{u}$, σ_θ and σ_ϕ), and of the dimensionless release height and buoyancy parameters (e.g., H/h and F^*).

In response to the recommendations of the American Meteorological Society workshop on stability classification schemes and sigma curves, Irwin (1979a, 1979b) proposed a generalized plume-dispersion scheme. This scheme requires either indirect estimates or direct measurements of σ_θ and σ_ϕ at the effective release height. During convective conditions, the mixing height (h) and the convective velocity scale (W_*) are required as input to the scheme. During stable conditions, measurements or estimates of the mixing height and the Obukhov length are required. Provisions are also made for incorporating the additional dispersion resulting from plume buoyancy and the wind-direction shear in the vertical.

For the lateral dispersion parameter, Irwin's (1979a) scheme uses Eq. (9.18) with Pasquill's (1976) suggested empirical function $f(x)$, which can be approximated as

$$f(x) = \frac{1}{[1 + 0.308x^{0.455}]}, \quad \text{for } x \leq 10^4 \text{ m}$$

$$f(x) = 0.333\left(\frac{10{,}000}{x}\right)^{1/2}, \quad \text{for } x > 10^4 \text{ m} \quad (9.24)$$

where x is in meters. For use in Eq. (9.18), σ_θ is to be measured or estimated at the effective release height. Irwin (1979b) recommended the use of Eq. (9.16) in which different forms of the function f_y are recommended for different stability conditions. These are based on the predictions of the statistical theory, similarity scaling, and experimental diffusion data. With the inclusion of the effects of wind-direction shear and plume buoyancy on lateral dispersion, σ_y can be expressed as

$$\sigma_y^2 = (\sigma_v t f_y)^2 + (0.174x\Delta\bar{\theta})^2 + 0.1(\Delta H)^2 \quad (9.25)$$

For the vertical dispersion parameter, Irwin's (1979a) scheme uses Eq. (9.22) with f_z expressed as a function of $t^* = tW_*/h$ and H/h for unstable and convective conditions. The suggested forms of the function $f_z(t^*, H/h)$ are essentially based on the convection tank data of Willis and Deardorff (1976, 1978) and the numerical large-eddy simulation results of Lamb (1978).

For stable and neutral conditions, f_z is simply considered as a function of t and H, with two different f_z versus t curves suggested for $H < 50$ m and $H > 100$ m (logarithmic interpolation between the two curves is suggested for 50 m $< H <$ 100 m) on the basis of field experimental data and the statistical theory prediction for large diffusion times. Alternative forms of f_z as a function of the dimensionless travel time t/T_o are recommended by Irwin (1979b), using Draxler's (1976) formulas for surface and elevated releases. The time scale T_o is specified as a function of the effective release height. With the inclusion of the plume buoyancy-induced dispersion, σ_z can be expressed as

$$\sigma_z^2 = (\sigma_w t f_z)^2 + 0.1(\Delta H)^2 \qquad (9.26)$$

It may be noted that the use of Taylor's hypothesis or the simple transformation $t = x/\bar{u}$ allows the replacement of $\sigma_v t$ and $\sigma_w t$ in Eqs. (9.25) and (9.26) by $\sigma_\theta x$ and $\sigma_\phi x$, respectively, at least for moderate and strong winds. The above transformation becomes questionable under weak- and calm-wind conditions.

In the absence of turbulence measurements at the site where the above dispersion scheme is to be used, the various methods of parameterizing the desired boundary layer and turbulence parameters have been suggested in the literature (Golder, 1972; Irwin, 1979a; van Ulden and Holtslag, 1985; Hanna and Chang, 1992; Arya, 1995). Some of the proposed parameterizations may be site specific, while almost all are restricted to flat and homogeneous terrain and steady-state atmospheric conditions. These restrictions also apply to the so-called generalized dispersion scheme recommended for use in Gaussian plume models. The suggested forms of the functions f_y and f_z by Irwin (1979a, 1979b) must be considered tentative; these should be revised in light of the more recent diffusion experiments and numerical simulations.

9.5.4 Briggs' Dispersion Parameterization for the CBL

For the convective boundary layer (CBL), Briggs (1985, 1988, 1993) has proposed different empirical diffusion parameterizations, which are based on mixed-layer similarity scaling, experimental diffusion data, and numerical modeling results. Some of these have already been discussed in Chapter 8, section 8.4, in the context of the mixed-layer similarity theory. The main difference from the above mentioned generalized dispersion scheme is that turbulence parameters are not necessary as input to dispersion relations. Instead, the plume dispersion parameters are normalized by the CBL depth and the ratios σ_y/h and σ_z/h are expressed as universal similarity functions of $X^* = xW_*/\bar{u}h, H/h$, and the dimensionless initial plume buoyancy flux F^* (see e.g., Eqs. 8.62). Another major difference is in the definitions of σ_z in Eq. (8.70) and σ_z' in Eq. (8.64),

which do not necessarily imply a Gaussian distribution of concentrations or particle displacements in the vertical. Empirical forms of the similarity functions related to dispersion from nonbuoyant releases in the CBL are given in Chapter 8, section 8.4. Briggs' (1988, 1993) relations incorporate the data from the CONDORS (convective diffusion observed with remote sensors) field experiment and other field studies (see e.g., Figs. 8.11 and 8.12). These can be used for unstable and convective conditions whenever measurements or estimates of h and W_* are available.

9.6 THE MAXIMUM GROUND-LEVEL CONCENTRATION

Determination of the maximum ground-level concentration $\bar{c}_{o\max}$ and its location $x_{\max}$ downwind of an elevated source are of considerable interest in regulatory applications of dispersion models, because air-quality standards and regulations generally refer to the maximum g.l.c. For the Gaussian plume model with given meteorological and source data and a specified dispersion scheme, it is straightforward and easy to run the model and compute ground-level concentrations and, hence, $\bar{c}_{o\max}$, every hour or every day for a long enough period (usually 1–5 years), to obtain the highest or the second highest concentration or other desired statistics. Such calculations are necessary for determining the impact of existing and proposed sources on the air quality of the local area (usually, within tens of kilometers of the source).

Here we will discuss the various factors that affect the maximum g.l.c. due to an elevated continuous point source, using the Gaussian plume-dispersion model with some analytical dispersion formulas or empirical dispersion curves. The starting point for this is the Gaussian formula for the g.l.c. along the plume centerline:

$$\bar{c}_o(x) = \frac{Q}{\pi \bar{u} \sigma_y \sigma_z} \exp\left(-\frac{H^2}{2\sigma_z^2}\right) \qquad (9.27)$$

in which H is the effective release height and $\bar{u}$ is the effective transport velocity across the depth of the plume. For release heights greater than 10 m, $\bar{u}$ is often approximated by the mean wind speed at H.

9.6.1 Nonbuoyant Releases

For nonbuoyant releases, H is also constant, independent of distance from the source, ignoring any plume rise due to momentum effects of release. For these conditions, $\bar{c}_o$ varies essentially in response to increasing values of σ_y and σ_z with distance from the source.

When σ_y and σ_z can be approximated by power laws of the form

$$\sigma_y = ax^p \qquad (9.28)$$

$$\sigma_z = bx^q \qquad (9.29)$$

as for example, in the BNL and the TVA schemes, substituting for them in Eq. (9.27), and using the con-

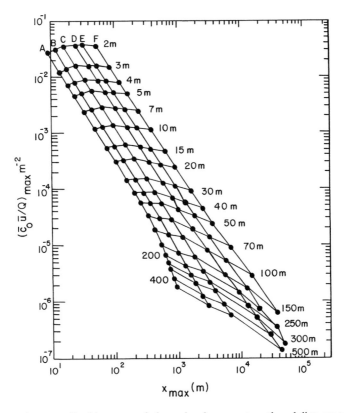

Figure 9.6 Maximum g.l.c. normalized by mean wind speed and source strength and distance to the maximum g.l.c. as functions of the Pasquill stability class and effective source height. From Turner, 1994.

dition $\partial \bar{c}_o/\partial x = 0$ or $\partial \ln \bar{c}_o/\partial x = 0$ for the maximum g.l.c., one can obtain

$$x_{\max} = \left[\frac{H^2 q}{b(p+q)}\right]^{1/2q} \quad (9.30)$$

$$\bar{c}_{o\max} = \frac{Q}{\pi \bar{u} a b}\left[\frac{b^2(p+q)}{qH^2}\right]^{(p+q)/2q}$$

$$\exp\left(-\frac{p+q}{2q}\right) \quad (9.31)$$

Note that for the particular case of $p = q$ (e.g., $p = q = 1/2$ according to the Fickian diffusion theory, and $p = q = 1$ according to the statistical and similarity theories), one can show that the maximum g.l.c. occurs where $\sigma_z = H/\sqrt{2}$ and

$$\bar{c}_{o\max} = \left(\frac{2b}{\pi a e}\right)\frac{Q}{\bar{u}H^2} \quad (9.32)$$

A similar result was derived in Chapter 6, using the Fickian diffusion theory. It is shown here to be more generally valid, provided σ_y and σ_z have similar variations with distance from the source.

For more general but different variations of dispersion parameters, as implied in the P–G curves and Briggs' interpolation formulas, one can calculate $\bar{c}_{o\max}$ and $x_{\max}$ numerically for different stability classes and release heights. The curves in Figure 9.6 are the result of such calculations, using the P–G dispersion scheme. These were obtained from the repeated solutions to Eq. (9.27) at a number of distances for the various pairs of Pasquill stability class and the effective release height (Turner, 1994). Note that for a fixed stability class, $\bar{c}_{o\max}$ and $x_{\max}$ are approximately inversely related to each other. As expected, with increasing release height the distance to the maximum g.l.c. increases, whereas $\bar{c}_{o\max}$ decreases. The benefit of very large release heights (tall stacks) is more realized during very stable conditions, compared with convective conditions. For most elevated releases, the maximum g.l.c. is expected to occur under unstable and convective conditions. Only for near-surface sources (say, $H < 5$ m) would the maximum g.l.c. be expected to occur under stable or near-neutral conditions.

More detailed graphs of the centerline g.l.c. as a function of distance from the source and release height can be obtained from the repeated use of the Gaussian plume formula (9.27) for each stability class and different release heights. Turner (1970, 1994) has presented these for fairly wide ranges of release heights and mixing heights. The values of $\bar{c}_{o\max}$ and $x_{\max}$ can easily be picked from these graphs to obtain the results of Figure 9.6. The same graphs can also be used to calculate and plot isopleths of constant g.l.c. For

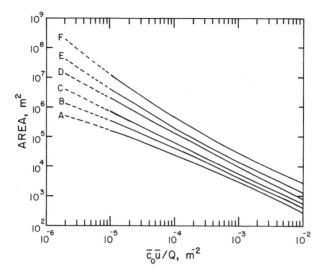

Figure 9.7 Area within a ground-level concentration isopleth as a function of the normalized g.l.c. ($\bar{c}_o\bar{u}/Q$) and stability for a surface source. From Turner, 1994.

this, the relationship between the centerline g.l.c. and off-axis g.l.c. can be used, that is,

$$\bar{c}(x,y,0) = \bar{c}_o(x,0,0)\exp\left(-\frac{y^2}{2\sigma_y^2}\right) \quad (9.33)$$

from which the off-axis coordinates of the isopleth can be determined as

$$y = \pm\left[2\ln\frac{\bar{c}_o}{\bar{c}}\right]^{1/2}\sigma_y \quad (9.36)$$

where $\bar{c}$ is the desired value of concentration represented by the isopleth. Turner (1970, 1994) has plotted graphs of g.l.c. isopleths for each stability class for a ground-level release ($H = 0$), as well as for $H = 100$ m. From these graphs, areas within the various isopleths can be determined. Figure 9.7 presents the results of computations for a surface release. As expected, the impacted area in which a specified concentration is exceeded decreases with the increase in the concentration value of the isopleth, but it increases with increasing stability. The dependence on stability is expected to be different for elevated releases.

9.6.2 Buoyant Release with Plume Rise

Although plume rise due to buoyancy and momentum effects of release will be discussed in more detail in the next chapter, we can consider here, at least qualitatively, the possible effects of plume rise on the maximum g.l.c. In the presence of a finite wind, the initially vertical plume from a stack or chimney is bent over and the plume rise z_c' above the top of the stack is an increasing function of x. Under stable atmospheric conditions, the rising plume tends to level off at some height ΔH above the stack top, which depends on the mean wind speed, atmospheric stability, and initial momentum and buoyancy fluxes. Under unstable and convective conditions, on the other hand, a buoyant plume may keep on rising with distance until it is restricted by the capping inversion layer at the top of the mixed layer.

In Gaussian plume models, the effective source height is considered as the sum of the stack height and the plume rise, that is,

$$H = H_s + z_c' \quad (9.37)$$

where z_c' is an increasing function of distance near the source but attains some constant value ΔH as a result of stable stratification or an elevated inversion. With the effective source height as an increasing function of distance, the maximum g.l.c. and its location are no longer given by Eqs. (9.30) through (9.32). However, $\bar{c}_{o\max}$ and $x_{\max}$ can still be determined easily by computing $\bar{c}_o(x)$ from Eq. (9.27) for a wide range of environmental and source conditions. An interesting result of such computations is that there is a "critical" finite wind speed at which $\bar{c}_{o\max}$ itself is a maximum (Briggs, 1984). This is due to the observed fact that the plume rise due to both the momentum and buoyancy effects of release decreases with increasing wind speed and vice versa. Without plume rise, Eq. (9.31) indicates that $\bar{c}_{o\max}$ should vary inversely proportional to wind speed and there may not be a "critical" wind speed, other than zero, at which $\bar{c}_{o\max}$ would be maximum. The Gaussian plume formulas (9.27) and (9.31) should not be used, however, under low wind-speed conditions, including the calm case.

9.7 MODEL EVALUATIONS AND UNCERTAINTIES

Use of Gaussian dispersion models has become very widespread, particularly in regulatory applications, as a result of legal requirements to analyze environmental impacts of existing and proposed sources of air pollution. The limitations of Gaussian models are also well known. Consequently, there has been considerable interest in determining the accuracy of such models and uncertainties associated with the same. The latter include the errors and uncertainties in the various meteorological and source parameters that are provided as input to dispersion models, uncertainties associated with the simplified model physics, and the inherent uncertainties due to the stochastic nature of atmospheric dispersion.

9.7.1 A Theoretical Estimate of Model Uncertainty

A rough estimate of the reducible uncertainty in model-predicted ground-level concentrations, due to errors or uncertainties of model input parameters, can be obtained by taking the logarithm of Eq. (9.27) and differentiating,

$$\frac{d\bar{c}_o}{\bar{c}_o} = \frac{dQ}{Q} - \frac{d\bar{u}}{\bar{u}} - \frac{d\sigma_y}{\sigma_y} - \frac{d\sigma_z}{\sigma_z} - d\left(\frac{H^2}{2\sigma_z^2}\right) \quad (9.38)$$

Further, identifying the differentials with finite differences or errors in estimates of the various input parameters, the corresponding error in model-computed concentrations can be estimated as

$$\frac{\Delta\bar{c}_o}{\bar{c}_o} = \frac{\Delta Q}{Q} - \frac{\Delta\bar{u}}{\bar{u}} - \frac{\Delta\sigma_y}{\sigma_y} - \frac{\Delta\sigma_z}{\sigma_z} - \Delta\left(\frac{H^2}{2\sigma_z^2}\right) \quad (9.39)$$

Assuming that errors associated with model input parameters are all random rather than systematic, the worst-case estimate of the percentage error in concentration would be the sum of the magnitudes of percentage errors in Q, $\bar{u}$, σ_y, and σ_z, and the absolute error in the estimate of $H^2/2\sigma_z^2$. Using a rather low estimate of ± 0.10 for each term on the right-hand side of Eq. (9.39), the estimated error in the model-predicted g.l.c. could easily be ± 50 percent for elevated sources and somewhat smaller for nonbuoyant surface sources.

To the above estimate one should add the inherent errors associated with the stochastic nature of turbulence and diffusion processes and the systematic error or bias introduced due to simplified physics and transport and diffusion parameterizations used in the Gaussian model. It is rather difficult to estimate these model uncertainties, but an overall assessment of the same can be made from model evaluation studies using extensive experimental data of good quality.

9.7.2 Empirical Estimates of Model Uncertainty

Before any systematic model evaluation studies were conducted, a short position paper on the accuracy of dispersion models was published by the American Meteorological Society's Committee on Atmospheric Turbulence and Diffusion (American Meteorological Society, 1978). They gave somewhat lower estimates (20–40%) of model uncertainties in concentration estimates for elevated sources in ideal circumstances (i.e., uniform terrain, steady meteorology, source and ambient parameters measured carefully by research-grade instruments). But, when dispersion modeling is applied in most real-world applications with input parameters subject to larger errors, and in circumstances that are different from the carefully controlled and idealized situation, model accuracy is estimated to be within a factor of 2 (see also Islitzer and Slade, 1968; Pasquill, 1974). This estimate still excludes those exceptional circumstances in which the use of the Gaussian dispersion models becomes highly questionable and model estimates may not even have the right order of magnitude.

The exceptional circumstances in which model uncertainties are expected to become very large are (Gifford, 1976; American Meteorological Society, 1978):

1. Aerodynamic wake flows of all kinds, including stack downwash, building wakes, highway vehicle wakes, and wakes generated by terrain obstacles
2. Flows over surfaces markedly different from those represented in the basic experiments for diffusion parameterization, including dispersion over forests, cities, water, and complex terrain
3. Highly buoyant emissions, including those from power-plant stacks and accidental releases of heavy and toxic gases
4. Extremely stable and unstable conditions with weak and calm winds

Success can be obtained in some of these exceptional cases only by developing specialized site-specific models based on observations of meteorological and dispersion parameters at the particular location and under specified conditions. Such site-specific models can have accuracies as good as a factor of 2.

There have been several workshops organized by the American Meteorological Society focused on performance evaluation and uncertainties of air quality models used in regulatory applications (Fox, 1981, 1984; Smith, 1984; Weil et al., 1992). A general review of model errors and uncertainties is given by Fox (1981). These can be divided into the following types:

1. Meteorological data errors, which result in part from an inability to measure the atmosphere with sufficient

spatial resolution. This problem is magnified in areas of irregular topography, and where both land and water surfaces are present, because limited observations may not be representative of the whole area over which dispersion is modeled. Another source of error is introduced by discretizing continuously varying meteorology, for example, characterizing stability in terms of Pasquill's stability classes A through F. A third source of error arises from the inability of sensors to measure the meteorological variables with sufficient accuracy.

2. Emissions data errors, which result from the variations in the fuel supply characteristics, as well as from variations in the facility's production level. Emissions input to models are generally not measured, but rather are estimated from conservative emission factors.

3. Model errors or uncertainties resulting from inadequate or incorrect representation of physics. Such errors can be categorized further on the basis of model components or elements. One source of error is related to modeling the type and height of release. Due to the various source aerodynamic interactions, such as stack downwash, entrainment in building wakes, and others, emissions are often distributed over an irregular volume, whereas the model assumes an ideal point or line source. The effective source height for buoyant emissions is determined by plume rise calculations in the model. Considerable error is often associated with plume rise estimates from highly parameterized or empirical plume rise formulas, which are used in regulatory air quality models. Another major source of uncertainty is oversimplified treatment of transport and diffusion in Gaussian-type models. Simple models assume straight-line transport with an effective transport velocity, which may be characterized by observations at one point. The presence of vertical and horizontal wind shears in the atmosphere introduces significant transport errors. Diffusion by random motions (turbulent velocity fluctuations) is simply parameterized in terms of dispersion parameters that are specified by empirical sigma schemes. There are significant errors associated with these dispersion parameters, as well as with the specification of the mixing height that limits the vertical dispersion.

4. Inherent uncertainty, which arises due to the stochastic nature of turbulence and diffusion processes. While other types of errors due to imperfect data and imperfect physics can be reduced, to some extent, by improved measurements and improved model physics, there will always be an irreducible, inherent uncertainty associated with atmospheric dispersion models. Most dispersion models are deterministic and can approximately represent only ensemble-averaged concentrations. Observed concentrations represent at most a few realizations of a theoretically large sample. These will always differ from model-predicted average concentrations. Inherent uncertainty in predicted concentrations is expected to be proportional to the variance of concentration fluctuations.

Considering all the above mentioned errors and uncertainties, our previous estimate of ± 50 percent uncertainty in model-predicted concentrations might be on the lower side. Even a more conservative estimate of the predicted concentrations being within a factor of 2 of the observed values may not be good enough. A more accurate assessment of the overall model uncertainty can be obtained only after comparing many sets of predicted and observed concentrations, paired in time and space. This is done as an integral part of model evaluation and validation studies.

9.7.3 Model Performance Evaluations

There have been many performance evaluations of air quality models in recent years, following the 1980 AMS Workshop on Dispersion Model Performance (Fox, 1981). At that workshop, a number of statistical measures of model performance were suggested and recommendations were made on performance evaluation, especially as applied to point source dispersion models. Three groups of both predicted and observed data for which performance measures may be used have been identified:

1. Observed and predicted concentrations paired for a particular location in space at a particular time
2. A peak concentration datum set with various degrees of time and space pairing
3. Cumulative frequency distributions of observed and predicted concentrations

The first group would represent the most stringent test of a model, because the model is required to simulate all concentrations at a particular place and time. Analysis of peak concentration data provides a less stringent test, because the model is tested only under worst-case conditions, which represent a subset of all conditions.

Two general types of performance measures have been identified: (1) statistical measures of difference and (2) measures of correlation between predicted and observed data. Difference measures are found to be better suited to quantitative performance evaluation. The following measures of difference have been suggested:

1. The average of the difference (observed − predicted), which is a measure of the bias
2. The variance or standard deviation of the difference, which is a measure of the noise or inherent uncertainty
3. The gross error of the difference

These are actually related measures, because the sum of the square of the bias and the variance equals the square of the gross variability. The suggested measures of correlation are time, space, and combined time–space correlations. These measures of difference and correlation can be applied to all three groups of data identified earlier. The performance measures recommended by the 1980 AMS workshop are now routinely used in model evaluation. A more recent review and update of these measures and their application to model evaluations is given by Weil et al. (1992).

An important application of the above procedure to a comparison of several dispersion parameterization schemes used in the Gaussian plume model was reported by Irwin (1983). He used meteorological and dispersion data from seventeen atmospheric diffusion experiments in which necessary information on mean transport winds, stability, and turbulence (σ_θ and/or σ_ϕ) was available, in addition to the data on ground-level concentrations and σ_y. Ten of the experiments involved near-surface releases ($H = 0.3$–3 m), while elevated releases ($H = 7.5$–255 m) were used in the seven other experiments. Direct estimates of σ_y could be made from concentration measurements in all the experiments, but σ_z could be measured only in a few cases of elevated releases. The latter was, therefore, indirectly estimated from the Gaussian formula (9.7) or (9.8), assuming no loss of tracer due to deposition at the surface. Some of these estimates may have contained large errors (more than 100%).

Irwin (1983) compared the observed and predicted dispersion parameter estimates, centerline ground-level concentrations, and maximum g.l.c. from elevated releases for five different parameterizations (models) of σ_y and σ_z. With the exception of one case in which the Pasquill–Gifford dispersion scheme was used for σ_z, dispersion formulas (9.16) and (9.21) based on turbulence parameters σ_v and σ_w (alternatively, σ_θ and σ_ϕ) were used. Models or parameterizations differed primarily in the expressions for the functions f_y and f_z. Comparisons involved simple statistics (e.g., mean and standard deviation) of the fractional bias or error

$$\text{FB} = \frac{2(P - O)}{(P + O)} \quad (9.40)$$

where P is the model-predicted estimate and O is the observed value, as well as the statistics of the ratio P/O.

A comparison of dispersion parameter estimates from several different parameterizations for both surface and elevated releases shows an overall superiority of Draxler's (1976) scheme over those of Cramer (1976) and Pasquill (1976). The statistical results support the general conclusions that there is greater uncertainty associated with the σ_z estimates in comparison with the σ_y estimates, and that there is greater uncertainty associated with the estimates for the elevated releases than for the surface releases. Also, the models employing turbulence data resulted in more estimates within a factor of 2 of the "measured" dispersion parameters, in comparison with the results of the P–G scheme. The mean fractional errors and their standard deviations are also found to be smaller for the more sophisticated turbulence-based schemes compared with the P–G scheme. A comparison of model-predicted and -observed centerline ground-level concentrations within a distance of 5.4 km from the source yielded similar results, which are summarized in Figure 9.8 Note that the Gaussian model with the Pasquill–Gifford (P–G) scheme predicts concentrations within a factor of 2 of the observed values for only about half of the cases. The accuracy and precision of the model are significantly improved if an appropriate dispersion parameterization scheme utilizing turbulence data is used (e.g., in model 1 based on Draxler's 1976 scheme).

A comprehensive evaluation and assessment of performance of an air quality/dispersion model should be based on extensive meteorological and diffusion data sets covering a wide range of wind flow, stability, and terrain roughness conditions. Limited concentration data from a few monitoring stations at one site may not be sufficient to assess the comparative performance of different Gaussian diffusion models with similar dispersion parameterization schemes. This was demonstrated by an earlier evaluation of several rural diffusion models by the U.S. Environmental Protection Agency and American Meteorological Society (Smith, 1984). The reviewers pointed out the various limitations of the Gaussian plume formulation, the Pasquill–Turner stability classification system, the Pasquill–Gifford dispersion scheme, and other commonly used attributes of those models. They showed that the correlation coefficients between observations and predictions, paired in space and time, over short time periods were nearly zero, with none of the models explaining more than 10 percent of the variance. This was stressed as a basic weakness that rendered other model performance measures relatively meaningless. One of the fundamental causes of the poor correlation is the inherent uncertainty in air quality/dispersion models. The other important contributors to the poor correlation were identified as inadequacy of the data base used and scientific flaws in the models.

A comprehensive review of the uncertainties associated with air quality models and performance measures for evaluating them has been given by Weil et al. (1992). They discuss the natural variability in concentration in terms of the variance and the probability density function of concentration fluctuations. Current practice in performance evaluation of air quality models is also described and a simplification of performance measures in terms of the fractional bias FB defined earlier and the fractional scatter

$$\text{FS} = \frac{2(\sigma_{co} - \sigma_{cp})}{(\sigma_{co} + \sigma_{cp})} \quad (9.41)$$

where σ_{co} is the standard deviation of observed concentrations and σ_{cp} is the standard deviation of predicted concentrations is recommended. The FB and FS indicate how well the model produces the average values and the spread (standard deviation) around the average values of observed concentrations. The ideal values of these measures are zero, but they can range from -2 to 2; a value of ± 0.67 corresponds to a prediction within a factor of 2 of the observation.

Weil et al. (1992) also describe several approaches for separating model error from natural variability or inherent uncertainty. The first step in evaluating a model should be an assessment of the model physics

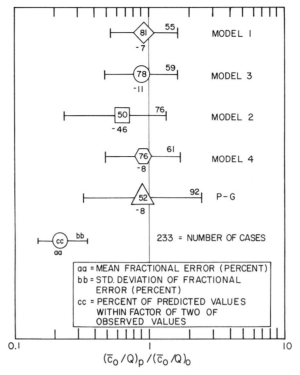

Figure 9.8 The ratio between the predicted and observed values of the ground-level concentration normalized by source strength for all stability conditions and downwind distances. P–G, Pasquill–Gifford. From Irwin, 1983.

in terms of the scientific basis and formulation of the model. The second step should be an evaluation of model predictions using available observations from laboratory and field experiments, as well as monitoring data from major source locations.

9.8 LIMITATIONS OF GAUSSIAN DIFFUSION MODELS

Limitations of most operational air quality and dispersion models arise from the simplifying assumptions implied in their formulation, simplified physics, and parameterization of complex turbulence and diffusion processes in these models. In particular, Gaussian diffusion models have limited applicability for relatively flat and homogeneous surfaces, reasonably steady and moderate to strong winds, moderately stable and unstable conditions, neutrally buoyant or slightly buoyant emissions, and relatively short distances (<50 km) from simple source configurations. Even under these idealized conditions, Gaussian diffusion models have large uncertainties due to natural variability and simplified model physics. Significant deviations from the idealized conditions introduce further limitations on the validity of such models, because model uncertainties might become too large to be acceptable. Gifford (1976) considered these as exceptional conditions, which include near-calm, extremely stable, and convective conditions; very rough (e.g., cities and forests) or very smooth (e.g., water) surfaces for which suggested diffusion parameterization may not be applicable; irregular and rugged terrain; and the presence of obstacles (e.g., hills and buildings) to the flow. To these one might add highly buoyant and dense gas emissions, stack downwash, and emissions of particulate matter and highly reactive gases. More sophisticated numerical models must be used to simulate and predict dispersion in these complex flow conditions. Such models will be discussed later in Chapter 11.

9.9 PRACTICAL APPLICATIONS OF GAUSSIAN DIFFUSION MODELS

Gaussian diffusion models have been extensively used in regulatory applications involving environmental impact assessments of existing and proposed sources of air pollution. In the United States, in particular, these models have received some sort of official blessing from state and federal regulatory agencies, since their use has been recommended in official regulatory guidelines (see e.g., U.S. Nuclear Regulatory Commission, 1979; U.S. Environmental Protection Agency, 1996). For example, Guideline on Air Quality Models by EPA (1996) recommends the use of nine standard air quality models, which are to be preferred for specific regulatory applications. Most of these models are based on the Gaussian formulation with empirical dispersion parameterization schemes. Another set of approved alternative refined models may be used for regulatory applications only after case-specific justifications are provided. The judicious use of air quality models has been incorporated in the U.S. Clean Air Act and its amendments. The U.S. EPA guideline recommends that air quality modeling techniques should be applied to state implementation plan (SIP) revisions for existing sources and new source reviews, including the prevention of significant deterioration. The guideline is intended for use by EPA regional offices, state and local agencies, and industry. A brief review of this guideline and the recommended air quality models is given here; for more details the reader should consult to the original reference.

9.9.1 Types of Air Quality Models

Air quality models are categorized into four generic classes: Gaussian, numerical, statistical, and physical. Within these classes, especially Gaussian and numerical models, a large number of individual algorithms may exist, each with its specific applications.

Gaussian models are the most widely used for estimating the impact of nonreactive pollutants from specified point and line sources in simple (flat and homogeneous) terrain. Numerical models may be more appropriate for applications in complex terrain and complex flow situations, as well as multiple point, line, and area sources emitting reactive pollutants on local (urban), regional, and global scales. Statistical or empirical modeling techniques are frequently employed in situations where incomplete scientific understanding of physical and chemical processes or lack of the required data bases render the use of Gaussian or numerical models impractical. Physical modeling involves the use of meteorological wind tunnels, convection tanks, or other fluid modeling facilities. It has been utilized for better understanding of flow and dispersion processes in complex flow and complex terrain situations. Here, our emphasis is primarily on Gaussian models.

On the basis of the level of sophistication of models, these are further subdivided into screening models and refined models. The former consist of relatively simple techniques that provide conservative estimates of the air quality impact of a specific source or source category. If the simpler analysis indicates a potential air pollution problem, an appropriate refined model is used for a more detailed and accurate estimate of source impact.

The modeling guideline identifies two classes of refined models. The so-called preferred models, listed in Table 9.6, are specifically recommended for regulatory applications involving different types of sources, land use, and terrain. In addition, a large number of alternative air quality models are listed and briefly described. But their use is considerably restricted by the additional requirement of providing justification in specific situations for which there is no suitable preferred model.

Table 9.6 Preferred Air Quality Models

1. Buoyant Line and Point Source Dispersion Model (BLP)
2. Line Source Dispersion Model (CALINE-3)
3. Climatological Dispersion Model (CDM 2.0)
4. Gaussian-Plume Multiple Source Air Quality Algorithm (RAM)
5. Industrial Source Complex Model (ISC3)
6. Urban Airshed Model (UAM)
7. Offshore and Coastal Dispersion Model (OCD)
8. Emissions and Dispersion Modeling System (EDMS)
9. Complex Terrain Dispersion Model Plus Algorithms for Unstable Situations (CTDMPLUS)

Source: U.S. EPA (1996).

9.9.2 Simple-terrain Gaussian Models

Simple terrain is considered to be an area where terrain features are all lower in elevation than the release height. The top five of the preferred air quality models listed in Table 9.6 are Gaussian models, which are recommended for use in simple terrain. A brief description of each of these is given in Appendix A of the revised guideline (U.S. Environmental Protection Agency, 1996). Information on their availability, users' guides, and other relevant references are also provided in the same. Model descriptions include input requirements, types of outputs available, model and source types, pollutant types, source-receptor relationships, plume behavior, specification of mean winds and horizontal and vertical dispersion parameters, chemical transformation, and physical removal processes.

The basic formulation of all the standard Gaussian models is the same and so are the specifications of horizontal transport winds and dispersion parameters. These models differ primarily in their specification of sources and receptors and treatments of chemical transformation and physical removal processes. The Pasquill–Gifford dispersion parameterization scheme and Briggs' urban dispersion formulas (Gifford, 1976) are recommended for rural and urban applications, respectively. Buoyancy-induced dispersion (Pasquill, 1976), plume rise (Briggs, 1969, 1971, 1975), and simple chemical transformation using an exponential decay are included in most of the models. Physical removal by gravitational settling and dry deposition is treated only in a few models dealing with particulate dispersion.

PROBLEMS AND EXERCISES

1. a. Show that the Gaussian plume formula (9.3) satisfies the mass conservation equation

$$\int_0^\infty \int_{-\infty}^\infty \bar{c}\bar{u}\, dy\, dz = Q$$

b. Also derive the expression for the cross-wind integrated g.l.c. from the same formula.

2. In an atmospheric diffusion experiment, a passive tracer was released continuously near the surface at the rate of 0.1 g s^{-1} and mean concentrations were measured using ground samplers spread along cross-wind arcs at various downwind distances from the source. From these measurements the plume centerline g.l.c. ($\bar{c}_o$) and the cross-wind integrated g.l.c. ($\bar{c}_{yo}$) could be estimated as follows:

x (m)	$\bar{c}_o$ (μg m^{-3})	$\bar{c}_y$ (μg m^{-2})
500	2.10	420
1,000	0.55	209
1,500	0.25	140

The observed mean wind speed at 10 m height was 3.5 m s^{-1} at the time of the experiment. Estimate the following at the three downwind locations: (a) the lateral dispersion parameter σ_y; (b) the vertical dispersion parameter σ_z and the plume height, z_p, assuming that the wind at 10 m is also the average transport velocity across the plume; (c) the average transport velocity over the plume height, using a power-law wind profile with an exponent of 0.2; (d) the revised estimates of σ_z, using the above estimated average transport velocities.

3. A burning dump emits 5 g s^{-1} of nitrogen oxides (NO$_x$). Estimate the mean concentrations of NO$_x$ at the distances of 1 km and 10 km downwind from the source on a clear night when the mean wind at 10 m height is 2.5 m s^{-1} and the wind profile can be represented by a power law, with an exponent of 0.5. You may use Briggs' interpolation formulas for dispersion over an open country and assume the dump to be a ground-level point source.

4. An expressway carries 10,000 vehicles per hour during the morning rush hour when the average speed of vehicles is 70 km hr^{-1}. At this speed, the average vehicle is estimated to emit 0.02 g s^{-1} of total hydrocarbons. Estimate the 10-min average concentration of total hydrocarbons at a distance of 500 m downwind of the expressway when the mean wind speed at 10 m height is 4.5 m s^{-1} and the wind profile exponent is estimated to be 0.15. You may use the Pasquill–Gifford dispersion curves.

5. Consider a continuous, ground-level, cross-wind line source of finite length b and strength Q_{cl}. Taking the origin of the coordinate system at the center of the line source and using the principle of superposition and the Gaussian plume formula for the point source, show that: (a) the mean g.l.c. at any point downwind of the source is given by

$$\bar{c}(x,y,0) = \frac{Q_{cl}}{(2\pi^{1/2}\bar{u}\sigma_z)}\left[\text{erf}\left(\frac{0.5b - y}{\sqrt{2}\sigma_y}\right) + \text{erf}\left(\frac{0.5b + y}{\sqrt{2}\sigma_y}\right)\right]$$

where

$$\text{erf}(t) = \frac{2}{\sqrt{\pi}}\int_0^t \exp(-t'^2)dt'$$

(b) the g.l.c. at large distances along the plume axis ($y = 0$) may be approximated by

$$\bar{c}_o(x,0,0) = \frac{Q}{\pi\bar{u}\sigma_y\sigma_z}$$

where Q is the strength of an equivalent point source.

6. A line of burning agricultural waste can be considered a finite line source 200 m long, which emits organics at the rate of 0.5 g m^{-1} s^{-1}. The mean transport velocity on this thinly overcast evening is 2.5 m s^{-1} perpendicular to the line source. Using the appropriate Briggs' dispersion formulas for open country, estimate the short-term average concentrations of organics at downwind distances of 500 m and 5000 m from the center of the line source. Compare these with the simpler, but rougher, estimates based on replacing the finite line source with an equivalent point source of strength 100 g s^{-1} and discuss your results.

7. a. Using the Gaussian plume diffusion formula for a continuous point release at an effective height H above the reflecting surface, derive the expressions for the maximum g.l.c. and its distance from the source, assuming that at short distances the plume dispersion parameters can be approximated by power-law relations $\sigma_y = ax^p$ and $\sigma_z = bx^q$.

b. Discuss the dependence of the maximum g.l.c. and its location on release height and atmospheric stability.

8. A 700-MW coal-fired power plant located in a flat countryside emits SO$_2$ at the rate of 4 kg s^{-1} with an effective release height of 180 m. Using the Briggs' dispersion formulas, estimate the following for a sunny afternoon when the mixing height is 1000 m, the mean wind speed at 10 m is 3.9 m s^{-1}, and the solar insolation may be considered strong: (a) the distance x_h at which the plume would extend to the top of the PBL; (b) mean transport velocity, assuming it to be equal to the wind speed at the effective release height and using a power-law wind profile with an exponent of 0.15; (c) the ground-level concentrations of SO$_2$ beneath the plume centerline as a function of downwind distance x from the source for x up to 30 km. Plot this on a log-log graph and estimate the maximum g.l.c. and the distance to the maximum x_{max}. Assume a uniform concentration in the vertical for $x \geq 2x_h$.

9. For the power plant given in Prob. 8, calculate and plot on linear graphs the following: (a) the vertical concentration profiles through the center of plume at distances of x_{max} and $2x_{max}$ and, hence, estimate the plume thicknesses at these locations; (b) the 100-μg m^{-3} g.l.c. isopleth and the surface area over which g.l.c. exceeds 100 μg m^{-3}.

10. For the power plant given in Prob. 8, estimate the maximum g.l.c. and its downwind distance from the source later in the afternoon when the solar insolation may be considered (a) moderate and (b) slight. Compare your results with those for the strong insolation and discuss the effect of discrete changes in the Pasquill stability class on the maximum g.l.c.

11. For the power plant in Prob. 8, estimate the maximum g.l.c. and its downwind distance from the source when the solar insolation is strong but the mean wind speed at 10 m height (a) increases to 6.5 m s^{-1} and (b) decreases to 1.9 m s^{-1}. Compare your results with those for Prob. 8 (c) and discuss the possible influence of wind speed on stability and the maximum g.l.c.

12. a. For the power plant in Prob. 8, estimate the maximum g.l.c. and its distance from the plant on a clear night when the wind speed at 10 m height is 2.4 m s^{-1} and the power-law profile exponent during the moderately stable conditions is estimated to be 0.5.

b. What would be the fumigation concentration at 1 km downwind of the plant next morning when the convective mixed layer extends to entrain all of the nighttime or early morning plume and wind speed at the plume height remains unchanged? Assume that, during fumigation, the concentration profile is approximately uniform in the vertical and $\sigma_{yf} = \sigma_y$ (stable) $+ H/8$.

13. A 15-m-tall industrial stack releases 2.5 g s^{-1} of a toxic gas. A sonic anemometer mounted on the top of the stack measured the following values of hourly averaged meteorological parameters at 4 AM and 4 PM:

Time	$\bar{u}$(m s^{-1})	σ_θ (deg.)	σ_ϕ (deg.)
4 AM	1.5	12.5	1.5
4 PM	3.0	25.0	12.5

a. For both the morning and the afternoon periods, estimate the dispersion parameters σ_y and σ_z as functions of x from 0.1 to 20 km, using the above measurements and the statistical theory relations (9.17) and (9.22) with the following empirical functions:

$$f_y^{-1} = 1 + 0.9\left(\frac{t}{1000}\right)^{0.5}$$

$$f_z^{-1} = 1 + 0.9\left(\frac{t}{50}\right)^{0.5}, \quad \text{for stable PBL}$$

$$f_z^{-1} = 1 + 0.9\left(\frac{t}{500}\right)^{0.5}, \quad \text{for unstable PBL}$$

where $t = x/\bar{u}$ in seconds.

b. Calculate and plot the g.l.c. as a function of distance for the two times.

c. Estimate and compare the maximum g.l.c. for the morning and afternoon periods.

14. For the industrial stack source in Prob. 13, the additional data on wind direction shear and buoyant plume rise during the early morning stable conditions are as follows:

Wind direction at 10 m = 180°
Wind direction at 200 m = 230°
Final plume rise $\Delta H = 50$ m

a. Estimate σ_y and σ_z at distances of 20 and 50 km from the stack, including the approximate contributions of buoyancy-induced dispersion and wind direction shear and compare these with their values due to turbulence alone.

b. Calculate the maximum g.l.c. and discuss the possible effects of buoyancy-induced dispersion and wind direction shear on the same.

15. The following hourly averaged meteorological data were collected at the site of a power plant whose impact is to be determined using the Gaussian plume model:

Maximum surface heat flux = 450 Wm^{-2}
Maximum mixing height = 2000 m
Mean wind speed in the mixed layer = 2.5 m s^{-1}
Mean temperature at 10 m = 26.8°C
Effective release height = 500 m

Compare different estimates of the lateral dispersion parameter by plotting σ_y as a function of x up to 50 km, using the following dispersion formulas proposed by Briggs (1974, 1988, 1993) for convective conditions:

$$\sigma_y = 0.22\, x(1 + 0.0001x)^{-1/2}$$
$$\sigma_y/h = 0.6X^*/(1 + 2X^*)^{1/2}$$
$$\sigma_y/h = 0.6X^*/(1 + 2X^{*2})^{1/6}$$

where $X^* = xW_*/h\bar{u}$.

16. a. Discuss the various types of uncertainties involved when the Gaussian plume model is used to estimate ground-level concentrations for a continuous point source at the surface.

b. The maximum estimated errors in the model input parameters are as follows:

Source strength: ±20%
Wind speed: ±10%
Lateral dispersion parameter: ±15%
Vertical dispersion parameter: ±15%

Estimate the maximum uncertainty in the calculated g.l.c. due to errors in model input parameters.

10

Plume Rise, Settling, and Deposition

10.1 MOMENTUM AND BUOYANCY EFFECTS OF RELEASE

In most of the diffusion theories and models discussed earlier in this book, the diffusing material has been assumed to be released in a passive manner. The momentum and buoyancy effects of release have not been considered except in a superficial way by including plume rise in the definition of effective source height. In reality, many sources of air pollution release material in the atmosphere with some initial vertical momentum and positive buoyancy, which make the released puff or plume rise gradually while dispersing under the influence of both the source-generated and naturally occurring turbulence. In some cases of cold dense gas releases and coarse particulate emissions, the trajectory of puff or plume actually falls with downwind distance from the release point. In general, the material released into the atmosphere may often have velocity and density different from those of the ambient air, and its behavior is likely to differ from that of a passive release. For a better understanding of the dispersion of such materials, we need to understand the possible effects of initial momentum and buoyancy of release on subsequent rise or fall, that is, on the mean puff or plume trajectory.

10.1.1 Rise of Hot Plumes

Most of the anthropogenic emissions into the atmosphere, especially from power generation, process emission, and industrial facilities, have much higher temperatures compared with the ambient air and frequently have high stack exit velocities. Although some of the major gaseous pollutants, such as CO_2 and SO_2, from these sources may have higher molecular weights than air, the fact that they are hot and much diluted by air and water vapor before their release means that their emissions are highly buoyant. The combined effects of both the momentum and buoyancy of emissions at the stack exit are to raise the plume center considerably above the initial height of release and to enhance near-source dispersion due to source- or stack-generated turbulence, as shown schematically in Figure 10.1.

The rise of plume is impeded by horizontal winds, which make it bend over; by the entrainment of ambient air into the plume due to effluent momentum- and buoyancy-generated turbulence; by mixing with ambient air due to ambient turbulence; and by decreasing density of ambient air with increasing height (stable stratification). The entrainment process is initially controlled by turbulence generated within the buoyant plume itself, but in later stages by atmospheric turbulence as the plume-generated turbulence dies out. The effect of ambient density stratification depends on whether the stratification is stable or unstable. In a stable atmosphere with increasing potential temperature and decreasing density with height, the rising plume eventually reaches an equilibrium level where the average density in the plume becomes equal to that of the environment. The plume may overshoot this level due to its upward momentum, but eventually settles down to its equilibrium height (see e.g., Fig. 10.1a). In an unstable atmosphere, on the other hand, the environmental density does not decrease as rapidly with height as that in the rising plume. The plume could rise to unlimited heights if it were not for the effect of entrainment of the ambient air into the plume and the presence of capping inversion on the top of the unstable mixed layer. The plume rise in an unstable PBL is essentially limited by the height of the inversion base above the stack top (see e.g., Fig. 10.1b).

For a well-defined point source, the plume rise is often determined from the visual or photographic observations of the visible smoke plume at various distances from the stack. During stable conditions,

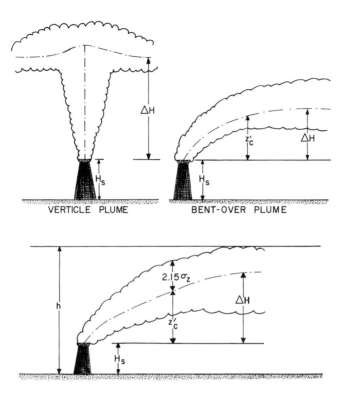

Figure 10.1 Schematic of plume rise in the atmospheric boundary layer in (a) stable stratification and (b) neutral and unstable conditions with a capping inversion.

plumes are observed to rise and level off not too far from the stack. From photographic observations, one can easily estimate the final plume rise. However, in near-neutral conditions, plumes are rarely seen to level off, and these may become invisible before final rise is reached. In unstable conditions, plumes are even more elusive as they swing up and down (looping) in response to convective updrafts and downdrafts and disperse rapidly. Sufficiently long time exposure may be used to estimate the trajectory of the rising plume as a function of distance for relatively short distances from the stack.

10.1.2 Fall of Dense Gas Plumes

Dispersion of cold, dense gases is found to be quite different from that of hot and light effluents. The gravitational slumping of very cold and dense gases released from their storage tanks can quickly bring these down to the ground level. Thus, for all practical purposes, the effective height of such releases can be assumed to be near ground level. But, when the release is in the form of a high-momentum jet from a controlled release valve of a pressurized storage tank, the jet can rise to considerable heights before falling off to the ground level under the influence of negative buoyancy. Trajectories of such jets under different environmental conditions have been systematically studied only in a few wind tunnel experiments (see e.g., Hoot et al., 1973; Schatzmann et al., 1993). These have shown that in early stages, the jet rise is governed by initial momentum and internal turbulence in the jet, but environmental wind shear and turbulence play greater roles in bending and dispersion of the jet and its eventual touchdown in later stages.

10.1.3 Stack Downwash

When effluents come out of a vertical stack at low momentum or low mean vertical velocity, and the horizontal flow around the stack is sufficiently strong, the effluent plume may be drawn down in the low-pressure region in the near wake of the stack. This phenomenon is referred to as stack downwash. A similar phenomenon, called building downwash, occurs when effluents are released from the rooftop or downwind side of a building and are drawn down in the low-pressure region of the building wake. Stack or building downwash essentially reduces the effective height of release, thereby increasing the ground level concentrations. Stack downwash can be avoided if the average exit velocity ($\overline{w}_s$) in the plume is more than one and one-half times the mean wind speed ($\overline{u}$) at the stack top. For $\overline{w}_s/\overline{u} < 1.5$, an empirical correction to effective release height, which accounts for the plume fall due to stack downwash, is often used (Briggs, 1984).

To avoid the adverse effect of building downwash, the height of the stack or chimney should be 1.5 to 2.5 times the height of the nearby building whose wake may cause the plume downwash and its entrainment into the recirculating flow region in the lee of the building. This so-called "good engineering practice stack height" also depends on the building aspect ratio. The traditional "2.5 times rule," that is, the effective source height being 2.5 times the building height, has been shown to be quite effective in reducing the adverse effect of building wake on ground-level concentrations (Huber and Snyder, 1982). It may be relaxed, however, for very tall and thin buildings, for which the necessary stack height for avoiding building wake effects might be as low as 1.5 building heights (Snyder and Lawson, 1976).

10.2 PLUME-RISE THEORY AND OBSERVATIONS

The calculation of plume rise is of crucial importance to the prediction of dispersion of buoyant effluents emitted from an industrial stack or chimney. In Gaussian dispersion models, plume rise is added to the physical stack height to estimate the effective source height. Except in strong winds, plume rise may increase the effective source height to several (2 to 10) times the actual stack height. Since the maximum g.l.c. ($\bar{c}_{o\max}$) is found to be approximately proportional to the inverse square of the effective source height, the importance of plume rise in reducing $\bar{c}_{o\max}$ by a factor of 4 to 100 or even more cannot be overstated. Unfortunately, plume rise may not be estimated to a great degree of accuracy, because most of the plume-rise theory and models are semiempirical and contain empirical constants or coefficients with large uncertainties in their values.

Plume-rise theory and models have been proposed in the literature since the mid-fifties. Extensive reviews of these have been given by Briggs (1969, 1971, 1975, 1984), Pasquill and Smith (1983), and Weil (1988a), among others. Early reviews lamented about too many different methods and formulas proposed in the literature for calculating plume rise, which did not agree with each other or with new observations of plume rise (Briggs, 1969, 1975).

Much of the confusion may have been due to the different definitions of plume rise adopted according to the convenience of the observer and limitations of the observing technique. For example, one common definition of the effective source height is the height of the plume at the largest distance where it is still detectable. This would obviously depend on the method of detection, as well as on atmospheric conditions. It also ignores the fact that in unstable conditions, buoyant plumes may keep on rising well beyond the distance they cease to be detectable by the observer's eyes or camera and that the plume rise is, in fact, a function of distance from the stack.

Briggs (1975) has proposed a more satisfactory definition of the effective source height H, which would yield the correct value of $\bar{c}_{o\max}$. He argues that, in dispersion modeling, one is usually most concerned with predicting $\bar{c}_{o\max}$ and the distance $x_{\max}$ from the source where the maximum g.l.c. occurs. At much shorter distances, where the plume is still rising, the g.l.c. is practically zero. At distances well beyond $x_{\max}$, additional plume rise might make g.l.c. fall off a little more rapidly with x, but not much. Therefore, according to Briggs, the final plume rise (ΔH) may simply be taken as the difference between the effective source height for yielding the correct $\bar{c}_{o\max}$ and the physical stack height. This is only practical if one has good measurements of $\bar{c}_{o\max}$ and a satisfactory dispersion model to obtain a set of plume rise formulas of sufficient general validity.

There are many factors that could possibly influence plume rise, such as the stack height and its internal cross-section area or diameter, efflux velocity, effluent temperature or density, atmospheric stability, temperature and wind profiles, turbulence intensity, and others. It is no wonder that so many plume rise formulas have been proposed, some purely empirical and dimensionally nonhomogeneous.

10.2.1 Different Approaches

Different approaches to express and calculate plume rise can be broadly divided into three categories. Prior to the formulation of a satisfactory theory, purely empirical plume-rise formulas were developed on the basis of the statistical correlation and regression of observed plume rise with a few of the relevant effluent and environmental variables. A brief review of the early empirical relations for the rise of buoyant plumes and momentum jets is given by Briggs (1969). Most of these have dimensional constants that restricted their application to within the narrow range of observational or experimental conditions for which they were derived. Such purely empirical formulas became obsolete following the development of more rational plume rise theory and models.

The second approach is also empirical, but based on sounder principles of dimensional analysis and similarity. Briggs (1968) used this approach to derive a simple set of plume-rise formulas for different effluent and environmental conditions, each containing a dimensionless empirical constant. Most of these turned out to be quite consistent with plume-rise data, as well as with the plume-rise theory based on the conservation equations of mass, momentum, and energy.

The best theoretical approach is based on the fundamental fluid dynamical equations expressing the conservation of mass, momentum, and buoyancy or thermodynamic energy for buoyant plumes and momentum jets. These equations are well known and have been reviewed by Briggs (1975, 1984), Weil (1988), and others. For the sake of simplicity, the conservation equations are integrated across the horizontal or vertical cross section of a plume or a jet, depending on whether the plume is vertical or bent over under the influence of horizontal cross-winds. Like the conservation equations for any turbulent flow, the equations for plumes and jets have the usual closure problems, which are surmounted only by certain ad hoc or empirical closure assumptions. The most commonly used closure assumptions involve the specification of the rate of growth of the plume radius or entrainment velocity. Different closure assumptions lead to different plume-rise formulas, which probably explains the proliferation of many plume and jet rise formulas that have been proposed in the literature (Briggs, 1969, 1975, 1984; Weil, 1988). Only some of the most commonly used relations will be given here together with their simpler derivation based on dimensional analysis. For more rigorous derivations using fluid dynamical conservation equations, the reader should refer to the above mentioned references.

10.2.2 Dimensional Analysis

Some aspects of dimensional analysis have been discussed in Chapter 4 in the context of similarity theories. For more details, the reader should refer to books on fluid mechanics/dynamics. For application to plume rise, we first consider the most relevant variables on which plume rise is likely to depend:

1. The initial momentum flux parameter, F_m, which for any arbitrary cross-section shape at the stack exit is traditionally defined as

$$F_m = \left(\frac{\bar{\rho}_s}{\bar{\rho}}\right) r_s^2 \bar{w}_s^2 \quad (10.1)$$

in which $\bar{\rho}_s$ is the average density and $\bar{w}_s$ is the average vertical velocity of effluent at the stack exit of equivalent radius r_s of a circular cross section. The kinematic momentum flux parameter F_m actually represents the total effluent momentum at the stack exit divided by $\pi\bar{\rho}$, where $\bar{\rho}$ is the density of the surrounding environment at the stack top.

2. The initial buoyancy flux parameter, F_b, which is traditionally defined as

$$F_b = \left(1 - \frac{\bar{\rho}_s}{\bar{\rho}}\right) g r_s^2 \bar{w}_s \quad (10.2)$$

It is also a kinematic parameter that actually represents the total buoyancy flux at the stack exit divided by $\pi\bar{\rho}$.

3. The static stability parameter

$$s = \frac{g}{T_v} \frac{\partial \bar{\theta}_v}{\partial z} = \frac{g}{T_v}\left(\frac{\partial \bar{T}_v}{\partial z} + \Gamma\right) \quad (10.3)$$

which is a measure of density stratification of the environment. This parameter is considered to be relevant to the determination of final plume rise only in a stable environment.

4. The mean horizontal wind speed, $\bar{u}$, at the stack top. The variation of wind speed with height, or wind shear, is generally ignored in the dimensional analysis of plume rise.

5. The time of travel, t, of the effluent parcel along the plume trajectory. This is considered to be a relevant variable in the dimensional analysis only of transitional rise of a plume or jet; it is ignored in the final rise. Note that the time of travel is considered more appropriate than the horizontal distance, x, from the stack, because the former can be used for vertical as well as bent-over plumes and jets. The conversion to $x = \bar{u}t$, using Taylor's translation hypothesis, may be done for bent-over plumes and jets only after the appropriate analysis.

6. Turbulence intensities in the effluent at the stack exit and those in the environment. These parameters are usually ignored in simplified dimensional analyses of plume rise, because their inclusion would lead to too many dimensionless parameters affecting plume rise.

Ignoring the effects of mean wind shear and turbulence intensities, the plume rise can be expressed as

$$\text{Plume rise} = f(F_m, F_b, \bar{u}, s, t) \quad (10.4)$$

which contains a total of six variables involving two fundamental dimensions (L and T). Dimensional analysis can be used to convert the dimensional expression (10.4) into a similarity relationship involving four dimensionless groups. Since the available observations and experiments on plume rise are not of sufficient accuracy and discrimination to determine empirically the forms of the similarity function involving more than one or two dimensionless parameters, attempts are made to simplify the analysis, so that only a few variables or parameters are considered at one time. Thus, instead of a generalized similarity relationship, different plume rise expressions are sought for different sets of conditions.

This strategy was originally used by Briggs (1968) in his derivation of plume rise formulas based on dimensional analysis. For each set of specified source and meteorological conditions, Briggs selected only the two most relevant independent variables on which plume rise might be expected to depend. This yielded only one dimensionless parameter, which was taken to be a universal constant for that set of restrictive conditions. Thus, definitive expressions of transitional or final plume rise could be obtained from dimensional considerations alone. Experimental data are then used to verify these relations and also to estimate the values of the dimensionless constants involved.

Several assumptions and arguments are made in order to reduce the number of variables to a bare minimum for each set of conditions. First, either the momentum flux or the buoyancy flux is considered as the most relevant parameter (independent variable), depending on whether the plume is momentum dominated (e.g., a nonbuoyant jet) or buoyancy dominated (e.g., a thermal plume). Second, the effect of static stability, s, is ignored in considerations of transitional plume rise under all stability conditions (note that $s = 0$ in a neutral PBL, as well as in a convective mixed layer). However, s is considered to be an important parameter (independent variable) in dimensional considera-

Table 10.1 Plume-rise Predictions Based on Dimensional Analysis

Stability/Wind Conditions	Type of Rise/Plume	Buoyancy-dominated Plume	Momentum-dominated Plume
Unstable and neutral/windy	Transitional/bent over	$z'_c = 1.6(F_b/\bar{u})^{1/3} t^{2/3}$	$z'_c = 2.0(F_m/\bar{u})^{1/3} t^{1/3}$
Stable/windy	Final/bent over	$\Delta H = 2.6(F_b/\bar{u})^{1/3} s^{-1/3}$	$\Delta H = 1.5(F_m/\bar{u})^{1/3} s^{-1/6}$
Stable/calm	Final/vertical	$\Delta H = 5.3 F_b^{1/4} s^{-3/8}$	$\Delta H = 2.4 F_m^{1/4} s^{-1/4}$
Neutral/windy	Final/bent over	$\Delta H = 400 F_b(\bar{u})^{-3}$	$\Delta H = 3.0 F_m^{-1/2}(\bar{u})^{-1}$

Source: Modified after Briggs, 1968.

tions of final rise in a stably stratified environment. Third, the mean wind speed, $\bar{u}$, is not a relevant variable under calm or near-calm conditions. In windy conditions also, when the wind is strong enough to bend the plume, $\bar{u}$ is not considered as a separate independent variable but is combined with the momentum or buoyancy flux parameter to form the ratio $F_m/\bar{u}$ or $F_b/\bar{u}$, which is then considered as the most relevant parameter in the dimensional analysis. The main argument for doing this is that, in a bent-over plume, the initial effluent momentum and buoyancy fluxes are diluted (reduced) along the plume length inversely proportional to $\bar{u}$. An exception is made, however, in consideration of the possible final rise under neutral conditions, for which $\bar{u}$ must be considered as an independent variable in order to form at least one dimensionless group.

Using the above assumptions and arguments, the plume rise formulas derived from dimensional considerations for different sets of conditions are given in Table 10.1. The values of the constants in the same may be considered as the best estimates obtained from analyses of plume-rise observations in the atmosphere as well as in laboratory facilities by various investigators (Briggs, 1968, 1969, 1975, 1984; Weil, 1988).

When both the effluent momentum and buoyancy contribute to plume rise, as occurs in most industrial stack plumes, the two effects may be considered additive and the total plume rise may be obtained by the summation of plume rise predicted for the momentum- and buoyancy-dominated plumes. This is only an approximation, which may result in an overestimation of the total plume rise, because the effluent momentum and buoyancy actually act together (simultaneously) to raise the plume. For more conservative estimates of plume rise, only the effect of buoyancy may be considered.

10.2.3 Transitional Plume Rise

Perhaps the best known plume-rise formulas are the so-called one-third and two-thirds laws for the transitional rise of momentum- and buoyancy-dominated plumes, respectively. Using Taylor's transformation, $t = x/\bar{u}$, these are also often expressed as

$$z'_c = 2.0 F_m^{1/3} (\bar{u})^{-2/3} x^{1/3} \tag{10.5}$$

$$z'_c = 1.6 F_b^{1/3} (\bar{u})^{-1} x^{2/3} \tag{10.6}$$

which can also be expressed in the dimensionless forms

$$\frac{z'_c}{L_m} = 2.0 \left(\frac{x}{L_m}\right)^{1/3} \tag{10.7}$$

$$\frac{z'_c}{L_b} = 1.6 \left(\frac{x}{L_b}\right)^{2/3} \tag{10.8}$$

in which the momentum and buoyancy length scales are defined as

$$L_m = \frac{F_m^{1/2}}{\bar{u}} \tag{10.9}$$

$$L_b = \frac{F_b}{\bar{u}^3} \tag{10.10}$$

The experimental plume-rise data in support of the one-third and two-thirds laws have been reviewed by Briggs (1975, 1984) and Weil (1988), who also give best estimates of the constants involved in the same. In particular, an experimental confirmation of Eq. (10.8) against field and laboratory data is shown in Figure 10.2. Experimental confirmation of Eq. (10.7) against laboratory data on momentum jet rise has been given by Hoult and Weil (1972) and others (Briggs, 1975, 1984). There is some question whether the constant involved may in fact be a weak function of the ratio $R = \bar{w}_s/\bar{u}$, as Briggs has suggested.

Further theoretical support for the above mentioned laws has been provided by their derivations based on the conservation equations of mass, momentum, and energy, all integrated across the plume cross section. For the bent-over plume rising under the influence of both the effluent momentum and buoyancy, the transitional plume rise is given by (Briggs, 1975, 1984)

$$z'_c = \left[\frac{3}{\beta_m^2} \frac{F_m x}{(\bar{u})^2} + \frac{3}{2\beta_b^2} \frac{F_b x^2}{(\bar{u})^3}\right]^{1/3} \tag{10.11}$$

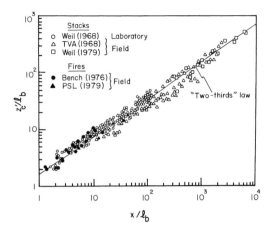

Figure 10.2 Observed trajectories of buoyancy-dominated plumes in the transient stage compared with the two-thirds law, Eq. (10.8). TVA, Tennessee Valley Authority; PSL, Physical Sciences Laboratory. From Weil, 1988.

or in the dimensionless form

$$\frac{z'_c}{L_b} = \left[\frac{3}{\beta_m^2} \left(\frac{L_m}{L_b} \right)^2 \frac{x}{L_b} + \frac{3}{2\beta_b^2} \left(\frac{x}{L_b} \right)^2 \right]^{1/3} \quad (10.12)$$

in which $\beta_b \simeq 0.6$ and $\beta_m \simeq 0.4 + 1.2R^{-1}$ are entrainment coefficients. Other investigators have found that $\beta_m \simeq 0.6$, independent of R (Hoult and Weil, 1972; Overcamp and Ku, 1986). Note that Eq. (10.11) or (10.12) reduces to the one-third and two-thirds laws when either the buoyancy flux term or the momentum flux term is neglected. It may be considered as an interpolation relation between the two laws, in which the momentum term is more significant near the stack (for $x < 2 \; \overline{u} F_m / F_b$), while the buoyancy term becomes more important and even the dominant term at distances much larger than $\frac{2\overline{u} F_m}{F_b}$.

In the presence of a strong enough inversion at the top of a neutral or unstable PBL, the plume rise will eventually be limited by the inversion (see e.g., Fig. 10.1b). In fact, there would be no further rise of plume after the top of the plume reaches the inversion base, that is, when

$$z'_c + 2.15\sigma_z = h - H_s \quad (10.13)$$

Thus, a strong inversion at the top of the PBL imposes a restriction on the transient plume rise, such that $z'_c \leq h - H_s - 2.15\sigma_z$, or the effective source height $H \leq h - 2.15\sigma_z$. Eventually, the plume axis will fall with further increase in distance, as a result of the downward spread of material after the plume impacts the inversion base. Here we ignore the more controversial matter of the partial or full penetration of weaker elevated inversions by highly buoyant plumes (Briggs, 1984; Weil, 1988). Detailed information on inversion thickness and strength is usually not available for such considerations. Also, the maximum g.l.c. will be expected only in the presence of a fully reflecting inversion capping the PBL.

10.2.4 Final Rise in a Stable Environment

Most of the observations of final plume rise in the atmosphere pertain to stably stratified conditions prevailing at night over land surfaces and even in daytime over colder-than-air water surfaces. Under such conditions, winds may be light near the surface, but are strong enough to bend over the elevated plumes from most industrial stacks and chimneys. Since such plumes are also often buoyancy dominated, there are plenty of data available to test the validity of the final plume rise formula in Table 10.1. This has indeed been done by Briggs (1969, 1975, 1984), Weil (1988), and others. The transient rise in stable, windy conditions follows the two-thirds law of the buoyancy-dominant plume discussed earlier. Experimental data to confirm the validity of the final rise formula for momentum-dominated plumes are lacking.

The final plume-rise formulas in Table 10.1 for stable and calm conditions have been confirmed or validated by experiments conducted in laboratory facilities (Morton et al., 1956; Morton and Middleton, 1973; Briggs, 1984). These can also be derived from the theoretical model proposed by Morton et al. (1956). The empirical values of the constants are based on Briggs' experiments in a saltwater-stratified channel. Here, ΔH represents the difference between the equilibrium plume height and the stack height. The applicability of these plume-rise formulas to the atmosphere is somewhat questionable, because calm conditions may not occur in the atmosphere except very close to the surface, and there is a substantial mean wind shear in the nocturnal stable boundary layer.

10.2.5 Final Rise in Neutral and Unstable Conditions

Strictly neutral conditions rarely occur in the atmospheric boundary layer. A near-neutral PBL may be realized, however, under strong winds and overcast conditions. The buoyancy-dominated as well as the momentum-dominated plumes are expected to acquire bent-over shapes whose trajectories may be described by the two-thirds and one-third laws, respectively. It is not clear up to what distance these transient plume-rise predictions would be applicable. Plume-rise observations in the atmosphere under near-neutral windy conditions do not extend to far enough distances from the stack to indicate a definite leveling of plumes or their deviation from the two-thirds law for buoyant plumes. However, both theoretical and physical considerations suggest that vigorous mechanical turbulence in the near-neutral PBL is capable of limiting the plume rise. The vigorous mixing of the plume with ambient air effectively dilutes the excess buoyancy and momentum of the plume. This is due to the process of "detrainment" or erosion of the plume by ambient turbulence. Different models have been proposed in the literature to describe plume rise limited by ambient turbulence (Briggs, 1975, 1984; Weil, 1988). In particular, the plume rise limited by mechanical turbulence in the lower part of the neutral PBL is given by the plume breakup model of Briggs as

$$\Delta H = 1.2 \left(\frac{F_b}{\bar{u}u_*^2} \right)^{3/5} (H_s + \Delta H)^{2/5}$$

$$= 1.3 \left(\frac{F_b}{\bar{u}u_*^2} \right) \left(1 + \frac{H_s}{\Delta H} \right)^{2/3} \quad (10.14)$$

This is an implicit equation for ΔH, which can be approximated to

$$\Delta H \simeq 1.3 \frac{F_b}{\bar{u}u_*^2} \quad (10.15)$$

for large final plume rise compared with the stack height ($\Delta H \gg H_s$).

Note that Eq. (10.15) is quite consistent with the final plume-rise prediction in Table 10.1 based on simple dimensional analysis, if one recognizes that in the neutral case

$$u_*^2 = C_D \bar{u}^2 \quad (10.16)$$

where the drag coefficient $C_D = 0.002$ to 0.004 for the neutral PBL, depending on the surface roughness. The above prediction of $\Delta H \propto F_b/\bar{u}^3$ for neutral conditions has been made by many investigators. The proportionality coefficient in this relationship may vary between 300 and 600 depending on the surface roughness. The plume-rise formula (10.15) should be preferred if an estimate of friction velocity is available.

For the momentum jet, Briggs' (1975) breakup model yields the following approximate formula for the final rise in neutral conditions:

$$\Delta H \simeq 1.5 \left(\frac{F_m}{\bar{u}u_*} \right)^{1/2} \quad (10.17)$$

which is not inconsistent with the formula given in Table 10.1 except for the difference in numerical coefficients. Again, Eq. (10.17) should be preferred if an estimate of u_* is available.

Verification of the above final rise formulas is still lacking, because plume-rise observations must be made at large distances, and neutral or near-neutral conditions in the PBL are rarely encountered. Even then, it is difficult to find situations when plume levels off because of ambient turbulence and not because of an elevated inversion layer. As a general guideline, a final rise formula should be used only when it gives lower plume rise than that given by the appropriate transient rise formula at the location of the maximum g.l.c. In neutral conditions, the plume rise probably terminates before the maximum g.l.c. occurs (Briggs, 1975).

Briggs (1971) considered the influence of ambient turbulence on buoyant plume rise in neutral and unstable conditions and suggested that the two-thirds law can be used only up to a finite distance x_f, beyond which the final rise can be assumed to have been reached, so that

$$\Delta H = 1.6 F_b^{1/3} (\bar{u})^{-1} x_f^{2/3}, \text{ for } x \geq x_f \quad (10.18)$$

Briggs showed that x_f essentially depends on the initial buoyancy flux F_b and proposed the following empirical relationships:

$$x_f = 49 F_b^{5/8}, \quad \text{for } F_b < 55 \text{ m}^4 \text{ s}^{-3}$$

$$x_f = 119 F_b^{2/5}, \quad \text{for } F_b \geq 55 \text{ m}^4 \text{ s}^{-3} \quad (10.19)$$

in which x_f is expressed in meters. The above final rise formula is often recommended and used in regulatory applications of the Gaussian plume models.

10.2.6 Final Rise in Convective Conditions

If a buoyant plume is rising in a deep convective boundary layer (CBL), large-scale convective eddies, especially downdrafts, would bring portions of the plume down to the ground level, thus causing the maximum g.l.c. relatively close to the stack. Briggs (1975) has derived the following formula for the final rise in the CBL, based on his "touchdown model":

$$\Delta H = 6.25 \left(\frac{F_b}{\bar{u} W_*^2} \right) \left(1 + \frac{2H_s}{\Delta H} \right)^2 \quad (10.20)$$

which is an implicit equation for ΔH. An explicit and simpler formula has been obtained by Briggs (1975, 1984) from his "plume breakup model"

$$\Delta H = 3.0 \left(\frac{F_b}{\bar{u}} \right)^{3/5} H_*^{-2/5} \quad (10.21)$$

in which $H_* = (g/T_o)(\overline{w'\theta_v'})_0$ is the buoyancy flux at the surface due to the combined effects of heating and evaporation.

A more sophisticated model of plume rise in the CBL has been described by Weil (1988). It is based on a specified probability density function (p.d.f.) of vertical velocity in the CBL. The model can be used for computing the mean plume trajectory, as well as vertical dispersion. The p.d.f. model provides a much better representation of the action of large-scale turbulent eddies (updrafts and downdrafts) on vertical dispersion in the CBL than the Gaussian model. Also, the p.d.f. approach does not require a final plume rise, which is a difficult quantity to predict and observe.

10.3 GRAVITATIONAL SETTLING OF PARTICLES

In considering the dispersion of particulate matter, one must take into account the gravitational settling of particles and the possible influence of the same on their mean trajectory. Gravitational settling is an important mechanism for the removal of particles from the effluent plume or puff and their deposition on the surface. An expression for the settling velocity can be derived from the basic fluid dynamical equations of motion around a small bluff body (particle) of a given shape (e.g., a sphere) and the resulting fluid drag due to viscous and pressure forces (see e.g., Kundu, 1990). These equations and their solutions are simple only so far as the flow around the body stays laminar. They become very complicated and analytically intractable, however, when the near-surface flow is separated and becomes turbulent in the wake of the body. However, an expression for the terminal settling velocity of particles falling in air or some other fluid can easily be derived from the simpler balance of gravitational and drag forces acting on the particle.

10.3.1 Terminal Fall Velocity

For the sake of simplicity and convenience, let us consider a spherical particle of diameter d and density ρ_p, falling at a constant velocity, v_g, in a fluid of density ρ.

When the particle is not accelerating or decelerating at all, that is, when it is falling at a constant (terminal) velocity, v_g, it is acted on by only two forces:

1. The gravitational force, including the effect of fluid buoyancy, which is given by

$$F_g = \frac{1}{6}\pi d^3(\rho_p - \rho)g \qquad (10.22)$$

2. The drag force due to the effects of viscous stresses and pressure around the body,

$$F_D = \frac{1}{2}C_D\rho A v_g^2 = \frac{\pi}{8}C_D\rho d^2 v_g^2 \qquad (10.23)$$

in which C_D is the drag coefficient of the particle and $A = \pi d^2/4$ is its cross-sectional area normal to the flow.

These two forces must be in balance (equal or opposite) when the particle is not accelerating, which yields

$$v_g = \left[\frac{4}{3}\frac{dg}{C_D}\left(\frac{\rho_p}{\rho} - 1\right)\right]^{1/2} \qquad (10.24)$$

Thus, the gravitational settling velocity depends on the particle diameter, the density ratio, and the drag coefficient. Equation (10.24) is applicable, irrespective of the flow conditions around the particle, so long as the flow can be considered as continuum. The nature of the flow, whether it is laminar or turbulent and separated or not separated, determines the drag coefficient. Standard empirical functions or curves of C_D as a function of Reynolds number, $\text{Re} = v_g d/\nu$, are available in the literature for different body shapes (Hoerner, 1958; Batchelor, 1967; White, 1974; Kundu, 1990). The one for a smooth sphere is given in Figure 10.3, covering a very wide range of Re values.

In particular, for very small spherical particles for which $\text{Re} < 1$, the drag coefficient is given by Stokes' solution for the creeping flow around a sphere as

$$C_D = \frac{24}{\text{Re}} = 24\nu/v_g d \qquad (10.25)$$

Substituting from Eq. (10.25) into (10.24) gives the well-known Stokes' equation

$$v_g = \frac{1}{18}\frac{gd^2}{\nu}\left(\frac{\rho_p}{\rho} - 1\right) \simeq \frac{1}{18}\frac{gd^2\rho_p}{\mu} \qquad (10.26)$$

The last approximation is particularly valid for heavy particles falling in air; here $\mu = \rho\nu$ is the dynamic viscosity. The implied assumption of $\text{Re} < 1$ would make the Stokes' law invalid for coarse particles of diameter greater than about 20 μm.

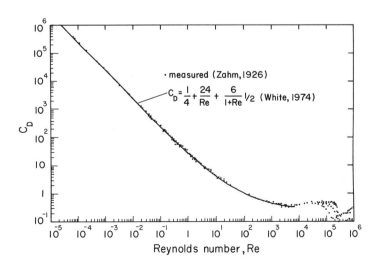

Figure 10.3 The measured drag coefficient as a function of Reynolds number for a smooth sphere, compared with an empirical relationship proposed by White (1974). From Stout et al., 1995.

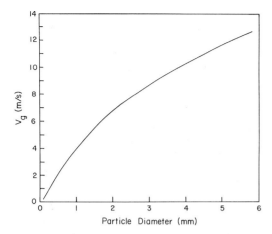

Figure 10.4 The gravitational settling velocity as a function of the particle diameter for particles of density $\rho_p = 10^3$ kg m^{-3}.

For large particles, one should use the more general expression (10.24) in which C_D is specified as a function of Re empirically. Then Eq. (10.24) becomes an implicit equation for v_g that can easily be solved by trial and error. The result for such a solution for smooth spherical particles with a density of 10^3 kg m^{-3} is shown in Figure 10.4. The settling speed for particles with different densities (ρ_p) can be obtained by multiplying the value of v_g from Figure 10.4 by $(\rho_p/1000)^{1/2}$, where ρ_p is expressed in kilograms per cubic meter.

The settling speed for nonspherical particles can be estimated by dividing the settling speed of an equivalent spherical particle having the same volume by a dynamical shape factor α. Typical values of α for different shape particles are given in Table 10.2, based on Chamberlain (1975). Note that the settling speed for nonspherical particles is less than that of an equivalent spherical particle, because the latter experiences the least amount of drag.

10.3.2 Mean Trajectory of Particles

If the mean shear can be ignored, as is done in Gaussian plume models, then the mean trajectory of particles of a given size (diameter) is expected to be a straight line with a downward slope of $\tan^{-1}(v_g/\overline{u})$ from the horizontal. The tilted plume axis would intersect the ground surface at a distance of $H_s\overline{u}/v_g$ from the source. The effective source height in the tilted Gaussian plume model is a decreasing function of x

$$H = H_s - \frac{v_g x}{\overline{u}}, \quad \text{for } x \leq \frac{H_s \overline{u}}{v_g}$$

$$H = 0, \quad \text{for } x > \frac{H_s \overline{u}}{v_g} \quad (10.27)$$

In the presence of wind shear, the mean trajectory of particles can be calculated from the dynamical equations of mean particle motion (see e.g., Stout and Arya, 1994). Sample trajectories of particles falling through a power-law wind profile are shown in Figure 10.5. These are based on the simplifying assumption of immediate adjustment of particles to fluid motion. Here m is the exponent of the power-law wind profile. Note that with increasing shear (increasing m), the normalized mean particle trajectory becomes increasingly steeper. For a given m, particle trajectory becomes increasingly steeper as the surface is approached. Thus, the effect of wind shear is to bend the mean particle trajectory or the plume axis such that the touchdown would occur closer to the source.

10.4 DRY DEPOSITION

Dry deposition refers to the transfer of airborne material, both gaseous and particulate, to the earth's surface, including soil, vegetation, and water, where it is removed (Seinfeld, 1986). The transfer processes leading to dry deposition at a surface can be divided into three stages, depending on the transfer mechanism involved. The first stage involves the transport of the material through the atmospheric surface layer to the immediate vicinity of the surface. The mechanism of transfer is the surface-layer turbulence. This stage is also frequently referred to as the aerodynamic component of transfer. The second stage involves the diffusion of material through the molecular sublayer (its characterization as the laminar sublayer may not be right, because smooth, streamlined laminar flow cannot establish on an irregular aerodynamically rough surface) adjacent to the surface itself. The primary mechanism of transfer through this very thin (order of millimeter) sublayer is molecular, although turbulent transfer also occurs intermittently. This stage is called the surface component of the transport. The third stage, called the substrate transfer component, involves the absorption of the material and its removal from the surface through chemical and biological processes. The solubility or absorptivity of the diffusing matter at the surface determines how much of the material is actually removed in this final stage of the substrate transfer and deposition process. Nonreactive gases, such as argon and helium, are not removed at all by

Table 10.2 The Dynamical Shape Factor α for Particles of Different Shapes

Shape	Ratio of Axes*	α
Ellipsoid	4	1.28
Cylinder	1	1.06
Cylinder	2	1.14
Cylinder	3	1.24
Cylinder	4	1.32

*In all cases preferential motion is perpendicular to the long axis.
Source: Chamberlain, 1975.

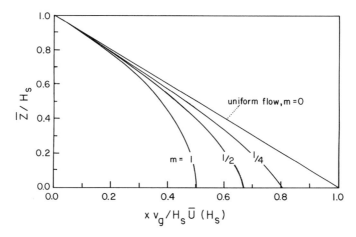

Figure 10.5 Calculated mean trajectories of particles falling through a power-law wind profile in the boundary layer.

dry deposition. Other gases might be removed through dry deposition on vegetation, wet soil, and water surfaces, but not on dry, bare surfaces.

10.4.1 Factors Influencing Dry Deposition

The various transfer mechanisms leading to dry deposition are complex and involve micrometeorological characteristics of the atmospheric surface layer through which turbulent transport occurs and the physical and chemical properties of the diffusing species and those of the surface material. Table 10.3 lists the various factors that are likely to influence dry deposition and removal rates of gaseous and particulate materials.

Micrometeorological variables influence the turbulent transfer through the surface layer. Of these, the friction velocity is the most important variable, with which dry deposition rates are found to be well correlated. Friction velocity is a measure of mean wind shear and shear-generated turbulence near the surface in both the above-canopy homogeneous surface layer and the canopy layer. Atmospheric stability also has a strong influence on mean wind distribution and turbulence in the surface layer and, hence, on dry deposition. Air temperature and humidity may influence chemical reactivity of diffusing gases.

Size, shape, and density of diffusing particles determine their settling velocities, which affect dry deposition. Other properties, such as hygroscopicity and solubility, determine whether particles will stick to a wet surface and not become resuspended. Electrostatic properties are also important in determining whether particles are attracted to the surface. Brownian diffusion becomes important in the diffusion of small (<1 μm) particles very close to the surface.

For gases, most important properties are their chemical reactivity with surface materials, including water and vegetation. Partial pressure near the surface and molecular diffusivity determine their diffusion rate through the molecular sublayer. But dry deposition will occur only as a result of absorption at the sur-

Table 10.3 Factors Influencing Dry Deposition and Removal Rates of Gases and Particles

Micrometeorological Variables	Properties of Depositing Material		Surface Variables
	Particles	Gases	
Wind speed at 10 m	Size (diameter)	Chemical reactivity	Aerodynamic roughness
Friction velocity	Shape	Molecular diffusivity	Wetness
Atmospheric stability	Density	Partial pressure	Canopy height
Turbulence	Diffusion coefficient	Solubility	Canopy structure
Air temperature	Hygroscopicity		Absorptivity
Relative humidity	Solubility		Porosity
Surface heat flux	Electrostatic effects		Electrostatic properties
	Impaction		Topography
	Interception		

Source: Modified from Sehmel, 1984.

face, molecular transfer through the substrate material, and biological processes taking place within the stomata of plant leaves. Thus, the surface properties are also very important in the final stage of the dry deposition process. The surface roughness may also be considered as a micrometeorological variable, but is more intimately related to the mean height of roughness elements and their shapes and distribution. For a vegetation canopy, the leaf area index (LAI) is a good measure of the canopy structure. Particulate material would be trapped more readily in a closed canopy than in an open canopy.

10.4.2 Quantitative Measures of Dry Deposition

Different quantitative measures have been used to represent and characterize dry deposition rates of gaseous or particulate material on a surface. The most fundamental of these is the vertical flux of the material near the surface, which can be expressed as

$$F_c = \overline{w'c'} + \overline{w}\,\overline{c} - v_g c - D\frac{\partial \overline{c}}{\partial z} \qquad (10.28)$$

Here, the vertical flux F_c is considered positive when it is directed upward. It consists of four terms, each representing a different transfer mechanism. The first term, $\overline{w'c'}$, represents the turbulent or eddy transport, which is the major contribution to the total flux outside the molecular sublayer. The second term, $\overline{w}\,\overline{c}$, represents the transport of material by mean vertical motion, which may not always be negligible, even near a flat, homogeneous surface (see e.g., Businger, 1986). There is some mean vertical velocity associated with the vertical fluxes of heat and water vapor at or near the surface. The mean vertical transport is further enhanced along sloping surfaces. However, it may be more appropriate to consider deposition and related fluxes normal to the sloping surface. In that case, $\overline{w}$ and w' will represent mean and fluctuating velocities in the direction of normal to the surface.

The third term in Eq. (10.28) represents the downward flux due to the mean settling speed of particles; it is important only for coarse and heavy particles ($d > 1$ μm). The last term is the contribution to the total flux due to molecular diffusion. It is negligible outside the thin molecular sublayer, but may become the dominant term within the molecular sublayer. Here, D represents the molecular diffusivity for the diffusing gas, or the Brownian diffusion coefficient for fine particles.

In the horizontally homogeneous turbulent surface layer, the expression (10.28) for the vertical flux may perhaps be simplified to

$$F_c \simeq \overline{w'c'} - v_g \overline{c} \qquad (10.29)$$

although one should also estimate $\overline{w}$ and, hence, $\overline{w}\,\overline{c}$, due to sensible and latent heat fluxes to make sure that the latter is indeed negligible. Equation (10.29) is the basis for the eddy correlation technique of measuring the vertical fluxes of diffusing gases ($v_g = 0$). This and other micrometeorological techniques of estimating mass fluxes and, hence, dry deposition have been reviewed by Businger (1986).

Direct measurements of downward fluxes of material near the surface for the purpose of characterizing dry deposition rates at the surface are not easy, particularly when deposition rates (fluxes at the surface) are not uniform under the dispersing plume. Even the indirect micrometeorological techniques of estimating fluxes over a large area would be too cumbersome to use routinely. These are best suited for estimating fluxes from uniform area sources, such as soil, water, and vegetation. In practical applications, dry deposition rates or fluxes are parameterized in terms of a simpler parameter, the dry deposition velocity, v_d, or the transfer velocity, v_t.

The dry deposition velocity is most commonly used in the parameterization of dry deposition rate (downward flux) and has been defined as (Chamberlain, 1953, 1966)

$$v_d = -\frac{F_c}{\overline{c}} \qquad (10.30)$$

where $\overline{c}$ is the mean concentration at some reference height above the surface and usually above the canopy or roughness layer. The transfer velocity is a somewhat related parameter that is frequently used for parameterizing transfer processes in engineering practice and is defined as

$$v_t = -\frac{F_c}{(\overline{c} - \overline{c}_o)} \qquad (10.31)$$

where $\overline{c}_o$ is the surface equilibrium concentration.

The distinction between the deposition velocity, defined in Eq. (10.30), and the transfer velocity, defined in Eq. (10.31), disappears only when $\overline{c}_o = 0$. This would be the case for a highly reactive species that would be readily absorbed on reaching the surface. The particulate material also would readily deposit and remain when it encounters a water surface, a wet soil surface, or a closed vegetation canopy, so that it would be reasonable to assume $\overline{c}_o = 0$. In most other cases of partial or no absorption, however, deposition and transfer velocities would differ due to the finite value of the surface equilibrium concentration. For these the concept and definition of deposition velocity become somewhat ambiguous and may even lead to some unphysical results (see e.g., Businger, 1986). For example, when water vapor is considered as a transferable species, the usual definition of deposition velocity, $v_{dq} = -\overline{w'q'}/\overline{q}$, makes it meaningless, because v_{dq} can be negative, positive, or zero and there is no relationship between vapor flux ($\overline{w'q'}$) and $\overline{q}$ at some reference height. The concept and definition of transfer velocity, on the other hand, are more meaningful and lead to more rational and physically plausible relationships, for example,

$$v_{tq} = -\frac{\overline{w'q'}}{(\overline{q} - \overline{q}_o)}$$

$$v_{t\theta} = -\frac{\overline{w'\theta'}}{(\overline{\theta} - \overline{\theta}_o)} \quad (10.32)$$

Note that Eqs. (10.32) imply a plausible bulk transfer or gradient transport relationship between the flux and the difference or gradient of the property being transferred. In the presence of significant gradients of specific humidity, temperature, and other scalar concentrations in the surface layer, the corresponding transfer velocities are all expected to be positive and directly related to the mean velocity at the reference height, friction velocity, or some other turbulence scaling velocity. Transfer velocities are useful only for parameterizing the fluxes near or at the surface when the surface equilibrium values $\overline{q}_o$, $\overline{\theta}_o$, and $\overline{c}_o$ can be determined. For water surfaces, θ_o and $\overline{q}_o$ can easily be measured or estimated, but for land surfaces these are often difficult to obtain. Also, the surface equilibrium values of concentrations of diffusing and depositing gases on natural surfaces cannot be measured or otherwise estimated. Therefore, the concept of transfer velocity has not been found to be practically useful for parameterizing dry deposition of airborne gases and particles on natural surfaces. Instead, dry deposition velocity has been extensively used in practical applications of atmospheric dispersion and deposition modeling. In these, deposition velocities are specified on the basis of empirical observations or theoretical estimates based on the approximate analogy with the momentum transfer.

10.4.3 Theoretical Relations for v_d

The transfer of material through the atmospheric surface layer, molecular sublayer, and ultimately to the surface and substrate can be viewed as analogous to the electric flow through a series of resistances. The resistance approach to the transfers of momentum, heat, and water vapor over bare soil and vegetative canopies is frequently used in agricultural and forest meteorology. It is also widely used in the parameterization of atmosphere–land surface exchange processes in climate models. The transfer velocity turns out to be the inverse of the total "resistance," which is considered as the sum of the resistances of the various components of the system (e.g., air, canopy, soil surface, etc.). When applied to dry deposition velocity for gaseous species, the resistance approach yields the relationship (Seinfeld, 1986)

$$v_d = (r_a + r_s + r_t)^{-1} \quad (10.33)$$

in which r_a denotes the aerodynamic resistance of the turbulent surface layer, r_s is the resistance of the molecular sublayer, and r_t represents the combined resistance of the surface and transfer to substrate. Thus, the above three resistances correspond to the three stages of the transfer/deposition process described earlier. Considering the different transfer mechanisms involved in the three stages, one can identify the various factors or properties on which r_a, r_s, and r_t might be expected to depend. Since aerodynamic resistance accounts for the turbulent diffusion of the material through the surface layer, r_a depends on the usual micrometeorological parameters, such as wind speed, friction velocity, stability, and surface roughness. The sublayer resistance, r_s, is expected to depend on the molecular properties as well as on the friction velocity and the roughness parameter. Finally, the transfer resistance, r_t, depends on the physicochemical interaction between the material and the surface, which is probably the most difficult to express quantitatively.

For dry deposition of particles, Eq. (10.33) must be modified to account for the gravitational settling velocity. Furthermore, for large and heavy particles, which may be considered to have deposited once they encounter the surface, the transfer resistance $r_t \simeq 0$, and the modified expression for dry deposition velocity is (Seinfeld, 1986)

$$v_d = (r_a + r_s + r_a r_s v_g)^{-1} + v_g \quad (10.34)$$

10.4.4 Aerodynamic Resistance

An expression for r_a can easily be obtained if we consider the turbulent transfer of material through the atmospheric surface layer to be analogous to that of momentum. The transfer or deposition velocity for momentum, by definition, is given by

$$v_{dm} = \frac{u_*^2}{\overline{u}(z_r)} \quad (10.35)$$

It can also be expressed in terms of the drag coefficient, $c_D = u_*^2/\overline{u}^2$, as

$$v_{dm} = c_D \overline{u} = c_D^{1/2} u_* \quad (10.36)$$

which expresses direct dependence of transfer or deposition velocity on $\overline{u}$ or u_*. An alternative expression, involving surface roughness and stability parameters, can be given using the surface-layer similarity relation for the mean wind speed, that is,

$$v_{dm} = \frac{k^2 \overline{u}}{[\ln(z_r/z_o) - \psi_m(z_r/L)]^2} \quad (10.37)$$

The aerodynamic resistance for momentum is, then, just inverse of v_{dm},

$$r_a = \frac{[\ln(z_r/z_o) - \psi_m(z_r/L)]^2}{k^2 \overline{u}} \quad (10.38)$$

or, alternatively,

$$r_a = \frac{[\ln(z_r/z_o) - \psi_m(z_r/L)]}{k u_*} \quad (10.39)$$

From these, the dimensionless product $r_a\bar{u}$ or $r_a u_*$ can be determined as a function of z_r/z_o and z_r/L; it decreases with increasing surface roughness and decreasing stability.

10.4.5 Sublayer Resistance

The resistance due to molecular diffusion in the thin sublayer near a flat rough surface has been formulated by Owen and Thompson (1963) to have the form

$$r_s u_* = a \, \text{Re}_*^m \, \text{Sc}^n + b \quad (10.40)$$

where $\text{Re}_* = u_* z_o/\nu$ is the roughness Reynolds number, $\text{Sc} = \nu/D$ is the Schmidt number, and a, b, m, and n are empirical constants. For a fairly large range of Schmidt numbers, Fernandez de la Mora and Friedlander (1982) have estimated $a \simeq 0.17$, $b \simeq 3$, $m = 0.5$, and $n = 2/3$. It is clear from Eq. (10.40) that the molecular sublayer resistance increases with increasing surface roughness and decreasing diffusivity of the material (gaseous or particulate).

For particles, diffusivity is much smaller than that for gases. Consequently, r_s is larger than that for gases. For fine particles, Brownian diffusion coefficient increases and, consequently, r_s decreases with decreasing size of particles. Fine particles ($d < 1 \ \mu\text{m}$) are transferred through the surface sublayer by Brownian diffusion in much the same way as gases. For coarse particles, both gravitational settling and inertial effects become important. Theoretical models of dry deposition velocity of particles, including the mechanisms of Brownian diffusion, gravitational settling, and inertial impaction of particles on the surface, have been proposed in the literature (see e.g., Sehmel, 1980, 1984; National Center for Atmospheric Research, 1983).

A different formulation for the surface sublayer resistance, which does not explicitly include Reynolds number and Schmidt number dependence, but relates r_s to only the usual micrometeorological variables, is described by Seinfeld (1986). Essentially it implies that the product r_s is inversely proportional to u_*, that is,

$$r_s = \frac{A}{k u_*} \quad (10.41)$$

in which the empirical constant A is estimated to lie between 2 and 2.6 (Wesely and Hicks, 1977). Note that Eq. (10.41) is an attempt to express the sublayer resistance in a form similar to that of the aerodynamic resistance.

10.4.6 Transfer Resistance

Probably the most uncertain component of the total resistance to transfer and deposition is the surface and substrate transfer resistance, r_t. The efficiency with which a surface absorbs diffusing gas molecules, or captures particles that impact on it, is determined largely by its chemical composition and physical structure, as well as by the physicochemical properties of the diffusing material. For gases, surface uptake rates by vegetation are essentially determined by biological factors, such as opening and closing of stomata and the associated stomatal resistance. For water vapor, r_t is determined by the saturation vapor pressure inside the vegetation and therefore the leaf temperature. For other gases, r_t would depend on their absorptivity or reactivity with the surface material. Some gases may not react at all and not be absorbed at the surface; for these $r_t = \infty$. Other gases, such as O_3 and SO_2, are readily taken up and destroyed by stomata of leaves. For these it would be important to know whether stomata are open or closed. Surface wetness also plays an important role in determining the rate of deposition of particles and certain gases. On wet surfaces hygroscopic materials will be more readily retained or absorbed than hydrophobic ones. If the surface is not a perfect absorber or retainer, a finite transfer resistance r_t is associated with it, since not all the impinging gas molecules or particles would adhere to the surface (Seinfeld, 1986). However, it is not easy to specify this resistance in a rational manner. This is the weakest link in the determination of total resistance and, hence, deposition velocity from Eq. (10.33) or (10.34).

10.4.7 Nomograms for v_d

To examine how micrometeorological variables, such as the near-surface wind speed or friction velocity, surface roughness, and stability, affect the deposition velocity of a gaseous species, we substitute from Eqs. (10.39) and (10.41) into Eq. (10.33) to express the dimensionless deposition velocity as

$$\frac{v_d}{u_*} = \left[\frac{\ln(z_r/z_o) - \psi_m(z_r/L) + A}{k} + r_t u_* \right]^{-1} \quad (10.42)$$

Alternatively, in terms of the wind speed at the reference height, Eq. (10.42) can be written as

$$\frac{v_d}{u_*} = \left[\frac{\bar{u}}{u_*} + \frac{A}{k} + r_t u_* \right]^{-1} \quad (10.43)$$

In the above relations, u_* is chosen as the appropriate scaling velocity for v_d, $\bar{u}$, and r_t^{-1}. It can easily be determined from the observed wind speed and the surface-layer stability parameter, z_r/L, using standard micrometeorological relations discussed in Chapter 4.

For a graphical representation of the deposition velocity as a function of micrometeorological variables, Eq. (10.43) would provide the most concise plot because it involves only two dimensionless parameters—$\bar{u}/u_*$ and $r_t u_*$—on which the dimensionless deposition velocity v_d/u_* is predicted to depend. Note that both the effects of surface roughness and stability are implicitly included in $\bar{u}/u_*$.

Using the value of $A = 2.6$ for SO_2 transfer (Wesely and Hicks, 1977) in Eq. (10.43), v_d/u_* is plotted in Figure 10.6 as a function of $\bar{u}/u_*$ for different values of $r_t u_*$. As expected, the dimensionless depo-

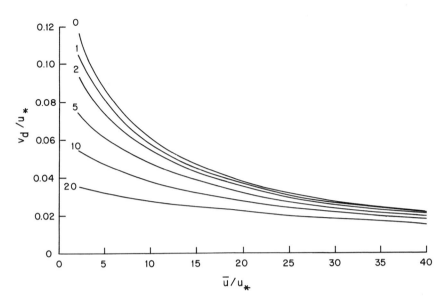

Figure 10.6 The dimensionless dry deposition velocity, v_d/u_*, as a function of normalized wind speed and the dimensionless surface transfer resistance, $r_t u_*$.

sition velocity decreases as the normalized wind speed, $\bar{u}/u_*$, and the dimensionless surface/substrate transfer resistance, $r_t u_*$, increase. Since the scaling velocity (u_*) itself would be expected to depend on the surface roughness and stability, it is not easy to draw conclusions from Figure 10.6 about the influence of these parameters on v_d.

The effects of the surface roughness parameter z_o and wind speed on dry deposition velocity for gases under neutral stability conditions were examined earlier by Seinfeld (1986), who plotted a series of graphs of v_d versus z_o for different wind speeds and transfer resistances. From these one can conclude that even for $r_t = 0$, implying a perfectly absorbing surface, the deposition velocity can be small if the wind speed is low and the surface is relatively smooth. The velocity increases with increasing wind speed and increasing roughness of the surface. As the transfer resistance is increased, the deposition velocity decreases rapidly. For $r_t > 0.01$ ms^{-1}, the deposition velocity is not very sensitive to changes in z_o. Thus, the transfer resistance corresponding to the final stage in the deposition process is a key parameter determining v_d (Seinfeld, 1986).

10.4.8 Measurements of Dry Deposition

Dry deposition velocities of gases and particles on natural surfaces have been measured or estimated using a variety of methods and techniques. Comprehensive reviews of these are given elsewhere (see e.g., Hicks and Wesely, 1978; McMahon and Denison, 1979; Droppo, 1980; Sehmel, 1980, 1984; National Center for Atmospheric Research, 1983; Businger, 1986). Only a brief mention of some of the techniques and a summary of experimental results are given here.

The deposition velocity is the ratio of the flux of the material at or near the surface to the mean concentration at a reference height of 1 to 10 m. Therefore, an experimental determination of v_d requires either direct measurement or indirect estimation of the vertical flux, in addition to the mean concentration measurement. The dry deposition flux of particulates can be determined from the measurement of the mass of collected particulate material over a specified surface sampling area and time. Filter sampling techniques are also used for this purpose. Dry deposition rates for gases are best determined from fast response concentration and vertical velocity measurements, using the eddy correlation and eddy accumulation techniques. Other micrometeorological methods used are gradient, Bowen ratio, variance, and dissipation methods (see e.g., Businger, 1986).

Figure 10.7 summarizes the experimental data on dry deposition velocity of particles as a function of particle radius and compares them with theoretical predictions based on the considerations of the different mechanisms, that is, Brownian diffusion, gravitational settling, and inertial impaction. Note that, in spite of the large scatter, the experimental data for coarse particles ($d > 1$ μm) clearly shows v_d to be increasing with increasing particle diameter and approaching the gravitational settling velocity for very large particles ($d > 20$ μm). The observed trend is in good agreement with theoretical predictions, which suggest a minimum deposition velocity for particles in the range 0.1 μm $< d \leq 1$ μm. For fine particles, the Brownian diffusion coefficient increases with decreasing parti-

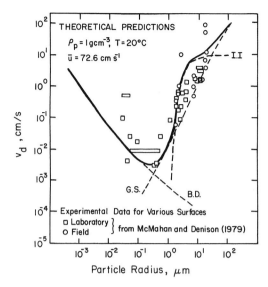

Figure 10.7 Experimental data on dry deposition velocities of particles compared with theoretical predictions based on the mechanisms of Brownian diffusion (B.D.), gravitational settling (G.S.), and inertial impaction (I.I.). From National Center for Atmospheric Research, 1983.

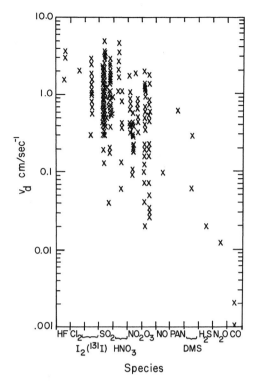

Figure 10.8 Experimental data on dry deposition velocities of the various gases ranked approximately in order of their reactivity. From National Center for Atmospheric Research, 1983.

cle size and, consequently, the deposition velocity also increases. Measurements of v_d for small particles have been made mostly in laboratory facilities, since field measurements are extremely difficult to make. Wesely et al. (1985) have reported measurements of dry deposition of particulate sulfur over various surfaces, using the eddy correlation techniques. Earlier field observations of v_d were characterized by large scatter; v_d values in any individual experiment range over several orders of magnitude (see e.g., Sehmel, 1980). Variables other than the particle diameter, such as wind speed and the vertical turbulence intensity, are also expected to influence dry deposition velocity.

Experimental data on dry deposition velocities of gases on different types of natural surfaces have been compiled and reviewed by McMahon and Denison (1979) and Sehmel (1980, 1984). The range of v_d values for each gas covers at least one order of magnitude, even for the same type of the surface. When different types of surfaces are considered, v_d values for a particular species range over several orders of magnitude. The most extensively measured are dry deposition velocities of sulfur dioxide (SO_2) and iodine (I_2); v_d values for the former range from 0.04 to 7.5 cm s^{-1} and those for the latter range from 0.02 to 26 cm s^{-1}. Even for uniform grass and water surfaces, the variability is quite large because of the expected dependence of v_d on micrometeorological variables. This should caution one against specifying a constant value of v_d, as is done in most atmospheric dispersion-deposition models. Although an average value of $v_d = 0.5$ cm s^{-1} is often used for SO_2, the value may vary over a range from 0.04 to 7.5 cm s^{-1}, depending on the type of surface, wind speed, stability, and possibly other factors.

Some of the experimental data on deposition velocities for the various gases ranked approximately in order of their reactivity are presented in Figure 10.8. Hydrogen fluoride and chlorine are among the most reactive species whose deposition velocities exceed 1 cm s^{-1}. Smaller range of values for these may only be an artifact of a few observations. More extensive measurements have been made of deposition velocities of iodine, sulfur dioxide, nitric acid, nitrogen dioxide, and ozone. For these, v_d values range over one to two orders of magnitude. Carbon monoxide is not very reactive and its deposition velocity is of the order of 10^{-3} cm s^{-1}.

10.5 DISPERSION-DEPOSITION MODELS

Dry deposition at the surface of any dispersing material from a surface or an elevated source can be considered in different ways. The deposition flux is usually parameterized in terms of deposition velocity, which is either specified empirically or estimated from appropriate theoretical relations.

Using the gradient transport (K)-theory, dry deposition is included by specifying the deposition flux as the surface boundary condition. Also, the integral mass conservation equation is modified to account for the amount of material deposited on the surface. Since the unknown surface or near-surface concentrations are involved in the modified boundary and integral mass-conservation conditions, analytical solutions of the diffusion equation are no longer possible, even for constant-eddy diffusivities. Numerical solutions to the advection-diffusion equation with variable-eddy diffusivities can be used to include the effects of dry deposition as well as gravitational settling for heavier particles. Such models will be discussed in Chapter 11.

Analytical dispersion-deposition models are essentially modifications of the basic Gaussian plume model for nonreactive or nonabsorbing surfaces. The proposed modifications fall into two broad categories of dispersion-deposition models, which are described below.

10.5.1 Source-depletion Models

Chamberlain (1953) proposed one of the simpler and most commonly used dispersion-deposition models. His source-depletion model accounts for the deposition of the airborne material by appropriately reducing the source strength as a function of downwind distance. For an elevated continuous point source, mean concentration and dry deposition flux at any point on the surface are given by

$$\bar{c}_o(x,y,0) = \frac{Q_x}{\pi \sigma_y \sigma_z \bar{u}} \exp\left(-\frac{y^2}{2\sigma_y^2}\right) \exp\left(-\frac{H^2}{2\sigma_z^2}\right) \quad (10.44)$$

$$F_d(x,y,0) = v_d \bar{c}_o(x,y,0) \quad (10.45)$$

Here Q_x is the depleted source strength at the downwind distance x, which is given by the consideration of mass conservation as

$$\frac{\partial Q_x}{\partial x} = -\int_{-\infty}^{\infty} F_d(x,y) dy \quad (10.46)$$

Substituting from Eqs. (10.44) and (10.45) into Eq. (10.46) yields, on integration,

$$\frac{\partial Q_x}{\partial x} = -\left(\frac{2}{\pi}\right)^{1/2} \frac{v_d Q_x}{\sigma_z \bar{u}} \exp\left(-\frac{H^2}{2\sigma_z^2}\right) \quad (10.47)$$

Equation (10.47) can also be written in the integral form as

$$\int_{Q_o}^{Q_x} \frac{dQ_x}{Q_x}$$

$$= -\left(\frac{2}{\pi}\right)^{1/2} \frac{v_d}{\bar{u}} \int_0^x \frac{dx}{\sigma_z \exp(H^2/2\sigma_z^2)}$$

or, $\ln \frac{Q_x}{Q_o}$

$$= -\left(\frac{2}{\pi}\right)^{1/2} \frac{v_d}{\bar{u}} \int_0^x \frac{dx}{\sigma_z \exp(H^2/2\sigma_z^2)} \quad (10.48)$$

The integral on the right-hand side of Eq. (10.48) and, hence, the ratio Q_x/Q_o can be evaluated for any specification of σ_z as a function of distance from the source. In particular, using the Pasquill–Gifford dispersion curves for open country, Van der Hoven (1968) determined the depleted source strength Q_x/Q_o as a function of distance, Pasquill's stability class, and effective source height for a constant ratio $v_d/\bar{u} = 10^{-2}$. His results for $v_d/\bar{u} = 0.01$ are shown in Figure 10.9; for other values of the ratio $v_d/\bar{u}$ the source depletion factor Q_x/Q_o may be obtained from Eq. (10.48)

Several conclusions can be drawn from the results in Figure 10.9. The source depletion factor for a fixed value of $v_d/\bar{u}$ is a decreasing function of x, which depends strongly on the source height as well as atmospheric stability. The effective source strength is depleted faster with increasing stability and decreasing height of release. It is also obvious from Eq. (10.48) that the source depletion factor is also strongly dependent on the ratio $v_d/\bar{u}$. From the graphs in Figure 10.9, Hanna et al. (1982) have estimated the downwind distance at which the source is depleted by 50 percent due to dry deposition, as a function of source height and Pasquill's stability class, for $v_d/\bar{u} = 0.01$. This distance exceeds 10 km under moderately and extremely unstable conditions, but it is only a few hundred meters under neutral and stable conditions for surface sources. For a given stability class, the 50-percent depletion distance increases with increasing height of release. Note that ground-level concentrations are also reduced due to dry deposition in proportion to the source depletion factor.

A serious limitation of Chamberlain's (1953) source-depletion model is that the depletion is assumed to occur over the whole depth of the plume rather than at the surface. The concentration gradient at the surface remains zero, as in the case of perfect reflection from the surface. In the case of absorption and deposition at the surface, one should have a finite positive concentration gradient. This is accomplished in the partial reflection model proposed by Csanady (1955) and further extended by Overcamp (1976). In this the contribution due to the image source is reduced by a factor α which is called the reflection coefficient ($\alpha = 1$ for perfect reflection and no deposition). The expression for concentration according to the partial reflection model is

$$\bar{c} = \frac{Q}{2\pi \bar{u} \sigma_y \sigma_z} \exp\left(-\frac{y^2}{2\sigma_y^2}\right) \left\{ \exp\left[-\frac{(z-H)^2}{2\sigma_z^2}\right] \right.$$

$$\left. + \alpha(x,z) \exp\left[-\frac{(z+H)^2}{2\sigma_z^2}\right] \right\} \quad (10.49)$$

in which the partial reflection coefficient is assumed to be a function of x and z. By equating the surface deposition rate, $v_d \bar{c}_o$, to the flux due to turbulent diffusion, Overcamp (1976) derived the following expression for $\alpha_o(x)$ at $z = 0$:

$$\alpha_o(x) = 1 - \frac{2v_d}{\left(v_d + \dfrac{\bar{u}H}{\sigma_z}\dfrac{d\sigma_z}{dx}\right)} \quad (10.50)$$

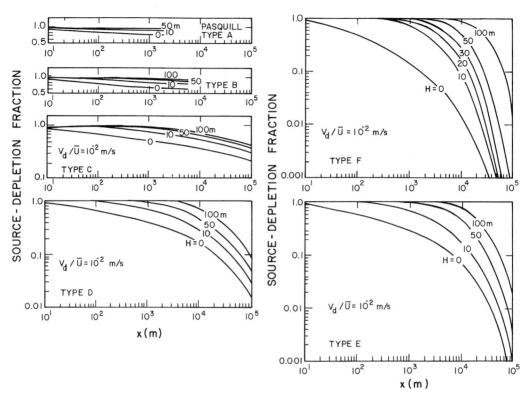

Figure 10.9 Ratio of the depleted source strength to the original strength or emission rate as a function of distance from the source and height of release for the various Pasquill's stability classes. From Van der Hoven, 1968.

Then, the ground-level concentration is given by

$$\bar{c}_o = \frac{Q[1 + \alpha_o(x)]}{2\pi\bar{u}\sigma_y\sigma_z}$$

$$\exp\left(-\frac{y^2}{2\sigma_y^2}\right)\exp\left(-\frac{H^2}{2\sigma_z^2}\right) \quad (10.51)$$

In the above partial reflection model dry deposition removes material from the lower part of the plume in which the concentration distribution becomes non-Gaussian. Also, the computations of $\bar{c}$ and $\bar{c}_o$ depend only on local plume dispersion parameters and do not require evaluation of an integral. Thus, the partial reflection model is simpler than Chamberlain's (1953) source-depletion model. However, it has been shown that the predictions of ground-level concentrations by the two models differ considerably, especially at large distances from the source (Horst, 1980). The partial reflection model greatly underpredicts g.l.c. at $x > 5$ km, while the Chamberlain model overpredicts g.l.c. Horst (1980b) has proposed a modified version of the source-depletion model, which removes most of the deficiencies of the Chamberlain model but it also makes it more complicated.

10.5.2 Surface-depletion Model

The source-depletion model assumes that the deposited material is a loss at the source rather than at the surface. The surface-depletion model proposed by Horst (1977) treats the loss of material to the surface more realistically. In this model the deposition flux $F_d(x,y)$ is represented as a material sink at a point (x,y) on the surface, that is, a ground-level source for diffusion of material deficit at that point. The mean concentration at any location is calculated as the sum of the nondepositing diffusion from the original source at $(0,0,H)$ and the diffusion from the distributed negative surface sources that account for the loss of material by deposition upwind of that location (Horst, 1980b). The resulting expression for the cross-wind-integrated concentration is

$$\bar{c}_y(x,z) = QD(x,z; H)$$

$$- \int_0^x v_d(z_*)\bar{c}_y(x',z_*)D(x - x',z; 0)dx' \quad (10.52)$$

where z_* is the reference height for deposition and the diffusion function is defined as

$$D(x,z; H) = \frac{1}{\sqrt{2\pi}\bar{u}\sigma_z}\left\{\exp\left[-\frac{(z - H)^2}{2\sigma_z^2}\right]\right.$$

$$\left. + \exp\left[-\frac{(z + H)^2}{2\sigma_z^2}\right]\right\} \quad (10.53)$$

The cross-wind-integrated concentration in the point-source plume is also related (proportional) to the above diffusion function; this is represented by the first term in the expression (10.52) for $\bar{c}_y$. The integrand in Eq. (10.52) is the cross-wind-integrated deposition flux, $v_d \bar{c}_y$, at location x' multiplied by the diffusion function for a source at the ground level and for a receptor at height z and downwind distance $x - x'$. Since the concentration at the reference height z_* appears in the integrand, Eq. (10.52) must initially be solved at $z = z_*$. This presents a problem if $z_* = 0$, because $D(x - x', 0; 0)$ is not well behaved at $x' \to x$. Therefore, in the source-depletion model the reference height for estimating deposition flux is assumed to be finite, but small, where surface concentration is commonly measured. A more serious limitation of the model is that the integral in Eq. (10.52) has to be evaluated numerically. This is particularly tedious since the integrand is also a function of x, the upper limit of integration.

The surface-depletion model is considered to be more accurate but also computationally more complex and expensive than the source-depletion models. The source-depletion models are easily extended to settling particles, whereas the surface-depletion model is applicable only to nonsettling contaminants. For these reasons, the former are more commonly used in regulatory applications.

10.6 APPLICATIONS

The importance of momentum and buoyancy effects of release has already been stressed in this chapter. An important practical application of the plume-rise theory and formulas reviewed here is in the estimation of the effective source height for use in simpler Gaussian dispersion models. The assumed separation of the processes of plume rise and turbulent dispersion is somewhat artificial; in reality the two processes occur simultaneously. These can be treated in an integral manner only in very sophisticated and computationally intensive models, such as large-eddy simulation, which will be discussed in the next chapter.

For modeling dispersion of large settling particles, it is important to consider their mean trajectory. Theoretical formulation of the drag and the terminal velocity of settling particles are used to determine mean trajectories of settling particles as they are carried by mean winds. The gravitational settling velocity is also used in the calculation of their deposition rate at the surface.

The process of dry deposition on natural surfaces has been treated only in an oversimplified and heuristic manner. The concept of dry deposition velocity has been extensively used for parameterizing dry deposition in short-range dispersion-deposition models, as well as in regional and long-range transport models. There are large uncertainties, however, in the specification of dry deposition velocities of reactive species on natural surfaces, especially vegetation.

PROBLEMS AND EXERCISES

1. Use the appropriate assumptions and dimensional analysis for deriving the plume-rise formulas for buoyancy-dominated plumes given in Table 10.1.

2. The following stack and effluent characteristics are measured for a 180-m power-plant stack:

Inside stack diameter at the exit = 6 m
Mean effluent exit velocity = 25 m s^{-1}
Mean temperature of effluent at exit = 140°C
Ambient air temperature at the stack height = 10°C

Calculate the effluent momentum and buoyancy flux parameters to be used for plume-rise calculations.

3. Using the appropriate plume-rise formulas, calculate the final plume rise under the following wind and stability conditions for the power-plant stack in Prob. 2: (a) nighttime stable and near-calm conditions with $\partial\bar{\theta}/\partial z = 0.05$°C m^{-1}; (b) early-morning stable conditions with $\partial\bar{\theta}/\partial z = 0.02$°C m^{-1} and $\bar{u} = 5.5$ m s^{-1} at the stack height.

4. For the power-plant stack in Prob. 2, calculate and plot plume rise as a function of downwind distance from the stack up to 2 km under the daytime unstable conditions with (a) $\bar{u} = 2.5$ m s^{-1} and (b) $\bar{u} = 5.0$ m s^{-1} in the mixed layer.

5. a. What would be the estimated final plume rise for the plumes in Prob. 4, according to Briggs' formula (10.18)?
   b. What are the other factors, besides the initial buoyancy flux and mean wind speed, that should be considered in the determination of the final plume rise?

6. A 700-MW coal-fired power plant located in a flat countryside emits SO$_2$ at the rate of 4 kg s^{-1} from a 180-m stack. At the exit of the 5-m by 5-m stack cross section the mean effluent velocity is 15 m s^{-1} and its temperature is 127°C. At the stack top, mean wind speed and ambient air temperature are observed to be 6.0 m s^{-1} and 10°C, respectively. Using Briggs' plume rise and dispersion formulas for the mid-day period of strong solar insulation when the mixing height is 1000 m, calculate the following: (a) plume rise as a function of downwind distance from the stack to a distance where the top of the plume just reaches the inversion base; (b) the g.l.c. of SO$_2$ beneath the plume centerline as a function of downwind distance until it attains its peak value. Compare the maximum g.l.c. and its location with those calculated in Prob. 8 of Chapter 9 without considering the plume rise.

7. For the power-plant stack in Prob. 2, estimate the final plume rise using Briggs' formulas for the following stability and wind conditions: (a) neutral with $\bar{u} = 10$ m s^{-1} and $u_* = 0.5$ m s^{-1}; (b) slightly stable with $\partial \bar{T}/\partial z = 0.01$ K m^{-1} and $\bar{u} = 7.5$ m s^{-1}; (c) very stable with $\partial \bar{T}/\partial z = 0.05$ K m^{-1} and $\bar{u} = 2.5$ m s^{-1}; Hence, discuss the effects of stability and wind speed on plume rise.

8. Calculate the gravitational settling velocity of spherical liquid droplets of density 10^3 times that of air at 20°C and diameter (a) $d = 5$ μm, (b) $d = 10$ μm, and (c) $d = 20$ μm. Also verify whether Stokes' law would be valid for the above droplets.

9. a. If the liquid droplets in Prob. 8 are released at a height of 20 m above the surface where the mean wind speed is 2 m s^{-1}, how far might each size droplets travel on the average before they fall on the surface?
   b. On the basis of your estimates of travel distance, would you consider it necessary to include gravitational settling in the calculation of the maximum g.l.c. for small

droplets of ≤ 20 μm? If so, what would be the effective source height at a distance of 1 km?

10. a. How would you calculate the gravitational settling velocity of very large particles and droplets that may not obey Stokes' law?

   b. Using the computational results of Figure 10.4, estimate the settling velocity of spherical particles of diameter 0.5 mm and density 2500 kg m^{-3} in air of density 1.25 kg m^{-3}. If these are released from a 200-m stack where mean wind speed is 10 m s^{-1}, what would be the effective source height for use in a tilted plume dispersion model?

11. a. Differentiate between the concepts of transfer velocity and dry deposition velocity used for expressing the deposition flux of a gaseous species.

   b. Calculate and plot the dry deposition velocity of SO_2 as a function of the transfer resistance (r_t) in the range of $r_t = 30$ to 1000 s m^{-1} under the following conditions:

Mean wind speed at 2 m = 8.0 m s^{-1}
Roughness parameter, $z_o = 0.01$ m
Neutral stability

12. Calculate and plot the dry deposition velocity of SO_2 as a function of the surface roughness (say, the roughness parameter z_o) and wind speed under neutral conditions for a specified transfer resistance of 100 s m^{-1}. You may consider the following ranges of parameter values:
Roughness parameter, $z_o = 10^{-3} - 10^{-1}$ m
Wind speed at 2 m = 0.5, 1.0, 1.5, and 2.0 m s^{-1}

13. Calculate and plot the dry deposition velocity of SO_2 as a function of stability (say, the Obukhov length L) for a specified transfer resistance of 100 s m^{-1} under the following conditions:

Mean wind speed at 2 m = 1.5 m s^{-1}
Roughness parameter, $z_o = 0.01$ m
Stable conditions with $L = 2, 5, 10, 20, 50$, and 100 m

Discuss the possible influence of stability on v_d.

14. Particles of different sizes are emitted in the neutral atmospheric surface layer above a vegetation canopy in which the mean wind speed at 10 m is 5.0 m s^{-1} and the roughness parameter is estimated to be 0.1 m.

   a. Calculate the aerodynamic resistance to momentum transfer at a reference height of 2 m.

   b. Estimate the surface (sublayer) resistance from the empirical relationship $r_s = 6.5/u_*$.

   c. Estimate the deposition velocities for particles whose gravitational settling velocities are 0.01, 0.05, and 0.25 m s^{-1}.

15. Consider the dispersion and deposition of particles released from an elevated source in the PBL. Using the Chamberlain source-depletion model, derive an expression for the ratio of the g.l.c. with deposition to that without deposition (perfect reflection) on the surface.

16. Using the graphical results of the source-depletion model given in Figure 10.9, determine the downwind distance from the source at which only 10 percent of the emitted material remains in the plume. Tabulate this for different release heights and Pasquill's stability classes and draw your conclusions about the influence of source height and stability on plume depletion due to deposition.

11

Numerical Dispersion Models

11.1 INTRODUCTION

So far in this book we have discussed the various theories of dispersion in turbulent flows, as well as some analytical models based on these theories for particular applications to atmospheric dispersion. The various limitations of the analytical dispersion models based on constant or variable eddy diffusivities have been mentioned in Chapter 6. The simpler Gaussian dispersion models with empirical dispersion parameters or coefficients, which are extensively used in regulatory applications, also have many limitations and uncertainties associated with them, as discussed in Chapter 9. All analytical dispersion models assume a flat and uniform surface that simply reflects, rather than absorbs, the pollutants reaching it, and an idealized (horizontally homogeneous and quasi-stationary) atmospheric boundary layer. Straightline uniform transport winds and constant diffusivities are assumed in most of the simpler models, while some allow for the power-law wind and diffusivity profiles. The limit to vertical diffusion by the capping inversion at the top of the PBL is considered in some of these models on an ad hoc basis. These analytical dispersion models may not be applicable beyond a downwind distance of the order of 10 km from a continuous point or line source. Even for shorter distances, transport and diffusion phenomena are greatly simplified and parameterized in these models.

To further improve model physics, incorporate more accurate boundary-layer and turbulence models or parameterizations, and consider inhomogeneities of the underlying surface including topography, numerical dispersion models should be used. Since atmospheric dispersion is a consequence of transport of material by mean winds and its diffusion by turbulence, a numerical dispersion model must either parameterize both the mean flow and turbulence in the PBL or model them separately, if not simultaneously. Depending on how turbulence is parameterized or modeled, and also on how the diffusion process is modeled, a variety of numerical dispersion models have been proposed in the literature. Here we will classify and describe them in terms of broad categories and not dwell much on the details of specific models and physical parameterizations used in the same.

11.2 SHORT-RANGE GRADIENT TRANSPORT MODELS

As in the case of analytical dispersion models described in Chapter 6, numerical K-theory models seek numerical solutions to the mean diffusion equation for specified initial and boundary conditions, mean transport winds, and eddy diffusivities. Unlike analytical solutions, however, numerical models permit much greater flexibility in the specifications of more realistic boundary conditions, variable-eddy diffusivities, and height-dependent winds. In short-range dispersion models, one can still assume an idealized PBL over a homogeneous surface, with realistic parameterizations for the vertical distributions of mean winds and turbulent diffusivities, as discussed in Chapter 4. In some situations, wind-direction shear can be ignored and one can use two-dimensional grid models for solving the finite-difference version of the cross-wind integrated diffusion equation. The influence of the Coriolis turning of wind with height can be considered only in a fully three-dimensional grid model (see e.g., Kao and Yeh, 1979).

11.2.1 Dispersion in the Surface Layer

By definition, there is no significant turning of wind with height in the surface layer, and the vertical fluxes of momentum and heat are nearly constant. However, mean wind speed and temperature vary rapidly with height. The distribution of mean winds and vertical-eddy diffusivities are best represented by Monin–Obukhov similarity relations (see section 4.8). Several numerical modeling studies of dispersion in the surface layer have been conducted using this similarity theory or scaling approach.

Yamamoto et al. (1970) modeled the diffusion of falling particles from a cross-wind line source in the stratified surface layer. They obtained numerical solutions to the governing steady-state diffusion equation

$$\bar{u}\frac{\partial \bar{c}}{\partial x} - v_g \frac{\partial \bar{c}}{\partial z} = \frac{\partial}{\partial z}\left(K_z \frac{\partial \bar{c}}{\partial z}\right) \quad (11.1)$$

after appropriately scaling and transforming it to a finite-difference form. The lower boundary (surface) was assumed to be a perfect sink where particle concentration must become zero. The concentration at the upper boundary, chosen well above the source height, was also assumed zero. Numerical computations were carried out for two values of the dimensionless terminal fall velocity ($kv_g/u_* = 0$ and 0.1), and for three stability conditions (unstable, neutral, and stable) with specified z_o/L. The results indicate strong effects of atmospheric stability and terminal velocity on particle concentration distributions and surface deposition rates. The effect of finite settling velocity is rather small on the maximum concentration but much larger on the height of maximum concentration. The height of maximum concentration initially decreases with downwind distance from the source, attains a minimum value, and then increases again with increasing distance. The latter is an obvious consequence of the perfectly absorbing surface. The rate of deposition of particles at the surface attains a maximum at some downwind distance; the maximum deposition rate decreases with increasing source height.

Nieuwstadt and van Ulden (1978) conducted a similar numerical study of the vertical dispersion of passive contaminants from a line source. They solved Eq. (11.1), but without the gravitational settling term, using the more appropriate reflecting or partially reflecting boundary conditions

$$K_z \frac{\partial \bar{c}}{\partial z} = v_d \bar{c}, \quad \text{for } z \to z_o$$

$$K_z \frac{\partial \bar{c}}{\partial z} = 0, \quad \text{for } z \to \infty \quad (11.2)$$

The lower boundary condition implies that the mass flux near the ground must be equal to the deposition rate. Nieuwstadt and van Ulden (1978) used two alternative specifications for the vertical-eddy diffusivity, $K_z = \bar{K}_h$ and $K_z = 1.35\, K_m$. The eddy diffusivities of heat and momentum as well as mean velocity were specified using the empirical M–O similarity expressions proposed by Businger (1973). Numerical solutions were obtained for different stability parameters, which corresponded with the selected runs of the Project Prairie Grass (Barad, 1958) and Porton experiments, so that the former could be compared with experimental results. To permit a comparison of the measured concentrations in continuous point source plumes with the solution of Eq. (11.1), the cross-wind integrated concentrations ($\bar{c}_y$) were estimated from the experimental data. Note that $\bar{c}_y$ for a point source is similar to $\bar{c}$ for the cross-wind line source, because they both follow similar diffusion equations. The numerical model results were shown to be in good agreement with experiments, both indicating non-Gaussian vertical concentration distributions of the form in Eq. (8.42). The results indicate that the profile exponent α is a function of stability: $\alpha \leq 1$ in very unstable or convective conditions, $\alpha \approx 1.3$ in near-neutral conditions, and $\alpha > 2$ in extremely stable conditions. A comparison of the computed and measured mean plume heights at 100 m from the source also showed good agreement. van Ulden (1978) further extended this model comparison study to include the PPG experimental data at farther downwind distances for many more runs covering a wider range of stability conditions. His results have already been discussed in Chapter 8, section 8.3

11.2.2 Dispersion in the Planetary Boundary Layer

A numerical model study of the dispersion from a continuous point source in the PBL with variable wind and diffusivity profiles was conducted by Ragland and Dennis (1975). They ignored the Coriolis turning of wind and assumed that the along-wind diffusion was negligible in comparison with advection. For an elevated point source, they solved the finite-difference form of the steady-state mean diffusion equation

$$\bar{u}\frac{\partial \bar{c}}{\partial x} = \frac{\partial}{\partial y}\left(K_y \frac{\partial \bar{c}}{\partial y}\right) + \frac{\partial}{\partial z}\left(K_z \frac{\partial \bar{c}}{\partial z}\right) \quad (11.3)$$

in which $\bar{u}$, K_y, and K_z were specified as functions of z and other parameters (e.g., geostrophic wind speed, surface heat flux, surface roughness, mixing height, etc.). The profiles in the surface layer were consistent with the Monin–Obukhov similarity theory. Above the surface layer, however, wind profile was assumed to be linear and eddy diffusivities were assumed constants. With a specified surface-layer height of 85 m and the mixing height of 275 m, the numerical model results were obtained for the various combinations of the surface roughness (z_o = 0.025, 0.25, and 2.5 m), the surface heat flux (H_o = 110, 0, and −40 W m^{-2}), and the upper-level wind speed (5, 10, and 15 m s^{-1}).

The presumed independence of mixing height on stability and the large ratio of the surface-layer height to the mixing height are some of the unrealistic features of the model proposed by Ragland and Dennis (1975). Nevertheless, the model results clearly show the expected strong effects of stability on concentration distributions in horizontal and vertical directions. When compared with the results of the Gaussian model with the Pasquill–Gifford dispersion scheme (Turner, 1970), the numerical model results show the additional influences of surface roughness, wind speed, and source height on plume dispersion. These effects are not explicitly considered in empirical dispersion parameterization schemes, although different schemes (e.g., PG, BNL, and TVA) implicitly contain the influence of different surface roughnesses and source heights used in the original experiments. The primary result of using height-dependent wind and diffusivity profiles is that higher ground-level concentrations occur over a larger area, while the downwind concentrations at the source level are reduced compared with the constant wind, constant diffusivity case. This can be explained by the fact that wind speed and eddy diffusivities are lower near the ground than at the source height. However, the maximum g.l.c. for a specified set of meteorological conditions does not differ much from that predicted by the Gaussian model. The ground-level concentrations predicted by the numerical model are sensitive to changes in the surface roughness. As the surface roughness is increased, the zone of high concentration shifts closer to the source and the maximum g.l.c. increases in much the same manner as increasing the heat flux. The normalized concentration, $\bar{c}\bar{u}/Q$, is independent of wind speed only for the neutral stability; for stable and unstable conditions, $\bar{c}\bar{u}/Q$ shows significant dependence on the upper-level wind speed. These effects of surface roughness and wind speed on plume dispersion parameters and ground-level concentrations are not included in simpler Gaussian dispersion models.

Kao and Yeh (1979) used a more sophisticated numerical model of wind flow and dispersion in an evolving daytime PBL. They solved the following set of time-dependent mean flow and diffusion equations:

$$\frac{\partial \bar{c}}{\partial t} + \bar{u}\frac{\partial \bar{c}}{\partial x} + \bar{v}\frac{\partial \bar{c}}{\partial y}$$
$$= K_y \frac{\partial^2 \bar{c}}{\partial y^2} + \frac{\partial}{\partial z}\left(K_z \frac{\partial \bar{c}}{\partial z}\right) \quad (11.4)$$

$$\frac{\partial \bar{u}}{\partial t} - f(\bar{v} - v_g) = \frac{\partial}{\partial z}\left(K_m \frac{\partial \bar{u}}{\partial z}\right) \quad (11.5)$$

$$\frac{\partial \bar{v}}{\partial t} + f(\bar{u} - u_g) = \frac{\partial}{\partial z}\left(K_m \frac{\partial \bar{v}}{\partial z}\right) \quad (11.6)$$

$$\frac{\partial \bar{\theta}}{\partial t} = \frac{\partial}{\partial z}\left(K_h \frac{\partial \bar{\theta}}{\partial z}\right) \quad (11.7)$$

The vertical-eddy diffusivities of momentum and heat in the surface layer were prescribed according to the Monin–Obukhov similarity relations (Businger, 1973) and those in the planetary boundary layer were specified as

$$K_m = \lambda^2 \left[\left(\frac{\partial \bar{u}}{\partial z}\right)^2 + \left(\frac{\partial \bar{v}}{\partial z}\right)^2 - \frac{g}{T_o}\frac{\partial \bar{\theta}}{\partial z}\frac{K_h}{K_m}\right]^{1/2} \quad (11.8)$$

where λ is the scale of energy-containing eddies. This is a mixing-length type of expression for K_m with the effect of stability explicitly included in terms of the flux Richardson number as

$$K_m = \lambda^2 \left|\frac{\partial \bar{V}}{\partial z}\right|(1 - \mathrm{Rf})^{1/2} \quad (11.9)$$

The ratio of eddy diffusivities K_h/K_m was estimated by Kao and Yeh (1979) from the surface layer similarity relations. They assumed $K_y = K_z = K_h$ throughout the PBL and specified them as functions of height and time for a typical day on the basis of observations taken during the Great Plains field program. The initial temperature profile, constant geostrophic winds ($u_g = 10$ m s^{-1} and $v_g = 0$), and the diurnally varying surface heat flux were also specified for numerically solving the above set of governing equations.

The computed mean velocity profiles appear to be reasonable in the sense that both the wind-speed and wind-direction shears in the middle part of the PBL decrease as the inversion height increases with increasing time. The vertical profiles of the cross-isobar angle show it to be decreasing with height and also with increasing time of the day. By mid-afternoon, the PBL grows up to 1100 m deep and wind-direction shears become small enough to be neglected. Still, the wind-speed and wind-direction shears appear to have been overestimated in the model due to the use of the gradient transport hypothesis.

For the computed velocity and eddy-diffusivity profiles, Kao and Yeh (1979) numerically integrated the mean diffusion equation (11.4) for a continuous point source of height $H = 100$ m above the surface with capping inversions located at 400 and 1100 m at 0835 and 1435 CST, respectively. Their results (concentrations relative to the source strength) for the morning time are shown in Figure 11.1. Figure 11.1(a) shows the concentration distribution on a vertical plane passing through the source and parallel to the x axis, which is aligned with the geostrophic wind, whereas Figure 11.1(b) shows the g.l.c. distribution. Note that the plume axis at the ground level is oriented with the near-surface winds, which are 21° to the left of the geostrophic wind or the x axis. The plume axis gradually shifts to the right with increasing height and is aligned with the x axis at the top of the PBL. The effect of the Coriolis turning of wind with height on plume dispersion is more clearly shown in figures 11.1(c) and (d), which represent the y–z cross sections

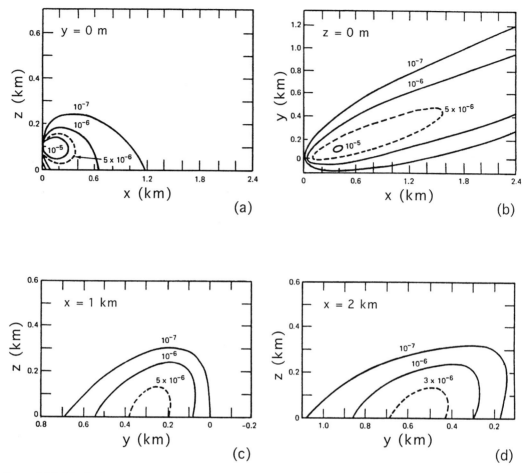

Figure 11.1 Distributions of concentration relative to the source strength at 0835 CST resulting from dispersion from a continuous point source located at height 100 m, with the inversion height at 400 m: (a) The x–z cross section through the source. (b) The x–y cross section at the ground level. (c) The y–z cross section through the plume at $x = 1$ km. (d) The y–z cross section at $x = 2$ km. From Kao, 1984.

through the plume at $x = 1$ and 2 km, respectively. Note that the maximum concentration shifts toward the right (in the northern hemisphere) with increasing height. This effect of wind-direction shear decreases with increasing time and becomes insignificant during convective conditions in the afternoon. The effect is expected to become much more important, however, in the nocturnal stable boundary layer, which was not modeled in the above mentioned study. Thus, the plume dispersion in the stratified PBL behaves very differently from that in the unidirectional, uniform flow assumed in simpler analytical models. In the former, even the lateral concentration distribution may not be Gaussian because of the asymmetrical transports caused by wind direction shears.

A variety of other first-order closure models of the PBL based on the gradient transport theory have been proposed in the literature; a comprehensive review of these is given by Holt and Raman (1988). In principle, computed wind field and eddy diffusivities from any PBL model can be used for solving the mean diffusion equation for the given source and boundary conditions. In practice, however, only a few gradient transport models have been utilized in dispersion applications (see e.g., Pasquill and Smith, 1983; Kao, 1984; Pielke, 1984).

11.2.3 Limitations of Gradient Transport Models

The various limitations of the K-theory, in general, and analytical gradient transport models, in particular, have been discussed in Chapter 6, section 6.7, and will not be repeated here. Numerical gradient transport or diffusion models also have the same limitations, but allow for more realistic and arbitrary variation of eddy diffusivities. They also permit the use of more realistic initial and boundary conditions and inclusion of source and sink terms representing chemical transformations and removal processes. A rational way of accurately specifying the various eddy diffusivities remains the major limitation, especially in nonhomogeneous PBL flows over complex terrain and around buildings. An additional limitation of numerical diffusion models is the pseudo numerical diffusion resulting from finite grid resolution and imperfect numerical advection and diffusion schemes used in these models. Simpler finite-difference schemes used with large grid spacing may result in large enough spurious or pseudo diffusion to overwhelm the real turbulent diffusion. More satisfactory higher order advection schemes with small grid spacing can minimize this spurious diffusion, but at increased computer costs. Another limitation of finite-grid numerical models is that near-source dispersion cannot be properly simulated. The pollutant puff or plume must be larger than the grid size in order to adequately resolve the same. For the same reason, the actual source height and geometry cannot be represented in a numerical model with a typical grid spacing much larger than the source dimensions.

11.3 TURBULENCE KINETIC ENERGY MODELS

In the simplest type of gradient transport models, the mean advection-diffusion equation is solved for the specified eddy diffusivities and the mean transport winds. Sometimes, mean winds are also computed from the equations of mean motion with a first-order closure scheme and specified eddy viscosity or mixing length. The explicit specification of eddy viscosity or diffusivity can be justified only in an idealized PBL for which adequate empirical data are available. In more complex boundary-layer and mesoscale flows, eddy diffusivities are expected to vary with time and space in response to the variations of turbulence structure. For a proper specification of eddy diffusivities, one must relate them to turbulence velocity and length scales and carry dynamical equations for the same. The most widely used approach utilizes the turbulence kinetic energy (TKE) equation

$$\frac{DE}{Dt} = -\left[\overline{u'w'}\frac{\partial \overline{u}}{\partial z} + \overline{v'w'}\frac{\partial \overline{v}}{\partial z}\right] + \frac{g}{T_{vo}}\overline{w'\theta'_v}$$
$$- \frac{\partial}{\partial z}\left(\overline{w'E'} + \frac{\overline{w'p'}}{\rho_o}\right) - \varepsilon \quad (11.10)$$

where E and E' are the mean and fluctuating turbulence kinetic energies per unit mass

$$E = \frac{1}{2}(\overline{u'^2} + \overline{v'^2} + \overline{w'^2}) \quad (11.11)$$

$$E' = \frac{1}{2}(u'^2 + v'^2 + w'^2) \quad (11.12)$$

The left-hand side of Eq. (11.10) represents the total derivative, which includes both the rate of change with time and the advection by mean flow, whereas the various terms on the right-hand side represent shear production, buoyancy production or destruction, turbulent transport, and the rate of dissipation of the TKE.

11.3.1 The Parameterized Length-Scale Model

In simpler TKE models, all the right-hand-side terms in Eq. (11.10) are parameterized. We have already discussed the K-theory parameterizations of momentum and heat fluxes. The turbulent transport term is parameterized in a similar manner, that is,

$$\overline{w'E'} + \frac{\overline{w'p'}}{\rho_o} = -K_E\frac{\partial E}{\partial z} \quad (11.13)$$

in which the diffusivity of TKE is assumed equal or proportional to K_m, that is, $K_E = K_m/\sigma_E$, where σ_E is the diffusivity ratio, similar to the turbulent Prandtl number $\sigma = K_m/K_h$. The rate of energy dissipation is usually expressed in terms of a dissipation length scale as

$$\varepsilon = \frac{C_\varepsilon E^{3/2}}{\ell_\varepsilon} \quad (11.14)$$

where C_ε is an empirical constant and the dissipation length ℓ_ε is related to the large-eddy length scale ℓ in a similar manner as the mixing length ℓ_m. In most of the TKE models it is assumed that $\ell_\varepsilon = \ell_m = \ell$, whereas some investigators have made a distinction between these length scales (Therry and Lacarrere, 1983). The most commonly used expression (parameterization) of the eddy viscosity is

$$K_m = C\ell E^{1/2} \quad (11.15)$$

which was originally suggested by Prandtl and Kolmogorov (Monin and Yaglom, 1971). Here C is an empirical constant that may depend on the definition of ℓ. The vertical diffusivities of heat and mass are generally assumed to be proportional to K_m. Sometimes, the ratios K_h/K_m and K_z/K_m are taken as functions of stability.

Thus, the TKE closure is also based on the gradient transport hypothesis, but eddy diffusivities are determined from physically realistic relationships involving stability and turbulence. Note that Eq. (11.15) implies that K_m is proportional to both the large-eddy length scale (ℓ) and the turbulence velocity scale ($E^{1/2}$). The TKE equation serves as the prognostic equation for the latter. Many diagnostic and some prognostic expressions for ℓ have been proposed in the literature (see e.g., Holt and Raman, 1988). Some of these are meant for only the neutral PBL, whereas others incorporate stability effects in different ways. Some investigators have also proposed different expressions for ℓ and ℓ_ε (Therry and Lacarrere, 1983).

11.3.2 The E-ε Model

In more sophisticated, but not necessarily more accurate, TKE models an additional dynamical equation for the rate of energy dissipation ε, or the large-eddy (mixing) length scale ℓ, is carried in the model. Using a prognostic equation for ε or ℓ eliminates the need for parameterizing the same. An equation for ε is more common because the concept of energy dissipation is easier to grasp than that of a mixing length. In terms of E and ε, eddy viscosity is usually expressed (parameterized) as

$$K_m = \frac{C_k E^2}{\varepsilon} \quad (11.16)$$

where C_k is an empirical constant that may depend on the type of flow. The derivation of the dissipation equation is quite involved (see e.g., Lumley, 1980) and will not be considered here. The ε equation contains many new unknown terms that have to be considerably simplified and parameterized. The highly parameterized form used in the so-called E-ε or k-ε models is given as

$$\frac{D\varepsilon}{Dt} = C_1 \frac{\varepsilon}{E}\left[-\overline{u'w'}\frac{\partial \overline{u}}{\partial z} - \overline{v'w'}\frac{\partial \overline{v}}{\partial z}\right.$$
$$\left. + \frac{g}{T_{vo}}\overline{\theta'_v w'}\right] - C_2 \frac{\varepsilon^2}{E} + \frac{\partial}{\partial z}\left(K_\varepsilon \frac{\partial \varepsilon}{\partial z}\right) \quad (11.17)$$

where C_1 and C_2 are empirical constants and K_ε is the vertical diffusivity of dissipation, which is usually linked to that of momentum as $K_\varepsilon = K_m/\sigma_\varepsilon$. Here, σ_ε is the inverse of the diffusivity ratio, similar to the turbulent Prandtl number.

The various terms on the right-hand side of Eq. (11.17) represent production, destruction, and turbulent transport of dissipation, respectively. The production and destruction terms in their parameterized forms have been assumed to be proportional to the production and dissipation of TKE. The various constants (C_k, C_1, C_2, etc.) in the so-called E-ε or k-ε model have been estimated from the laboratory experimental data. The "standard" values of the same may not be appropriate for the stratified, rotating atmospheric PBL (see e.g., Detering and Etling, 1985; Beljaars et al., 1987). Yet sufficiently accurate measurements of production, dissipation, and turbulent transport terms in the TKE and ε equations are not available for a direct empirical determination of these constants for the PBL. Different values have been suggested from comparisons of model-predicted and observed wind and eddy viscosity profiles by different investigators (see e.g., Holt and Raman, 1988).

11.3.3 The E-ℓ Model

In this model the eddy viscosity is determined by Eq. (11.15), with E and ℓ computed from their dynamical (prognostic) equations. Thus, instead of the ε equation, a prognostic equation for the length scale ℓ (more appropriately, the product $E\ell$) is included in the model. Mellor and Yamada (1974) proposed such a model as a compromise between their level 2 and 3 models. The parameterized form of the prognostic equation for $E\ell$ looks similar to that for the dissipation rate ε (see e.g., Mellor and Yamada, 1982). It also has the same limitations of highly uncertain closure constants.

A comprehensive review of the different types of TKE-closure models has been given by Holt and Raman (1988). They also tested some of the models with different parameterization schemes against limited field experimental data collected during the 1979 Monsoon experiment. One of their two general conclusions is that mean profiles of potential temperature, humidity, and horizontal wind components show little sensitivity to the type of closure scheme as long as the effects of turbulent mixing in the PBL are properly handled. Thus, in the determination of K_m, using a diagnostic formulation of ℓ or a prognostic determination of ε makes little difference. The second main conclusion is that the TKE closure shows closer agreement with the observed turbulence structure in the PBL than the first-order closure. Among the different TKE models, as a group, the E-ε model performs better than the parameterized ε model. This is because the former contains more physics of energy dissipation. Among E-ε parameterizations, the modified Detering and Etling (1985) scheme performed best. It should be emphasized that these conclusions are based on rather limited comparisons of model results with one day's observations in a convective marine boundary layer.

The utilization of the TKE closure in atmospheric dispersion models has been made only recently (Mellor and Yamada, 1982; Boybeyi et al., 1995; Zhang et al., 1996). The basic flow model yields mean velocities, turbulence kinetic energy, and the vertical eddy diffusivities of momentum and heat. These computed fields of variables are, then, used in a dispersion model that may be a first-order closure (Zhang et al., 1996), a second-order closure (Sun, 1986, 1989), or a Monte Carlo particle dispersion model (Boybeyi et al., 1995). Here we consider only the first-order closure or gradient diffusion model. In this, the vertical diffusivity of material is often assumed equal to that of heat, which is assumed proportional to eddy viscosity, that is, $K_z = K_h = \alpha K_m$, where α may be specified as a function of stability. The horizontal diffusivities must also be specified in order to solve the mean diffusion equation. These may be assumed proportional to $E^{1/2}$ and also proportional to the large-eddy length scale in the appropriate direction. Since information about the latter may not be available, horizontal diffusivities are usually assumed equal or proportional to the vertical diffusivity. Such an assumption appears to be reasonable for short-range dispersion, but becomes invalid for mesoscale and regional dispersion.

The TKE closure has a definite advantage over the simple gradient transport approach in that the eddy diffusivities are directly related to the turbulence kinetic energy and computed from the TKE equation. In the E-ℓ and E-ε models, even the large-eddy length scale is computed, rather than explicitly specified. Such models represent a reasonable compromise between the overly simplistic gradient transport models and much more complicated second- and higher order closure models. They are particularly useful for modeling dispersion in nonhomogeneous flows over complex terrain and around buildings. Most of the conceptual limitations of the K-theory remain in the TKE-closure models as well.

11.4 HIGHER ORDER CLOSURE MODELS

As mentioned earlier in Chapter 6, the gradient transport theory has some serious flaws and becomes invalid in certain situations, especially in unstable and convective boundary layers where turbulent transports are not necessarily down the mean gradients. The numerical first-order closure and the TKE-closure models based on the K-theory also fail to account for the significant transports by large convective eddies. In these cases, it becomes necessary to use higher order closure models based on the dynamical equations for turbulent transports or fluxes. We will not present these lengthy and complex equations here; the reader may refer to reviews by Monin and Yaglom (1971), Donaldson (1973), Kao (1984), Stull (1988), and Sorbjan (1989) for the derivation of the dynamical equations for the second moments (variances and covariances). A Cartesian tensor notation is invariably used to minimize the length and the number of equations.

The second-moment equations contain the unknown third moments, correlations involving pressure and velocity or other scalar fluctuations, and the molecular dissipation terms, which have to be parameterized to close the set of equations, that is, to reduce the number of unknowns to the number of equations. Thus, the closure problem shifts to the higher order second-moment equations. The models based on first- and second-moment equations are called second-order closure models. Certainly, the second-moment equations contain considerable amounts of physical information regarding the dynamics of turbulent fluctuations. Therefore, it has been argued that the second-order closure approach should be more general than the first-order closure. Similarly, a third-order closure model utilizing the dynamical equations for first, second, and third moments might be considered even more general, because it contains more physics of turbulence. Unfortunately, the governing equations also become more complex and numerous and their closure becomes much more difficult and ad hoc in the absence of empirical data on higher moments. Here we will give a brief review of only some of the second-order closure models that have been used for modeling atmospheric dispersion. But first we consider the different closure approaches used for modeling the turbulent transport, dissipation, and pressure terms that appear in the second-moment equations.

11.4.1 Second-order Closure Methods

Parameterization of Turbulent Transport The third-moment terms appearing in the second-moment equations cause the fundamental closure problem of the latter. Physically, they represent the turbulent transports of second moments and are found to be rather insignificant in stably stratified flows including the SBL. The vertical turbulent transports have been observed to become very important and often govern the dynamics of unstable and convective boundary layers. Their accurate parameterization is considered crucial to the proper modeling of the CBL using the second-order closure approach. The simplest and the most commonly used parameterization of turbulent transport is based on the gradient-diffusion concept, whereby third moments are expressed as the products of second moments and some form of turbulent diffusivity. The diffusivity is expressed either as a product of turbulence length and velocity scales, as in Eq. (11.15) (see e.g., Donaldson, 1973; Mellor and Yamada, 1974), or in terms of the turbulence kinetic energy (E) and the rate of energy dissipation, as in Eq. (11.16) (see e.g., Wyngaard et al., 1974; Wyngaard, 1975). In the latter approach, a turbulence diffusion time scale is defined as $\tau_D = E/\varepsilon$ and the diffusivity is expressed as $K \sim \tau_D E$, which is consistent with Eq. (11.16).

Both of the third-moment parameterization methods are based on the concept of down-gradient diffusion. The same concept, when applied to second moments in the first-order and TKE closure models, has been found to be inconsistent with observations in the convective boundary layer. Measurements of third moments in the CBL indicate that the buoyancy-driven turbulence can also cause countergradient transport of second moments. However, the partial failure of the gradient transport hypothesis for third moments does not appear to be as catastrophic as that for second moments. For this reason, the second-order closure models using the simpler gradient transport hypothesis have simulated most of the mean flow and turbulence structure reasonably well. A more complicated functional expansion approach, originally proposed by Lumley and Khajeh-Nouri (1974), can remove some of the limitations of the simpler approach, but it also introduces many more terms and empirical constants whose values cannot be determined from the available experimental data. Another approach is to carry the dynamical equations for third moments, but these have their own closure problems (Zeman and Lumley, 1976; Andre et al., 1978; Lumley, 1979).

Parameterization of Dissipation An accurate modeling or parameterization of the molecular dissipation terms in the second-moment equations, particularly the variance equations, is important because these terms determine the rate at which turbulence variances and kinetic energy are destroyed. In some models, ε also determines the turbulence length and time scales that are used in the parameterization of turbulent transport and pressure terms. In the simplest approach, ε, ε_θ, and so on are expressed in terms of the appropriate length and velocity scales, as in Eq. (11.14). Alternatively, dynamical equations are also carried for ε, ε_θ, etcetera. Such equations are very complex (see e.g., Lumley, 1979) and only highly simplified and parameterized forms of the same, such as Eq. (11.17), are used in second-order closure models (Wyngaard et al., 1974; Wyngaard, 1975). The molecular dissipation terms in the dynamical equations for covariances or fluxes are usually neglected, using Kolmogorov's local-isotropy hypothesis for large Reynolds number flows. The same hypothesis also implies that the rate of dissipation is the same for the energies associated with different velocity components, that is, $\varepsilon_u = \varepsilon_v = \varepsilon_w = \varepsilon/3$.

Parameterization of Return-to-Isotropy Terms The parameterization of the covariances between fluctuating pressure and fluctuating velocity, temperature, or other scalar gradients is probably the most difficult problem of higher order closure modeling. These pressure terms are also called the return-to-isotropy terms because they tend to equalize the different velocity variances through redistribution of energy between them and act to destroy any covariances in the flow. These are the primary destruction terms in the covariance equations, since molecular dissipation terms become negligibly small due to local isotropy. For this reason, the simplest way to parameterize these terms is to make them proportional to the excess variance or covariance to be destroyed and inversely proportional to the time scale of turbulence (Mellor and Yamada, 1974; Wyngaard et al., 1974). Thus, the return-to-isotropy term in the dynamical equation for the covariance $c'u'$ can simply be parameterized by $-c'u'/\tau_D$ multiplied by an empirical constant. This simplistic approach, originally due to Rotta (1951), has not been found to be entirely satisfactory and other more refined and also more complex parameterizations have been proposed (Gibson and Launder, 1978; Lumley, 1979; Wyngaard, 1982).

The fluctuating pressure field in a stratified shear flow, such as the PBL, can be expressed entirely in terms of the fluctuating velocity and temperature fields and the mean velocity shear. The resulting Poisson equation for p' clearly indicates that pressure fluctuations at any point depend on the velocity and temperature fields in a large neighborhood around that point (see e.g., Lumley, 1979; Wyngaard, 1982). From the expression of p' one can easily obtain expressions for the various return-to-isotropy terms. Formally, these terms constitute three distinct contributions due to (1) nonlinear turbulence-turbulence interactions, (2) mean shear-turbulence interactions, and (3) buoyancy-turbulence interactions. One can probably ignore the relatively minor contributions of rotational terms. The simple Rotta parameterization can account for only the nonlinear turbulence interactions. It leads to equal lateral and vertical velocity variances in the neutral PBL, an undesirable feature for air pollution dispersion modeling. More refined parameterizations can approximately account for all the interactions and, hence, explicitly include the effects of mean wind shear and buoyancy (Launder et al., 1975; Lumley, 1979; Wyngaard, 1982). An additional effect of the proximity to the ground surface, which limits the vertical extent of large eddies, has been considered in the parameterization proposed by Gibson and Launder (1978). An elaborate parameterization scheme including all these effects has been adopted in a second-order closure mesoscale model for air pollution dispersion applications (Andren, 1990). An evaluation of the scheme against the PBL measurements under different stability conditions suggests that the inclusion of wall effects leads to a significant improvement in the model performance. But this improvement comes at the expense of the increased complexity of the parameterization scheme, which involves six additional closure constants. More complex parameterizations based on Lumley's functional expansion approach also involve more empirical constants, some of which can be determined by applying symmetry, incompressibility, and integral constraints. Unfortunately, these are not always found to be consistent with the observed mean flow and turbulence structure of the PBL (Zeman, 1981).

In second-order closure models to be used for air pollution applications, it would be desirable to keep the various closure and parameterization schemes as simple as possible, without sacrificing their generality and the accuracy of the predicted mean velocity and turbulence quantities that have a direct effect on dispersion.

11.4.2 Dispersion Models Based on Higher Order Closure

Following the development of second-order closure models of the PBL in the mid-seventies, Lewellen and Teske (1976) reported one of the first applications of such a model to the dispersion of pollutants in the PBL. They used the simplest closure assumptions involving a large-eddy length scale ℓ, a turbulence velocity scale, and a minimum number of empirical closure constants. For the length scale, two different approaches are used and compared. The dynamical length-scale equation contains a great deal of arbitrariness and its accurate parameterization would contain a large number of additional empirical constants. A simplified dynamical equation is used in the model by Lewellen and Teske (1976). They also compared it with the simpler approach of explicitly specifying ℓ. Closure constants were determined from several critical flow experiments, including some in the homogeneous surface layer. Using the same set of constants, an extensive model verification was done by comparing model predictions with the experimental data obtained from the various turbulent flows in the laboratory and the atmosphere. The comparison with the atmospheric data was limited to only the surface layer.

Lewellen and Teske (1975, 1976) extended their second-order-closure turbulence model to study plume dispersion in the PBL. The model contains about twenty dynamical differential equations including those for the mean concentration, the variance of concentration fluctuations, and all the covariances or turbulent transports. The equations for mean velocity and turbulence fields are solved separately, and these computed flow fields are utilized in the solution of the species mean diffusion and covariance equations. For modeling dispersion near the source, where plume dimensions are smaller than the large-eddy length scale, both the length and velocity scales of plume diffusion are reduced from the corresponding turbulence length and velocity scales. The plume length scales are assumed proportional to the plume spread or dispersion parameter. At large distances, where the plume size becomes larger than the large-eddy length scale, plume diffusion scales are assumed equal to turbulence scales. Model results are presented for dispersion from continuous line and point sources located at the ground level, as well as at an elevated level. These are compared with those of the Gaussian plume-dispersion model with Pasquill's (1974) recommended dispersion parameterization. A number of interesting conclusions have been drawn from this comparison.

For a line source released at the ground level under neutral conditions, there is close agreement between the second-order closure model and Pasquill's Gaussian plume model. But the variation in the vertical spread with z_o in the latter should be interpreted as a variation with roughness Rossby number G/fz_o. Under the more general stratified atmospheric conditions, the specification of one stability parameter (e.g., Pasquill's stability class, or Richardson number) is not sufficient. Additional information on the time history of turbulence, as contained in the mixing height, must be provided. The influence of stability on vertical spread is much stronger than the influence of roughness Rossby number. For unstable conditions, the second-order closure model yields an upper bound on σ_z of about 0.6 h. The vertical concentration distributions for stable conditions remain close to the Gaussian, but show strong departures from the Gaussian for unstable conditions. At large enough distances from the source, the maximum concentration lifts off the ground, as previously observed by Willis and Deardorff (1976) in their convection tank experiments.

For a continuous point source at the surface, there is very little difference from the ground-level line source, so far as the vertical spread and concentration profiles are concerned. But the lateral plume spread and concentration profiles are strongly influenced by wind-direction shear under neutral and stable conditions. A strong departure from the Gaussian distribution is caused by the variation of wind direction with height above the surface. With increasing stability and increasing distance from the source, the plume is distorted asymmetrically by the Ekman turning of wind. This effect disappears in very unstable and convective conditions.

Yamada (1977) used a much simpler version of the second-order closure model for modeling dispersion in a diurnally evolving PBL. The model, designated as the level 2.5 model, is intermediate between the level 2 and 3 models proposed by Mellor and Yamada (1974). It includes prognostic differential equations for only the mean velocity components, potential temperature, concentration, and turbulence kinetic energy. The fluxes and variances are expressed through diagnostic algebraic equations, which are much easier to solve. This level 2.5 model may be considered similar to a TKE closure model with some added diagnostic relations for scalar fluxes and variances. These relations essentially imply down-gradient turbulent transport or diffusion. The model was used to simulate the dispersion of a nonbuoyant and inert pollutant from elevated sources ($H = 40$, 100, and 200 m) in the simulated PBL at different times during a continuous two-day and -night period of the Wangara Experiment. Since diffusion measurements were not made during that experiment, no comparisons could be made with observations. Simulated concentration distributions in the point-source plumes show some expected and interesting features. Large differences are observed between the daytime and nighttime dispersion patterns. Pollutants reach the top boundary (1200 m above the ground) in the afternoon, but very little material reaches above 300 m during the night. These differences are caused by large changes in mixing heights and turbulence intensities between daytime and night. The stack height strongly influences the ground-level concentrations, especially in nighttime stable conditions. For $H = 200$ m, very little pollutant reaches the surface within 20 km from the stack during the night. Simulated plumes show large horizontal spreading even during the night. This is largely due to wind shear and pseudo diffusion in the numerical model, since horizontal turbulent diffusion terms were actually neglected (these were estimated to be an order of magnitude smaller than the advection terms).

Mellor and Yamada (1982) have described their efforts in developing different levels (2–4) of turbulence-closure models, ranging from the simple TKE closure to the full second-order closure. They also show how the various closure constants could be determined from the available experimental data. The various geophysical applications of their models including free and forced convections, diurnal evolution of the PBL and pollutant dispersion in the same, two- and three-dimensional flows with topography, mesoscale thermal circulations, and boundary-layer parameterization in large-scale atmospheric and oceanic models have been discussed. Decreasing model levels represent decreasing complexity and decreasing requirements for computer time and storage. For one- and perhaps two-dimensional model simulations, higher level models can be used. However, for applications to three-dimensional mesoscale and large-scale atmospheric flows, a simpler 2- or 2.5-level model would be more desirable.

Sun and Chang (1986a, 1986b) used a TKE closure model with specified length scale for modeling dispersion in the convective condition. They assumed $\ell_\varepsilon = \ell = 0.25\lambda_m$, where λ_m is the wavelength corresponding to the peak in the spectrum of vertical velocity. Using the empirical expression of λ_m given by Caughey and Palmer (1979) and setting $C_\varepsilon = 0.1$ and $C = 0.41$, Sun and Chang (1986a) simulated the CBL for Day 33 of the Wangara Experiment. Then, diffusion from a continuous point source in this CBL was modeled by numerically solving the prognostic differential equations for the mean concentration $\bar{c}$, the vertical mass flux $\overline{c'w'}$, and the covariance between concentration and temperature $\overline{c'\theta'}$. Actually, to use the simpler two-dimensional model, the above equations were integrated along the cross-wind (y) direction. The numerical solution, then, yields the distributions of cross-wind integrated concentration ($\bar{c}_y$) for the various specified source heights. These are presented in the convenient dimensionless form, using the usual mixed-layer similarity scaling. The model results are compared with the laboratory measurements of Willis and Deardorff (1978, 1981), atmospheric diffusion experiments, and several other numerical models (Lamb, 1978, 1982). The simplified second-order closure model was shown to successfully simulate the main observed features of the dispersion in the CBL. In particular, for an elevated source in the mixed layer, the maximum cross-wind integrated concentration first descends, until it intercepts the ground, and then gradually ascends. For a near-surface source, the maximum concentration stays near the ground for a short distance from the source before it gradually moves upward in the mixed layer. These interesting dispersion phenomena are attributed to the integrated influence of updrafts and downdrafts on plume dispersion in the CBL.

The model of Sun and Chang (1986) was slightly modified by Sun (1989) to permit some countergradient heat flux in the manner suggested by Deardorff (1966):

$$\overline{w'\theta'} = -K_h\left(\frac{\partial \bar{\theta}}{\partial z} - \gamma_c\right) \quad (11.18)$$

This modification to the original gradient transport relationship, with $\gamma_c = 0.5 \times 10^{-3}$ K m^{-1}, resulted in a better agreement with the observed concentrations. In particular, the model-computed ground-level concentrations due to surface and elevated sources agree well with observations, as well as with other more sophisticated numerical simulations. Other plume statistics, such as the mean plume height and the standard deviation of the vertical concentration distribution, also agree with measurements.

Enger (1986) used a similar higher order closure model of dispersion from a continuous point source in the CBL. The mean wind and turbulence are computed from another model based on Mellor and Yamada's (1974) level 3 model, but with the change that the pressure–temperature interaction term is parameterized according to Lumley's (1979) suggestion. Expressions for the turbulence (dissipation) length scale for the surface layer and the transition layer are derived from the turbulence data. Empirical constants in the closure relations are also estimated from the laboratory and field experimental data. The model successfully simulated the observed features of dispersion in the CBL, as shown by the comparison of model results with the experimental data from Willis and Deardorff (1978, 1981) and the Project Prairie Grass. An improved parameterization of pressure–velocity correlation terms has been added in order to make the above model more useful for air pollution applications (Andren, 1990). The improved scheme allows for the model-predicted lateral and vertical velocity variances in the neutral PBL to be different. The scheme includes a very sophisticated closure for the pressure redistribution or return-to-isotropy terms, including the wall effects, as suggested by Gibson and Launder (1978). The proximity of the surface limits the vertical extent of the eddies and, hence, the transfer of energy from horizontal to vertical motions in the lower part of the PBL. This wall effect is included in the parameterization of pressure terms in the model. The improved parameterization is implemented in a mesoscale model containing dynamical prognostic equations for mean momentum, potential temperature, and turbulence kinetic energy. All other second moments are obtained through diagnostic expressions.

Andren (1990) has evaluated a one-dimensional version of the above mentioned model with the simplified closure scheme against turbulence data from stable, neutral, and convective atmospheric boundary layers. Model simulations are made with and without the inclusion of wall effects in the parameterization of pressure redistribution terms. These are compared with the measured variances, fluxes, and dissipations, using appropriate PBL similarity scaling. Inclusion of wall effects is shown to improve the model performance in all stability conditions. It gives a better prediction of ratios between different velocity variances, which is of particular importance in air pollution dispersion modeling.

11.4.3 Limitations of Higher Order Closure Models

In adopting the higher order closure approach, an implied assumption has been that the inclusion of second-moment equations should lead to a better prediction of the mean fields. This is still an article of faith rather than an established fact. However, even if there is no significant improvement in the prediction of mean fields by a second-order closure model, the model provides a great deal of additional information on turbulence variances and covariances that cannot be obtained from a first-order closure model. The information on turbulence velocity variances is particularly useful in dispersion applications. Higher order closure models also permit countergradient diffusion and have been shown to successfully simulate the various observed features of dispersion under convective conditions. However, the simpler parameterizations of the third-order turbulent transports often fail in the CBL, unless buoyancy effects are accounted for through additional terms. This points out a major limitation of the second-order closure approach in that the closure relations validated and evaluated by laboratory shear-flow experiments may not be applicable to buoyancy-driven geophysical flows, such as the CBL. Some investigators have proposed third-order closure models for these flows. But they have their own closure problems and limitations. Another major limitation of higher order closure models is that there are too many empirical closure constants of which only a few can be evaluated from the available experimental data. This limitation becomes more and more severe as the order of closure is increased. Only limited data are available on third and higher moments of turbulence in the atmosphere for a proper determination of closure constants and validation of models. Similar measurements of higher moments involving concentration fluctuations are not yet available.

A major limitation of all the ensemble-averaged turbulence closure models is that ensemble averaging forces one to parameterize the effects of the entire spectrum of eddy motions, including the largest energy-containing eddies (Wyngaard, 1982). These large eddies in a turbulent flow are very sensitive to the geometry of the flow, as well as to the dominant mechanisms of production and destruction of turbulence in that flow. Consequently, some researchers have expressed doubt whether any second-order closure model could be applied to the wide range of conditions encountered in the atmosphere. Reservations about the general validity and widespread use of higher order closure models for dispersion applications have also been expressed in the literature (Lamb, 1982). These concerns are minimized in a large-eddy simulation that explicitly resolves most of the energy-containing and transporting eddies.

11.5 LARGE-EDDY SIMULATIONS

As mentioned in Chapter 2, the three-dimensional large-eddy simulation (LES) is currently the best available approach to atmospheric turbulence and diffusion modeling. It is also computationally the most expensive approach, generally requiring the use of supercomputers. Although the origins of LES can be traced back to the general circulation models of the early sixties, its application to PBL modeling started from the pioneering works of Deardorff (1970a&b, 1972, 1973, 1974) in the early seventies. The LES attempts to resolve explicitly all the large-scale eddy motions that contain most of the turbulence kinetic energy and are responsible for almost all the turbulent transports of momentum, heat, and mass in the flow. The important effects of small subgrid scale (SGS) motions, such as viscous dissipation of kinetic energy and molecular dissipation of scalar variances, are parameterized. Relatively minor contributions of SGS motions to the TKE, variances, and covariances (fluxes) are also parameterized. These parameterizations constitute a SGS closure model, which may be a first-order or higher order type, similar to the ensemble-averaged turbulence closure models discussed in the previous sections. Unlike the full (ensemble-averaged) turbulence closure models, however, the SGS closure models usually provide only small corrections to the explicitly resolved LES fields. Therefore, their importance and desired accuracy depend on the expected fraction of the total TKE that is left in subgrid scale motions. If this fraction is relatively small (say, less than 0.1), even a crude, but simple, SGS closure model might be adequate. If a significant fraction (say, greater than 0.2) of the TKE is expected to reside in the unresolved subgrid scale motions, a more accurate SGS closure model would be desirable. Thus, the grid size and the SGS model should be chosen according to the expected size or scale of the most energetic eddies in the flow.

The grid size in the model often depends on the size of the model flow domain and the available computer capacity (speed, memory, etc.), which essentially determines the number of grid points (nodes) that can be handled while carrying out the numerical integration of the governing equations of motion and scalar dispersion. Using the currently available supercomputers with parallel-processing architecture, eddy sizes down to a few meters can be resolved over limited domains of a few hundred meters. More often, though, grid size is limited to several tens of meters in large-eddy simulations of the diurnally evolving PBL with its maximum depth of the order of 1 km. This grid resolution is hardly sufficient for the nocturnal stable boundary layer in which the characteristic size of the most energetic eddies becomes small. These and other aspects of LES will be briefly discussed here; for more comprehensive reviews the reader may refer to Deardorff (1973), Mason (1989, 1994), Nieuwstadt et al. (1992), Galperin and Orszag (1993), and Andren (1995).

11.5.1 Grid-volume-averaged Equations for LES

The set of equations of motion, potential temperature, and scalar diffusion used in LES is obtained by grid-volume averaging of the instantaneous equations. This operation is similar to ensemble averaging and utilizes the same Reynolds averaging conditions as discussed in Chapter 2. But there are some subtle and conceptual differences as discussed by Mason (1994). The grid-volume averaging amounts to applying a spatial filter which, for all practical purposes, is determined by the grid resolution used in the model. The optimum filter is not necessarily the one with sharp cutoffs at grid boundaries; such a filter was used by Deardorff (1972, 1973, 1974). Gaussian and other smooth spatial filters have been found to be more efficient for mathematical operations (Leonard, 1974). Leonard has defined this filtering operation quite generally (see also Wyngaard, 1982).

If we define the volume averaging or the spatial filter operation so that the average fields are denoted by an overbar, and let $u = \bar{u} + u'$, $v = \bar{v} + v'$, and so on, in which prime denotes the subgrid scale fluctuation, then the volume averaging of the instantaneous equations of motion, temperature, and so on, yields the appropriate equations for LES. For filters preserving the average values of variables upon second-filter operation (e.g., $\bar{\bar{u}} = \bar{u}$, etc.), the filtered LES equations have exactly the same forms as ensemble-averaged equations (2.45) through (2.51). There is, however, an important conceptual difference that relates to the nature of average variables and Reynolds stresses and fluxes. In the ensemble-averaged equations, all variables are well behaved and are at most gradually varying functions of time (for nonstationary flows) and space (for nonhomogeneous flows). The spatial variations occur only in the directions of inhomogeneity, such as the vertical direction in the atmosphere. In the grid-volume-averaged or filtered equations, on the other hand, all variables have random turbulent fluctuations in time and space superimposed on their ensemble averages. These fluctuations represent large-eddy motions with scales larger than the filter or grid size. The subgrid-scale fluxes also have this random variability with respect to time and space. But, on the average, these are much smaller than the ensemble-averaged Reynolds fluxes, because they represent only minor contributions of SGS motions. Thus, although the convenient use of the same notation for ensemble and grid-volume averaging leads to the same set of equations, the important differences between the nature of variables involved in the two cases must be recognized.

For the more general filter operations that may not preserve previously filtered average variables on a second-filter operation or averaging, that is, for which $\bar{\bar{u}} \neq \bar{u}$, and so on, additional terms are generated by the averaging or filter operation on the instantaneous equations. For these, each of the subgrid-scale Reynolds fluxes inside the parentheses on the right-hand sides of Eqs. (2.46) through (2.51) is replaced by two terms, for example,

$$\overline{u'^2} \to \overline{u^2} - \bar{u}^2$$
$$\overline{u'v'} \to \overline{uv} - \bar{u}\bar{v}$$
$$\overline{u'w'} \to \overline{uw} - \bar{u}\bar{w} \qquad (11.19)$$

in Eq. (2.46) and similarly in other equations, where $\to$ denotes that the left-hand term is to be replaced by the right-hand terms, but the two are not necessarily equal. Using the Cartesian tensor notation for velocity u_i, with $i = 1, 2$, and 3 representing the three components, one can write, in general,

$$\overline{u_i' u_j'} \to \overline{u_i u_j} - \bar{u}_i \bar{u}_j \qquad (11.20)$$

By substituting $u_i = \bar{u}_i + u_i'$ and $u_j = \bar{u}_j + u_j'$, and taking the average, it is easy to show that the two sides of arrow in Eq. (11.20) are different, as

$$\overline{u_i u_j} = \overline{\bar{u}_i \bar{u}_j} + \overline{u_i' \bar{u}_j} + \overline{\bar{u}_i u_j'} + \overline{u_i' u_j'} \qquad (11.21)$$

Only for the particular filters satisfying the conditions $\bar{\bar{u}}_i = \bar{u}_i$, etc., the two sides of Eq. (11.20) become exactly equal. Even otherwise, the RHS is often considered as a combined subgrid-scale Reynolds stress tensor, τ_{ij}, i.e.,

$$\tau_{ij} = \overline{u_i u_j} - \bar{u}_i \bar{u}_j \qquad (11.22)$$

For the sake of convenience, however, the grid-volume-averaged or filtered equations used in LES are often written in the same form as the ensemble-averaged equations (2.45) through (2.51), implying a strong correspondence between τ_{ij} and $\overline{u_i' u_j'}$. The difference between the two may be entirely obscured by the same subgrid-scale parameterization used for both by most of the investigators. Some have used separate treatments or parameterizations of the different terms resulting from the decomposition of the Reynolds stress in Eq. (11.21). The decomposition may have some advantages, but the separate treatment of the terms also increases computational costs. The combination of simplicity and computational efficiency makes the combined approach preferable to the piecemeal treatment (Ferziger, 1993).

11.5.2 Subgrid-scale Models

In order to close the set of grid-volume-averaged or filtered equations, the unknown SGS Reynolds stresses and fluxes must be parameterized in terms of the resolved variables for which the set of LES equations is numerically solved. The various SGS-closure approaches or models have been reviewed, among others, by Deardorff (1973), Ferziger (1993), Mason (1994), and Andren (1995).

The Smagorinski Model The simplest and the most widely used SGS model is that due to Smagorinski (1963), who originally used it in an atmospheric general circulation model. The Smagorinski model is essentially a first-order closure model based on the classical Boussinesq concept of eddy viscosity, with the difference that the gradient transport hypothesis is applied to small-scale SGS motions only. Using the usual eddy–viscosity relation, the SGS Reynolds stress is expressed as

$$\tau_{ij} = -K\left(\frac{\partial \bar{u}_i}{\partial x_j} + \frac{\partial \bar{u}_j}{\partial x_i}\right) = -2KS_{ij} \qquad (11.23)$$

where S_{ij} denotes the rate of SGS deformation, and K represents the SGS eddy viscosity, which should be distinguished from the overall eddy viscosity introduced in Chapter 4. Even when the ensemble-averaged fluxes may not be down the ensemble-averaged gradients, as often occurs in the convective mixed layer due to large-eddy convective motions, it is reasonable to expect that the gradient transport hypothesis might be more generally valid for small-scale (SGS) motions. Some of the models of small-scale turbulence based on scale-dependent eddy viscosity have successfully predicted the small-scale turbulence structure, such as the energy spectrum in the equilibrium range (Hinze, 1975).

It is interesting to note that Eq. (11.23) can be derived in several different ways: heuristically, as Smagorinsky (1963) did; from a production equals dissipation argument applied to the dynamical equations for the SGS Reynolds stress; and from the various turbulence theories (see, e.g., Ferziger, 1993). These derivations suggest that the SGS eddy viscosity can be expressed as

$$K = \ell_o^2 S \qquad (11.24)$$

$$S^2 = \frac{1}{2}\left(\frac{\partial \bar{u}_i}{\partial x_j} + \frac{\partial \bar{u}_j}{\partial \bar{u}_i}\right)^2 \qquad (11.25)$$

where ℓ_o is a mixing-length scale which is related to the filter scale ℓ_f, or the geometric mean grid size

$$\Delta = (\Delta x \, \Delta y \, \Delta z)^{1/3} \qquad (11.26)$$

Following the approaches suggested by Smagorinsky (1963) and Lilly (1967), Deardorff (1970a, 1972) used the simple proportionality relation

$$\ell_o = C_s \Delta \qquad (11.27)$$

in his original LES studies of the PBL and channel flows. However, he found that the proportionality coefficient C_s was not really universal, but depended, to some extent, on mean flow shear, stability, and possibly other factors. For example, the optimum values of $C_s = 0.13$ and 0.21 were found for the neutral and unstable PBLs, respectively. Deardorff also found it necessary to adjust C_s near the wall or surface. Later, other investigators also found that the theoretically predicted value of $C_s = 0.2$ for homogeneous and isotropic turbulence may not be applicable to inhomogeneous shear flows. For the latter, it is found that the above value of C_s must be reduced by half or more (Ferziger, 1993). The possible dependence of C_s on Reynolds number has also been recognized.

A different analysis of the SGS mixing-length scale by Mason (1994), using the original approach of Lilly (1967) but with a spherical filter of scale ℓ_f, yields the following expression for ℓ_o:

$$\ell_o = C_f \ell_f \qquad (11.28)$$

where C_f is related to the Kolmogorov constant α, for $\alpha = 1.5$, $C_f = 0.17 \simeq 0.2$ for the particular filter used in the analysis. The precise value of C_f, in general, must depend on the filter shape. It is clear from Eqs. (11.27) and (11.28) that only when the filter scale is equal to the average grid size does $C_s = C_f$. Mason (1994) has shown that $\ell_f = \Delta$, or $C_s = C_f \simeq 0.2$ is indeed the optimum choice of filter scale as well as of constants in SGS mixing-length relations, at least for neutral boundary-layer and channel flows. Large-eddy simulations with larger values of C_s will be numerically accurate but represent only a limited range of scales. Those with smaller values of C_s are likely to suffer badly from finite-difference errors and will be difficult to interpret. The adoption of the optimum value of $C_s \simeq 0.2$ requires the use of sound numerical methods free from spurious numerical diffusion. Larger values of C_s might be desirable to make the LES results insensitive to the numerical method, which may not be the best.

In large-eddy simulations of the atmospheric boundary layer and other mesoscale circulations, we should also be concerned about the possible effects of buoyancy or stability on the SGS model. If the filter scale or the mean grid size are actually within a long inertial subrange, then one would expect a negligible influence of buoyancy on the model. A SGS model would be required to express the SGS heat flux in the equation for $\bar{\theta}$, but this would be identical to the parameterization of the SGS scalar flux in the diffusion equation. With the Smagorinsky model this closure involves a diffusivity of heat or scalar that is proportional to that of momentum. For example, Deardorff (1972) assumed $K_h = 3 \, K_m$ in his simulation of the unstable PBL. A more rigorous estimate of the proportionality constant can be made by extending Lilly's (1967) analysis to the scalar spectrum; this yields $K_h \simeq 2 \, K_m$ (Mason, 1994). This procedure is well justified for simulating unstable and convective boundary layers and other flows with an expected long inertial subrange in their turbulence spectra, especially when the filter scale is far removed (much smaller) from the scales of most energetic and buoyant eddy motions.

In most practical simulations of stratified flows, the theoretical requirement of a small grid size lying within a long inertial subrange is unlikely to be achieved, and some direct effect of stability on the subgrid model must be incorporated. To date the only available approaches for incorporating the stability effects on eddy viscosity are those used in ensemble-averaged Reynolds flux closures. Deardorff (1973) used an equation for subgrid turbulence kinetic energy and derived an expression for the subgrid-scale eddy viscosity that is dependent on flux or gradient Richardson number. However, this amounts to a higher order TKE closure. The simpler approach of using a stability-dependent subgrid mixing length model seems equally successful and computationally less expensive. By equating shear and buoyant production to the rate of dissipation in the subgrid TKE equation, and using the usual parameterizations $\varepsilon \sim E^{3/2}/\ell$, and $K \sim E^{1/2}\ell$, it is easy to show that

$$K = \ell^2 S (1 - \text{Rf})^{1/2} \qquad (11.29)$$

Equation (11.29) may be considered as a generalization of the relation (11.24) for the neutral flow to buoyant or stratified flows. The Rf-dependent term in Eq. (11.29) represents the influence of stability on E only. The additional influence of stability on the mixing-length scale is implied by the relation $K \sim E^{1/2}\ell$, and should be considered separately. The value of ℓ may exceed ℓ_o by a finite ratio in unstable and convective conditions and should tend to zero at the critical value of Richardson number (Ri_c or Rf_c) in very stable conditions. The ratio of diffusivities of heat and momentum may also depend on Ri or Rf. Such a stability-dependent subgrid model is considered to be more crucial to accurately simulating the stable boundary layer (SBL) and also the near-surface and upper transition layers of the CBL. In these, significant energy may reside in subgrid or subfilter scales and turbulence spectra may have little or no inertial subranges. This is the reason the LES approach could not be used for modeling stable boundary layers until recently.

Dynamic Models The so-called dynamic subfilter models are based on the idea that the smallest resolved scales of the LES can provide information on the subgrid-scale stresses on which the subgrid or subfilter model can be based. This has been applied in two ways. In the first, called the scale similarity model, Bardina et al. (1980) applied a second filter operation to the resolved scales to isolate the stresses on the smallest resolved scales. They assumed that the principal scales that require modeling, that is, those lying just below the filter size, are similar in structure to the smallest resolved scales. Then they expressed the SGS stress tensor in the following form (see e.g., Bardina et al., 1980; Ferziger, 1993):

$$\tau_{ij} = \overline{\bar{u}_i \bar{u}_j} - \tilde{\bar{u}}_i \tilde{\bar{u}}_j \tag{11.30}$$

Here, $\tilde{\bar{u}}_i$ is the twice-filtered velocity field, which is different from $\bar{u}_i$. The second filter, denoted by $\sim$, need not be identical to the first, although in most simulations using this model the two filters have been taken as the same. This scale similarity model provides a simple parameterization of the subgrid-scale stress and reproduces the structure of τ_{ij} quite well. However, the model does not dissipate the subgrid-scale energy at a sufficiently high rate. To correct for this and have other desired properties, Bardina et al. (1980) added a mean stress, provided by the Smagorinsky model, to their above modeled stresses. This combination, called the mixed model, improved the values of the subfilter stresses in a diagnostic sense, but any improvement to the resolved motions is probably limited and hard to quantify (Mason, 1994).

The second approach in dynamic modeling uses the magnitude of stresses occurring on the smallest scales to determine the value of the mixing length in the Smagorinsky model (Germano, 1991; Germano et al., 1991). This provides a self-consistent method of determining the fundamental parameters C_s and Δ of the Smagorinsky model. In Germano's approach, the definition of the subgrid-scale stress in Eq. (11.22) is adopted and the filtered LES equations are filtered a second time to obtain the equations for $\tilde{\bar{u}}_i$. The equations obtained are similar to those produced by the first filter operation, except that τ_{ij} is replaced by a new subfilter stress:

$$T_{ij} = \overline{\tilde{u}_i \tilde{u}_j} - \tilde{\bar{u}}_i \tilde{\bar{u}}_j \tag{11.31}$$

The scale of the second filter should be greater than ℓ_f of the Smagorinsky model. The mixing-length scale associated with double filtration must also be greater than that associated with the first filter operation. The assumption that the Smagorinsky model with the same constant C_s should apply on both scales, and the use of an identity involving filter operations, led to the derivation of six values of mixing lengths (one for each component of the stress tensor) at each grid point. This variation in mixing length can also be viewed as a variation of the Smagorinsky constant C_s (Mason, 1994). In principle, according to this dynamic model, C_s can be different at every point in the flow at every instant. Large fluctuations of C_s in both time and space can result in numerical instability and nonrealizable fields, such as negative eddy viscosities or mixing lengths. Germano et al. suggested a spatial averaging procedure to ensure positive eddy viscosities and avoid computational instability. This method provides a slowly varying mixing-length scale for the Smagorinsky model.

The above dynamic SGS model has some distinct advantages over the original Smagorinsky model (see e.g., Ferziger, 1993; Mason, 1994). Because changing the value of C_s can compensate for an incorrect choice of the length scale, the latter need not be prescribed accurately. Furthermore, variations of C_s with nondimensional shear, Reynolds number, stability, and so on, are accounted for automatically in the model. In contrast to the Smagorinsky model, it has the capability of perhaps being able to deal with flow laminarization at small Reynolds number. Unfortunately, the dynamic model has not been applied to and tested in high Reynolds number flows, such as the PBL. Some reservations about its applicability to very large Reynolds-number flows have been expressed (Mason, 1994). For these another class of SGS models, called the stochastic backscatter models, has been found more promising.

Stochastic Subfilter Models The Smagorinsky subgrid model and other models discussed so far consider an ensemble-averaged subgrid stress conditional on the situation prevailing at each grid point. In reality, the subgrid stresses must differ from one realization to another in a random manner. The fluctuations in the subgrid stress imply an energy transfer from the subgrid scales to the resolved scales (Mason, 1994). These energy transfers were noted earlier by Leslie and Quarini (1979), who proposed a rigorous subgrid closure using the eddy-damped quasi-normal Markovianized (EDQNM) theory of turbulence. For high Reynolds numbers, however, the EDQNM-based closure approach is found to be an order of magnitude more expensive than the Smagorinsky model. Also, the EDQNM theory is strictly applicable only to homogeneous and isotropic turbulence, for which it is indeed the most complete and rigorous theory of turbulence available to date. If the filter scale is small enough to lie in the inertial subrange, all the subgrid scales would be expected to lie in the equilibrium range. For this situation, the subgrid closure based on the EDQNM theory can be used in the LES. But its much greater complexity and associated computer costs may not be justified if the LES results of the resolved fields are only weakly sensitive or not sensitive to the SGS model. A simple modification of the Smagorinsky model to account for the stochastic effects resulting in the backscatter of energy might be adequate. Such an approach has been used in stochastic backscatter subgrid or subfilter models proposed by Leith (1990) and Mason and Thomson (1992). Here we give only a brief and qualitative description of the Mason and Thomson model which has been used in large-eddy simulations of the PBL under different stability conditions (Mason, 1994).

The stochastic subfilter model proposed by Mason and Thomson (1992) involves several assumptions. The SGS stress tensor is assumed to have statistically isotropic Gaussian fluctuations. The length scale of these fluctuations is assumed to be that of the filter operation, ℓ_f, while the time scale and energy levels are derived from both this length scale, ℓ_f, and the local rate of energy dissipation ε. With these assumptions, the rate of energy backscatter is estimated to be $C_B\varepsilon$, where C_B is a constant of the order of 1 but actually depends on the filter shape ($C_B \simeq 1.4$ for the sharp spectral cutoff filter used by Mason and Thomson). Implementation of the backscatter model in LES involves the derivation of a random stress field that has been subjected to the filter operation and scaled so as to give the estimated energy backscatter rate of $C_B\varepsilon$. The two main uncertainties in the model are associated with the value of the backscatter coefficient C_B and the application of the filter operation to derive suitable random stresses. Both are linked to our ignorance of the correct form of the filter that corresponds to the Smagorinsky model (Mason, 1994).

The backscatter constant C_B determines the ratio, ℓ_o/ℓ_f, between the subgrid mixing-length scale and the filter scale. Following an analysis similar to that used by Lilly (1967), Mason and Thomson (1992) derived the relation

$$c_f = \frac{\ell_o}{\ell_f} = c_{fo}(1 + c_B)^{1/2} \quad (11.32)$$

where C_{fo} represents the ratio between the subgrid mixing length and filter length scales without backscatter. Thus, inclusion of backscatter has an effect of increasing the above length scales ratio. For their sharp cut filter, Mason and Thomson (1992) obtained a value of $C_f \simeq 0.22$.

Another backscatter constant, $C_{B\theta}$, is involved when one considers the fluctuations of a passive scalar θ. Mason and Thomson (1992) estimated $C_{B\theta} \simeq 0.32\ c_B$, which implies a value for the subgrid turbulence Prandtl number Pr $\simeq 0.77$ or the subgrid eddy-diffusivity ratio $K_h/K_m \simeq 1.3$, for their sharp spectral cut filter. The exact values of the above parameters and backscatter constants still depend on the form of the filter, but values of these critical constants depend on the presence of backscatter. Overall, the introduction of backscatter through the use of a stochastic subgrid or subfilter model seems to have a limited but beneficial effect on the large-eddy simulations of the flow interior, particularly when the grid resolution does not extend too far into the inertial subrange (Mason, 1994).

SGS Models Near Surfaces As a boundary in the form of a rigid surface (e.g., the earth's surface in the case of the PBL flow) is approached in a large Reynolds-number flow, the characteristic size of energy-containing eddies normal to the surface decreases in proportion to the distance from the surface. The near-surface flow also contains larger scale tangential (horizontal in the case of PBL) motions that arise as large eddies become elongated in the direction of mean flow. Thus, in a large-eddy simulation with a finite grid size, more and more energy and turbulent stresses are contained within the subgrid scales as the surface is approached. In the first grid adjacent to the surface, most of the energy is likely to remain in unresolved subgrid scale motions, while a smaller fraction may actually be resolved. The ratio of the resolved to unresolved energy can be increased by reducing the grid size near the surface, but there is always a limit imposed by the total number of grid points that can be carried in the model. The subgrid scale closure model becomes more important in accounting for the unresolved turbulence kinetic energy and turbulent transports near the surface.

None of the SGS models described so far is completely satisfactory close to the surface. The effectiveness of any subgrid model near the surface is usually measured by how closely the empirical surface-layer similarity relations are reproduced by the LES with the particular SGS model. In particular, for the neutral boundary layer over an aerodynamically rough surface, the mean velocity profile in the surface layer is expected to follow the log law and velocity variances must be appropriately related to the surface shear stress. It is well known that with the conventional Smagorinsky subgrid model the LES mean velocity field near the surface does not follow the log law. Instead, mean velocity gradients are considerably overestimated by the model, so that the dimensionless wind shear $\phi_m = (kz/u_*)\partial \bar{u}/\partial z$ is greater than 1 (Mason, 1994).

The above mentioned defect of the Smagorinsky model may be corrected if the mixing-length scale in the model is appropriately modified near the surface, so that $\ell = kz$ close to the surface and it approaches its constant value ℓ_o in the interior flow. An interpolation expression that smoothly matches with the expected behaviors in the near-surface and interior regions has been proposed by Mason and Thompson (1992):

$$\ell^{-n} = (kz)^{-n} + \ell_o^{-n} \quad (11.33)$$

An optimum value of $n = 2$ has been determined by Mason and Brown (1994) in conjunction with their stochastic backscatter model. To implement these matches smoothly, a refinement of the vertical mesh near the surface, so that Δz is sufficiently small closest to the surface and gradually increases to its constant interior value, has been recommended (Mason, 1994). Some investigators do not refine the vertical mesh near the surface, but specify an appropriate drag coefficient at the lowest grid point. Scalars such as potential temperature and specific humidity are treated in an analogous manner.

The stochastic subgrid model accounting for backscatter has also been extended to the near-surface layer (Mason and Thomson, 1992). There is, however, no exact theoretical basis for estimating the magnitude and details of backscatter. Very close to the surface the dominant energy-containing eddies are essentially subgrid or subfilter scale and virtually all turbulent transfers must be described by the subgrid model. Very little backscatter of energy is expected, and small-scale eddies should provide a deterministic value of stress on the grid or filter scale. No formal theory is available to provide the transition from the stochastic backscatter appropriate in the inertial subrange in the interior of the flow toward a deterministic empirical model very close to the surface. Mason and Thomson (1992) proposed a simple physical model of random eddies of length scale ℓ_e that are responsible for the stresses in the filter volume of scale ℓ_f, assuming that $\ell_e < \ell_f$. The values of ℓ_e and ℓ_f are not precisely known, but ℓ_e is assumed proportional to the mixing-length scale ℓ of the near-surface subgrid model and ℓ_f is assumed proportional to the mixing-length scale ℓ_o for the interior flow. Then, an expression for the energy backscatter rate is obtained as

$$\frac{\partial K}{\partial t} = c_B \left(\frac{\ell}{\ell_o}\right)^5 \varepsilon \qquad (11.34)$$

which implies a very strong sensitivity of the energy backscatter rate on the ratio ℓ/ℓ_o. With an empirical adjustment of either the backscatter coefficient C_B or the specification of ℓ close to the surface, Mason and Thomson's (1992) stochastic subfilter model can give the expected mean velocity and scalar gradients in the atmospheric surface layer. The results of this model indicate that backscatter is required to give a correct simulation of the poorly resolved near-surface regions of the neutral and unstable PBLs as well as the entire stably stratified boundary layer (Mason, 1994). However, the computational cost of the stochastic backscatter model is high and the resulting resolved flow is significantly agitated by stochastic excitations, particularly near the surface. The addition of SGS backscatter also requires a considerably larger eddy viscosity to dissipate the extra energy generated (Sullivan et al., 1994).

Higher Order SGS-Closure Models The TKE- and second-order closure approaches to subgrid closure modeling in LES were proposed and first used by Deardorff (1973, 1974). These are similar to the TKE- and higher order closure models based on the ensemble-averaged equations, which have been discussed earlier. In particular, a second-order SGS model contains many more constants compared with the one or two in the Smagorinsky model, and some of these have to be specified rather arbitrarily (see e.g., Deardorff, 1974; Schmidt and Schumann, 1989). The improvement in the overall simulation is found to be marginal and may not be justified by the added complexity of the model, as well as the increase in computational costs. It has been argued that additional computer resources could better be used for more accurate simulation by reducing the grid size or increasing the resolution of LES, while the subgrid model is kept relatively simple. Thus, second- or higher order subgrid closure approach is generally not recommended for LES. But the TKE approach with a simple length-scale prameterization might be a good compromise between the first-order and second-order subgrid closure models. This also appears to be a more rational approach for simulating stably stratified flows in which the effects of buoyancy on subgrid fluxes are likely to become more significant. But the relative benefits of an improved TKE closure with somewhat coarser grid resolution and a simple first-order closure with finer grid resolution, both using the same computer resources, have not been compared for large-eddy simulations of stably stratified flows including the SBL.

In the standard SGS model based on the subgrid TKE equation, the SGS eddy viscosity is expressed in terms of the TKE as

$$K = C\ell E^{1/2} \qquad (11.35)$$

In the prognostic equation for E the rate of energy dissipation is simply parameterized in terms of the mixing-length scale as

$$\varepsilon = \frac{C_\varepsilon E^{3/2}}{\ell} \qquad (11.36)$$

Such a model was first proposed by Deardorff (1973) and subsequently used by Deardorff (1980), Moeng (1984), and others. For unstable and neutral conditions, $\ell = \Delta$, while for stable stratification, ℓ is reduced as suggested by Deardorff (1980), that is,

$$\ell = \min\left(\Delta, \frac{0.76E^{1/2}}{N}\right) \qquad (11.37)$$

When the grid size falls within the inertial subrange, Moeng and Wyngaard (1988) estimated the values of constants $C = 0.1$ and $C_\varepsilon = 0.93$ from their spectral analysis.

A major modification of the standard TKE-closure SGS model to correct for the lack of sufficient resolution near the surface and to account for the effect of mean wind shear on subgrid stress has been proposed by Sullivan et al. (1994). Their proposed SGS model is based on an earlier idea of splitting the SGS stresses into two parts: (1) the isotropic part, which is proportional to the magnitude of the fluctuating strain rate, and (2) the inhomogeneous part, which is proportional to the magnitude of the mean strain rate (Schumann, 1975). The two parts are expressed in terms of the fluctuating and mean-field eddy viscosities, respectively.

The fluctuating-eddy viscosity is given by the standard SGS model with the modified TKE equation in which the mean-shear contribution is taken out from the shear production term. An expression for the mean-field eddy viscosity is derived from a match with the Monin–Obukhov similarity relations in the surface layer. The mean-field eddy viscosity is introduced mainly to account for the inadequate grid resolution and other shortcomings of the standard SGS model as the surface boundary is approached. Its specification ensures the consistency of the mean LES fields with the empirical similarity relations. Its contribution to the subgrid stresses dies down rapidly in going away from the surface. The proposed modification to the parameterization of the SGS stresses requires little additional computational cost. Further details of this model and the results of its application in large-eddy simulations of neutral and unstable PBL flows are given by Sullivan et al. (1994). A more recent application of the model to weakly stratified stable atmospheric boundary layers has been reported by Andren (1995), who also compared it with several other SGS models. A major conclusion of Andren's study is that the SGS model proposed by Sullivan et al. is about as successful in improving the near-surface behavior as a computationally more expensive stochastic SGS model.

11.5.3 Boundary Conditions

The numerical solution of the grid-volume–averaged equations of motion, potential temperature, and scalar concentrations depends not only on the finite-difference methods used, but also on the boundary conditions. These aspects of the LES modeling have been considered in some detail, among others by Deardorff (1973) and Mason (1994). Here we briefly review the lateral, upper, and lower boundary conditions that are commonly used in LES.

Lateral Boundary Conditions The use of cyclic or periodic boundary conditions at the vertical sides of the model domain has been very common. It has obvious merits and justification in flow problems with horizontal homogeneity. Such conditions are also imposed on the pressure fluctuations even though a large-scale horizontal gradient may exist in the mean pressure. There are many situations of interest to modelers, however, in which air is affected by the surface in such a manner that more momentum, heat, or mass of pollutants, for example, must pass out the downstream side than enters the upwind side. In this case, instead of cyclic boundary conditions being used, certain variables may have to be specified differently at the inflow and outflow boundaries. There may be sufficient information available on the specification of certain variables (e.g., velocity, temperature, etc.) on the upwind side (approach flow) but not on the downwind (outflow) side. In general, it is difficult to specify physically realistic and accurate lateral boundary conditions when the underlying surface is not homogeneous in temperature, roughness, or topography. In large-eddy simulations of flow over isolated hills and buildings, most investigators have used periodic lateral boundary conditions for lack of any better alternative. With such boundary conditions, effectively an array of hills or buildings is implied. To reduce sensitivity to the lateral boundary conditions, one must extend the domain much farther out from the region of one's primary interest.

Upper Boundary Conditions The upper boundary of the model domain at $z = H$ should be placed sufficiently high, preferably above the maximum mixing height, so that practically all the turbulence and diffusion processes would occur below it. The most commonly used upper boundary conditions specify zero gradients of horizontal velocity and potential temperature and zero vertical velocity (Deardorff, 1973). These imply zero shear stress at $z = H$ and a rigid lid or an impenetrable inversion at the top. More realistic upper boundary conditions allow for an inversion layer penetrable by convective elements and entrainment of nonturbulent fluid from above. An alternative boundary condition on the horizontal velocity at $z = H$ is to specify it as a constant known value (e.g., the geostrophic velocity). The boundary condition on the pressure at $z = H$ can be specified diagnostically. It is necessary to provide the correct pressure-gradient boundary condition in the solution of the Poisson equation for pressure (see e.g., Deardorff, 1973).

Lower Boundary Conditions These boundary conditions are required to specify the subgrid Reynolds stresses and heat flux at a height just above the roughness elements. For all practical purposes, this height is considered $z = 0$ where $\overline{w} = 0$ is specified. The subgrid stress or flux comprises the total stress or flux at $z = 0$ because $\overline{w} = 0$.

At the lowest interior grid point, z_1, the flow may be assumed to be directed along the direction of the surface stress. The angle α of the latter, as measured counterclockwise from the x axis, is therefore given by

$$\alpha = \tan^{-1}\left(\frac{\overline{v}}{\overline{u}}\right)_{z=z_1} \quad (11.38)$$

The two components of stress at the lower boundary can be specified in terms of the local friction velocity as

$$\overline{(u'w')}_o = -u_*^2 \cos \alpha$$
$$\overline{(v'w')}_o = -u_*^2 \sin \alpha \quad (11.39)$$

Similarly, the horizontal components of the heat flux can be specified in terms of the product, $u_* \theta_*$, of friction velocity and friction temperature scales, or alternatively in terms of the surface heat flux if the latter is known (Deardorff, 1973). The problem is then shifted to the specification or parameterization of u_* and possibly also θ_*/θ_*.

Starting from the original suggestion of Deardorff (1973, 1974), the Monin–Obukhov similarity relations have been used to specify u_* and θ_* in terms of computed velocities and potential temperatures at the first interior grid level (see e.g., Moeng, 1984; Mason, 1989; Schmidt and Schumann, 1989). This, of course, requires that z_1 be set sufficiently small so as to lie within the surface layer. As pointed out by Deardorff (1973), any specification of u_* and θ_* using the surface-layer similarity relations constitutes an assumption because those relations are approximately valid only for large time averages or ensemble averages of velocity, temperature, and the corresponding fluxes. In numerical calculations using LES, on the other hand, the same similarity relationships are assumed to hold at each grid point near the surface at each time step. This leads to several alternative choices for the lower boundary conditions, depending on whether local or spatially averaged values of certain variables are used in the similarity relations (Schowalter et al., 1996).

Usually, the local stress or friction velocity at the surface is related to the local velocity at the first interior grid point, that is,

$$u_*^2 = c_D(\overline{u}_1^2 + \overline{v}_1^2) \tag{11.40}$$

in which the drag coefficient c_D may also be a local variable or it may be constant for the whole domain but can vary with time. According to the M–O similarity theory,

$$c_D = k^2\left[\ln\frac{z_1}{z_o} - \psi_m\left(\frac{z_1}{L}\right)\right]^{-2} \tag{11.41}$$

in which L may be calculated either locally as

$$L = -\frac{T_{vo}u_*^3}{kg(\overline{w'\theta_v'})_o} \tag{11.42}$$

or from the horizontally averaged fluxes. The specification of the stress at the lower boundary also depends on whether the surface temperature or the surface heat flux is specified.

An alternative method of treating the lower boundary conditions is to specify first or second vertical derivatives of velocity and potential temperature at the lowest interior grid points such that the subgrid Reynolds stresses and heat fluxes obtained from the SGS closure model agree with values deduced from the surface-layer similarity relations. This method was used by Deardorff (1970, 1972) to specify values of variables on exterior grid points so as to produce the desired values of the vertical derivatives. The sensitivity of the LES results to different lower boundary conditions has not been studied thoroughly. But our recent preliminary investigations of this for the convective boundary layer reveal that the computed velocity variances and third moments are little affected by different boundary conditions (Schowalter et al., 1996). However, the horizontal anisotropy of turbulence, as indicated by the ratio $\langle v'v'\rangle/\langle u'u'\rangle$, where $\langle\ \rangle$ denotes spatial averaging over the horizontal domain, is found to be more sensitive to different lower boundary conditions. The specification of surface heat flux seems preferable to the specification of surface temperature as a function of time. The latter boundary condition gives rise to organized horizontal roll vortices, which are not observed in the convective boundary layer. These rolls appear to be more associated with the limited domain size; a sufficiently large horizontal domain is required to suppress the roll vortices in the model. In the specification of the surface stress or u_*^2, a local drag coefficient might be preferable to a global one based on horizontally averaged fields.

11.5.4 Applications of LES to Dispersion

In what might be the first application of LES to particle dispersion, Deardorff and Peskin (1970) reported the Lagrangian statistics of one- and two-particle displacements in a large-eddy simulated turbulent channel flow at large Reynolds number. In spite of the small numbers of grid points (6720) and tagged fluid particles used by them with the computer resources of that time, they obtained remarkably realistic Lagrangian and Eulerian autocorrelation coefficients as functions of dimensionless time lag and spatial separation along the mean flow direction. The computed mean-square particle displacements were found to be consistent with Taylor's (1921) theory.

A somewhat limited study of particle dispersion in neutral and unstable PBLs, using the LES mean and turbulent flow fields, was reported by Deardorff (1972). The primary focus of this was on the dimensionless mean particle height as a function of the dimensionless travel time, following the release of a large number (800) of particles from a horizontal plane near the surface. In the neutral PBL, the mean particle height increased almost linearly but slowly with time without reaching its equilibrium value within the simulation time. Under unstable and convective conditions, however, mean particle trajectories rose much more steeply, overshot the expected equilibrium level at one-half the height of capping inversion z_i, then decreased and reached the equilibrium level after small oscillations.

A more thorough investigation of particle dispersion in the convective boundary layer, using Deardorff's (1974) large-eddy simulation of the same, has been conducted by Lamb (1978, 1979, 1982). Lamb's approach is based on the Lagrangian counterpart of the instantaneous mass continuity equation of an inert nonbuoyant chemical species. His formulation of the Lagrangian diffusion model for calculating ensemble mean concentrations involves a probability density function p of particle displacements, similar to that introduced in Chapter 8. A procedure for evaluating this function from the computed velocity field in Deardorff's (1974) large-eddy simulation of the CBL is outlined by Lamb (1982). It requires only a description of the trajectory of each of the N particles released from a given source position, which is given by

$$\frac{dX_i^n}{dt} = \overline{u}_i[X_i^n(t), t] + u_i'[X_i^n(t), t] \tag{11.43}$$

Here X_i^n denotes the vector position of the particle n ($n = 1, 2, \ldots, N$) released at $(0,0,H)$, $\bar{u}_i$ represents the resolved velocity field given by the LES model, and u_i' represents the SGS velocity fluctuation. The process by which u_i' is determined is described in detail by Lamb (1981, 1982). Assuming that SGS fluctuations are approximately isotropic and lie within the inertial subrange, $u_i'(t)$ can be specified in terms of its value at the previous time step $u_i'(t - \Delta t)$ and a computer-generated random vector with zero mean and variance equal to $2E/3$, where E is the subgrid scale kinetic energy given by the LES model at the particle's position $X_i(t)$ at time t. The statistics of a large ensemble of particle releases were utilized to compute the probability density function of particle displacements. From this, Lamb calculated the ensemble mean concentrations, as well as plume dispersion parameters σ_y and σ_z (Lamb, 1982). These have been presented in dimensionless forms using the mixed-layer similarity scaling, as described in Chapter 8.

Figure 11.2 shows the calculated nondimensional cross-wind integrated concentration distributions for continuous point sources of different dimensionless heights ($H/h = 0.025$, 0.25, 0.5, and 0.75). On the basis of these and the calculated dispersion parameters as functions of the dimensionless distance from the source, two distinct dispersion regimes can be identified—one for sources located in the surface layer and the other for sources located in the mixed layer. For a near-surface source, particles released into the base of an updraft are immediately transported upward, while those emitted into a downdraft move approximately horizontally until eventually they too are swept into updrafts and are carried upward. Consequently, after a sufficiently large travel time (of the order of h/W_*) that a majority of particles have entered updrafts, the location of maximum particle concentration lifts off the ground and rises toward the inversion base. This unusual phenomenon was first observed in the convection tank diffusion experiment by Deardorff and Willis (1975). In their experiment, the maximum concentration due to a surface source in the simulated CBL was observed to lift off the surface after a travel time of about 0.5 h/W_* and it reached a height of about 0.75 h at a travel time of about h/W_*. For the elevated sources in the mixed layer, the numerical simulation results of Lamb (1978, 1982) showed the centerlines of plumes descending with the dimensionless travel time or distance from the source. These were later confirmed by further convection tank experiments by Willis and Deardorff (1978, 1981). Both the lift-off and the descending-plume phenomena have been explained in Chapter 8, where some results of atmospheric diffusion experiments are also discussed. Lamb's (1982) numerical model results are found to be in excellent agreement with observations.

Encouraged with his success on passive plume diffusion modeling, Lamb (1982) also extended his study to buoyant plumes. For this purpose, a component u_i'' representing the buoyancy-generated particle velocity was added to the basic trajectory equation (11.43). Values for u_i'' were obtained by solving the simplified,

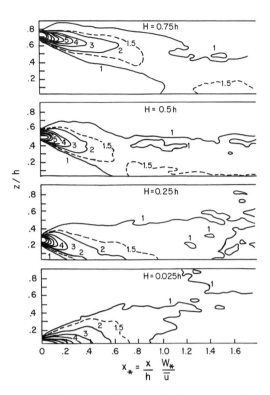

Figure 11.2 Calculated normalized cross-wind integrated concentration as a function of normalized distance from the source and the dimensionless height for continuous point sources of dimensionless heights $H/h = 0.025$, 0.25, 0.50, and 0.75 in the convective boundary layer. From Lamb, 1982.

spatially averaged, time-dependent momentum, heat, and mass conservation equations derived by Briggs (1975). A comparison of a simulated buoyant plume with a small buoyancy flux $F_b = 13$ m^4 s^{-3} with the corresponding simulation of a nonbuoyant plume indicated that buoyancy has two basic effects: an upward displacement of the effective release height and an upward rotation of the plume axis. The magnitude of the displacement (plume rise) and the angle of rotation are both directly proportional to F_b. The latter effect is not considered in the conventional plume-rise models. Lamb's (1982) first-order numerical model simulations of dispersion from buoyant sources were also considered somewhat unreliable due to some unrealistic model assumptions. Therefore, he proposed a more sophisticated (also more complicated) formulation for the component of particle velocity produced by the particle's buoyancy (Lamb, 1982). An important result of his numerical simulation is that even with a very small plume buoyancy flux, the material from an elevated plume in the mixed layer may not reach ground level until it has coursed at least once through the depth of the mixed layer. Consequently, the maximum g.l.c. is much smaller than that for the corresponding nonbuoyant plume.

An interesting application of LES to the study of diffusion from uniform area sources at the surface and the top of the CBL has been made by Wyngaard and Brost (1984). They used a dry version of the LES model used by Deardorff (1980), but with a different finite-difference scheme. Their subgrid scale parameterization was based on the subgrid TKE equation, which was solved in conjunction with the grid-volume–averaged equations for velocity components, potential temperature, and scalar concentration. Wyngaard and Brost (1984) investigated the two distinct problems of bottom-up and top-down diffusion in the CBL. In the top-down problem, a finite value of the scalar flux at the mixed-layer top and zero value at the bottom (surface) were specified. In the bottom-up problem, they maintained a surface flux by specifying a uniform value of mean concentration at the surface that was greater than that at the first interior grid level and forced the flux to zero at the mixed-layer top. The resulting profile of the bottom-up flux followed the expected linear shape, similar to that of the heat flux. But the top-down flux profile is distinctly curved; this curvature is due to time-varying mean concentration gradient in the vertical. Perhaps the most interesting and profound result of the LES study is that the top-down and bottom-up vertical-eddy diffusivities computed from the corresponding scalar flux and mean concentration profiles in the same CBL have quite different magnitudes and vertical distributions. The maximum value of the bottom-up diffusivity for the surface area source is nearly one and one-half times the maximum top-down diffusivity for the area source at the top of the mixed layer. This is consistent with an earlier finding of Lamb and Duran (1978) that K_z depends strongly on the source height. This dependence of diffusivity on the source height is perhaps related to the implied dependence of the apparent diffusivity on distance from a continuous source in a homogeneous field of turbulence, as discussed in Chapter 7.

Another important application of a LES model to the study of the behavior of passive and buoyant plumes in the CBL has been reported by Nieuwstadt and de Valk (1987) and van Haren and Nieuwstadt (1989). They simulated three CBLs with the same boundary layer height $h = 1500$ m and the same geostrophic wind speed of 5 m s^{-1}, but different values of the surface heat flux corresponding to different convective velocities ($W_* = 1.0, 1.9,$ and 2.4 m s^{-1}). With use of a 5-km by 5-km by 2-km domain with 40 by 40 by 40 grid points, their horizontal and vertical resolutions were rather coarse ($\Delta x = \Delta y = 125$ m and $\Delta z = 50$ m). Consequently, their simulation of continuous point source by an instantaneous line source represented by a horizontal bar running parallel to the x axis, with a vertical cross section of 125 m by 100 m, is also somewhat crude. With the x axis oriented along the mean flow direction, the diffusion from an instantaneous line source as a function of time can be interpreted in terms of the dispersion from a continuous point source as a function of the distance from the source by means of the transformation $t = x/\bar{u}$. Such a transformation is only valid if along-wind diffusion is negligible compared with the transport by mean velocity (Willis and Deardorff, 1976).

van Haren and Nieuwstadt (1989) performed simulated dispersion experiments at dimensionless release heights $H/h = 0.15$ and 0.48, which correspond to the mixed-layer source and dispersion regime. For their buoyant plume simulations, the line source was given an excess potential temperature with respect to the ambient air. Following the mixed-layer similarity scaling, they specified three different values for the dimensionless plume buoyancy flux parameter $F_* = 0, 0.01,$ and 0.02, where

$$F_* = \frac{F_b}{W_*^3 h}. \qquad (11.44)$$

was introduced earlier by Willis and Deardorff (1983) and F_b is the buoyancy flux parameter defined in Eq. (10.2). Obviously, $F_* = 0$ corresponds to the simpler nonbuoyant or passive plume case. The mean plume height $\bar{z}$ and dispersion statistics ($\sigma_y, \sigma_z,$ and $\bar{c}_y$) were obtained by averaging over a horizontal plane. Such an area average for the instantaneous line source is equivalent to a time average for a continuous point source. For the domain length of 5 km and $\bar{u} = 5$ m s^{-1}, an averaging time of 1000 s is implied. Since this was not considered enough to get statistically stable averages, three different realizations were performed for each experiment by placing the line source at different y locations. The results are presented in the mixed-layer similarity framework for easy comparison with the data from field and laboratory experiments.

For passive (nonbuoyant) plumes, the LES results of Nieuwstadt and deValk (1987) and van Haren and Nieuwstadt (1989) are shown to be in good agreement with the experimental data from Willis and Deardorff (1980) and CONDORS field experiment (Kaimal et al., 1986). For buoyant plumes with $F_* = 0.01$ and 0.02, however, the model underpredicted both the dimensionless mean plume height ($\bar{z}/h$) and the vertical dispersion parameter (σ_z/h) at short distances ($X^* < 0.5$) from the source, but well predicted the lateral dispersion parameter (σ_y/h). The discrepancies between the model-computed and observed parameters are explained in terms of the large initial dimension of the simulated source and initial momentum effects of release in experiments. The most important conclusions based on these large-eddy simulations of dispersion are that there is little influence of plume buoyancy on vertical and lateral dispersion parameters, but the mean plume height and concentrations are strongly influenced by plume buoyancy. Figure 11.3 shows the effect of F_* on dimensionless cross-wind integrated g.l.c. as a function of dimensionless distance from the source. Ground-level concentrations near the source including the maximum g.l.c. are significantly reduced with increasing F_*.

A detailed analysis of the computed mean plume height, effectively separating the plume rise due to

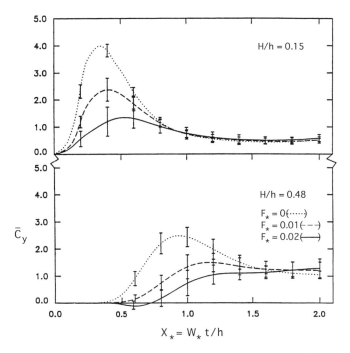

Figure 11.3 Dimensionless cross-wind integrated surface concentration as a function of dimensionless distance from the source and the plume buoyancy parameter for the elevated sources ($H_s/h = 0.15$ and 0.48) in a convective mixed layer. From van Haren and Nieuwstadt, 1989.

buoyancy from the plume motion due to convective turbulence, yields some interesting results (van Haren and Nieuwstadt, 1989). In particular, the plume rise due to buoyancy does not follow the well-known two-thirds law and is much smaller than that predicted by this law. The two-thirds law of plume rise is based on the assumption that a buoyant plume grows by self-generated turbulence. It has been verified by extensive plume-rise experiments and observations in both turbulent and nonturbulent ambient flows under mostly neutral and stably stratified conditions (see, e.g., Briggs, 1975, 1984). In the CBL, the plume is apparently torn apart by large-scale convective eddies instead of maintaining a double vortex structure, dominated by plume-induced turbulence. A LES model with a coarse grid is probably not capable of simulating the small-scale dynamics of a buoyant plume. However, the model results clearly indicate the strong influence of convective turbulence on the rise or fall of a buoyant plume, which considerably differs from that on a passive plume released from the same height. This is because buoyancy transports the plume to a higher level in the CBL. As a result, the buoyant plume from a relatively low release height near the surface may rise less, whereas the plume released at higher elevation in the mixed layer may descend faster. Therefore plume buoyancy strongly affects the contribution of ambient turbulence to the mean plume height. But one cannot simply add the plume rise due to buoyancy to the mean plume height for the nonbuoyant source.

According to the LES studies of buoyant plumes conducted so far, the processes of plume rise and dispersion are interactive and, hence, inseparable. The commonly used practice of adding plume rise to the release height to obtain an effective source height in regulatory applications of short-range dispersion models is not supported by these more sophisticated numerical models. A more detailed analysis of plume dispersion statistics for nonbuoyant elevated releases in the CBL is given by Nieuwstadt (1992a). The overall or total dispersion parameters σ_y and σ_z show only weak dependence on the release height in the mixed layer. They both show the expected behavior predicted by Taylor's (1921) theory for small diffusion times ($t_* = W_*t/h \ll 1$). At large diffusion times (say, $t_* > 1$), σ_y appears to be consistent with Taylor's result for homogeneous turbulence, but σ_z approaches a constant value as the concentration distribution in the vertical tends to become uniform and well mixed across the CBL.

The total dispersion parameters are further decomposed into two parts—meandering and relative diffusion—and both contributions are calculated as a function of dimensionless diffusion time and release height. It is found that initially there is a substantial contribution to dispersion by large turbulent eddies (convective updrafts and downdrafts) in the form of plume meandering. At large diffusion times the influence of meandering becomes smaller, and relative diffusion becomes more dominating. These results can be interpreted in terms of the well-known looping plume behavior in the CBL.

Further large-eddy simulations of dispersion from passive sources in neutral and convective boundary layers have been reported by Sykes and Henn (1992) and Henn and Sykes (1992). They used medium- and high-resolution models, both with a variable vertical grid spacing for a better resolution of the surface layer and the interfacial transition layer. They also used several different domain sizes relative to the mixing height. The small area source function used for dispersion in the CBL is Gaussian with a standard deviation of $0.04\,h$. This source size is relatively large compared with a typical stack exhaust diameter, but is more representative of an effective plume size following an initial spread due to source turbulence. The scalar diffusion equation is solved using an advection-diffusion scheme, which uses higher order moments of the concentration distribution within each grid cell to maintain an accurate description of the concentration field. The emphasis is placed on concentration fluctuations due to random turbulent velocity fluctuations. The standard deviation of concentration fluctuations, their temporal and spatial scales, and their probability distributions are examined. It is found that the intensity of concentration fluctuations, $\sigma_c/\bar{c}$, continually decreases with distance downstream from the source. The vertical distributions of both mean concentration and σ_c become nearly uniform at downstream distances corresponding to $X^* \gtrsim 2.5$. Concentration probability distributions show intermittency near the source, but they become nearly Gaussian farther downstream in the plume (Henn and Sykes, 1992). There are significant differences in the distributions of concentration fluctuations in neutral and convective boundary layers. The distribution in the former tends toward the log-normal distribution.

11.6 LAGRANGIAN STOCHASTIC MODELS

In the Lagrangian approach to dispersion modeling, one tracks particles through a turbulent environment using either a numerically computed Eulerian velocity field or a stochastic model for the Lagrangian particles velocities. Lamb (1978, 1982) followed the former approach using the velocity field calculated in Deardorff's (1974) large-eddy simulation of the CBL. The successful applications of this approach to atmospheric dispersion have already been discussed in the preceding section (see also Lamb, 1982; Weil, 1988b). The Lagrangian particle-dispersion modeling utilizing the LES for mean flow and turbulent velocity fields is the most computer intensive and expensive approach. The instantaneous velocities at all the grid points at each time step must be stored in the computer memory for calculation of the trajectories of thousands of particles, and, hence, obtaining of the desired dispersion statistics.

The second approach is much less expensive and is based on a stochastic model of the Lagrangian particle velocity. The model first separates the instantaneous velocity into a mean and a turbulent component, that is,

$$\frac{dX_i}{dt} = U_i = \overline{U}_i + U'_i \tag{11.45}$$

in which the overbar represents time averaging, and prime, as usual, denotes turbulent fluctuation. By definition, the Lagrangian velocity U_i corresponds to an individual particle. A small, passive, nonbuoyant particle can also be considered as a tagged fluid particle, because it faithfully follows the fluid motion. The choice of an appropriate time for averaging is not always clear, particularly in the atmosphere with its wide range of scales of motion. It should not be less than the travel time at which dispersion statistics are desired. For a stationary and homogeneous turbulent flow, the Lagrangian mean velocity $\overline{U}_i$ is expected to be identical to the Eulerian mean velocity $\bar{u}_i$. In stratified shear flows, such as the atmospheric surface layer, the relationship between the two is more complicated, but may be obtained from other theoretical and empirical considerations (e.g., the surface-layer similarity theory discussed in Chapter 8). The turbulent component of the Lagrangian velocity cannot be predicted in detail for any individual fluid particle; it is usually specified using a stochastic or random-walk model.

11.6.1 Random-walk Model for Homogeneous Turbulence

Smith (1968) proposed a random-walk hypothesis that states that the Lagrangian turbulent velocity a short time later can be expressed in terms of its previous velocity and a random velocity fluctuation, that is,

$$U'_i(t + \Delta t) = R_L(\Delta t)U'_i(t) + U''_i \tag{11.46}$$

Here, $R_L(\Delta t)$ is the value of the Lagrangian autocorrelation function $R_L(\xi)$ at $\xi = \Delta t$ (see Section 5.9), and $U''_i(t)$ is the purely random or the uncorrelated part of velocity fluctuation. Smith (1968) called Eq. (11.46) a linear regression equation, because he could fit a straight line through a plot of experimentally measured values of $V'(t + \Delta t)$ versus $V'(t)$ (see also Hanna, 1979). It also represents a Markov-chain process and, hence, belongs to the family of Markov equations.

To specify the Lagrangian particle velocity at each time step in the numerical model using Eq. (11.46), one must specify $R_L(\Delta t)$ as well as $U''_i(t)$. The former is easily obtained from the commonly used exponential form (5.88) of the Lagrangian autocorrelation function, that is,

$$R_L(\Delta t) = \exp\left(-\frac{\Delta t}{T_{iL}}\right) \tag{11.47}$$

provided the integral time scale can be specified (see section 5.10). The random fluctuation U_i'' is picked from a very large ensemble of random numbers whose mean is zero, and the variance is given by

$$\overline{U_i''^2} = \overline{U_i'^2}[1 - R_L^2(\Delta t)] \quad (11.48)$$

where the repeated index does not imply summation. Equation (11.48) is obtained by squaring and averaging Eq. (11.46) and assuming that the velocity variance does not change during the short time interval Δt. It is more convenient to express U_i'' in terms of its standard deviation, that is,

$$U_i'' = \sigma_i[1 - R_L^2(\Delta t)]^{1/2} \eta(t) \quad (11.49)$$

where σ_i is the standard deviation of the velocity fluctuations U_i' and $\eta(t)$ is a dimensionless random variable with zero mean and variance = 1. Substituting from Eqs. (11.47) and (11.49) into Eq. (11.46), one obtains

$$U_i'(t + \Delta t) = aU_i'(t) + \sigma_i(1 - a^2)^{1/2}\eta(t) \quad (11.50)$$

where

$$a = \exp\left(-\frac{\Delta t}{T_{iL}}\right) \quad (11.51)$$

or some other suitable expression or value for $R_L(\Delta t)$. For very large time steps with $\Delta t \gg T_{iL}$, $a \approx 0$, so that the velocity at any time is completely specified by $\sigma_i\eta(t)$ and is not correlated at all with its value at the previous time.

A Monte-Carlo-type random number generator is often used to specify η at any time step. For this reason, the random-walk model is also called the Monte-Carlo model. The distribution of η is often chosen to be Gaussian for simplicity; this is also the most appropriate choice for homogeneous turbulent flows.

The Markov-chain equation (11.50) is considered to be particularly suitable and valid for stationary and homogeneous turbulence. In such an idealized flow, not only is the mean Lagrangian velocity equal to the Eulerian mean velocity, but also the distributions of Lagrangian velocity fluctuations become identical to those of Eulerian velocity fluctuations. Consequently, σ_i in Eq. (11.50) can be replaced by more easily measured σ_u, σ_v, or σ_w.

Random-walk simulations of fluid-particle trajectories (Thompson, 1971; Hall, 1975; Reid, 1979), in which the particle velocities are represented by a Markov sequence such as Eq. (11.50), are based on an assumed equation of motion of a fluid particle. The equation used or implied implicitly in such models is the Langevin equation, which describes the motion of a particle subject to a retarding force and a random acceleration. Legg and Raupach (1982) show how the Langevin equation is related to Markov-chain models of Lagrangian velocity. They start with the one-dimensional equation for the vertical particle velocity

$$\frac{dW}{dt} = -\alpha W + \lambda \xi(t) \quad (11.52)$$

where α and λ are coefficients to be specified later and $\xi(t)$ is Gaussian white noise, which is a stationary stochastic process with a Gaussian probability density function and zero mean. The solution to the above stochastic differential equation is given by

$$W(t) = W(0)e^{-\alpha t} + \lambda \int_0^t e^{\alpha(s-t)}\xi(s)ds \quad (11.53)$$

which is then decomposed into mean and fluctuating parts

$$\overline{W}(t) = \overline{W}(0)e^{-\alpha t} \quad (11.54)$$

$$W'(t) = W'(0)e^{-\alpha t} + \lambda \int_0^t e^{\alpha(s-t)}\xi(s)ds \quad (11.55)$$

From Eq. (11.55), Legg and Raupach (1982) derived the following expressions for the variance and covariance of vertical velocity:

$$\overline{W'^2}(t) = \overline{W'^2}(0)e^{-2\alpha t} + \frac{\lambda^2}{2\alpha}(1 - e^{-2\alpha t}) \quad (11.56)$$

$$\overline{W'(0)W'(t)} = \overline{W'^2}(0)e^{-\alpha t} \quad (11.57)$$

If the Lagrangian integral time scale for the particle velocity is defined as

$$T_{iL} = \int_0^\infty \frac{\overline{W'(0)W'(t)}dt}{\overline{W'^2}(0)} \quad (11.58)$$

expresssions (11.56) and (11.57) for the variance and covariance enable the coefficients in the Langevin equation (11.52) to be determined as

$$\alpha = \frac{1}{T_{iL}} \quad (11.59)$$

$$\lambda = \sigma_w\sqrt{2\alpha} = \sigma_w\left(\frac{2}{T_{iL}}\right)^{1/2} \quad (11.60)$$

Thus, if σ_w and T_{iL} are known, Eq. (11.53) determines the velocities, and therefore trajectories, of an ensemble of tagged fluid particles with a prescribed distribution of initial velocity $W(0)$.

Legg and Raupach (1982) argue that although the ensemble of particle velocities determined by the Langevin equation constitutes a Markov process, no Markov process can exactly represent fluid particle velocities in a turbulent flow. Thus, Eqs. (11.52) and (11.53) cannot describe exactly the velocities of marked particles in a turbulent flow. They become relevant to turbulent dispersion only when one considers the particle velocities at discrete times, t_0, $t_0 + \Delta t$, $t_0 + 2\Delta t$, and so on. If Δt is chosen much greater than the time scale over which particle acceleration remains correlated, then the sequence of particle velocities will be a Markov sequence. Then it can be shown that the Langevin equation is equivalent to the Markov-chain equation (11.50). It can also be argued that the time step Δt to be used in particle velocity and trajectory calculations should be much smaller than the Lagrangian integral time scale T_{iL}.

11.6.2 Random-walk Models for Inhomogeneous Turbulence

In any realistic dispersion model of the atmospheric boundary layer, one must take into account the inhomogeneity of mean flow and turbulence in the vertical direction, even when the PBL might be stationary and horizontally homogeneous. It has been shown in Chapter 8, through the use of the statistical and similarity theories of dispersion, that in the stratified surface layer $\overline{U}$ and $\overline{W}$ differ from $\overline{u}$ and $\overline{w} = 0$, respectively, and also that $\overline{W'^2}$ is different from $\overline{w'^2}$. Approximate empirical relations between Lagrangian and Eulerian mean velocities and variances have been derived from the same theories. These can be used in the random-walk model described above. Some investigators have also suggested modifications to the Markov-chain equation (11.50) to include explicitly a term representing inhomogeneity of turbulence (see e.g., Wilson et al., 1981a, 1981b; Legg and Raupach, 1982).

Wilson et al. (1981a, 1981b) simulated dispersion in the lower part of the atmospheric boundary layer with σ_w increasing with height, as happens in a canopy layer and also in the CBL. They found that the use of Eq. (11.50) resulted in far too may particles collecting in the near-surface region of low turbulence. They reasoned that, for $\partial \sigma_w / \partial z > 0$, the particles crossing an x–y plane going downward would have higher velocity on the average than those traveling upward. To satisfy the mass continuity across the x–y plane, the vertical Lagrangian velocity distribution must be skewed, with a preference for upward motion. Therefore, Wilson et al. (1981a) added a bias velocity to the Markov equation for the Lagrangian vertical velocity:

$$W(t + \Delta t) = aW(t)$$
$$+ \sigma_w (1 - a^2)^{1/2}\, \eta(t) + \Lambda_L \frac{\partial \sigma_w}{\partial z} \quad (11.61)$$

Here Λ_L is a length scale whose values were found by comparing the model-predicted mean concentrations with those measured during the Project Prairie Grass field experiment (Wilson et al., 1981b). It is found that Λ_L is function of height above the ground and stability. It should be noted that the vertical Lagrangian velocity W in Eq. (11.52) includes both the mean and fluctuating parts. The last term amounts to a positive bias to only the mean vertical velocity. It is assumed to be a function of height and stability, but independent of travel time.

The proposed modification of the Markov-chain equation for the vertical particle velocity by Wilson et al. (1981a) was based on heuristic arguments. A theoretically more sound and rigorous approach based on the Langevin equation has been used by Legg and Raupach (1982). They argue that when there is a gradient of vertical velocity variance, the equation of vertical motion for a fluid particle must include a mean force due to the action of the mean pressure gradient on the particle. Hence, the appropriate Langevin equation for this case of inhomogeneous turbulence is

$$\frac{dW}{dt} = -\alpha W + \lambda \xi(t) + F \quad (11.62)$$

where

$$F = -\frac{1}{\rho_o} \frac{\partial \overline{p}_1}{\partial z} = \frac{\partial \overline{w'^2}}{\partial z} \quad (11.63)$$

Here $\overline{p}_1$ denotes the deviation of the mean pressure from the reference hydrostatic pressure p_o and ρ_o is the density of the reference atmosphere. Following an analysis of the Langevin equation similar to that for the case of stationary and homogeneous turbulence, but with some additional implied assumptions, Legg and Raupach (1982) derived an expression for the mean drift velocity

$$\overline{W}(t) = T_{iL} \frac{\partial \sigma_w^2}{\partial z} \quad (11.64)$$

The assumptions of stationary turbulence and $\overline{W'^2} = \overline{w'^2} = \sigma_w^2$, implied in the derivation of Eq. (11.64), are somewhat questionable in the case of nonhomogeneous turbulence. However, the above result is consistent with the following, more general expression of the mean drift velocity of particles obtained from the Fokker–Planck form of diffusion equation (see Monin and Yaglom, 1971, p. 610):

$$\overline{W}(t) = T_{iL} \frac{\partial \sigma_w^2}{\partial z} + \sigma_w^2 \frac{\partial T_{iL}}{\partial z} \quad (11.65)$$

Since the questionable assumption of T_{iL} = constant, implied by constant α, was made in the derivation of Eq. (11.64), the use of Eq. (11.65) should be preferable. With the addition of mean drift velocity, the appropriate Markov-chain equation for the vertical particle velocity is

$$W(t + \Delta t) = aW(t) + \sigma_w (1 - a^2)^{1/2}\, \eta(t)$$
$$+ T_{iL} \frac{\partial \sigma_w^2}{\partial z} + \sigma_w^2 \frac{\partial T_{iL}}{\partial z} \quad (11.66)$$

A more general form of the nonlinear stochastic differential equation for the vertical velocity has been used in more recent random-walk models of dispersion (Thomson, 1987; Luhar and Britter, 1989; Hurley and Physick, 1993; Sawford, 1993; Wilson and Sawford, 1996):

$$dW = A(Z,W)dt + B(Z,W)d\eta \quad (11.67)$$

Here A and B are functions of particle position (Z) and vertical velocity (W), and $d\eta$ is the random velocity increment. For the more general vector or tensor form of the stochastic differential equation, the reader should refer to Thomson (1987) and Sawford (1993). In high Reynolds number flows the evolution of the particle velocity $W(t)$ can be assumed to be a Markov process because the particle acceleration is significantly correlated only for times much shorter than the lifetime of the energetic eddies. Hence the derivative dW can be represented by a nonlinear stochastic equation of the form (11.67). If $P(Z,W)$ is the joint probability density function of all the tracer or marked fluid particles in the ensemble of flows, then away from any source it should satisfy the Fokker–Planck equation (Thomson, 1987; Sawford, 1993)

$$W\frac{\partial P}{\partial Z} + \frac{\partial(AP)}{\partial W} = \frac{\partial^2(B^2 P)}{2\partial W^2} \qquad (11.68)$$

Equation (11.68), in conjunction with the well-mixed condition, can be used to determine A and B, which have also been called the drift and diffusion terms, respectively.

For the simplest problem of one-particle dispersion in stationary and homogeneous turbulence, $A = -W/T_{iL}$ and $B = \sqrt{2}\sigma_w/T_{iL}^{1/2} = (c_o\varepsilon)^{1/2}$, where $c_o \approx 2.0$ is Kolmogorov's structure function constant and ε is the rate of energy dissipation. For certain inhomogeneous turbulent flows, such as the surface layer and CBL, A and B have been determined uniquely for the one-dimensional case (see e.g., Thomson, 1987; Luhar and Britter, 1989). For the more general three-dimensional dispersion problems, however, the solutions to Eq. (11.68) may not be unique (Sawford, 1993). Wilson and Sawford (1996) have reviewed the theoretical basis for and the advantages of the Lagrangian stochastic or random-walk models for dispersion in the PBL.

11.6.3 Computation of Particle Trajectories and Mean Concentrations

The trajectories of marked fluid particles, or neutrally buoyant passive particles released from a specified source geometry and height, are computed by numerically integrating the Lagrangian equations (11.45) of particle motion. In these the mean velocities are usually specified in terms of an empirical mean wind profile or are otherwise calculated from a separate PBL or mesoscale atmospheric model. Fluctuating velocities are computed at each time step using the random-walk models or Markov-chain expressions discussed in the preceding subsection. For example, the vertical displacement of the particle is given by

$$dZ = W(t)dt \qquad (11.69)$$

or

$$Z(t + 1) = Z(t) + W(t)\Delta t \qquad (11.70)$$

in which Δt is chosen as a small fraction (10–20%) of T_{iL}. The vertical inhomogeneity of the PBL turbulence requires small time steps (of the order of a few seconds) in random-walk models of particle dispersion. Failure to observe the restrictions on the time step can lead to an accumulation or deficit of particles in the boundary regions (Thomson, 1987). Thousands of individual particle trajectories are generated in order to calculate the statistics of particle displacements such as mean height and standard deviations. The computation of the mean concentration field due to a simulated continuous source requires an accounting of the total or average time spent by particles in specified grid volumes or bins.

In practice, particle displacements are calculated for short time increments (Δt) and the number of increments (n_j) that all particles spend in a particular grid cell or volume $\Delta x_j \Delta y_j \Delta z_j$. If the continuous point source strength is Q and the total number of trajectories is N, then the mean concentration in the grid cell j is (Reid, 1979)

$$\bar{c}_j = \frac{n_j \Delta t Q}{N \Delta x_j \Delta y_j \Delta z_j} \qquad (11.71)$$

This method gives a minimum definable concentration for $n_j = 1$ and also a maximum definable concentration corresponding to $n_j = N\Delta x_j/\bar{u}\Delta t$. It is necessary to make a suitable choice of model parameters to ensure that the concentration range of interest is covered. Since the ratio of the maximum to the minimum definable concentration is $N\Delta x_j/\bar{u}\Delta t$, to obtain a wide range of concentration estimates large numbers of trajectories must be computed with small time steps relative to the time taken to advect a particle through a cell. Both of these require large compute time. In practice, the grid cells are allowed to expand in size with increasing downwind distance, so that decreasing concentrations may be resolved.

An alternative method of computing mean particle concentrations is based on the expression

$$\bar{c}(x_i, t) = Q \int_{-\infty}^{t} p(x_i, t | x_i', t') dt' \qquad (11.72)$$

where p is the probability density that a particle released at x_i' at time t' will be found at x_i at time t. The probability density function can be estimated from a large ensemble of particle trajectories (Lamb, 1978, 1982).

11.6.4 Applications of Random-walk Models

Thompson (1971) was probably the first to publish results obtained from a random-walk model of atmospheric dispersion. He simplified Eq. (11.50) by assuming $a = 0$, so that the particle velocity fluctuations at any time were independent of their values at any earlier time and were completely specified by a random number generator. Using 1000 particle trajectories in a homogeneous flow, he showed that the computed mean and the variance of particle displacements generally agreed (within a few percent) with their expected values predicted by the statistical theory (Taylor, 1921). Satisfied with the model for the simple (homogeneous) case, Thompson (1971) simulated two-dimensional dispersion on the upstream side of a hill. For this, he used Eq. (11.50) with an exponential form of the autocorrelation coefficient (a) and varying turbulence levels with height. Without the explicit accounting of the inhomogeneity of mean flow and turbulence over the hill, the calculated dispersion pattern is not likely to be right.

Dispersion in the Surface Layer Hall (1975) modeled two-dimensional dispersion in the neutral surface layer using the Markov-chain equation (11.50) for the vertical velocity fluctuation. He assumed that $\overline{W} = 0, \overline{U}(z)$ is given by the log law, and $U(t) = \overline{U}(z) \pm \sigma_u$, in which the plus or minus sign was determined by a probability density such that $\overline{U'W'}/\sigma_u\sigma_w = -0.25$. Using the neutral surface-layer similarity relations, Hall (1975) specified $\sigma_u = 2.2u_*$, $\sigma_w = 1.3u_*$, and $T_{iL} = 2.4z/\overline{u}$. His random-walk simulations of dispersion in the neutral surface layer showed good agreement with field experimental data.

Reid (1979) used the same formulation for the vertical velocity as Hall's, but ignored longitudinal velocity fluctuations. He estimated $T_{iL} \simeq 0.26z/u_*$ using the alternative expressions for eddy diffusivity $K = \sigma_w^2 T_{iL} = kzu_*$ in the neutral surface layer in conjunction with $\sigma_w^2 \simeq 1.56u_*^2$. Another theoretical approach based on an empirical form of the spectrum of vertical velocity yielded $T_{iL} \simeq 0.60z/u_*$, but the best agreement with field experimental data was obtained with $T_{iL} = 0.4\ z/u_*$. Two-dimensional random-walk simulations of dispersion in the neutral surface layer were made for surface and elevated releases. For each simulation, 10,000 particle trajectories were computed. From these the vertical dispersion parameter σ_z and the mean plume height $\overline{Z}$ were obtained as functions of distance from the source and the release height.

Random-walk simulation of particle trajectories and dispersion in vertically inhomogeneous turbulence have been reported by Wilson et al. (1981a, 1981b, 1981c). As pointed out in the previous section, Wilson et al. (1981b) found it necessary to add a mean drift velocity $\overline{W}$ to the Markov-chain equation for the vertical velocity. Their proposed empirical expression $\overline{W} = \Lambda_L \partial\sigma_w/\partial z$, which is included in Eq. (11.61), is found to be quite consistent with the more rigorous theoretical derivation of the same by Legg and Raupach (1982). Wilson et al. (1981c) specified σ_w as a function of height and stability, while T_{iL} was calculated from the eddy-diffusivity relationship $K_w = \sigma_w^2 T_{iL}$, where K_w is the diffusivity of water vapor which was specified using the appropriate surface-layer similarity relation. The length scale Λ_L was essentially treated as an adjustable empirical parameter, which was determined as a function of height and stability by best fitting of the numerical model results with the diffusion data from the Project Prairie Grass under different stability conditions.

Longitudinal velocity fluctuations have been included in random-walk models of atmospheric dispersion used by some investigators (see e.g., Ley, 1982; Legg, 1983; Zannetti, 1990). Both Ley (1982) and Legg (1983) found an increase in the mean parti-cle height and plume depth of between 4 and 20 percent when longitudinal fluctuations were included. They simulated dispersion in the neutral surface layer with σ_w constant with height. They also assumed that the Lagrangian turbulence statistics are equivalent to the corresponding Eulerian statistics, that is, that $\overline{W'^2} = \overline{w'^2} = \sigma_w^2$, $\overline{U'^2} = \overline{u'^2} = \sigma_u^2$, and $\overline{U'W'} = \overline{u'w'} = -u_*^2$. The expected correlation between U' and W' makes the Markov-chain equation of the former more complicated because it must also include the latter. Ley (1982) proposed two different formulations for the horizontal particle velocity. The first formulation is Markovian and simpler, the second is more complicated. The two formulations produced almost identical results that are in good agreement with experimental data. A further simplification neglecting correlation between U' and W' did not change the results by more than 10 percent. A comparison of the numerical model simulations with the Lagrangian similarity theory for the neutral surface layer (see Chapter 8) yielded values of the two empirical constants $b \simeq 0.39$ and $c \simeq 0.50$ appearing in the similarity relations (8.1) through (8.3).

Dispersion in the Neutral PBL Davis (1983) extended the formulation of random-walk modeling to simulate dispersion from elevated source in the whole PBL under near-neutral conditions. He used a Markov-chain expression for the vertical velocity very similar to Eq. (11.66), but with the difference that the last term representing drift velocity was reduced by a factor $(1 - a)$. For the longitudinal velocity fluctuation U', a modified Markov relation accounting for the desired correlation between U' and W' was utilized, although longitudinal velocity fluctuations were found to have a relatively small effect on model predictions. The mean wind and stress profiles in the neutral PBL were specified using an obscure analytical model, while turbulence parameters σ_u, σ_w, T_{Lu}, and T_{Lw} were specified on the basis of empirical data on PBL turbulence. Some of the empirical constants were optimized from a comparison of the observed and predicted mean concentration profiles and σ_z as a function of distance from the source in the surface layer. Then, model simulations of dispersion in the PBL were made and compared with the empirical dispersion curves obtained by Singer and Smith (1966), Briggs (1973), and Draxler (1976), each with appropriate values of the release height (H), the surface roughness (z_o), and the friction velocity (u_*).

Following the demonstrated good agreement with empirical dispersion curves, Davis (1983) systematically investigated the effects of H, z_o and u_* on the mean plume height ($\overline{z}$), the σ_z, and the normalized g.l.c. It is found that, at all downwind distances, σ_z decreases slightly with an increase in release height. For a fixed H and z_o, an increase in u_* results in larger values of $\overline{z}$ and σ_z. The strong sensitivity of σ_z to u_* or $\overline{u}$ has not been considered in widely used empirical dispersion curves or formulas. For a fixed geostrophic wind speed and release height, σ_z is also found to increase with increasing surface roughness. This dependence is, of course, also implied in the empirical dispersion parameters derived from experimental diffusion data from different sites.

Dispersion in the CBL Measurements and large-eddy simulations of vertical velocity fluctuations in the convective boundary layer (CBL) have revealed that the corresponding p.d.f. is positively skewed, but has a negative mode. This, of course, is a consequence of the organized convective circulation in the CBL, characterized by updrafts and downdrafts. Convection tank experiments by Willis and Deardorff (1976, 1978, 1981), as well as numerical simulations of dispersion in the CBL by Lamb (1978, 1982), using Deardorff's (1974) LES model, demonstrated the surprising effects of a skewed p.d.f. of vertical velocity on dispersion from surface and elevated sources in the CBL. For a surface source, the maximum concentration in the plume stayed at the ground level only for a relatively short dimensionless travel time or distance X_* from the source, and then it lifted off the ground and slowly moved to the upper part of the CBL. On the other hand, the height of maximum concentration due to an elevated source decreased with downwind distance, touched the surface, and then lifted off. This has been attributed to larger areas of slower moving downdrafts and smaller areas of faster moving updrafts, that is, to the skewness of the vertical velocity distribution.

Baerentsen and Berkowicz (1984) performed Monte-Carlo simulations of plume dispersion in the CBL and compared their results with those of experiments by Willis and Deardorff (1976, 1978, 1981). They found very good agreement. They expressed the p.d.f. of vertical velocity as a sum of the two Gaussian distributions with different statistics, one for updrafts and the other for downdrafts. The height dependence of the p.d.f. parameters is characterized by specifying the normalized variance and skewness of the vertical velocity as functions of the dimensionless height z/h on the basis of experimental data. The Markov-chain expressions for the vertical velocity of particles in updraft and those in downdraft are similar but with different statistics (mean, standard deviation, and autocorrelation coefficient). At the time of release, each particle is assigned a positive or a negative phase depending on whether the particle is released in an updraft or a downdraft, according to its release height, while a fixed ratio of the number of particles released in updrafts to that in downdrafts is maintained. Particles are assumed to be reflected perfectly when they hit the top or the bottom of the CBL. The overall Lagrangian time scale of vertical velocity is also expressed as the weighted sum of two time scales, one for particles in updrafts and one for those in downdrafts. The probability of a particle reversing its phase at each time step is simply expressed in terms of the lifetime of convective eddies (updrafts and downdrafts). Corrections for the effects of vertical gradients in variance and mean velocity in updrafts and downdrafts are also used. With only one adjustable parameter in their random-walk model, Baerentsen and Berkowicz (1984) showed that the calculated dimensionless cross-wind-integrated concentrations, mean plume height, and the vertical-dispersion parameter as functions of normalized downwind distance compare well with the corresponding results from Willis and Deardorff's convection tank experiments.

Similar random-walk models of dispersion in the CBL using non-Gaussian random forcing have also been proposed by de Bass et al. (1986) and Sawford and Guest (1987). However, Thomson (1987) showed that these models do not satisfy some basic physical requirements such as the well-mixed condition, which requires that if particles of a passive tracer are initially well mixed in the CBL they must remain so at subsequent times. This is a general condition that leads to the fulfillment of the other criteria discussed by Thomson (1987). It has also been shown that the random forcing in a random-walk model must be Gaussian.

Luhar and Britter (1989) have developed a nonlinear random-walk model with Gaussian random forcing based on the approach of Thomson (1987). Thomson developed a random-walk model of dispersion in the free convective surface layer, using a skewed PDF of vertical velocities. Luhar and Britter (1989) extended his approach to the whole CBL using the skewed PDF of Baerentsen and Berkowicz (1984), which was constructed from two Gaussian distributions. They also verified the validity of the closure assumptions used by the former investigators and derived the Markov-chain equation for the vertical velocity which satisfies the well-mixed condition. Their model simulation results showed excellent agreement with those from Willis and Deardorff's (1976, 1978, 1981) convection tank experiments.

Hurley and Physick (1993) investigated the effects of certain simplifying assumptions that permit the use of large time steps in random-walk simulation of dispersion in the CBL. Thomson (1987) derived the restrictive conditions on the time step Δt needed to satisfy the well-mixed condition. Inhomogeneity of turbulence near the lower and upper boundaries generally requires small time steps. This restriction can be eased significantly if turbulence is assumed to be homogeneous over the whole depth of the CBL. Hurley and Physick (1993) found the assumption of homogeneous turbulence quite adequate, provided one uses a more realistic PDF for the vertical velocity.

Luhar et al. (1996) have made a comparison of the several closure assumptions that have been used with the bi-Gaussian form of the PDF of vertical velocity. After pointing out the shortcomings of these closure assumptions, they proposed a new closure using new laboratory measurements of PDF in the CBL. The proposed PDF model has the desirable property of reducing to a simple Gaussian form in the limit of zero skewness. Significant differences are found between the values of the PDF parameters obtained from the various closures and the saline water tank data of Hibberd and Sawford (1994). The results of the random-walk model with the new PDF showed excellent agreement with experimental data.

Dispersion in the Stable Boundary Layer and Katabatic Flows Compared with their use for the neutral and convective boundary layers, random-walk models, along with other aspects of turbulence and dispersion, are not so well developed for the stable boundary layer (SBL). Luhar and Rao (1993) developed and tested three different formulations of dispersion in stably stratified katabatic flows in which turbulence was assumed to be inhomogeneous with a Gaussian PDF. The parameters influencing dispersion in a katabatic flow are the layer thickness, mean velocities, and turbulence, all of which vary with downwind distance. Luhar and Rao (1993) used the two-dimensional katabatic flow model of Nappo and Rao (1987), which is based on the TKE closure. Their simplest random-walk formulation assumes that particle velocities are not correlated at all over the duration of the time step. This leads to a pure random-walk model in which the increment in particle height $Z(t)$ is expressed in terms of the time step, the height-dependent vertical-eddy diffusivity, the mean Eulerian vertical velocity, and a random forcing function. Another more accurate and also more complex formulation is based on the Langevin equation and is similar to that proposed by Thomson (1987). Only minor differences in the predictions of dispersion parameters by different model formulations are found, so that one can as well use the simpler formulation for simulating dispersion in stable flows. Except for the peak concentrations, which were generally well predicted by the model, the simulated concentrations differed considerably from the observed concentrations as a function of time after a finite-duration release. At all the sampling sites, predicted concentrations decayed much more rapidly than the observed concentrations following their respective peak values. This discrepancy is attributed to pooling of the drainage air in the valley basin that could not be simulated in the two-dimensional katabatic flow model.

Dispersion in Mesoscale and Complex Terrain Flows Random-walk Lagrangian transport and dispersion models have been combined with mesoscale atmospheric models to simulate dispersion in mesoscale and complex terrain flows (see e.g., Etling et al., 1986; Gross et al., 1987; McNider et al., 1988; Segal et al., 1988; Yamada and Bunker, 1988; Andren, 1990; Physick and Abbs, 1991; Pielke et al., 1992; Physick et al., 1993; Uliasz, 1994; Boybeyi et al., 1995; Moran and Pielke, 1996a, 1996b). These models are typically applied over domain sizes of 50 to 500 km, with a horizontal resolution of 1 to 10 km. Particle velocity consists of two components: (1) the resolved mean velocity from the mesoscale model and (2) the turbulent fluctuating velocity from the Lagrangian stochastic model.

The mesoscale model provides the various parameters needed in the dispersion model, such as the mixing height, the surface fluxes of momentum, and heat, and in some cases the turbulence kinetic energy. From these, turbulence variables (σ_u, σ_v, σ_w, T_L, etc.) are estimated from diagnostic relations based on empirical similarity relations. The widespread use of diagnostic similarity relations developed from similarity theories and experimental data for the horizontally homogeneous atmospheric boundary layer in complex terrain flows is a questionable practice. But there is no alternative choice in the absence of systematic experimental studies of turbulence and diffusion in such flows.

Three-dimensional mesoscale Lagrangian dispersion models generally require large computer resources including supercomputers and work stations, because a small grid size is needed to adequately resolve complex flows around individual topographical features and a large number of particles must be released to simulate dispersion from each source. One way of reducing the computing resources needed for running such a model is to maximize the time step used for simulating particle trajectories. For this it may be necessary to simplify the Lagrangian dispersion model. For example, the assumption of local homogeneity of turbulence in all three directions permits the use of a long time step in the CSIRO model (Physick et al., 1993). Hurley and Physick (1993) have shown that neglecting inhomogeneity of turbulence in the CBL is a reasonable approximation, but the skewness of the PDF cannot be neglected.

In spite of the above mentioned limitations, the random-walk modeling technique is particularly suited to study and simulate dispersion in complex terrain flows where other dispersion theories and models, such as the gradient transport, statistical, and similarity theories; higher order closure models; and large-eddy simulation are inappropriate, impractical, or invalid. Successful applications of the Lagrangian stochastic models and their validation against the recent data from mesoscale tracer field experiments have been reported in the literature (Eastman et al., 1995; Moran and Pielke, 1996a, 1996b).

In general, the Lagrangian stochastic modeling approach is soundly based in scientific principle, and provides an excellent description of one-particle or absolute dispersion in the PBL, as judged by the agreement with field observations (Wilson and Sawford, 1996). It is a simple and natural model of the dispersion process that utilizes the available information on a turbulent flow. It is also a flexible technique for simulating ensemble-averaged dispersion and calculating puff or plume dispersion parameters. Random-walk dispersion models do not suffer from numerical problems, such as grid resolution, artificial diffusion, and negative concentrations, that are frequently encountered by gradient transport and higher order closure models. Such models are commonly used, along with mesoscale flow models, to simulate dispersion in complex terrain and other complex flow situations. In contrast, Eulerian closure models, irrespective of the level of closure, are fundamentally flawed due to the need for assumptions on the joint concentration-velocity statistics, especially for the description of the near-source dispersion. They are also burdened by serious problems of inadequate grid resolution and numerical diffusion, in addition to the more fundamental closure problem.

Many interesting challenges still remain in further development of Lagrangian stochastic models for two-particle, relative dispersion that can describe concentration fluctuation statistics, the dispersion of heavy particles and buoyant gases, dispersion in canopy layers, and dispersion in very stable flows with sporadic or intermittent generation of turbulence (Wilson and Sawford, 1996).

PROBLEMS AND EXERCISES

1. Compare the merits and limitations of the analytical gradient transport models with those of the numerical K-theory models of dispersion in the PBL.

2. How would you parameterize the vertical-eddy diffusivity for numerical modeling of dispersion in (a) the surface layer and (b) the PBL under different stability conditions. Suggest some specific expressions for K_z with appropriate experimental or theoretical justifications.

3. Qualitatively discuss the effect of wind-direction shear on the lateral dispersion of a continuous point source plume in the stable boundary layer. How would you study this using a numerical K-theory model?

4. a. Write down the Navier–Stokes equations of instantaneous motion u_i in the atmospheric boundary layer in the Cartesian coordinate system fixed to the earth's surface, using the usual Boussinesq assumptions or approximations.

   b. Expressing $u_i = \bar{u}_i + u_i'$ and using Reynolds' averaging rules, derive the equations for the mean and fluctuating motions.

5. a. Starting from the equation for u_i' derived in Prob. 4(b), derive the TKE equation and simplify it by neglecting the viscous diffusion terms in comparison with turbulent diffusion or transport terms.

   b. Simplify the TKE equation for the stationary, horizontally homogeneous PBL and discuss the significance of the various terms in the same.

6. a. Starting with the equation for the concentration fluctuation c', derive an equation for the variance of concentration fluctuations in a turbulent atmosphere.

   b. Simplify the above variance equation for the horizontally homogeneous PBL, keeping in mind that the scalar field may not be horizontally homogeneous.

7. a. Utilizing the equations for u_i' and c', derive the dynamical equation for the covariance $\overline{u_i'c'}$.

   b. Discuss the physical significance of the various terms in the $\overline{u_i'c'}$ equation and simplify it for the horizontally homogeneous PBL.

8. a. Define the flux Richardson number Rf as the ratio of buoyancy destruction to shear production terms in the TKE equation (11.10) for a stably stratified flow.

   b. Using the gradient transport relations for turbulent fluxes show that $Rf = Ri K_h/K_m$.

9. a. Using the surface-layer similarity relations for E, ε, K_m, and $\ell_\varepsilon = \ell$ under neutral stability condition, estimate the values of constants C_ε, C, and C_k in the parametric relations (11.14) through (11.16) used in TKE-closure models.

   b. Examine the sensitivity of the above estimated values of constants to the ratio E/u_*^2 which has been observed to range from 4 to 6 in the neutral surface layers of laboratory and atmospheric flows.

10. a. What are the merits and limitations of the TKE closure approach over the first-order closure approach to modeling turbulence and dispersion?

    b. What type of turbulence closure model would you use for modeling dispersion in a hilly terrain and why?

11. a. Write down the general parametric relation for the third moments $\overline{u_i'u_k'c'}$ using either of the gradient transport approaches.

    b. Discuss the limitations of the gradient transport hypothesis for parameterizing third moments in the PBL.

12. a. How would you parameterize the molecular dissipation terms in the variance and covariance equations?

    b. Discuss the usefulness and applications of Kolmogorov's local isotropy hypothesis in higher order closure modeling.

13. a. Starting from the equation for u_i', derive the Poisson equation for p'.

    b. Show that pressure fluctuations at any point in a turbulent flow depend on the velocity and temperature fields in a large neighborhood around that point.

14. Discuss the merits and limitations of the second-order closure approach and compare it with the first-order closure or gradient transport model.

15. a. Distinguish between the subgrid scale closure models used in large-eddy simulation and the similar full turbulence closure models.

    b. How would you select optimum grid spacings in a LES?

16. a. Starting from the instantaneous equations of motion u_i, derive the grid-volume-averaged equations, using an appropriate spatial filter, for use in LES.

    b. Distinguish between the Reynolds stresses appearing in the ensemble-averaged and the grid-volume-averaged equations.

17. Discuss the merits as well as limitations of the Smagorinski SGS closure model.

18. a. How would you include the buoyancy effect on the SGS eddy diffusivity in the stable boundary layer?

    b. Derive Eq. (11.29) from the subgrid TKE equation after using appropriate assumptions and parameterizations for ε and K.

19. a. Explain why the constants C and C_ε in the parametric relations (11.35) and (11.36) for the subgrid K and ε might be different from those in the similar relations for the overall eddy diffusivity and the rate of energy dissipation.

    b. Discuss the possible effect of stable stratification on subgrid mixing length.

20. a. Discuss the limitations of SGS models in approaching the surface boundary.

b. Why is the TKE-closure approach more appropriate in parameterizing the effects of stable stratification and proximity to the surface? Which SGS model would you use in the large-eddy simulation of the stable boundary layer?

21. a. Derive an expression for the random velocity fluctuation appearing in the Markov-chain equation (11.46).

b. How would you specify the parameters a and σ_i in a random-walk model of the neutral surface layer?

22. a. Under what conditions does the Langevin equation (11.52) become equivalent to the Markov-chain equation (11.46)?

b. How would you consider the effects of the vertical inhomogeneity of turbulence velocity σ_w and Lagrangian length scale Λ_L in the random-walk model of dispersion in the free convective surface layer?

23. a. Discuss the various advantages of the random-walk modeling approach over other numerical models of atmospheric dispersion.

b. How would you specify the mean velocity field and turbulence variances for use in a random-walk model?

12

Urban and Regional Air Quality Models

12.1 INTRODUCTION

Historically, air pollution came to be regarded as a serious problem only for large cities and commercial centers. With the advent of the industrial revolution and later coming of the automobile, air quality of most large urban and industrial areas took a nosedive. It is no wonder that early efforts in air pollution control were primarily directed toward improvement of air quality in those areas. For implementing efficient control measures and strategies, various urban air quality models have been developed. These differ from other local or short-range dispersion models mainly in the variety and number of sources that are distributed over the whole urban area and also in that they deal with the peculiar effects of nonhomogeneous urban boundary layer (UBL) and urban heat island on dispersion.

Urban air pollution in the form of urban plumes is transported by atmospheric winds to rural, pristine, and wilderness areas far away from its source region. In addition, many rural and agricultural sources as well as natural sources located within and outside the region contribute to the regional air pollution. Transport distances and travel times involved for these pollutant species are large enough for some chemical and physical transformations to take place and for dry and wet removal processes to occur as pollutants are mixed through increasing depths of the troposphere. The terrain over any large region is generally nonuniform in its roughness and thermal characteristics; it may also contain complex topography. Consequently, meteorological parameters affecting transport, diffusion, transformation, and removal processes are functions of both time and space. Therefore, a regional air quality model must include detailed information on all the anthropogenic and natural emissions (estimated for each horizontal grid area), advected pollutants, land use and topography, and regional meteorology including dispersion characteristics. In this chapter we will discuss the various components of urban and regional air quality models, review different types of air quality models, and give some examples of mathematical models that are used to simulate and study urban and regional air quality.

12.2 COMPONENTS OF AN AIR QUALITY MODEL

The various components of a comprehensive urban airshed or regional air quality model are schematically shown in Figure 12.1. A simpler analytical urban air quality model for a particular pollutant species may contain only a few of these, while a regional photochemical oxidant or acidic precipitation model would have most, if not all, of the components depicted in Figure 12.1.

12.2.1 Sources and Emissions Inventory

Sources of air pollutants emitted in the particular urban or regional area as well as those advected by mean winds from outside the area constitute the most important component. A comprehensive emissions inventory of anthropogenic and natural sources is necessary.

An urban area contains thousands, or even millions, of individual sources that range from small process emission sources, such as incinerators and backyard grills, to large sources, such as fossil-fuel power plants and other industrial facilities. The application of a dispersion model to each of these sources is impractical. Even the basic assumption implied in such an approach, that contributions of individual sources to the total concentration at a receptor point are additive, is

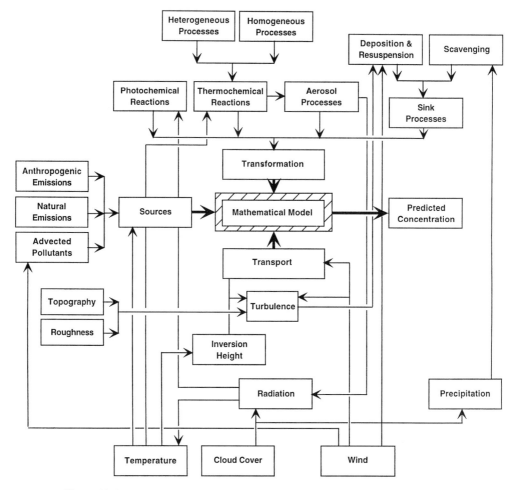

Figure 12.1 Components of a comprehensive air quality model. From Seinfeld, 1986.

questionable when pollutants from different sources may react with each other. Consequently, emissions from all the small sources in each grid area are combined together into a larger area source, assuming that emissions are uniform over that particular area. Since vehicular emissions constitute a major portion of urban area sources, their effective height of release is assumed to be the ground level. Large stack emissions are usually considered as point sources whose effective release heights are calculated from the given source and meteorological parameters, using appropriate plume-rise formulas. Dispersion from each of these major point sources is calculated and the resulting concentrations at a receptor point are added to the contribution from area sources. The number of point sources treated this way may vary from tens to hundreds, depending on the size and industrial character of the urban area. An accurate emissions inventory, including a knowledge of the diurnal, weekly, and seasonal variations of emissions, is essential for successful urban air quality modeling.

An emissions inventory is a list of the amount of pollutants emitted by all the sources within the fixed boundaries of a particular area in a given time period. It is usually not practical and economically feasible to monitor continually every emission source in the area. It is also not feasible to monitor continually the pollutant concentrations in ambient air at many points and apply the appropriate diffusion equations to calculate the emissions. In practice, the method used to develop the emission inventory utilizes actual monitoring of emissions from major sources and estimating of emissions from smaller sources scattered over the area from emission factors developed for each source type. To develop an emission inventory for an area, one must (1) list the types of sources, such as automobiles, home fireplaces, and so on, (2) determine the types of air pollutants emitted from each source type, such as SO_2, CO, and particulates, (3) find the appropriate emission factors (see e.g., U.S. Environmental Protection Agency, 1973–1983) for each of the pollutants of concern, and (4) estimate the number and size of specific sources in the area (Stern et al., 1984). Valid emission factors for each source of air pollution are the key to an accurate emission inventory. Actual emission data are available from many handbooks, government publications, and journal articles. Emission factors must be critically examined to make sure that the tests from which they were obtained are representative of the actual (prevailing) operational conditions.

Data on emissions are available from many sources including city, county, and state planning and industrial development commissions; chambers of commerce; census bureaus; local and national associations of manufacturers or distributors; individual dealers or distributors of coal, oil, gasoline, and so on; local utility companies; insurance companies; air pollution control agencies; and state transportation departments.

A comprehensive inventory of anthropogenic emissions in the eastern half of the United States was prepared for the 1985 National Acid Precipitation Assessment Program (NAPAP) (Saeger et al., 1989). Major point, area, and mobile source emissions in the inventory varied according to the day of the week (Saturday, Sunday, or weekday). Mobile source emissions were adjusted for daily average temperature and diurnal temperature range. Some stationary sources of volatile organic compounds were also adjusted for the daily average temperature. The anthropogenic emissions from the 1985 NAPAP inventory have also been projected to the year 2005 using emissions' category-specific growth factors and existing emission controls and federal and state regulations (Roselle, 1991; Roselle et al., 1991).

Natural emissions include the biogenic emissions of hydrocarbons as well as nitrogen- and sulfur-containing compounds. Biogenic emissions constitute the most uncertain component of emission inventory for use in regional air quality models. For the United States, Pierce et al. (1990) developed the Biogenic Emissions Inventory System (BEIS) from the biogenic emission inventories of Lamb et al. (1987) and others. BEIS gives estimates of isoprene, α-pinene, monoterpenes, and unidentified hydrocarbons from trees and crops and also NO_x emissions from grassland soils. Emission rates were determined from land-use data; measured or estimated emission fluxes, standardized to 30°C and full sunlight, for twenty-five different varieties of trees and plants; and environmental correction factors. A simple forest canopy model is used to obtain vertical profiles of sunlight intensity and leaf surface temperature, which are then used to determine environmental correction factors. There are large uncertainties involved in the various components of BEIS. In particular, biogenic nonmethane hydrocarbon (NMHC) emissions estimated from BEIS may have an uncertainty factor of up to 3.

12.2.2 Transport Winds

All air quality models require information on mean transport winds. In the simpler urban air quality models based on Gaussian plume-dispersion parameterization, the mean transport wind is assumed to be uniform in both the horizontal and vertical directions within the urban boundary layer. It is usually specified on the basis of meteorological observations at a nearby airport or from a mast or tower within the urban area. An appropriate power-law wind profile can be used in conjunction with the observed data at some reference height to estimate the transport wind at any desired height of release or the layer-averaged transport wind. If meteorological data are available from different sites spread over an urban area, different transport winds (speeds and directions) can be assigned to different grid squares through an objective interpolation scheme. A more sophisticated dispersion model would be more appropriate with the spatially variable wind field. With only limited meteorological data available from a few sites in a complex urban terrain, a suitable mesoscale model of the urban heat island and boundary layer would be more appropriate and desirable (see e.g., Seaman et al., 1989; Byun and Arya, 1990). The available near-surface observations and upper-air data can be used to specify initial and boundary conditions in the model. Such a mesoscale model of the urban boundary layer can also give more realistic spatially varying mixing height and turbulence (fluxes and velocity variances) as necessary inputs to a sophisticated urban air quality model.

In regional air quality models, transport winds are necessarily considered spatially variable and have to be specified at all grid points at each time step. With a large domain size and horizontal grid spacings and time steps that are commonly used, both the mesoscale and synoptic variations of transport winds are accounted for in these models. Some long-term averaged regional air quality models utilize only the climatological data on wind speed and wind direction at representative surface and upper-air stations in the region. Most other models, especially those directed to short-term calculations or predictions, use wind observations of the period considered. In some of these, winds at the 850-mb level (about 1500 m height), which are routinely available from upper-air soundings every 6 or 12 hours, are used. Several models require vertical wind profiles derived from upper-air wind observations or from surface and tower observations as input (Zwerver and van Ham, 1985).

Ideally, the prescribed wind field in a mesoscale or regional model should be divergence free. If the wind field is not divergence free, unwanted loss or gain of pollutant mass may occur. Also numerical problems may arise when nondivergent wind field is used. For example, in a region of strong convergence very sharp gradients in pollutant concentration may occur, which is likely to result in large numerical errors. However, simple procedures or techniques have been developed to transform a divergent wind field into a divergence-free field (see e.g., Sherman, 1978).

Regional air quality models utilizing observed winds from surface and upper-air synoptic stations cannot be used to forecast air quality of next day or so because meteorological fields are not updated from a weather-forecasting model. The coarse resolution of the synoptic network also does not permit the influence of mesoscale circulation systems to be taken into account. Ideally, a regional air quality model should have input from another meteorological model of the same or larger domain. But simultaneous running of meteorological and air quality models for forecasting purposes would require huge computational resources. The Regional Acid Deposition Model (RADM), which has been used to calculate episodic pollutant concentrations and dry and wet deposition of acids in North America, utilizes the meteorological fields computed by a mesoscale meteorological model (Chang et al., 1987). A more comprehensive regional modeling system under development by U.S. Environmental Protection Agency will also have the flexibility of a mesoscale meteorological model providing the necessary meteorological input to the air quality model (Ching et al., 1995; Byun et al., 1996; Dennis et al., 1996).

12.2.3 Diffusion Parameterization

In simpler urban air quality models, the parameterization of horizontal and vertical diffusion is usually based on the empirical dispersion parameters obtained from tracer experiments over cities. In the United States, the plume-dispersion parameters derived from the St. Louis Dispersion Study by McElroy and Pooler (1968), or Briggs' (1973) interpolation formulas for σ_y and σ_z based on the same data, are most widely used as recommended in the U.S. EPA's Guideline on Air Quality Models (U.S. Environmental Protection Agency, 1986, 1996). In more sophisticated numerical grid models, such as the Urban Airshed Model (UAM), horizontal and vertical-eddy diffusivities in the mean diffusion equations for pollutant species are specified. Horizontal diffusivity ($K_x = K_y$) is set to a user-specified constant value (nominally 50 m² s⁻¹). Vertical-eddy diffusivity is specified using mixed-layer and the PBL similarity relations.

Most of the regional dispersion and air quality models use a deterministic description of atmospheric transport and diffusion as represented by the mean advection-diffusion equation for any species i

$$\frac{\partial \overline{c}_i}{\partial t} + \overline{u}\frac{\partial \overline{c}_i}{\partial x} + \overline{v}\frac{\partial \overline{c}_i}{\partial y} + \overline{w}\frac{\partial \overline{c}_i}{\partial z}$$
$$= \frac{\partial}{\partial x}\left(K_x \frac{\partial \overline{c}_i}{\partial x}\right) + \frac{\partial}{\partial y}\left(K_y \frac{\partial \overline{c}_i}{\partial y}\right)$$
$$+ \frac{\partial}{\partial z}\left(K_z \frac{\partial \overline{c}_i}{\partial z}\right) + R_i + S_i \quad (12.1)$$

where R_i and S_i represent chemical reaction and source/sink terms, respectively. In this equation, turbulent diffusion is represented by the eddy diffusivities K_x, K_y, and K_z, which have to be specified or parameterized. Several regional models use the above equation to represent mean transports and diffusion in all directions. In most cases, however, simplifying assumptions are made.

In a region with essentially flat terrain, often the mean vertical velocity is small enough to be neglected. When observed (interpolated) winds are used, these are made divergence free in order to conserve pollutant mass and avoid numerical problems associated with divergent or convergent flow fields. Also the simplifying assumption that horizontal diffusion can be neglected in comparison with horizontal transport can be justified in some cases. However, this assumption appears to be more drastic than the slender-plume approximation used in short-range dispersion theories and models in which only the longitudinal diffusion in the mean flow direction is neglected, while the lateral diffusion normal to the plume axis is considered as important as the vertical diffusion. The neglect of horizontal diffusion in regional models with horizontal grid sizes of more than 10 km can be justified because horizontal diffusion terms are typically an order of magnitude smaller than horizontal advection and certainly smaller than the spurious numerical diffusion introduced by commonly used finite-difference schemes in the numerical integration of the advection equations (McRae et al., 1982; Chang et al., 1987). Then, Eq. (12.1) simplifies to

$$\frac{\partial \overline{c}_i}{\partial t} + \overline{u}\frac{\partial \overline{c}_i}{\partial x} + \overline{v}\frac{\partial \overline{c}_i}{\partial y}$$
$$= \frac{\partial}{\partial z}\left(K_z \frac{\partial \overline{c}_i}{\partial z}\right) + R_i + S_i \quad (12.2)$$

In a mixed Eulerian-Lagrangian (hybrid) formulation used in some models, Eq. (12.2) is split into two equations—one representing the rate of change of concentration due to advection only and the other representing the effect of turbulent diffusion (see e.g., Zwerver and Ham, 1985). The former is solved by determining Lagrangian trajectories. If the vertical wind shear can also be neglected, the transport can be represented as a column of air that travels along trajectories. The vertical diffusion in the moving column is represented by the Eulerian diffusion equation

$$\frac{\partial \overline{c}_i}{\partial t} = \frac{\partial}{\partial z}\left(K_z \frac{\partial \overline{c}_i}{\partial z}\right) \quad (12.3)$$

in which the eddy diffusivity K_z is continuously adapted to the existing conditions along the trajectory. The vertical-eddy diffusivity may be parameterized by using the mixed-layer and PBL similarity relations, similar to those used in the Urban Airshed Model, or other empirical expressions (see e.g., Louis, 1979; Holt and Raman, 1988). Horizontal-eddy diffusivity, if not ignored, is usually assigned a constant value, more or less arbitrarily. Several regional models use empirical Gaussian dispersion coefficients, even though their validity is generally limited to distances of 50 km or less. In some Lagrangian trajectory models, uniform mixing is assumed in the vertical up to a specified mixing height, so that K_z becomes irrelevant.

12.2.4 Chemical Transformations

Transport distances and travel times involved for pollutants even over large cities are usually not long enough for most of the pollutant species to undergo substantial chemical transformations. Consequently, such transformations are usually ignored in urban air quality models, except for those specifically dealing with the photochemistry of ozone and smog formation over large urban complexes during warm (usually summer) sunny days. For example, in most of the preferred air quality models in the U.S. EPA's Guideline (U.S. Environmental Protection Agency, 1996), chemical transformations are very simply treated using either a linear or an exponential decay in which the decay rate or the half-life of the pollutant species is specified as input to the model. Obvious exceptions are the various photochemical box models and urban airshed models that employ sophisticated chemistry submodels, including both the kinetics and mechanisms of the reactions converting primary pollutants into secondary pollutants (see e.g., Finlayson-Pitts and Pitts, 1986, Ch. 9).

Our basic knowledge of tropospheric chemistry increased rapidly in the late 1960s and early 1970s, following the discovery of the importance of OH radical as well as the development of many specific, sensitive, and accurate techniques for studying reaction kinetics and mechanisms. A number of explicit chemical mechanisms, including all reactions between primary and secondary pollutant species as well as their reactive intermediate products, have been proposed (see e.g., Leone and Seinfeld, 1985; Seinfeld, 1986). However, these involve hundreds of reactions and corresponding rate constants. Many of the rate constants and intermediate and final products of the individual reactions are not known. Even if these were known and could be specified, the amount of computer time required for numerical integration of the rate equations associated with each of the hundreds of individual hydrocarbons found in the ambient air would be prohibitive. For this reason, explicit chemical mechanisms are not directly utilized in urban air quality models. However, they have been used indirectly in the development of more practical condensed or lumped chemical mechanisms for use in photochemical urban air quality models, such as the urban airshed models.

The lumped chemical mechanisms are used to reduce the numbers of reactions and species, so that they can be reasonably handled in the model simultaneously with other components (e.g., emissions, meteorology, etc.). The inorganic chemistry is usually treated in terms of explicit reactions because a relatively small number of inorganic species and reactions are involved.

The organic chemistry involving nonmethane hydrocarbons (NMHC) is treated by lumping together a number of reactions and chemical species. For example, in the most commonly used carbon bond mechanism (CBM), the reactions are lumped not in terms of the usual classification of organics, but in terms of the bonding of the atoms in reactants (Whitten et al., 1980). The CBM-IV developed by Gery et al. (1989), which is currently used in the U.S. EPA's (1996) Urban Airshed Model and other air quality models, accommodates thirty-four species and eighty-two reactions. In some models, a combination of an explicit mechanism with a relatively few "lumped" reactions is used (Finlayson-Pitts and Pitts, 1986).

Chemical mechanisms or submodels are usually tested and validated against experimental data obtained in smog chambers. There are a number of adjustable parameters involved, even in explicit chemical mechanisms, which reflects uncertainties either in the specific reactions or in the smog chamber experiments. The modeler must exercise sufficient care both in the interpretation of the experimental data and in the use of adjustable parameters in any chemical submodel.

Most of the regional air quality models deal with sulfur and nitrogen oxides and their reaction products sulphates, nitrates, and sulfuric and nitric acids. In industrial regions, anthropogenic emissions dominate over the natural emissions and the latter are often neglected. The anthropogenic contribution to NO_x emissions is primarily NO (about 85%) and, to a lesser extent, NO_2. These gases are chemically more reactive than SO_2. The chemical transformation of NO to NO_2 depends on the ozone concentration, while the reverse transformation of NO_2 to NO is governed by a photolytic process that strongly depends on the intensity of solar radiation. Photochemical models are used to describe these transformations, resulting in many photochemical products, especially in the presence of reactive anthropogenic and natural hydrocarbons or VOCs. The same chemical kinetic mechanisms as discussed earlier in the context of urban air quality models are also used in regional photochemical models. Again, the carbon bond mechanism is probably the most widely used mechanism.

Earlier regional models considered only the gas-phase oxidation of SO_2 and/or NO_2 by the hydroxyl (OH) radical. Such a treatment has been found to oversimplify the influence of NO_x and VOCs on SO_2 and NO_2 oxidation rates. Much more sophisticated chemical kinetic mechanisms have been used in three-dimensional numerical regional models (see e.g., Lamb, 1985; Chang et al., 1987; Young et al., 1989). For example, the CBM-IV is used in the revised version of the U.S. EPA's Regional Oxidant Model (ROM), whereas a hybrid or combination mechanism (Stockwell, 1986) containing seventy-seven reactions among thirty-six species is utilized in the Regional Acid Deposition Model (RADM). These simulate reactions among nitrogen oxides, sulfur oxides, HO radical, hydrogen peroxide, organic peroxides, ozone, and numerous other species. For details of the chemical reactions and transformations the reader should refer to the original publications and books on atmospheric chemistry (e.g., Whitten et al., 1980; Leone and Seinfeld, 1985; Finlayson-Pitts and Pitts, 1986; Seinfeld, 1986; Stockwell, 1986; Singh, 1995).

The formation of sulfuric and nitric acids, which are the primary components of acidic deposition, involves many homogeneous gas phase and aqueous phase reactions, as well as heterogeneous reactions on the surfaces of solid particles (see e.g., Finlayson-Pitts and Pitts, 1986, Ch. 11). In power-plant plumes, the observed rates of oxidation of SO_2 are typically less than 10 percent per hour. But much higher rates are often observed when the plume passes through a cloud or fog bank. Rates of conversion appear to be higher at noon than at nighttime and generally higher in summer compared with the winter season. These diurnal and seasonal effects indicate the importance of photochemistry and, perhaps, temperature, in the oxidation of SO_2 (Finlayson-Pitts and Pitts, 1986). The rates of oxidation of SO_2, both in plumes and in ambient air, are generally higher when hydrocarbons, nitrogen oxides, and the host of secondary pollutants they produce are also present. The presence of liquid water in aerosols, fog, and clouds is also an important factor in determining the overall rate of conversion of SO_2 into sulfates including H_2SO_4. The possible reactions involved in the gas phase, liquid phase, and on aerosol surfaces, which may contribute to the observed rates of conversion of SO_2, are reviewed elsewhere (Finlayson-Pitts and Pitts, 1986; Seinfeld, 1986).

The oxidation of NO_x in power-plant plumes and in ambient air into nitrates and nitric acid has not received as much attention with regard to acidic deposition, because SO_2 is considered to play a dominant role in most regions of concern in Europe and the northeast United States. However, in power-plant plumes, rates of conversion of NO_x between 0.2 and 12 percent per hour have been observed. These have similar diurnal and seasonal variations as SO_2 conversion rates. Conversion rates in the so-called urban plumes, that is, the ambient air that has passed over urban areas, are also high (up to 24% per hour). Most of the NO_x is converted to HNO_3 and PAN; the various homogeneous gas phase and aqueous phase reactions involved in conversion are reviewed by Finlayson-Pitts and Pitts (1986). Which phase is most important depends on the particular meteorological conditions, such as temperature, relative humidity, intensity of solar radiation, and presence of clouds.

Chemical submodels used in long-range transport and regional acid deposition models range from overly simplistic models using only one lumped reaction representing the conversion of SO_2 into H_2SO_4 with an arbitrarily specified conversion rate constant to highly sophisticated carbon bond mechanism and other chemical kinetic mechanisms (see e.g., Zwerver and van Ham, 1985; Finlayson-Pitts and Pitts, 1986; Chang et al., 1987). Rodhe et al. (1981) first incorporated the simplified reactive hydrocarbon and NO_x chemistry in their long-range transport model for sulfuric acid formation and deposition. Since then, a number of long-range transport and regional acidic deposition models using sophisticated chemistry submodels have been proposed (Seigneur and Saxena, 1984; Venkatram et al., 1984; Chang et al., 1987). For example, the chemistry submodel of Seigneur and Saxena (1984) includes eighty gas phase reactions, twenty-seven liquid phase reactions, and twelve equilibria describing the gas-liquid partitioning.

12.2.5 Removal Processes

Due to the limited transport distances and travel times involved in urban air quality models, removal of pollutants by dry and wet deposition processes is usually ignored. In some models, dry deposition of particulate material including sulphate aerosols and some gases like SO_2 and NO_x is considered. Both dry and wet removal processes become extremely important in long-range transport and regional air quality and must be incorporated in numerical models of the same.

Dry Deposition Dry deposition is included by using the concept of deposition velocity v_d introduced earlier in Chapter 10. The net downward flux of depositing material is expressed as the product of the deposition velocity and the average concentration at some low reference height of the order of 1 m. The deposition velocity in principle is dependent on the reference height, because mean concentration also depends on the height above the surface. The parameterization or specification of v_d is highly uncertain, however, because it depends on many factors including micrometeorological parameters (e.g., stability, u_*, etc.) and properties of the surface and the depositing material. Most models use a constant value for the deposition velocity of a species that is assumed to be representative of the whole modeling region. In some models v_d is simply taken proportional to the friction velocity, whereas more surface and meteorological characteristics, such as roughness, stability, and wind speed, are included in the specification of v_d in other models (see e.g., Zwerver and Ham, 1985; Seinfeld, 1986).

The most sophisticated parameterizations of v_d in regional air quality models are based on the resistance approach described in Chapter 10. The resulting expression (10.33) for v_d involves aerodynamic and viscous sublayer resistances, which are easily expressed in terms of the friction velocity, the roughness parameter, and stability. However, the surface and substrate transfer resistance, which represents the ability of the surface to uptake and absorb diffusing gas molecules or to capture particles impacting upon it, is the most uncertain component. It is not generally possible to specify this resistance in a completely objective and rational manner. Since deposition is dependent on physical, chemical, and biological characteristics of the surface, the surface resistance to transfer should in principle be determined for each type of the surface considered in a regional model. It may also show considerable variation with time due to changes in meteorological variables. In most regional air quality models, however, only average resistances or average deposition velocities are often used (see e.g., Zwerver and van Ham, 1985).

Measurements of dry deposition velocities for SO_2 and NO_x on different types of surfaces and estimated surface resistances for the same have been reviewed by van Aalst and Diederen (1985). The reported values of dry deposition velocity range from 0.1 to 2 cm s^{-1} for SO_2, from 0.001 to 0.1 cm s^{-1} for NO, and from 0.003 to 1.9 cm s^{-1} for NO_2. These cover a wide variety of surfaces, such as dry snow, water, soil, grass, crops, and forests. Some field studies of NO_x deposition to vegetation have also been conducted; these suggest deposition velocities ranging from <0.1 to 1.5 cm s^{-1}.

van Aalst and Diederen (1985) have also reviewed the limited data on dry deposition of sulphate and nitrate aerosols. The deposition velocity of aerosol particles depends on many variables, the most important of which are particle diameter and density, friction velocity, and surface characteristics. For the smallest particles ($d < 0.1$ μm). Brownian diffusion plays a dominant role; the deposition velocity decreases with increasing particle diameter. For large particles ($d > 20$ μm), gravitational settling becomes more important and dominant; deposition velocity increases with increasing particle diameter ($v_d \propto d^2$). For intermediate-size particles ($d = 0.1$–20 μm), inertial impaction on and interception by the surface also play a role. Deposition by impaction increases with particle size and wind speed, and it is also strongly influenced by surface structure. Several theoretical models of particulate deposition on water, snow, and vegetation surfaces have been formulated. However, validation of such models is often limited due to the lack of suitable field data.

From the strong dependence of deposition velocity on particle size, it is clear that size distribution of sulfate or nitrate aerosol will determine the range of v_d as well as its average value for the aerosol. For example, the deposition velocity of coarse sulphate particles in marine aerosol may be an order of magnitude larger than that of fine continental sulfate aerosol. Thus, even a small fraction of large particles in the sulfate aerosol may have large influence on the average deposition velocity. Reported values of the deposition velocity for sulfate aerosol range from <0.1 to 2 cm s^{-1}. For nitrate aerosol, estimates of v_d are much more uncertain because direct flux measurements are lacking.

Wet Deposition Wet deposition is defined as the composite removal process by which airborne pollutants (gases and particulates) are absorbed into precipitation elements (water droplets, ice particles, and snowflakes) and thus deposit onto the earth's surface during precipitation. This process includes the attachment of pollutants to cloud droplets by the various physical mechanisms during cloud formation, in-cloud scavenging (washout), and attachment to the falling hydrometeors (rainout). A number of different terms are used more or less synonymously with wet deposition, including precipitation scavenging, wet removal, washout, and rainout. However, washout is sometimes used to represent in-cloud scavenging, whereas rainout generally refers to below-cloud scavenging by falling precipitation.

A comprehensive review of theories, models, and measurements of precipitation scavenging of pollutant gases and particles has been given by Slinn (1984). A more condensed theoretical treatment is presented by Seinfeld (1986, Ch. 16). Here we give only a brief description of how precipitation scavenging and wet deposition are represented or parameterized in regional air quality models.

In many cases it is possible to represent the local rate of removal of gaseous or particulate pollutants by wet deposition as a first-order process, that is,

$$\text{Local removal rate} = \Lambda_w(z,t)\bar{c}(x,y,z,t) \quad (12.4)$$

where $\Lambda_w(z,t)$ is the local washout coefficient that has dimensions of $[T^{-1}]$ and is considered a function only of the height above the surface and time. This simple first-order representation can be used when the scavenging is irreversible, such as for aerosol particles and highly soluble gases (Seinfeld, 1986, Ch. 16). The wet deposition flux to the surface is the integral of wet removal from all volume elements aloft, that is,

$$F_w = \int_0^\infty \Lambda_w(z,t)\bar{c}(x,y,z,t)dz \quad (12.5)$$

From this one can define the wet deposition velocity

$$v_w = \frac{F_w}{\bar{c}_o} \quad (12.6)$$

where $\bar{c}_o(x,y,t)$ is the ground-level concentration of the gaseous or particulate pollutant being scavenged and deposited at the surface with precipitation.

When the material being scavenged is uniformly distributed vertically, for example, in convective mixed layer of depth h, then the wet deposition velocity is given by

$$v_w = \int_0^h \Lambda_w(z,t)dz = \overline{\Lambda}_w h \quad (12.7)$$

where $\overline{\Lambda}_w$ is the average value of the washout coefficient for the whole layer. For the typical values of $h = 1000$ m and $\overline{\Lambda}_w = 10^{-4}$ s^{-1}, $v_w = 0.1$ m s^{-1}.

For certain practical applications, the scavenging or washout ratio is defined as

$$w_r = \frac{\bar{c}_w}{\bar{c}_o} \quad (12.8)$$

where $\bar{c}_w$ is the mean concentration of the material in precipitation at the surface and $\bar{c}_o$ is near-surface concentration in air. The wet deposition flux can be expressed in terms of $\bar{c}_w$ and the intensity or rate of precipitation, p_r, that is,

$$F_w = \bar{c}_w p_r \qquad (12.9)$$

From equations (12.6) through (12.9), the relationship between the wet deposition velocity v_w and the washout ratio w_r is

$$v_w = w_r p_r \qquad (12.10)$$

Slin (1984) has estimated the order of magnitude of washout ratio $w_r \sim 10^6$ to 10^7. To calculate the rate of wet deposition it is necessary to determine the washout or scavenging coefficient Λ_w. For detailed calculations of precipitation scavenging rates of particles and gases and the corresponding scavenging coefficients using theoretical and mathematical models, the reader should refer to Slin (1984) and Seinfeld (1986). Generally, the model predictions are accurate to an order of magnitude or so only.

Quantitative information about precipitation scavenging obtained through field measurements also has large uncertainties. Many meteorological parameters and specific air pollution characteristics of the air parcel in which precipitation scavenging is studied influence the mechanisms and the rate of wet deposition. Systematic field studies must be conducted to isolate and understand the effects of different parameters. Field measurements are also required to validate model predictions. In most models of mesoscale and long-range transport and also in most field measurements of precipitation scavenging and wet deposition, simplified treatments of the wet deposition process are often used (see e.g., van Aalst and Diederen, 1985). Wet deposition or removal process is often parameterized as a first-order decay, with the decay constant depending on precipitation intensity. Alternatively, the wet deposition velocity is specified as in Eq. (12.10).

12.3 URBAN DIFFUSION AND AIR QUALITY MODELS

A variety of urban diffusion models are used to predict concentrations of primary pollutants resulting from urban emissions. Most urban air quality models also predict concentrations of secondary pollutants resulting from photochemical reactions in urban air during transport and dispersion. These range from simple empirical models to complex three-dimensional urban airshed models. Here we review only certain classes of mathematical models of urban diffusion and air quality.

12.3.1 Gaussian Diffusion Models

The basic formulation of the Gaussian plume-dispersion model for a continuous point source has been discussed in Chapter 9. Recommended empirical schemes for specifying the dispersion parameters over urban areas are also reviewed there. Simpler urban diffusion models apply the same Gaussian plume formulation to multiple point, area, and line sources of pollution in an urban area. Due to the distributed nature of these sources, however, a numerical model becomes necessary even when an analytical formulation is used for each individual point source or each grid element of the area source emissions.

Following Lucas' (1958) early study, a number of simple urban air pollution models, each based on the simple Gaussian point source diffusion model, were proposed in the sixties and early seventies (Turner, 1964; Gifford and Hanna, 1973). Differences are only in the details of how the area source summation is carried out and how various meteorological and dispersion parameters are included. Some models incorporate computations or analyses of temporally and spatially varying meteorological fields; other simpler models assume uniform winds and other meteorological parameters. Here we give a brief description of two simple models as examples of Gaussian urban diffusion models.

Turner (1964) proposed a Gaussian diffusion model for the city of Nashville, Tennessee, where SO_2 concentration measurements were made earlier at thirty-two sampling stations during a one-year study. He chose 24-hr concentrations for his model verification. An emission inventory of sulfur dioxide sources in mile-square areas of Nashville and surrounding environs was used to calculate source strengths for each square-mile area of the 17 mile by 16 mile rectangular domain at 2-hr intervals. All sources were assumed to have the same effective height of 20 m. Average wind direction, wind speed, and stability were evaluated from observations at a 10-m height at four sites within the urban area. In the model, wind velocity and stability were assumed to be uniform in both the horizontal and vertical dimensions and to vary temporally only at the 2-hr intervals. The Gaussian dispersion parameters σ_y and σ_z were specified as functions of travel time $t = x/\bar{u}$ and Turner's (1964) stability classes 1 through 5. These correspond closely to the Pasquill–Gifford dispersion curves. An exponential decay of SO_2 with a half-life of 4 hours was assumed to account for its chemical transformation and removal.

The computations of 2-hourly ground-level concentrations at each receptor (sampling) point in the model were facilitated by the calculation of relative concentrations $\bar{c}_o/Q$, representing the contributions of all the upwind sources to a given receptor point. For this it is convenient to use the well-known reciprocal theorem, which states that the relative g.l.c. at the receptor located at (0,0,0) from a source located at (x,y,H), where the x axis is oriented upwind, is the same as the relative g.l.c. at the receptor at $(x,y,0)$ from an elevated source at $(0,0,H)$, where the x axis is oriented downwind. The receptor has the same orientation and distance from the source in either case

(Gifford, 1959b). The orientation of the x axis is determined by the wind direction. Since the emission source-strength grid is fixed in the geographical (east-west and north-south) coordinates, it is easy to transform them into the along-wind and cross-wind coordinates for dispersion calculations (see e.g., Turner, 1964). The concentration at any receptor point can be obtained by repetitive multiplication of source strength and relative concentration and summation for all the upwind grid-area sources. These 2-hourly concentrations were further averaged to obtain 24-hourly average concentrations for comparison with the corresponding measurements at thirty-two sampling sites. Comparisons of model calculated and observed concentrations for thirty-five test periods showed that 58 percent of all calculated values were within ±10 ppb of the observed concentrations. Excluding zero values of both calculated and observed concentrations, 70 percent of the calculated values were within a factor of 2 of the observed values. But a general tendency toward overestimation by the model is noted, especially downwind of major sources.

The Atmospheric Turbulence and Diffusion Laboratory (ATDL) model described by Hanna (1971) and Gifford and Hanna (1973) is another example of Gaussian-based urban diffusion models. The ATDL model's formulation for area sources has been incorporated in the U.S. EPA's (1986, 1996) multiple-source urban air quality model RAM. The mean concentration at any receptor point where the origin of the coordinates is located is given by an integration over the upwind half-plane

$$\bar{c}_o = \int_0^\infty \int_{-\infty}^\infty \frac{Q_a}{\pi \bar{u} \sigma_y \sigma_z} \exp\left(-\frac{y^2}{2\sigma_y^2}\right) dy\, dx \quad (12.11)$$

where sources are assumed to be at the ground level. For further simplification, using a narrow plume hypothesis, the area source strength Q_a is assumed to be a function only of x. Then, Eq. (12.11) is simplified to

$$\bar{c}_o = \frac{(2/\pi)^{1/2}}{\bar{u}} \int_0^\infty \frac{Q_a}{\sigma_z} dx \quad (12.12)$$

in which the effective transport velocity $\bar{u}$ is assumed to be spatially uniform. Next, Eq. (12.12) is written in a form consistent with the typical square grid that is used in urban emission inventory. If the receptor point is located in the center of the grid square "0" with its area source strength Q_{a0}, and upwind grid squares $1, 2, \ldots, n$, spaced at intervals of Δx, have area source strengths $Q_{a1}, Q_{a2}, \ldots, Q_{an}$, the piecewise integration of Eq. (12.12) yields

$$\bar{c}_o = \frac{(2/\pi)^{1/2}}{\bar{u}} \left[\int_0^{\Delta x/2} \left(\frac{Q_{a0}}{\sigma_z}\right) dx \right.$$
$$\left. + \int_{\Delta x/2}^{3\Delta x/2} \left(\frac{Q_{a1}}{\sigma_z}\right) dx + \ldots \right] \quad (12.13)$$

To perform the above integrations analytically, the vertical dispersion parameter is assumed to have the power-law form

$$\sigma_z = ax^b \quad (12.14)$$

in which the parameters a and b are estimated from the urban dispersion curves for σ_z versus x for each stability class (Hanna et al., 1982). Substituting from Eq. (12.14) into (12.13) and carrying out the integrations, one obtains a simple analytical expression for the g.l.c.

$$\bar{c}_o = \frac{(2/\pi)^{1/2} (\Delta x/2)^{1-b}}{\bar{u} a (1 - b)} \left\{ Q_{a0} + \sum_{i=1}^n Q_{ai} \right.$$
$$\left. [(2i + 1)^{1-b} - (2i - 1)^{1-b}] \right\} \quad (12.15)$$

Generally, n is the number of grid squares necessary to reach the upwind edge of the urban area. Equation (12.15) is best valid for an averaging time of the order of 1 hour for which the assumed dispersion parameterization is applicable. Extension to longer averaging times is also easily made by following a simple procedure outlined by Hanna et al. (1982).

It has been noted from the applications of the ATDL model to several urban areas that the calculated g.l.c. at any receptor point was usually proportional to the emissions Q_{a0} in the grid square in which the receptor is located. The reason for this is that the distribution of urban area emission is quite smooth and the coefficients of the upwind source terms in Eq. (12.15) are generally small. For most applications, it may be sufficient to use the simpler version of the ATDL model, which is obtained by approximating the various Q_{ai} terms by Q_{a0} in Eq. (12.15):

$$\bar{c}_o = A \frac{Q_{a0}}{\bar{u}} \quad (12.16)$$

where

$$A = (2/\pi)^{1/2} \frac{[\Delta x (2n + 1)/2]^{1-b}}{a(1 - b)} \quad (12.17)$$

Note that the dimensionless coefficient A depends on the upwind distance from the receptor to the edge of city, $\Delta x (2n + 1)/2$, and stability. It increases with increasing distance, as well as with increasing stability. For distances between 5 and 20 km, the estimated values of A range between 50 and 800 (Hanna et al., 1982). This simple model has been verified against measured concentrations of particulate matter in several cities. However, a comparison of the observed and model-computed concentrations of SO_2 suggest considerable overestimation (by a factor of 4) by the model. This may be due to the fact that most of the sources of SO_2 in cities are elevated whereas the model assumes ground-level sources. To properly account for the elevated release, it is recommended that the coefficient A should be reduced by a factor of 4 (Hanna et al., 1982).

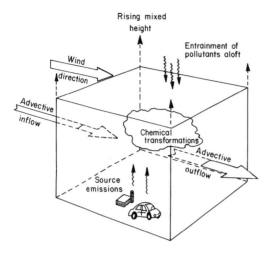

Figure 12.2 Schematic diagram showing basic elements of a box model. From Schere and Demerjian, 1978.

12.3.2 Box Models

The second category of simple urban air quality models has been called box models. Figure 12.2 is a schematic diagram of a box model including source emissions near the lower boundary (surface), advective inflow and outflow to and from the sides, entrainment of pollutant from aloft due to increasing mixing height, and chemical transformations. Since uniform mixing is assumed to occur within the box whose horizontal boundaries enclose the urban area of interest, the model can predict only the volume-averaged concentration as a function of time. Diffusion from individual sources is not considered, but all sources are considered in estimating source emissions into the box. With the simplified treatment of meteorology in terms of the effective transport winds and mixing height, one can use a sophisticated chemical and photochemical module.

For the simplest box model without chemical transformations, one can derive a simple differential equation for the average concentration $\bar{c}$ within the box from the consideration of mass conservation within the box or a slice of the box with a unit width normal to the wind flow, length L in the direction of flow, and height h equal to the mixing depth. The rate of change of mass within the box must be equal to the sum of the rates at which the pollutant mass is added by all the emission sources in the box, the change due to horizontal advection, and the change due to entrainment from the top resulting from the growth in mixing height. This can be expressed mathematically as (see e.g., Hanna et al., 1982)

$$Lh\frac{d\bar{c}}{dt} = LQ_a + \bar{u}h(\bar{c}_b - \bar{c})$$
$$+ L\frac{\partial h}{\partial t}(\bar{c}_a - \bar{c}) \quad (12.18)$$

where $\bar{c}_a$ is the average concentration aloft ($z > h$) over the city and $\bar{c}_b$ is the average background concentration upwind of the city. Equation (12.18) can be rewritten in the form

$$\frac{d\bar{c}}{dt} + \left(\frac{\bar{u}}{L} + \frac{1}{h}\frac{\partial h}{\partial t}\right)\bar{c}$$
$$= \frac{Q_a}{h} + \frac{\bar{u}\bar{c}_b}{L} + \frac{1}{h}\frac{\partial h}{\partial t}\bar{c}_a \quad (12.19)$$

which can be solved easily for the specified values of Q_a, $\bar{c}_a$, $\bar{c}_b$, $\bar{u}$, h, and $\partial h/\partial t$. Further simplifications can be made for the negligible background concentrations ($\bar{c}_a = 0$, $\bar{c}_b = 0$). Furthermore, if conditions become steady state ($\partial \bar{c}/\partial t = 0$, $\partial h/\partial t = 0$), say late in the afternoon, then the equilibrium concentration is given simply by

$$\bar{c}_e = \frac{L}{h}\frac{Q_a}{\bar{u}} \quad (12.20)$$

This expression is similar to Eq. (12.16) resulting from the simplified ATDL model. The equilibrium or steady-state average concentration within the urban environment is directly proportional to the total rate of emission from the urban area and inversely proportional to the product of mean wind speed and mixing height, also known as the ventilation factor.

If one defines a time scale $L/\bar{u}$ as the flushing time required for the air to pass completely over the urban area and a dimensionless time $t_* = t\bar{u}/L$, then Eq. (12.19) without the background concentration terms can be written as

$$\frac{d\bar{c}}{dt_*} = \bar{c}_e - \bar{c} \quad (12.21)$$

The solution to Eq. (12.21) is

$$\bar{c} = \bar{c}_e + (\bar{c}_o - \bar{c}_e)e^{-t_*} \quad (12.22)$$

where $\bar{c}_o$ is the initial value of concentration at $t_* = 0$, which is larger than the equilibrium concentration. According to the above simple box model, first proposed by Lettau (1970), the concentration decreases exponentially with increasing time and approaches its equilibrium value given by Eq. (12.20) after a time two to three times the flushing time. Because of its simplicity, this type of box model is often used as a screening model in regulatory applications.

With appropriate accounting for chemical and photochemical transformations, photochemical box models have also been used for predicting concentrations of ozone and other photochemical pollutants in urban areas. For example, Demerjian and Schere (1979) used such a model for predicting the ozone air quality of Houston, Texas. They used Eq. (12.18) plus a source term representing chemical transformations. The vertical entrainment term is found to be important, because ozone is often trapped above the inversion layer over urban areas at night and is mixed down to the surface by the growing unstable or convective mixed

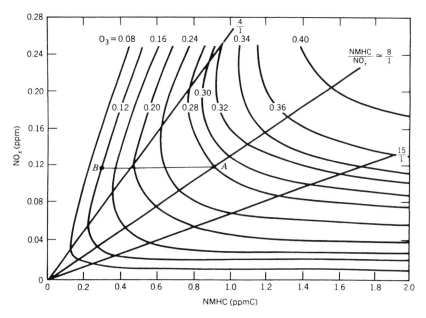

Figure 12.3 Ozone isopleths used in the EKMA (empirical kinetic modeling approach) approach. NMHC, non-methane hydrocarbons. From Dodge, 1977.

layer the next morning. The most complicated part of the photochemical box model is the multistep chemical kinetic mechanism. Demerjian and Schere (1979) used a thirty-six-step mechanism and found that predicted hourly averaged concentrations of hydrocarbons, NO_x, CO, and ozone were within a factor of 2 of observed concentrations.

One variant of the photochemical box model that has been widely used for estimating ozone concentration in urban areas is the so-called EKMA (empirical kinetic modeling approach) technique (Dodge, 1977). This approach is based on the generation of ozone concentration isopleths using an ozone isopleth plotting package (OZIPP). These isopleths are three-dimensional plots of daily maximum hourly average ozone concentrations that are generated in mixtures with various initial NMHC and NO_x concentrations. Figure 12.3 shows a set of such isopleths that were calculated using the EKMA approach analogous to that used in photochemical box models. The following assumptions were made in obtaining the isopleths in Figure 12.3 (Finlayson-Pitts and Pitts, 1986):

1. The hydrocarbon mix consists of propane and butane in a 1:3 ratio (as carbon), while aldehydes account for 5 percent (as carbon) of the initial NMHC.
2. Initial NO_2 concentrations were 25 percent of the initial NO_x concentration.
3. No addition of fresh pollutants occurred during the simulation period of 8:00 AM to 5:00 PM (local time) for the summer solstice at 34°N latitude.
4. Dilution is assumed to occur at a rate of 3 percent per hour until 3:00 PM with no dilution afterward.
5. Background or transported ozone is assumed to be negligible.

The chemical mechanism currently used in EKMA is the carbon-bond mechanism, but other mechanisms using a lumped approach can also be incorporated. These mechanisms are usually validated against the smog chamber data. The ozone isopleths in Figure 12.3 illustrate the importance of both NMHC and NO_x concentrations as well as of their ratio in the formulation of ozone control strategies. Ozone isopleths show a ridge along the diagonal running from the lower left to the upper right of the diagram. The predicted ozone maximum concentration increases with simultaneous increases in the initial concentrations of NMHC and NO_x provided their ratio is consistent with the ridge line; for the conditions of Figure 12.3 this ratio is about 6. The characteristic NMHC/NO_x ratios of different urban areas will, in general, not coincide with the ridge line but will lie on one or the other side of it. The NMHC/NO_x ratio has important implications for ozone control in different urban areas.

If an urban area is characterized by large NMHC emissions such that the NMHC/NO_x ratio is greater than the ridge line value, the ozone isopleths run almost parallel to the NMHC axis. Thus, the maximum O_3 concentration is not very sensitive to the hydrocarbon control if NO_x is kept constant. Ozone formation under these conditions is limited by the available NO_x. Therefore, reducing NO_x while keeping NMHC levels constant or reducing both NMHC and NOx simultaneously while keeping their ratio constant would reduce the maximum ozone concentration.

A different control strategy might be more appropriate for an urban area with smaller NMHC/NO$_x$ ratio than the ridge line value in Figure 12.3. Here reducing NMHC with constant NO$_x$ would lead to significant reductions in the ozone maximum. The same can be accomplished by decreasing both NMHC and NO$_x$ simultaneously while keeping their ratio constant. However, in contrast to the situation of greater NMHC/NO$_x$ ratios, reducing NO$_x$ while keeping NMHC constant would lead to higher ozone concentrations until the ridge line is reached. Therefore, the best overall control strategy, if applied uniformly to all urban areas with significant ozone pollution, is to control both NMHC and NO$_x$ emissions. The EKMA modeling approach can play a very useful role in developing optimum control strategies for different urban areas. City-specific versions of EKMA in which ozone isopleth diagrams are generated using data more representative of particular locations and transport of pollutants from outside is also taken into account are found to be more appropriate and useful for this purpose. Some of the limitations of this approach have been discussed by Finlayson-Pitts and Pitts (1986).

12.3.3 Three-dimensional Grid Models

The most comprehensive and sophisticated urban air quality models are the so-called urban airshed models that incorporate realistic descriptions of emissions, meteorology, chemistry, and removal processes occurring in an air basin. Here we consider only the Eulerian grid models in which pollutant concentrations are calculated at fixed grid locations at different times or time steps. These are based on the ensemble-averaged advection-diffusion equations including chemical transformation terms for all the chemical species of interest. A sophisticated chemistry module is also an essential part of any photochemical airshed model.

In most airshed models, meteorology is specified as an input from measurements. In a few cases, a mesoscale model of the urban boundary layer and the surrounding environment is also included. Such a model can simulate sea–land breezes in coastal cities, as well as other mesoscale circulations associated with the urban heat island and topography. The mesoscale model can provide much more detailed and accurate information on spatially inhomogeneous transport winds in a complex urban area. Available observations from a few sites in and around the city can be used to specify initial and boundary conditions for the mesoscale model (see e.g, Byun, 1987; Seaman et al., 1989).

The combination of transport, diffusion, and chemistry submodels in an urban airshed air quality model makes it very much like a regional air quality model, albeit for the smaller domain and smaller grid sizes. It is not surprising that the U.S. EPA's Urban Airshed Model (UAM) is also frequently used for studies of ozone pollution over regions extending over hundreds of kilometers and including several cities. A brief description of this model is given here as an example of an urban air quality model designed for regulatory applications (Morris and Myers, 1990; U.S. Environmental Protection Agency, 1996).

The UAM is an urban or small-regional scale, three-dimensional grid type numerical simulation model. It incorporates a condensed photochemical kinetics mechanism for urban atmospheres. In particular, a simplified version of the carbon-bond mechanism (CBM-IV) developed by Gery et al. (1989) is employed. The CBM-IV utilizes an updated simulation of PAN chemistry that includes a peroxy-peroxy radical termination reaction, which is significant when the urban atmosphere is NO$_x$ limited. The mechanism accommodates thirty-four species and eighty-two reactions. The UAM is designed for computing ozone concentrations under short-term, episodic conditions lasting 1 to 2 days resulting from emissions of oxides of nitrogen (NO$_x$), volatile organic compounds (VOCs), and carbon monoxide. The model treats VOC emissions as their carbon-bond surrogates.

Gridded, hourly emissions of CO, NO, NO$_2$, and ten categories of VOCs from low-level sources are required as input. For major elevated point sources, in addition to their hourly emissions, stack parameters such as stack height, diameter, exit velocity, and exit temperature are also required for calculating effective release heights. Initial concentrations of all the CBM-IV species and hourly concentrations of each pollutant at each vertical grid level along the inflow boundaries as well as at the top of the model domain must also be specified.

Meteorological data required as input to the UAM are: hourly, gridded, horizontal wind ($\bar{u}$ and $\bar{v}$) components for each vertical grid level; hourly, gridded surface temperatures, water vapor mixing ratios, exposure classes, and mixing heights; hourly vertical potential temperature gradients above and below the mixing height; hourly surface pressures; and gridded roughness lengths. Mean vertical velocity at each vertical grid cell interface is calculated from the continuity equation using the input horizontal wind field. The model has a maximum of eight vertical layers of which five are below the mixing height and three are above the mixing height. In a simplified version of the UAM with five layers, only two are below the mixing height. Thus, the vertical resolution in the model varies with the diurnally varying mixing height.

The horizontal and vertical dispersion are modeled through the mean advection-diffusion equations for the various pollutant species. In these the horizontal-eddy diffusivity is set to a user-specified constant value (nominally 50 m^2 s^{-1}). The vertical-eddy diffusivity is calculated using different surface-layer and PBL similarity relations for stable, neutral, and unstable conditions. For the stable boundary layer, the formulation of Businger and Arya (1974) is employed. For the daytime unstable conditions, however, the parameterization of eddy diffusivity in the model is somewhat questionable. A recent sensitivity study of the UAM by Nowacki et al. (1996) suggests that vertical diffusion is considerably overestimated in the model during daytime unstable conditions. Calculated air pollutant concentrations during several photochemical smog episodes in Atlanta, Georgia, are found to depend strongly on the parameterization of vertical diffusivity. Nowacki et al. (1996) have suggested several alternative parameterization schemes to alleviate the problem of overestimating vertical diffusivity in the model.

12.4 REGIONAL AIR QUALITY MODELS

Regional transport and air quality models can be classified according to their objectives, spatial and temporal scales, and mathematical approach and framework (van den Hout and van Dop, 1985). The objectives of a model are strongly related to the types of pollutants (e.g., inert trace gases, reactive pollutants, or photochemical pollutants) whose concentrations are to be computed or predicted by the model. Sometimes, concentrations are related to other quantities such as deposition (e.g., in acidic deposition models) or dosage (e.g., in radioactive dosage models).

Regional models usually cover horizontal scales ranging from 50 to 5000 km. These can further be subdivided into mesoscale (50–500 km) models and synoptic scale (500–5000 km) models, depending on the area of primary interest and the density of observational network utilized for specifying meteorology and initial concentrations in the model. With increasing spatial scales, the observational frequency usually decreases, so that data become sparse in time. For example, in synoptic scale models, upper-air radiosonde and rawindsonde data, usually collected at 12 hourly intervals, are utilized. In mesoscale models, on the other hand, more frequently (say, hourly) obtained surface synoptic data as well as data from any special mesoscale observation system or network can be used.

Another subdivision of air quality models can be made on the basis of averaging time. Episodic models deal with short-term averages of the order of an hour, over limited periods of days to a week. In these the short-term variations of meteorological conditions in time and space must also be taken into account. In long-term models with averaging times of the order of months to a year, however, meteorological input and other data are often simplified.

Finally, regional models can be subdivided or classified according to their mathematical framework. Eulerian models describe all relevant variables and physical properties at fixed points in space. Lagrangian or trajectory models, on the other hand, describe the variables and physical and chemical processes in a coordinate system moving along with the wind. Some models have both Eulerian and Lagrangian features and are therefore called hybrid models. The degree of sophistication with which the various model components, such as emissions, transports, chemical transformations, and removal processes, are treated varies considerably amongst the models. Eulerian models can treat these in a most detailed and sophisticated manner, but they also require more computational resources.

A comprehensive review of thirty-five regional transport and air quality models proposed up to 1985 has been given by van den Hout and van Dop (1985). Most of these are no longer in use because more sophisticated models have become available. Some are still utilized, often with updated model physics and parameterizations. By way of examples, a few of the better known regional air quality models are summarized in the following.

12.4.1 ADOM

The Acid Deposition and Oxidant Model (ADOM) is a continental (3500–7000 km)-scale Eulerian grid model for describing transports and chemical transformations of hydrocarbons, NO_x, and SO_x. It is an episodic model with horizontal grid resolution of several tens of kilometers. One can use up to twenty layers in the vertical, giving a fine resolution of the PBL. The wind field is derived from a PBL model. The vertical diffusivity is specified as a function of normalized height and stability using the PBL similarity expression of Brost and Wyngaard (1978). Horizontal diffusivity is parameterized using Smagorinski's expression for eddy viscosity or diffusivity as a function of grid size and velocity deformation. The model contains sophisticated parameterizations of dry and wet deposition. Interchangeable chemical kinetic mechanisms describing sulphate and nitrate aerosol formation or photochemical oxidants can be used. The model requires emission inventories of hydrocarbons, NO_x, and SO_x as well as concentration distributions of some gaseous and aerosol reactive species.

12.4.2 CIT Model

The California Institute of Technology (CIT) model is an Eulerian airshed model for photochemical air pollution (McRae et al., 1982; Harley et al., 1993). It is an episodic mesoscale model for a domain size up to 300 km and horizontal resolution of a few kilometers. The model contains several layers in the vertical with a specified variable mixing height. The wind field is specified through an interpolation of surface and upper-air observations. It is then smoothed and made divergence free. Alternatively, a mesoscale meteorological model can be used to provide the wind field. The vertical-eddy diffusivity is given by different expressions for unstable, neutral, and stable conditions based on similarity scaling. Horizontal and vertical dispersion from point source emissions is assumed Gaussian close to the source. Dry deposition velocity is expressed in terms of surface-layer parameters (e.g., z_o, u_*, and L); wet deposition is not considered. The photochemistry is based on the LCC/SAPRC lumped kinetic mechanism, which includes 106 chemical reactions among fifty species. Input requirements for the model include meteorological data, estimated emissions from point and area sources, and initial and boundary values of concentrations of various chemical species.

12.4.3 STEM

STEM is an episodic Eulerian grid model for describing the regional transports and transformations of SO_x and NO_x compounds (Carmichael et al., 1984; Carmichael et al., 1991). The horizontal domain may be varied from 100 km to the continental scale, while the vertical domain extends up to 8 km. Wind field is specified from interpolation of surface wind data and vertical extrapolation through the use of power-law wind profile. Vertical eddy diffusivity and mixing height are specified on the basis of Yamada and Mellor's (1975) PBL model. Horizontal diffusion is ignored. An extensive chemical kinetic mechanism, including photochemistry and cloud physics, and removal at the earth's surface, are utilized.

12.4.4 ROM

The Regional Oxidant Model (ROM), developed originally by Lamb (1983) and later revised by Young et al. (1989), has been used extensively for regulatory applications by the U.S. Environmental Protection Agency. It is designed for simulating most of the physical and chemical processes responsible for the formation and transport of ozone and other photochemical products on regional or continental scales. Most of the northeastern United States is included in the model domain. A curvilinear coordinate system with a horizontal grid resolution of $(1/4)°$ longitude and $(1/6)°$ latitude is used. The PBL and capping inversion or cloud layer are simulated using three dynamic layers with spatially and temporally varying thicknesses. During the day, the two bottom layers one and two represent the unstable or convective boundary layer; layer three represents the inversion or cloud layer. During the nighttime, layer 1 represents the stable boundary layer and layer two becomes the residual layer. Ozone and the remnants of other photochemical reaction products may remain in layers two and three during the night and can be transported long distances downwind. The dynamic three-layer system accounts for both the spatial and temporal variations of the PBL over the modeling domain, while it minimizes computational costs.

Horizontal advection, diffusion, and gas-phase chemistry for specified pollutants are modeled prognostically in the three layers. The ensemble-averaged mass conservation equations for all chemical species are solved simultaneously with specified initial and boundary conditions, transport winds, and eddy diffusivities. Wind fields are obtained from interpolation of surface observations and upper-air soundings. A prognostic submodel is used for the stable boundary layer flow (layer 1) when surface inversion is present over most of the model domain.

Horizontal diffusivity is parameterized in terms of convective velocity scale and mixing height during unstable conditions and is neglected during stable conditions. Vertical fluxes of pollutants are determined at both the top and bottom surfaces of each layer. Several physical processes can cause mass exchanges between adjacent layers. These are horizontal and vertical advections, turbulent eddies, changes in layer height, terrain forcing, and cumulus cloud updrafts. Each of these processes and the associated material fluxes are simulated in the ROM. Surface emissions are specified as a mass flux through the bottom of layer one; the dry deposition flux is also combined with the same. The model simulates dry deposition fluxes through empirical parameterization of deposition velocities and near surface concentrations. Wet deposition is not considered in the model.

The chemical kinetic mechanism embedded in ROM is the carbon-bond mechanism (CBM-IV) of Gery et al. (1989). The mechanism consists of eighty-two reactions involving thirty-four chemical species. Chemical time steps vary depending on chemical reaction rates, but typically range from 10 to 60 s. A solar radiation algorithm is used to compute photolytic rate constants; it considers the effect of cloud cover.

Anthropogenic emissions are specified from a comprehensive emissions inventory, such as the 1985 National Acid Precipitation Assessment Program inventory. Emissions from anthropogenic sources of VOC, NO_x, and CO over the northeast United States were provided by the NAPAP inventory. Major point, area, and mobile source emissions were also included; these varied hourly, depending on the day of the week (weekday, Saturday, or Sunday). Biogenic emissions are obtained from the Biogenic Emissions Inventory System (Pierce et al., 1990). Biogenic emissions are considered to have large uncertainties. Roselle et al. (1991) have investigated the sensitivity of the model ozone concentrations to uncertainties in biogenic hydrocarbon emissions. When biogenic emissions were decreased or increased by a factor of 3, predicted ozone concentrations also decreased or increased, depending on the availability of NO_x. The model has been extensively used for studying the impact of various control strategies on ozone concentrations in different parts of the northeastern United States.

12.4.5 RADM

The Regional Acid Deposition Model (RADM) is a comprehensive Eulerian model designed to study acidic deposition on regional and continental scales (Chang et al., 1987). It simulates the chemistry and meteorology of the entire depth of the troposphere and follows the development and movement of precipitation systems that generate acidic deposition. The model was originally designed to study acidic deposition in the northeastern United States. It uses a horizontal grid size of approximately 80 km over a 2400 km by 2400 km domain. In the vertical the chemical model is divided into six layers of which the lowest three levels are typically in the PBL during the day. The number of vertical layers and their thicknesses are flexible in the model.

RADM uses hourly, three-dimensional fields of horizontal winds, temperature, and water vapor mixing ratio. In addition, two-dimensional hourly fields of surface temperature, surface pressure, and precipitation rate are required over the model domain. A hydrostatic, mesoscale meteorological model (MM-IV) is currently used to generate these meteorological fields. The use of synoptic surface observations and upper-air sounding data does not provide adequate spatial and temporal resolution of transport winds, mixing height, and precipitation for use in RADM. The meteorological model provides better resolved, smoothly varying, mass-consistent fields for the chemistry model. The former has the same horizontal domain and grid size but fifteen layers in the vertical. Both models use the terrain-following σ vertical coordinate system. Meteorological initial and boundary conditions are provided by the National Meteorological Center global analysis, upgraded by incorporating rawinsonde data. Surface energy and moisture budgets are parameterized using a slab model that incorporates land use data on the various surface characteristics. A high-resolution PBL submodel (Zhang and Anthes, 1982) is utilized to quantify subgrid scale vertical transports or fluxes of momentum, heat, and moisture.

RADM solves a set of twenty-four chemical species conservation equations which include the usual advection, diffusion, chemical transformation, source, and sink terms. Both the dry and wet removal processes are included in the model in highly parameterized forms. At each grid point, eleven additional species are computed, assuming that a local chemical steady state exists for these species. Methane is held constant at 1.7 ppm. Subgrid vertical turbulent transports of trace gases are parameterized through the use of vertical-eddy diffusivity following Louis (1979). The mixing-length expressions for K_z for unstable and stable conditions include the effect of stability in terms of the bulk Richardson number for the particular layer. The horizontal subgrid scale diffusion is neglected, in comparison with horizontal advection and numerical diffusion introduced by the integrated upstream finite-difference scheme used in the model.

The RADM2 chemistry model simulates the complex gas phase chemistry responsible for the formation of acidic compounds. Using 158 reactions among sixty-three species, the RADM gas phase chemical reaction mechanism (Stockwell et al., 1990) simulates concentrations and conversion rates of nitrogen oxides, sulfur oxides, HO radical, hydrogen peroxide, organic peroxides, ozone, and numerous other species. The RADM chemical reaction mechanism condenses the explicit mechanism of Leone and Seinfeld (1985) and the carbon bond mechanism of Whitten et al. (1985). Further details of gas phase chemistry, rate constants, and numerical solution technique used in RADM are given by Chang et al. (1987) and Stockwell et al. (1990).

Dry deposition to the surface is parameterized in terms of spatially and temporally varying deposition velocity, which is expressed as the inverse of the sum of the aerodynamic resistance, molecular sublayer resistance, and surface resistance. The aerodynamic and sublayer resistances are parameterized in terms of friction velocity and surface roughness over each land-use type in a particular grid area. The surface resistance is a seasonally and diurnally varying species-specific parameter that depends on the type of surface. Surface resistances have been estimated from the experimental data for only a limited number of species, such as SO_2. For other species, surface resistances are qualitatively scaled to the SO_2 value using different scaling factors over land and water surfaces. Deposition velocities are computed for each hour by taking an area-weighted average of the deposition velocities over all land types encountered within each grid area.

The effects of all clouds present above each grid point are parameterized in RADM by modeling the chemical lifetime of a representative cloud and imposing the resulting changes in trace gas concentration on the grid-averaged concentration. Each representative cloud has a depth and fractional area of coverage determined by large-scale environmental parameters. A one-dimensional diagnostic cloud model is used to specify vertical distributions of cloud microphysical and dynamical properties. Chemical interactions in the cloud are modeled using a box aqueous phase chemical and scavenging submodel, which calculates time-dependent chemical composition of cloud water and rainwater. For further details of the characterization of cloud fields, aqueous phase chemistry, and scavenging of soluble species in RADM, the reader should refer to Chang et al. (1987).

RADM uses hourly emissions of SO_2, sulfate, NO, NO_2, NH_3, CO, and ten classes of volatile organic compounds (VOC). For the northeastern United States, these were derived from the two comprehensive emission inventories developed by NAPAP and the Electric Power Research Institute (EPRI). These inventories specify seasonally averaged emissions for weekday or weekend periods. Both the NAPAP and EPRI inventories were analyzed and modified before being incorporation into the model. All emissions were spatially aggregated into appropriate model grids. Significant differences (up to 30–60%) were found in the emissions of SO_x, NO_x, and VOC given by the two earlier inventories, reflecting uncertainties in emissions or real differences between the two base years (1980 for EPRI and 1982 for NAPAP). Later, the 1985 NAPAP inventory helped resolve some of these differences. In RADM, all areas sources are considered emitting into the lowest model layer, but point sources may emit into higher model layers, depending on their effective source heights after the buoyant plume rise is accounted for.

To solve numerically the various species' mass conservation and chemical equations in a regional episodic model such as RADM, the initial and boundary conditions for concentrations must be specified. Ideally, these should be based on observations. This is not practically feasible, however, because of the lack of suitable observational networks for large regional domains. To minimize the sensitivity of model results to lateral boundary conditions, domain boundaries are placed sufficiently far from the major emission sources. Then concentrations at the domain boundaries may be specified as ambient levels for an unpolluted atmosphere. The regional model itself can be used to generate reasonable initial conditions over the model domain. For example, following the initialization with horizontally uniform, vertically varying, clean atmospheric conditions, RADM is run for 2 days, with realistic meteorology, prior to the simulation period. During this spin-up time a quasi steady state is attained among the emissions, transport, and dry deposition of the various trace gases in the model domain. For the sake of simplicity, cloud and precipitation processes may not be activated during the initialization period, unless prevailing weather conditions prior to the simulation period indicate otherwise.

There are many computational, physical, and chemical sources of uncertainty in any comprehensive regional model such as RADM. Several parameterizations of gas and aqueous phase chemistry, dry deposition, and cloud and precipitation processes are difficult to verify because of a lack of appropriate data and probably contain large uncertainties. Other uncertainties are introduced by meteorological models or observations. In addition, large errors can arise because of uncertainties in emission rates and initial and boundary conditions involving the various trace gases. An Eulerian model using finite-difference numerical schemes also contains uncertainties associated with numerical diffusion and unresolved chemical interactions. Considerable research effort and model sensitivity studies will be required to quantify the effects of these errors and uncertainties on model calculations. Validation of a large regional model against experimental data presents an even greater challenge.

12.4.6 Model System KAMM/DRAIS

For application to smaller regions (say, 200 km × 200 km) with complex topography, a nonhydrostatic mesoscale model with smaller grid spacing of about 3 km in the horizontal has been developed and used in Germany (Adrian and Fiedler, 1992; Vogel et al., 1995). The meteorological driver of this system is called KAMM, while transport and diffusion of air pollutants are handled by the submodule DRAIS. The RADM2 chemical mechanism (Stockwell et al., 1990) has recently been incorporated in the KAMM/DRAIS model system to study the interaction of the various physical and chemical processes determining the temporal and spatial distributions of ozone and other reactive species on a small regional scale (Vogel et al., 1992, 1995).

The mesoscale meteorological model KAMM solves the ensemble-averaged equations for meteorological variables (velocities, potential temperature, and specific humidity) in a terrain-following coordinate system. The basic thermodynamic and dynamic forcing for KAMM is derived from either an operational weather prediction model or from observations. Important boundary conditions for the model are those of temperature and specific humidity at the lower boundary. They are both modified by the land use. A soil vegetation model is used to obtain the lower boundary conditions for temperature and humidity for KAMM at each time step.

The transport and diffusion submodule DRAIS numerically solves the mass conservation equations for all the pollutant species of interest. These include advection, diffusion, chemical production, removal, and source terms. Horizontal-eddy diffusivities are assumed to be 2.3 times the vertical-eddy diffusivity of material, which is assumed equal to that of heat. Under neutral and stable conditions in the PBL the vertical-eddy diffusivities of momentum and heat are parameterized using the surface-layer and PBL similarity relations. For unstable and convective conditions, eddy diffusivities are determined using a nonlocal closure scheme.

Vogel et al. (1995) have reported on a specific application of the KAMM/DRIAS modeling system to study the influence of topography and biogenic VOC emissions on ozone concentrations during high-temperature episodes in the state of Baden-Wurttemberg, Germany, which includes several large cities and industrial areas. They also give details of topography, land use, and anthropogenic and biogenic emissions in their model domain. Comparisons of simulated and observed concentrations of NO_x and ozone show a good agreement during the daytime hours. But large differences are found during the night; these are caused by local effects that cannot be resolved by the model even with a grid size of 3 km. During the daytime, enhanced mixing reduces the influence of local sources on measurements. Well-known features of the diurnal variation of ozone are reproduced by the model. Model results show a good correlation between temperature and ozone concentration. This is partly due to the strong dependence of biogenic VOC emissions on temperature. The neglect of biogenic emissions results in significantly lower ozone concentrations.

12.4.7 Next-generation Models

The U.S. Environmental Protection Agency is developing a third-generation modeling system called Models-3 (Ching et al., 1995; Byun et al., 1996; Dennis et al., 1996). The Models-3 is intended to be a flexible and general air quality modeling system with multiscale and multipollutant capabilities. It is designed to take advantage of the enhanced computational capabilities provided by high-performance computing and communication (HPCC) architectures. The system framework consists of six processing layers including the user interface, a system manager, a UNIX environment, the computational programs, data access, and data storage. The air quality modeling system resides in the computational programs layer and consists of several models or submodels dealing with emissions, chemical transport and transformations, and analysis and visualization (Ching et al., 1995).

The standard synoptic meteorological observational network was designed to detect synoptic-scale weather systems; it does not provide adequate information to characterize transport and diffusion at the finer scales desired in urban and regional air quality models. For this reason a prognostic mesoscale meteorological model with capabilities of grid nesting to resolve finer scales and estimate micrometeorological parameters such as mixing height, friction velocity, and surface heat flux is used as a driver for the chemistry transport model. In Models-3, the latest version of the Mesoscale Meteorological Model (MM5) is recommended to be used, although the modular design would permit its substitution by any other meteorological model. This model includes a four-dimensional data assimilation of observations, nesting capability, the choice of a hydrostatic or nonhydrostatic set of equations, and adequate boundary-layer physics and parameterization.

The emissions model to be used in Models-3 must be capable of (1) speciating for different chemical mechanisms; (2) providing temporal and spatial variations of emissions from the various point and area sources; (3) developing control strategy projections; and (4) generating appropriate emissions input to different urban and regional air quality models. It should be flexible enough to allow for easy updates and modifications as the science and methodologies of emissions modeling change. An adaptation of the Geocoded Emissions Modeling and Projection System (GEMAP) is proposed to be utilized in Models-3 (Ching et al., 1995; Dennis et al., 1996).

The chemistry transport model (CTM) is essentially the driving engine for any regional air quality model. The design attributes required for the CTM of Models-3 are: (1) multipollutant capability to handle different air pollution problems related to oxidants, acidic deposition, visibility, and particulate matter; (2) multiscale capability with flexibility to operate at a wide range of spatial scales; and (3) easy extension to be able to incorporate addition pollutants and chemical transformation and removal processes. It solves a large set of species mass conservation equations coupling chemistry with advection and diffusion processes. The various atmospheric processes being modeled include gas and aqueous phase chemistry, dry deposition, cloud and precipitation scavenging, and transport and diffusion. Models-3 CTM is designed to have the flexibility of choice among a short list of chemical mechanisms, specifically CBM-IV (Gery et al., 1989), RADM-2 (Stockwell et al., 1990), and SAPRC-90 (Carter, 1990). It has flexible and generalized chemistry readers and solvers. Several optional numerical solution methodologies are provided in the model (see e.g., Byun et al., 1996).

In order to have the multiscale modeling capability, Models-3 utilizes a generalized coordinate system that allows easy transformation among different horizontal and vertical coordinates by simple changes in a few scaling parameters, boundary conditions, and origin and orientation of axes. Grid spacing is uniform in the horizontal with flexibility of choice from a few to 100 km for a domain size of 100 to 3000 km. A variable vertical grid spacing allows for a better resolution of the PBL. The grid-nesting capability in the model provides a means of resolving fine scale features such as found in urban areas and over and around complex terrain. Two-way nesting procedures permit significant feedbacks from nested fine scales to coarse scales of the larger domain. A plume-in-grid approach in combination with grid nesting may address some subgrid-scale reactive plume issues. Key processes embedded in the plume submodel are plume rise, plume transport and dispersion, plume chemistry, and surface deposition.

The numerical differential equation solver is isolated from the chemical mechanism to achieve a high level of modularity and allow updating or replacement of the chemical mechanism. The Models-3 also incorporates the chemistry and physics (dynamics) of aerosols that are either emitted as primary pollutants or formed from complex photochemical processes in the atmosphere. For more detailed descriptions of the design requirements, conceptual framework, and targeted capabilities of the U.S. EPA's Models-3 the reader should refer to Ching et al. (1995), Byun et al (1996), and Dennis et al. (1996). Several prototypes of varying complexity and functions were made available for verification, testing, and evaluation in 1997, and technical transfer to the public domain is projected to take place in June 1998, particularly for the Models-3 Community Multi-scale Air Quality (CMAQ) system.

Another third-generation model, called the Comprehensive Modeling System (CMS), is being developed in concert with and complementary to Models-3. This is being initiated by the Consortium for Advanced Modeling of Regional Air Quality funded by EPRI (Hansen et al., 1994). Other next generation air quality models are being developed in Canada and Europe.

12.5 APPLICATIONS OF AIR QUALITY MODELS

Urban air quality models such as UAM-IV are extensively used for running "what-if" scenarios and developing optimum control strategies for many urban areas that have been identified for nonattainment of ambient air quality standards for certain criteria pollutants. For example, when Charlotte, North Carolina, was designated a nonattainment area for ozone, state environmental scientists ran many runs of UAM-IV with different control strategies, such as reducing point source emissions of NO_x, and came up with a scientifically sound and cost-effective state implementation plan. Earlier, Scheffe (1990) reported on the UAM study of five cities including Atlanta, Georgia. A more recent model study of ozone formation in Atlanta is described by Sillman et al. (1995). They also compared their model results with measurements made during a field study. Different model scenarios created with alternative emissions and meteorology are evaluated. The California Institute of Technology model has been used to study photochemical smog in southern California, particularly in Los Angeles (Seinfeld, 1986; Harley et al., 1993), and also in Athens, Greece (Giovannoni and Russell, 1995).

Urban air quality models are also used in predicting possible impacts of proposed plants or facilities on urban air quality. The use of local and urban air quality models in preparing the environmental impact assessments of existing or proposed sources of air pollution has been more or less mandated by the Clean Air Act Amendments of 1990 in the United States.

When the pollution problem is more widespread over an urban-industrial complex or corridor such as southern California and the northeastern United States, or over an extensive region including rural areas, a regional air quality model can be used for developing cost-effective control strategies. In the United States, a variety of regional air quality models have been used for studying regional air quality and the effects of various control strategies on the same. Regional smog or haze caused by ozone and photochemical oxidants is the most serious problem in many parts of the United States, especially southern California, the southeastern Atlantic states, and the northeastern corridor, as well as in other parts of the world with high traffic density, heavy industrialization, and warm climates. The Regional Oxidant Model (ROM) has been used for studying different control strategies for the northeastern United States (Lamb, 1986). It has also been used for studying the effects of uncertainties in biogenic emissions of NMHC on ozone and other photochemical oxidants (Roselle et al., 1991). A more recent version of the Urban Airshed Model-V (UAM-V) uses a regional-to-urban grid that can model ozone and other photochemical oxidants on a much wider range of scales necessary to study both the urban and regional air pollution problems. The CIT model has been extensively used in southern California (Harley et al., 1993). For regional acidic deposition modeling, RADM and ADOM have been extensively used in the United States and Canada. Different regional acid deposition and oxidant models have been utilized in Europe (Zwerver and van Ham, 1985; Vogel et al., 1995). More comprehensive reviews of the various regional and long-range transport models and their applications to specific air pollution problems are given elsewhere (Zwerver and van Ham, 1985; Air and Waste Management Association, 1993). Recent developments in the formulation of more general and comprehensive modeling systems with modular and flexible design are likely to make their use more widespread for solving our urban and regional air quality problems.

PROBLEMS AND EXERCISES

1. Differentiate between urban and regional air quality models and discuss the importance of chemical transformation and removal processes in these models.

2. Draw a schematic of the various components of an urban air quality model for carbon monoxide and discuss their importance to predicting concentrations of CO in an urban area.

3. Discuss different ways of specifying transport winds in a regional air quality model, stating their merits as well as limitations.

4. Distinguish between the two different methods of parameterizing diffusion in Eulerian urban air quality models and discuss their relative merits and limitations.

5. Distinguish between explicit and lumped chemical mechanisms. Which of these is most commonly used in urban and regional air quality models and why?

6. Which factors are likely to determine the wet deposition velocity of a particular gaseous species? How is it commonly specified in regional air quality models?

7. A city extends 20 km along the mean wind direction. The mean transport speed in the daytime unstable boundary layer is 4.5 m s^{-1}. If the stability in the late afternoon may be characterized by Pasquill's class C and Briggs' urban dispersion scheme is used for parameterizing σ_z, calculate the ground-level concentration of SO_2 at the downwind edge of the city where the area source strength is estimated to be 1 μg km^{-2} s^{-1}. You may use the simplified version of the ATDL model.

8. For Prob. 7, if the mixing height over the city in late afternoon is 500 m when steady state may be assumed to occur and background concentrations of SO_2 are negligible, calculate the average concentration of SO_2 over the city using the simple box model. Discuss the differences between the ATDL and box models, as well as their limitations.

9. For Prob. 8, if the initial concentration of SO_2 at noon was twice the equilibrium or steady-state concentration, calculate and plot concentration as a function of time using the box model. State all the simplifying assumptions you make.

10. What are the limitations of parameterizing diffusion through the use of the K-theory in three-dimensional urban air quality models? How would you specify eddy diffusivities in such models?

11. Distinguish between mesoscale and continental scale regional air quality models. How would you specify transport winds and other meteorological fields in these models?

12. Discuss the various sources of uncertainties in regional air quality models. Describe the various difficulties that may be encountered in satisfactory validation of such models.

13. What are the primary distinguishing features of U.S. EPA's regional air quality models ROM and RADM?

14. Describe the basic concept and primary objectives of the Models-3 being developed by U.S. EPA. In what respects might such a comprehensive air quality modeling system be superior to existing urban and regional air quality models?

15. Discuss the various regulatory applications of urban and regional air quality models.

References

Abramowitz, M., and Stegun, I.A. (1965). *Handbook of Mathematical Functions*. Dover, NY.

Adrian, G., and Fiedler, F. (1992). Simulation of unstationary wind and temperature fields over complex terrain and comparison with observations. *Beitr. Phys. Atmos.*, 64, 27–48.

Adrian, R.J. (1991). Particle-imaging techniques for experimental fluid mechanics. *Ameri. Rev. Fluid Mech.*, 23, 261–304.

Ahrens, C.D. (1994). *Meteorology Today*. 5th ed. West Publishing Co., St. Paul, MN.

Air and Waste Management Association (1991). RCRA and the Clean Air Act in the '90s. P*roc. 19th Annual Government Affairs Seminar*, Air and Waste Management Association, Pittsburgh.

Air and Waste Management Association (1993). *Regional Photochemical Measurement and Modeling Studies*. Air and Waste Management Association, Pittsburgh.

American Meteorological Society (1978). Accuracy of dispersion models. *Bull. Amer. Meteorol. Soc.*, 59, 1025–1026.

Andre, J.C., de Moor, G., Lacarrere, P., Therry, G., and du Vachat, R. (1978). Modeling the 24–hour evolution of the mean turbulent structures of the planetary boundary layer. *J. Atmos. Sci.*, 35, 1861–1883.

Andren, A. (1990). Evaluation of a turbulence closure scheme suitable for air-pollution applications. *J. Appl. Meteorol.*, 29, 224–239.

Andren, A. (1995). The structure of stably stratified atmospheric boundary layers: A large-eddy simulation study. *Quart. J. Roy. Met. Soc.*, 121, 961–985.

Andren, A., Brown, A., Graf, J., Moeng, C.H., Mason, P.J., Nieuwstadt, F.T.M., and Schumann, U. (1994). Large-eddy simulation of neutrally stratified boundary layer: A comparison of four computer codes. *Quart. J. Roy. Met. Soc.*, 120, 1457–1484.

Angell, J.K. (1964). Measurements of Lagrangian and Euleria properties of turbulence at a height of 2500 feet. *Quart. J. Roy. Meteorolol. Soc.*, 90, 57–71.

Arya, S.P.S. (1972). The critical condition for the maintenance of turbulence in stratified flows. *Quart. J. Roy. Meteorol. Soc.*, 98, 264–273.

Arya, S.P.S. (1975). Buoyancy effects in a horizontal flat-plate boundary layer. *J. Fluid Mech.*, 68, 321–343.

Arya, S.P. (1988). *Introduction to Micrometeorology*. Academic Press, San Diego, CA.

Arya, S.P. (1991). Finite-difference errors in estimation of gradients in the atmospheric surface layer: *J. Appl. Meteorol.*, 30, 251–253.

Arya, S.P. (1995). Atmospheric boundary layer and its parameterization. In *Wind Climate in Cities* (J.C. Cermak, A.G. Davenport, E.J. Plate, and D.X. Viegas, eds.), pp. 41–66. Kluwer Acad. Pub. Dordrecht, The Netherlands.

Arya, S.P.S., Capuano, M.E., and Fagen, L.C. (1987). Some fluid modeling studies of flow and dispersion over two-dimensional low hills. *Atmos. Environ.*, 21, 753–764.

Arya, S.P.S., and Shipman, M.S. (1981). An experimental investigation of flow and diffusion in the disturbed boundary layer over a ridge. I. Mean flow and turbulence structure. *Atmos. Environ.*, 15, 1173–1184.

Arya, S.P.S. Shipman, M.S., and Courtney, L.Y. (1981). An experimental investigation of flow and diffusion in the disturbed boundary layer over a ridge. II. Diffusion from a continuous point source. *Atmos. Environ.*, 15, 1185–1194.

Atkinson, B.W. (1981). *Meso-scale Atmospheric Circulations*. Academic Press, New York.

Baerentsen, J.H., and Berkowicz, R. (1984). Monte Carlo simulation of plume dispersion in the convective boundary layer. *Atmos. Environ.*, 18, 701–712.

Bains, P.G. (1995). *Topographic Effects in Stratified Flows*. Cambridge University Press, New York.

Barad, M.L. (1958). Project Prairie Grass, a field program in diffusion. *Geophys. Res. Papers*, No. 59, Vols. I and II, Report AFCRC-TR-58–235, U.S. Air Force Cambridge Research Center.

Bardina, J. Ferziger, J.H., and Reynolds, W.C. (1980). Improved sub-grid models for large eddy simulation. AIAA Paper 80–1357. American Institute of Aeronautics and Astronautics.

Batchelor, G.K. (1949). Diffusion in a field of homogeneous turbulence, I. Eulerian analysis. *Aust. J. Sci. Res.*, 2, 437–450.

Batchelor, G.K. (1950). The application of the similarity theory of turbulence of atmospheric diffusion. *Quart. J. Roy. Meteorol. Soc.*, 76, 133–146.

Batchelor, G.K. (1952). Diffusion in a field of homogeneous turbulence. II. The relative motion of particles. *Proc. Camb. Phil. Soc.*, 48, 345–362.

Batchelor, G.K. (1953). *The Theory of Homogeneous Turbulence*. Cambridge University Press, Cambridge, England.

Batchelor, G.K. (1957). Diffusion in free turbulent shear flows. *J. Fluid Mech.*, 3, 67–80.

Batchelor, G.K. (1959). Note on diffusion from sources in a turbulent boundary layer. Unpublished manuscript.

Batchelor, G.K. (1964). Diffusion from sources in a turbulent boundary layer. *Arch. Mech. Stosowanej.*, 16, 661–670.

Batchelor, G.K. (1967). *An Introduction to Fluid Dynamics*. Cambridge University Press, London.

Batchelor, G.K., and Townsend, A.A. (1956). Turbulent diffusion. In *"Surveys in Mechanics"* (G.K. Batchelor and R.M. Davies, ed.), pp. 352–399. Cambridge University Press, Cambridge, England.

Baumgartner, A. Enders, G., Kirchner, M., and Mayer H. (1982). Global Climatology. In *Engineering Meteorology* (E.J. Plate, ed.), pp. 125–177. Elsevier, Amsterdam.

Beljaars, A.C.M., Walmsley, J.L., and Taylor, P.A. (1987). A mixed spectral finite-difference model for neutrally stratified boundary layer flow over roughness changes and topography. *Boundary Layer Meteorol.*, 38, 273–303.

Bendat, J.S., and Piersol, A.G. (1986). *Random Data Analysis and Measurement Procedures*, 2d ed. Wiley, New York.

Berlyand, M.E. (1975). *Prediction and Regulation of Air Pollution*. Kluwer Academic Publ., Dordrecht, Holland.

Blackman, R.B., and Tukey, J.W. (1958). *The Measurement of Power Spectra*. Dover, New York.

Bosanquet, C-H., and Pearson, J.L. (1936). The spread of smoke and gases from chimneys. *Trans. Faraday Soc.*, 32, 1249–1263.

Boussinesq, J. (1877). Essai sur la theorie des eaux courantes. *Mem. pres. par div. savants a l'Acad. Sci., Paris*, 23, 1–680.

Boybeyi, Z., Raman, S., and Zannetti, P. (1995). Numerical investigation of possible role of local meteorology in Bhopal gas accident. *Atmos. Environ.*, 29, 479–496.

Briggs, G.A. (1968). Momentum and buoyancy effects. In *Meteorology and Atomic Energy—1968* (D.H. Slade, ed.), pp. 189–202. U.S. Department of Energy, Technical Information Center, Oak Ridge, TN.

Briggs, G.A. (1969). *Plume Rise*. AEC Critical Review Series, U.S. AEC Technical Information Center, Oak Ridge, TN.

Briggs, G.A. (1971). Some recent analyses of plume rise observations. In *Proceedings of the Second International Clean Air Congress* (H.M. Englund and W.T. Berry, eds.), pp. 1029–1032. Academic Press, New York.

Briggs, G.A. (1973). Diffusion estimation of small emissions. Contribution No. 79, Atmospheric Turbulence and Diffusion Laboratory, Oak Ridge, TN.

Briggs, G.A. (1975). Plume rise predictions. In *Lectures on Air Pollution and Environmental Impact Analyses* (D.A. Haugen, ed.), pp. 59–111. American Meteorological Society, Boston.

Briggs, G.A. (1984). Plume rise and buoyancy effects. In *Atmospheric Science and Power Production* (D. Randerson, ed.), pp. 327–366. U.S. Department of Energy, Technical Information Center, Oak Ridge, TN.

Briggs, G.A. (1985). Analytical parameterization of diffusion: The convective boundary layer. *J. Climate Appl. Meteorol.*, 24, 1167–1186.

Briggs, G.A. (1988). Analysis of diffusion field experiments. In *Lectures on Air Pollution Modeling* (A Venkatram and J.C. Wyngaard, eds.), pp. 63–117. American Meteorological Society, Boston.

Briggs, G.A. (1993). Plume dispersion in the convective boundary layer. Part II: Analysis of CONDORS field experimental data. *J. Appl. Meteorol.*, 32, 1388–1425.

Brost, R.A., and Wyngaard, J.C. (1978). A model study of the stably stratified planetary boundary layer. *J. Atmos. Sci.*, 35, 1427–1440.

Brown, M.J., Arya, S.P., and Snyder, W.H. (1993). Vertical dispersion from surface and elevated releases: An investigation of a non-Gaussian plume model. *J. Appl. Meteorol.*, 32, 490–505.

Brunt, D. (1927). The period of simple vertical oscillations in the atmosphere. *Quart. J. Roy. Meteorol. Soc.*, 53, 30–31.

Businger, J.A. (1973). Turbulent transfer in the atmospheric surface layer. In *Workshop on Micrometeorology* (D.A. Haugen, ed.), pp. 67–100. American Meteorological Society, Boston.

Businger, J.A. (1986). Evaluation of the accuracy with which dry deposition can be measured with current micrometeorological techniques. *J. Climate Appl. Meteor.*, 25, 1100–1124.

Businger, J.A., and Arya, S.P.S. (1974). Height of the mixed layer in the stably stratified planetary boundary layer. *Advan. Geophys.*, 18A, 73–92.

Businger, J.A., Wyngaard, J.C., Izumi, Y., and Bradley, E.F. (1971). Flux-profile relationships in the atmospheric surface layer. *J. Atmos. Sci.*, 28, 181–189.

Byers, H.R. (1974). *General Meteorology*. McGraw-Hill, New York.

Byun, D.W. (1987). A two-dimensional mesoscale numerical model of St. Louis urban mixed layer. Ph.D. dissertation, Department of Marine, Earth and Atmospheric Sciences, North Carolina State University, Raleigh.

Byun, D.W., and Arya, S.P.S. (1990). A two-dimensional mesoscale numerical model of an urban mixed layer. I. Model formulation, surface energy budget, and mixed-layer dynamics. *Atmos. Environ.*, 24A, 829–844.

Byun, D.W., Dabdub, D., Fine, S., Hanna, A.F., Mathur, R., Odman, T. Russell, A., Segall, E.J., Seinfeld, J.H., Steenkiste, P., and Young, J. (1996). Emerging air quality modeling technology for high performance computing and communication environments. In *Air Pollution Modeling and its Applications*, XI, 491–502.

Calder, K.L. (1952). Some recent British work on the problem of diffusion in the lower atmosphere. In *Air Pollution*, Proceedings of the U.S. Technical Conference on Air Pollution, pp. 787–792, McGraw-Hill, New York.

Carmichael, G.R., Kitada, T., and Peters, L.K. (1984). A second generation combined transport/chemistry model for the regional transport of SO_x and NO_x compounds. In *Air Pollution Modelling and Its Applications III* (C. de Wispelaere, ed.), pp. 525–528. Plenum Press, New York.

Carmichael, G.R., Peters, L.K., and Saylor, R.D. (1991). The STEM-II regional scale acid deposition and photochemical oxidant model. I. An overivew of model development and application. *Atmos. Environ.*, 25A, 2077–2090.

Carpenter, S.B., Montgomery, T.L., Leavitt, J.M., Colbaugh, W.C., and Thomas, F.W. (1971). Principal plume dispersion models: TVA power plants. *J. Air Pollut. Cont. Assoc.*, 21, 491–495.

Carslaw, H.S., and Jaeger, J.C. (1959) *Conduction of Heat in Solids*. Oxford University Press, Oxford, England.

Carter, W.P.L. (1990). A detailed mechanism for the gas

phase atmospheric reactions of organic compounds. *Atmos. Environ.*, *24A*, 481–518.

Caughey, S.J., Kitchen, M., and Leighton, J.R. (1983). Turbulence structure in convective boundary layers and implications for diffusion. *Boundary-Layer Meteorol.*, *25*, 345–352.

Caughey, S.J., and Palmer, S.G. (1979). Some aspects of turbulence through the depth of the convective boundary layer. *Quart J. Roy. Meteorol. Soc.*, *105*, 811–827.

Caughey, S.J. Wyngaard, J.C. and Kaimal, J.C. (1979). Turbulence in the evolving stable boundary layer. *J. Atmos. Sci.*, *36*, 1041–1052.

Cermak, J.E. (1963). Lagrangian similarity hypothesis applied to diffusion in turbulent shear flow. *J. Fluid Mech.*, *15*, 49–64.

Chamberlain, A.C. (1953). Aspects of travel and deposition of aerosol and vapor clouds. Report No. A.E.R.E. HP/R1261. Her Majesty's Stationery Office, London.

Chamberlain, A.C. (1966). Transport of gases to and from grass and grass-like surfaces. *Proc. Roy. Soc. London*, *A290*, 236–265.

Chamberlain, A.C. (1975). The movement of particles in plant communities. In *Vegetation and the Atmosphere* (J.L. Monteith, ed.), Vol. 1, pp. 155–203. Academic Press, London.

Chandrasekhar, S. (1943). Stochastic problems in physics and astronomy. *Rev. Modern Phys.*, *15*, 1–89.

Chang, J.S., Brost, R.A., Isaksen, I.S.A., Madronich, S., Middleton, P., Stockwell, W.R., and Walcek, C.J. (1987). A three-dimensional Eulerian acid deposition model: Physical concepts and formulation. *J. Geophys. Res.*, *92*, 14,681–14,700.

Chatwin, P.C. (1968). The dispersion of a puff of a passive substance in the constnt stress region. *Quart. J. Roy. Meteorol. Soc.*, *94*, 401–411.

Chaudhry, F.H., and Meroney, R.N. (1973). Similarity theory of diffusion and the observed vertical spread in the diabatic surface layer. *Boundary-Layer Meteorol. 3*, 405–415.

Ching, J.K.S., Byun, D.W., Hanna, A., Odman, T., Mathur, R., Jang, C., McHenry, J., and Galluppi, K. (1995). Design requirements for multiscale air quality models. In *Proc. Mission Earth Symposium on High Performance Computing*, Phoenix, 533–538.

Cleugh, H.A. and Oke, T.R. (1986). Suburban-rural energy balance comparisons in summer for Vancouver, B.C. *Boundary-Layer Meteorol.*, *36*, 351–369.

Cramer, H.E. (1957). A practical method of estimating the dispersal of atmospheric contaminants. In *Proceedings of the Conference on Applied Meteorology*, American Meteorological Society, Boston.

Cramer, H.E. (1976). Improved techniques for modeling the dispersion of tall stack plumes. In *Proceedings of the 7th International Technical Meeting on Air Pollution Modeling and Its Applications, No. 51*, NATO/CCMS, 731–780.

Cramer, H.E., Record, F.A., and Vaughan, H.C. (1958). The study of the diffusion of gases or aerosols in the lower atmosphere. Report AFCRC-TR-58-239, Massachusetts Institute of Technology, Cambridge.

Csanady, G.T. (1955). Dispersal of dust particles from elevated sources. *Aust. J. Phys.*, *8*, 545–550.

Csanady, G.T. (1973). *Turbulent Diffusion in the Environment*. D. Reidel Pub. Co., Dordrecht, Holland.

Davis, P.A. (1983). Markov-chain simulations of vertical dispersion from elevated sources into the natural planetary boundary layer. *Boundary-Layer Meteorol. 26*, 355–376.

Deardorff, J.W. (1966). The countergradient heat flux in the lower atmosphere and in the laboratory. *J. Atmos. Sci.*, *23*, 503–506.

Deardorff, J.W. (1970a). A three-dimensional numerical investigation of the idealized planetary boundary layer. *Geophys. Fluid Dyn.*, *1*, 377–410.

Deardorff, J.W. (1970b). Convective velocity and temperature scales for the unstable planetary boundary layer and for Rayleigh convection. *J. Atmos. Sci.*, *27*, 1211–1213.

Deardorff, J.W. (1972). Numerical investigation of neutral and unstable planetary boundary layers. *J. Atmos. Sci.*, *29*, 91–115.

Deardorff, J.W. (1973). Three-dimensional numerical modeling of the planetary boundary layer. In *Workshop on Micrometeorology* (D.A. Haugen, ed.), pp. 271–311. American Meteorological Society, Boston.

Deardorff, J.W. (1974). Three-dimensional numerical study of the height and mean structure of a heated planetary boundary layer. *Boundary-Layer Meteorol. 7*, 81–106.

Deardorff, J.W. (1978). Observed characteristics of the outer layer. In *Short Course on the Planetary Boundary Layer* (A.K. Blackadar, ed.). Amer. Meteorological Society, Boston.

Deardorff, J.W. (1980). Stratocumulus-capped mixed layers derived from a three-dimensional model. *Boundary-Layer Meteorol.*, *18*, 495–527.

Deardorff, J.W., and Peskin, R.L. (1970). Lagrangian statistics from numerically integrated turbulent shear flow. *Phys. Fluids*, *13*, 584–595.

Deardorff, J.W., and Willis, G.E. (1974). Computer and laboratory modeling of vertical diffusion of non-buoyant particles in the mixed layer. *Adv. Geophys.*, *18B*, 187–200.

Deardorff, J.W., and Willis, G.E. (1975). A parameterization of diffusion in the mixed layer. *J. Appl. Meteorol.*, *14*, 1451–1458.

Deardorff, J.W., and Willis, G.E. (1982). Ground-level concentrations due to fumigation into an entraining mixed layer. *Atmos. Environ.*, *16*, 1159–1170.

de Bass, A.F., van Dop, H., and Nieuwstadt, F.T.M. (1986). An application of the Langevin equation for inhomogeneous conditions to dispersion in a convective boundary layer. *Quart. J. Roy. Meteorol. Soc.*, *112*, 165–180.

Demerjian, K.L., and Schere, K.L. (1979). Applications of a photochemical box model for O_3 air quality in Houston, Texas. *Proceedings of Specialty Conference on Ozone/Oxidants: Interactions with the Total Environment*, U.S. Environmental Protection Agency, Research Triangle Park, NC.

Demuth, C. (1978). A contribution to the analytic steady solution of the diffusion equation for line source. *Atmos. Environ.*, *12*, 1255–1258.

Dennis, R.L., Byun, D.W., Novak, J.H., Galluppi, K.J., and Coats, C.J. (1996). The next generation of integrated air quality modeling: EPA Models-3. *Atmos. Environ.*, *30*, 1925–1938.

Detering, H.W., and Etling, D. (1985). Application of the E-ε turbulence model to the atmospheric boundary layer. *Boundary-Layer Meteorol.*, *33*, 113–133.

Diehl, S.R., Smith, D.T., and Sydor, M. (1982). Random-walk simulation of gradient-transfer processes applied to dispersion of stack emission from coal-fired power plants. *J. Appl. Meteorol.*, *21*, 69–83.

Dodge, M.C. (1977). Combined use of modeling techniques and smog chamber data to derive ozone precursor relationships. In *Proceedings of the International Conference on Photochemical Oxidant Pollution and its Control, Vol. II* (Dimitriades, ed.), pp. 881–889. EPA-

600/3-77-001b, U.S. Environmental Protection Agency, Research Triangle Park, NC.

Donaldson, C. du P. (1973). Construction of a dynamic model of the production of atmospheric turbulence and the dispersal of atmospheric pollutants. In *Workshop on Micrometeorology* (D.A. Haugen, ed.), pp. 313–392. American Meteorological Society, Boston.

Draxler, R.R. (1976). Determination of atmospheric diffusion parameters. *Atmos. Environ.*, 10, 99–105.

Draxler, R.R. (1984). Diffusion and transport experiments. In *Atmospheric Science and Power Production* (D. Randerson, ed.) pp. 367–422. U.S. Department of Energy, Technical Information Center, Oak Ridge, TN.

Droppo, J.G. (1980). Experimental techniques for dry deposition measurements. In *Atmospheric Sulfur Deposition* (D.S. Shriner, C.R. Richmond, and S.E. Lindberg, eds.), pp. 209–221. Ann Arbor Press, Ann Arbor, MI.

Dutton, J.A. (1995). *Dynamics of Atmospheric Motion*. Dover Pub., New York.

Dyer, A.J. (1965). The flux-gradient relation for turbulent heat transfer in the lower atmosphere. *Quart. J. Roy. Meteorol. Soc.*, 91, 151–157.

Dyer, A.J. (1967). The turbulent transport of heat and water vapor in unstable atmosphere. *Quart. J. Roy. Meteorol. Soc.*, 93, 501–508.

Eastman, J.L., Pielke, R.A., and Lyons, W.A. (1995). Comparison of lake-breeze model simulations with tracer data. *J. Appl. Meteorol.*, 34, 1398–1418.

Eberhard, W.L., Moninger, W.R., and Briggs, G.A. (1988). Plume dispersion in the convective boundary layer. Part I. CONDORS field experiment and example measurements. *J. Appl. Meteorol.*, 27, 599–616.

Einstein, A. (1905). Uber die von molekularkinetischen Theorie der warme geforderte Bewegung von in ruhenden Flussigkeiten suspendierten Teilchen. *Ann. Phys.*, 17, 549.

Eisenbud, M. (1973). *Environmental Radioactivity*, 2 ed. Academic Press, New York.

Eisenbud, M. (1976). The primary air pollutants—radioactive; their occurrence, sources, and effects. In *Air Pollution*, 3rd ed., Vol. I (A.C. Stern, ed.), pp. 197–231. Academic Press, New York.

Ellison, T.H. (1959). Turbulent diffusion. *Sci. Progress*, 47, 495–506.

Enger, L. (1986). A higher order closure model applied to dispersion in a convective PBL. *Atmos. Environ.*, 20, 879–894.

Etling, D., Preuss, J., and Wamser, M. (1986). Application of a random walk model to turbulent diffusion in complex terrain. *Atmos. Environ.*, 20, 741–747.

Fernandez, de la Mora, J., and Friedlander, S.K. (1982). Aerosol and gas deposition to fully rough surfaces: Filteration model for blade shaped elements. *Int. J. Heat Mass Transfer*, 25, 1725–1735.

Ferziger, J.H. (1993). Subgrid-scale modeling. In *Large Eddy Simulation of Complex Engineering and Geophysical Flows* (B. Galperin and S.A. Orszag, eds.), pp. 37–54. Cambridge University Press, Cambridge, England.

Finlayson-Pitts, B.J., and Pitts, J.N. (1986). *Atmospheric Chemistry*. Wiley-Interscience Pub., New York.

Fleagle, R.G., and Businger, J.A. (1980). *An Introduction to Atmospheric Physics*, 2d ed. Academic Press, New York.

Flohn, H. (1969). *Climate and Weather*. McGraw-Hill, New York.

Fox, D.G. (1981). Judging air quality model performance. *Bull. Amer. Meteorol. Soc.*, 62, 599–609.

Fox, D.G. (1984). Uncertainty in air quality modeling. *Bull. Amer. Meteorol. Soc.*, 65, 27–36.

Frenkiel, F.N., and Katz, I. (1956). Studies of small-scale turbulent diffusion in the atmosphere. *J. Meteorol.*, 13, 388–394.

Fuquay, J., Simpson, C.L., and Hinds, W.T. (1964). Prediction of environmental exposures from sources near the ground based on Hanford experimental data. *J. Appl. Meteorol.*, 3, 761–770.

Galperin, B., and Orszag, S.A. (1993). *Large Eddy Simulation of Complex Engineering and Geophysical Flows*. Cambridge University Press, Cambridge, England.

Garratt, J.R. (1992). *The Atmospheric Boundary Layer*. Cambridge University Press, Cambridge, England.

Garratt, J.R., Wyngaard, J.C., and Francey, R.J. (1982). Winds in the atmospheric boundary layer—prediction and observation. *J. Atmos. Sci.*, 39, 1307–1316.

Gates, D.M. (1980). *Biophysical Ecology*. Springer-Verlag, Berlin.

Gates, O.M. (1972). *Climate*. Harper and Row, New York.

Germano, M. (1991). Turbulence: The filtering approach. *J. Fluid Mech.*, 238, 325–336.

Germano, M., Piomelli, U., Moin, P., and Cabot, W.H. (1991). A dynamic subgrid-scale eddy viscosity model. *Phys. Fluids*, A3, 1760–1765.

Gery, M., Whitten, G., Killus, J., and Dodge, M. (1989). A photochemical kinetics mechanism for urban and regional scale computer modeling. *J. Geophys. Res.*, 94, 12,925–12,956.

Gibson, M.M., and Launder, B.E. (1978). Ground effects on pressure fluctuations in the atmospheric boundary layer. *J. Fluid Mech.*, 86, 491–511.

Gifford, F.A. (1953). A study of low level air trajectories at Oak Ridge, Tennessee. *Mon. Weath. Rev.*, 81, 179–192.

Gifford, F.A. (1955). A simultaneous Lagrangian-Eulerian turbulence experiment. *Mon. Weath. Rev.*, 83, 293–301.

Gifford, F.A. (1957a). Relative atmospheric diffusion of smoke puffs. *J. Meteorol.*, 14, 410–414.

Gifford, F.A. (1957b). Further data on relative atmospheric diffusion. *J. Meteorol.*, 14, 475.

Gifford, F.A. (1959a). Statistical properties of a fluctuating plume dispersion model. *Advan. Geophys.*, 6, 117–137.

Gifford, F.A. (1959b). Computation of pollution from several sources. *Int. J. Air Water Pollut.*, 2, 109–110.

Gifford, F.A. (1961). Use of routine meteorological observations for estimating atmospheric dispersion. *Nuclear Safety*, 2, 47–51.

Gifford, F.A. (1962). Diffusion in the diabatic surface layer. *J. Geophys. Res.*, 67, 3207–3212.

Gifford, F.A. (1968). An outline of theories of diffusion in the lower layers of the atmosphere. In *Meteorology and Atomic Energy—1968* (D.H. Slade, ed.), pp. 65–116. U.S. Department of Energy, Technical Information Center, Oak Ridge, TN.

Gifford, F.A. (1976). Turbulent diffusion-typing schemes: A review. *Nuclear Safety*, 17, 68–85.

Gifford, F. (1977). Tropospheric relative diffusion observations. *J. Appl. Meteorol.*, 16, 311–313.

Gifford, F.A. (1981). Estimating ground-level concentration patterns from isolated air-pollution sources: A brief summary. *Environ. Res.*, 25, 126–138.

Gifford, F.A., and Hanna, S.R. (1973). Modeling urban air pollution. *Atmos. Environ.*, 7, 131–136.

Giovannoni, J.-M. and Russell, A. (1995). Impact of using prognostic and objective wind fields on the photochemical modeling of Athens, Greece. *Atmos. Environ.*, 29, 3633–3653.

Godish, T. (1991). *Air Quality*, 2d ed. Lewis Pub., Chelsea, MI.

Golder, D. (1972). Relations among stability parameters in the surface layer. *Boundary-Layer Meteorol.*, 3, 47–58.

Goldsmith, J.R., and Friberg, L.T. (1977). Effects of air pollution on human health. In *Air Pollution*, 3rd ed., Vol. II (A.C. Stern, ed.). Academic Press, New York.

Goody, R.M. and Yung, Y.L. (1990). *Atmospheric Radiation: Theoretical Basis*, 2d ed. Oxford University Press, New York.

Griffin, R.D. (1994). *Principles of Air Quality Management*. Lewis Pub., Ann Arbor, MI.

Griffiths, R.F. (1994). Errors in the use of Briggs parameterization for atmospheric dispersion coefficients. *Atmos. Environ.*, 28, 2861–2865.

Gross, G., Vogel, H., and Wipperman, F. (1987). Dispersion over and around a steep obstacle for varying thermal stratification—numerical simulations. *Atmos. Environ.*, 21, 483–490.

Hage, K.D. (1964). Particle fallout and dispersion below 30 km in the atmosphere, Report SC-DC-64–1463, Sandia Corp., Natl. Tech. Info. Serv., Springfield, VA.

Halitsky, J. (1968). Gas diffusion near buildings. In *Meteorology and Atomic Energy—1968* (D.H. Slade, ed.), pp. 221–255. U.S. Department of Energy, Technical Information Center, Oak Ridge, TN.

Hall, C.D. (1975). The simulation of particle motion in the atmosphere by a numerical random-walk model. *Quart. J. Roy. Meteorol. Soc.*, 101, 235–244.

Haltiner, G.J., and Martin, F.L. (1957). *Dynamical and Physical Meteorology*. McGraw-Hill, New York.

Hanna, S.R. (1971). A simple method of calculating dispersion from urban area sources. *J. Air Pollut. Cont. Assoc.*, 21, 774–777.

Hanna, S.R. (1979). Some statistics of Lagrangian and Eulerian wind fluctuations. *J. Appl. Meteorol.*, 18, 518–525.

Hanna, S.R. (1981). Lagrangian and Eulerian time-scale relations in the day-time boundary layer. *J. Appl. Meteorol.*, 20, 242–249.

Hanna, S.R. (1982a). Applications in air pollution modeling. In *Atmospheric Turbulence and Air Pollution Modeling* (F.T.M. Nieuwstadt and H. van Dop, eds.), pp. 275–310. D. Reidel Pub. Co., Dordrecht, Holland.

Hanna, S.R. (1982b). A review of atmospheric diffusion models for regulatory applications. Technical Note No. 177, WMO-No. 581, World Meteorological Organization, Geneva.

Hanna, S.R., Briggs, G.A., Deardorff, J., Egan, B.A., Gifford, F.A., and Pasquill, F. (1977). AMS Workshop on Stability Classification Schemes and Sigma Curves—summary of recommendations. *Bull. Amer. Meteorol. Soc.*, 58, 1305–1309.

Hanna, S.R., Briggs, G.A., and Hosker, R.P. (1982). *Handbook on Atmospheric Diffusion*. U.S. Department of Energy, Technical Information Center, Oak Ridge, TN.

Hanna, S.R., and Chang, J.C. (1992). Boundary-layer parameterizations for applied dispersion modeling over urban areas. *Boundary-Layer Meteorol.*, 58, 229–259.

Hansen, D.A., Dennis, R.L., Ebel, A., Hanna, S.R., Kaye, J., and Thuillier, R. (1994). The quest for an advanced regional air quality model. *Environ. Sci. Tech.*, 28, 71A–77A.

Harley, R.A., Russell, A.G., McRae, G.J., Cass, G.R., and Seinfeld, J.H. (1993). Photochemical modeling of the Southern California air quality study. *Environ. Sci. Tech.*, 27, 378–388.

Haugen, D.A. (1959). Project Prairie Grass, a field program in diffusion. *Geophys. Res. Papers, No. 59, Vol. III*, Report AFCRC-TR-58–235, U.S. Air Force Cambridge Research Center.

Hay, J.S., and Pasquill, F. (1957). Diffusion from a fixed source at a height of a few hundred feet in the atmosphere. *J. Fluid Mech.*, 2, 299–310.

Hay, J.S., and Pasquill, F. (1959). Diffusion from a continuous source in relation to the spectrum and scale of turbulence. In *Atmospheric Diffusion and Air Pollution* (F.N. Frenkiel and P.A. Sheppard, eds.), *Advances in Geophys.*, 6, 345–365. Academic Press, New York.

Heffter, J.L. (1965). The variation of horizontal diffusion parameters with time for travel periods of one hour or longer. *J. Appl. Meteorol.*, 4, 153–156.

Henn, D.S., and Sykes, R.I. (1992). Large-eddy simulation of dispersion in the convective boundary layer. *Atmos. Environ.*, 26A, 3145–3159.

Hess, S.L. (1959). *Introduction to Theoretical Meteorology*. Holt, Rinehart and Winston, New York.

Hibberd, M.F., and Sawford, B.L. (1994). A saline laboratory model of the planetary covective boundary layer. *Boundary-Layer Meteorol.*, 67, 229–250.

Hicks, B.B., and Wesely, M.L. (1978). An examination of some micrometeorological methods for measuring dry deposition. Report No. EPA-600/7-78-160, U.S. Environmental Protection Agency, Research Triangle Park, NC.

Hinze, J.O. (1975). *Turbulence*, 2d ed. McGraw-Hill, New York.

Hoerner, S.F. (1958). *Fluid Dynamic Drag*. Published by the author, New York.

Hogstrom, U. (1964). An experimental study on atmospheric diffusion. *Tellus*, 16, 205–251.

Hojstrup, J. (1982). Velocity spectra in the unstable boundary layer. *J. Atmos. Sci.*, 39, 2239–2248.

Holt, T., and Raman, S. (1988). A review and comparative evaluation of multilevel boundary layer parameterizations for first-order and turbulent kinetic energy closure schemes. *Rev. Geophys.*, 26, 761–780.

Holton, J.R. (1992). *An Introduction to Dynamic Meteorology*, 3rd ed. Academic Press, San Diego.

Holzworth, G.C. (1964). Estimates of mean maximum mixing depths in the contiguous United States. *Mon. Weath. Rev.*, 92, 235–242.

Hoot, T.G., Meroney, R.N., and Peterka, J.A. (1973). Wind tunnel tests of negatively buoyant plumes. Report No. EPA-660/3-74-003, U.S. Environmental Protection Agency, Research Triangle Park, NC.

Horst, T.W. (1977). A surface depletion model for deposition from a Gaussian plume. *Atmos. Environ.*, 11, 41–46.

Horst, T.W. (1979). Lagrangian similarity modeling of vertical diffusion from a ground-level source. *J. Appl. Meteorol.*, 18, 733–740.

Horst, T.W. (1980a). Discussion on "a numerical study of the vertical dispersion of passive contaminants from a continuous source in the atmospheric surface layer." *Atmos. Environ.*, 14, 267–269.

Horst, T.W. (1980b). A review of Gaussian diffusion-deposition models. In *Atmospheric Sulfur Deposition* (D.S. Shriner, C.R. Richmond, S.E. Lindberg, eds.), pp. 275–283, Ann Arbor Science, Ann Arbor, MI.

Hosker, R.P. (1984). Flow and diffusion near obstacles. In *Atmospheric Science and Power Production* (D. Randerson, ed.), pp. 241–326. U.S. Department of Energy, Technical Information Center, Oak Ridge, TN.

Houghton, J.T. (1977). *The Physics of Atmospheres*. Cambridge University Press, Cambridge, England.

Hoult, D.P., and Weil, J.C. (1972). A turbulent plume in a laminar crossflow. *Atmos. Environ.*, 6, 513–531.

Huang, C.H. (1979). A theory of dispersion in turbulent shear flow. *Atmos. Environ.*, 13, 453–463.

Huber, A.H., and Snyder, W.H. (1982). Wind tunnel investigation of the effects of a rectangular-shaped building on dispersion of effluents from short adjacent stacks. *Atmos. Environ.*, 16, 2837–2848.

Hunt, J.C.R., and Simpson, J.E. (1982). Atmospheric boundary layer over nonhomogeneous terrain. In *Engineering Meteorology* (E.J. Plate, ed.), pp. 269–318. Elsevier, Amsterdam.

Hunt, J.C.R., and Snyder, W.H. (1980). Experiments on stably and neutrally stratified flow over a model three-dimensional hill. *J. Fluid Mech.*, 96, 671–704.

Hunt, J.C.R., and Weber, A.H. (1979). A Lagrangian statistical analysis of diffusion from a ground-level source in a turbulent boundary layer. *Quart. J. Roy. Meteorol. Soc.*, 105, 423–443.

Hurley, P., and Physick, W. (1993). A skewed homogeneous Lagrangian particle model for convective conditions. *Atmos. Environ.*, 27A, 619–624.

Ide, Y., Okamoto, H., and Okabayashi, K. (1988). On the parameterization of diffusion in the turbulent boundary layer. *Atmos. Environ.*, 22, 1863–1870.

Intergovernmental Panel on Climate Change (1995). *Climate Change 1995: The Science of Climate Change.* Cambridge University Press, Cambridge, England.

Irwin, J.S. (1979a). Estimating plume dispersion—a recommended generalized scheme. *Preprints, Fourth Symposium on Turbulence, Diffusion, and Air Pollution*, pp. 62–69. American Meteorological Society, Boston.

Irwin, J.S. (1979b). Scheme for estimating dispersion parameters as a function of release height. Report No. EPA 600/4-79-062, U.S. Environmental Protection Agency, Research Triangle Park, NC.

Irwin, J.S. (1980). Dispersion estimate suggestion #8: Estimation of Pasquill stability categories. Docket Reference No. II-B-10. U.S. Environmental Protection Agency, Research Triangle Park, NC.

Irwin, J.S. (1983). Estimating plume dispersion—a comparison of several sigma schemes. *J. Climate Appl. Meteorol.*, 22, 92–114.

Irwin, J.S., and Binkowski, F.S. (1981). Estimating of the Monin-Obukhov scaling length using on-site instrumentation. *Atmos. Environ.*, 15, 1091–1094.

Islitzer, N.F., and Slade, D.H. (1968). Diffusion and transport experiments. In *Meteorology and Atomic Energy—1968* (D.H. Slade, ed.), pp. 117–188. U.S. Department of Energy, Technical Information Center, Oak Ridge, TN.

Izumi, Y. and Caughey, S.J. (1976). Minnesota 1973 atmospheric boundary layer experiment data report. AFGL Environmental Research Paper No. 547, National Technical Information Service, Springfield, Virginia.

Jacobson, A.R., and Morris, S.C. (1976). The primary air pollutants—viable particulates; their occurrence, sources and effects. In *Air Pollution*, 3rd ed., *Vol. I* (A.C. Stern, ed.), pp. 169–196. Academic Press, New York.

Kaimal, J.C., Eberhard, W.L., Moninger, W.M., Gaynor, J.E., Troxel, S.W., Uttal, T., Briggs, G.A., and Start, G.E. (1986). Project CONDORS: Convective diffusion observed by remote sensors. Boulder Atmospheric Observatory Report No. 7, NOAA Wave Propagation Laboratory, Boulder, CO.

Kaimal, J.C., Eversole, R.A., Lenschow, D.H., Stankov, B.B., Kahn, P.H., and Businger, J.A. (1982). Spectral characteristics of the convective boundary layer over uneven terrain. *J. Atmos. Sci.*, 39, 1098–1114.

Kaimal, J.C. and Finnigan, J.J. (1994). *Atmospheric Boundary Layer Flows.* Oxford University Press, Oxford, England.

Kaimal, J.C., Wyngaard, J.C., Haugen, D.A. Cote, O.R., Izumi, Y., Caughey, S.J., and Readings, C.J. (1976). Turbulence structure in the convective boundary layer. *J. Atmos. Sci.*, 33, 2152–2169.

Kaimal, J.C., Wyngaard, J.C., Izumi, Y., and Cote, O.R. (1972). Spectral characteristics of surface layer turbulence. *Quart. J. Roy. Meteorol. Soc.*, 98, 563–589.

Kampe de Feriet, M.J. (1939). Les fonctions aleatoires stationnaires et la theorie statistique de la turbulence homogene. *Ann. Soc. Sci. Brux.*, 59, 145–194.

Kao, S.K. (1984). Theories of atmospheric transport and diffusion. In *Atmospheric Science and Power Production* (D. Randerson, ed.), pp. 189–239. U.S. Department of Energy, Technical Information Center, Oak Ridge, TN.

Kao, S.K., and Yeh, Y.N. (1979). A note on the effects of stack and inversion heights on diffusion of pollutants in the planetary boundary layer. *Atmos. Environ.*, 13, 873–878.

Karacostas, T.S., and Marwitz, J.D. (1980). Turbulent kinetic energy budgets over mountainous terrain. *J. Appl. Meteorol.*, 19, 163–174.

Kazanski, A.B., and Monin, A.S. (1957). Shape of smoke plumes. *Izv. Acad. Nauk. SSSR, Geofiz. Ser.*, No. 8, 1020–1033.

Kellog, W.W. (1956). Diffusion of smoke in the stratosphere. *J. Meteorol.*, 13, 241–250.

Kiehl, J.T., and Trenberth, K.E. (1977). Earth's annual global mean energy budget. *Bull. Amer. Meteorol. Soc.*, 78, 197–208.

King, J.C. (1988). Some measurements of turbulence length scale in the stably-stratified surface layer. In *Stably Stratified Flow and Dense Gas Dispersion* (J.S. Puttock, ed.), pp. 39–53. Clarendon Press, Oxford.

Klemp, J.B., and Lilly, D.K. (1975). The dynamics of wave-induced downslope winds. *J. Atmos. Sci.*, 32, 320–339.

Kolmogorov, A.N. (1941). Local structure of turbulence in an incompressible fluid at very high Reynolds numbers. *Doklady Akad. Nauk., SSSR*, 30, 299–303.

Koppen, W., and Geiger, R. (1955). *Climate of the Earth*, 3rd ed. Wallmap and Perthes, Darmstadt, Germany.

Kundu, P.K. (1990). *Fluid Mechanics.* Academic Press, New York.

Lamb, B., Guenther, A., Gay, B., and Westberg, H. (1987). A national inventory of biogenic hydrocarbon emissions. *Atmos. Environ.*, 21, 1695–1705.

Lamb, R.G. (1978). A numerical simulation of dispersion from an elevated point source in the convective boundary layer. *Atmos. Environ.*, 12, 1297–1304.

Lamb, R.G. (1979). The effects of release height on material dispersion in the convective planetary boundary layer. *Preprint Vol., Fourth Symposium on Turbulence, Diffusion, and Air Pollution*, pp. 27–33. American Meteorological Society, Boston.

Lamb, R.G. (1982). Diffusion in the convective boundary layer. In *Atmospheric Turbulence and Air Pollution Modelling* (F.T.M. Nieuwstadt and H. van Dop, eds.), pp. 159–229. D. Reidel Pub. Co., Dordrecht, Holland.

Lamb, R.G. (1983). A regional scale (1000 km) model of photochemical air pollution: Part 1. Theoretical formulation. Report no. EPA-600/3-83-035, U.S. Environmental Protection Agency, Research Triangle Park, NC.

Lamb, R.G. (1985). A regional-scale (1000 km) model of photochemical air pollution, Part 3. Tests of the numerical algorithms. Report no. EPA-600/3-85-037, U.S. Environmental Protection Agency, Research Triangle Park, NC.

Lamb, R.G. (1986). Numerical simulations of photochemical air pollution in the northeastern United States: ROM1 applications. Report No. EPA-600/3-86-038, U.S. Environmental Protection Agency, Research Triangle Park, NC.

Lamb, R.G., and Duran, D.R. (1978). Eddy diffusivities derived from a numerical model of the convective planetary boundary layer. *Nuovo Cimento, 1C,* 1–17.

Landsberg, H.E., and Essenwanger, O., *World Survey of Climatology, Vol. 1A: General Climatology.* Elsevier, Amsterdam.

Launder, B.E., Reece, G.J., and Rodi, W. (1975). Progress in the development of a Reynolds-stress turbulence closure. *J. Fluid Mech., 68,* 537–566.

Legg, B.J. (1983). Turbulent dispersion from an elevated line source: Markov-chain simulations of concentration and flux profiles. *Quart. J. Roy. Meteorol. Soc., 109,* 645–660.

Legg, B.J., and Raupach, M.R. (1982). Markov-chain simulation of particle dispersion in inhomogeneous flows: The mean drift velocity induced by a gradient in Eulerian velocity variance. *Boundary-Layer Meteorol., 24,* 3–13.

Leith, C.E. (1990). Stochastic backscatter in a subgrid-scale model: Plane shear mixing layer. *Phys. Fluids, A2,* 297–299.

Lenschow, D.H., Li, X.S., Zhu, C.J., and Stankov B.B. (1988a). The stably stratified boundary layer over the Great Plains. I. Mean and turbulence structure. *Boundary-Layer Meteorol., 42,* 95–121.

Lenschow, D.H., Zhang, S.F., and Stankov, B.B. (1988b). The stably stratified boundary layer over the Great Plains. II. Horizontal variations and spectra. *Boundary-Layer Meteorol., 42,* 123–135.

Leonard, A. (1974). Energy cascade in large eddy simulations of turbulent fluid flows. *Adv. Geophys., 18A,* 237–248.

Leone, J.A., and Seinfeld, J.H. (1985). Comparative analysis of chemical reaction mechanisms for photochemical smog. *Atmos. Environ., 19,* 437–464.

Leslie, D.C., and Quarini, G.L. (1979). The application of turbulence theory to the formulation of subgrid modelling procedures. *J. Fluid Mech., 91,* 65–91.

Lettau, H. (1970). Physical and meteorological basis for mathematical models of urban diffusion. In *Proceedings of Symposium on Multiple Source Urban Diffusion Models,* Air Pollution Control Official Publication No. AP86, U.S. Environmental Protection Ageny.

Lettau, H.H., and Davidson, B. (1957). *Exploring the Atmosphere's First Mile.* Pergamon Press, Oxford, England.

Lewellen, W.S., and Teske, M. (1975). Turbulence modeling and its application to atmospheric diffusion. Part I: Recent program development, verification, and application. Report No. EPA-600/4-75-016A, U.S. Environmental Protection Agency, Research Triangle Park, NC.

Lewellen, W.S., and Teske, M.E. (1976). Second-order closure modeling of diffusion in the atmospheric boundary layer. *Boundary-Layer Meteorol., 10,* 69–90.

Ley, A.J. (1982). A random walk simulation of two-dimensional turbulent diffusion in the neutral surface layer. *Atmos. Environ., 16,* 2799–2808.

Lilly, D.K. (1967). The representation of small-scale turbulence in numerical simulation experiments. In *Proceedings of IBM Scientific Symposium on Environmental Sciences,* pp. 195–210. Thomas J. Watson Research Center, Yorktown Heights, NY.

Lin, C.C. (1960). On a theory of dispersion by continuous movements. *Proc. Natl. Acad. Sci., 46,* 566–570.

Lin, J.T., and Pao, Y.H. (1979). Wakes in stratified fluids. *Ann. Rev. Fluid Mech., 11,* 317–338.

Liou, K.N. (1980). *An Introduction to Atmospheric Radiation.* Academic Press, New York.

Louis, J.-F. (1979). A parametric model of vertical eddy fluxes in the atmosphere. *Boundary-Layer Meteorol., 71,* 187–202.

Lovelock, J.E., Maggs, R.J., and Wade, R.J. (1973). Halogenated hydrocarbons in and over Atlantic. *Nature, 241,* 194–196.

Lowry, W.P. (1970). *Weather and Life: An Introduction to Biometeorology.* Academic Press, New York.

Lu, J., and Arya, S.P. (1995). A laboratory simulation of urban heat-island-induced circulation in a stratified environment. In *Wind Climate in Cities* (J.E. Cermak, A.G. Davenport, E.J. Plate, and D.X. Viegas, eds.), pp. 405–429. Kluwer, Dordrecht, Holland.

Lucas, D.H. (1958). The atmospheric polluton of cities. *Int. J. Air Pollut., 1,* 71–86.

Luhar, A.K., and Britter, R.E. (1989). A random walk model for dispersion in inhomogeneous turbulence in a convective boundary layer. *Atmos. Environ., 23,* 1911–1924.

Luhar, A.K., Hibbard, M.F., and Hurley, P.J. (1996). Comparison of closure schemes used to specify the velocity PDF in Lagrangian stochastic dispersion models for convective conditions. *Atmos. Environ., 30,* 1407–1418.

Luhar, A.K., and Rao, K.S. (1993). Random-walk model studies of the transport and diffusion of pollutants in katabatic flows. *Boundary-Layer Meteorol., 66,* 395–412.

Lumley, J.L. (1979). Computational modeling of turbulent flows. *Adv. Appl. Mech., 18,* 123–176.

Lumley, J.L. (1980). Second-order modeling of turbulent flows. In *Prediction Methods for Turbulent Flows* (W. Kollmann, ed.), pp. 1–31. Hemisphere Pub., London.

Lumley, J.L., and Khajeh-Nouri, B. (1974). Computational modeling of turbulent transport. *Adv. Geophys., 18A,* 169–192.

Lumley, J.L., and Panofsky, H.A. (1964). *The Structure of Atmospheric Turbulence.* Wiley-Interscience, New York.

Lyons, T.J., and Scott, W.D. (1990). *Principles of Air Pollution Meteorology.* CRC Press, Boca Raton, FL.

MacCready, P.B., Smith, T.B., and Wolf, M.A. (1961). Vertical diffusion from a low altitude line source, Vols. I and II. Meteorology Research, Inc., Altadena, CA.

Machta, L. (1958). Global scale dispersion by the atmosphere. In *Proceedings of the Second United Nations Conference on the Peaceful Uses of Atomic Energy,* Geneva, *Vol. 18,* p. 519.

Mason, P.J. (1989). Large-eddy simulation of the convective atmospheric boundary layer. *J. Atmos. Sci., 46,* 1492–1516.

Mason, P.J. (1994). Large-eddy simulation: A critical review of the technique. *Quart. J. Roy. Meteorol. Soc., 120,* 1–26.

Mason, P.J., and Brown, A.R. (1994). The sensitivity of large-eddy simulations of turbulent shear flow to sub-grid models. *Bound.-Layer Meteorol., 70,* 133–150.

Mason, P.J., and King, J.C. (1984). Atmospheric boundary layer over a succession of two-dimensional ridges and valleys. *Quart. J. Roy. Meteorol. Soc., 110,* 821–845.

Mason, P.J., and Sykes, S.I. (1979). Flow over an isolated hill of moderate slope. *Quart. J. Roy. Meteorol. Soc., 105,* 383–395.

Mason, P.J., and Thompson, D.J. (1992). Stochastic backscat-

ter in large-eddy simulations of boundary layers. *J. Fluid Mech.*, *242*, 51–78.
McElroy, J.L., and Pooler, F. (1968). St. Louis dispersion study. U.S. Public Health Service, National Air Pollution Control Administration, Report AP-53.
McMahon, T.A., and Denison, P.J. (1979). Empirical atmospheric deposition parameters—a survey. *Atmos. Environ.*, *13*, 571–585.
McNider, R.T., Moran, M.D., and Pielke, R.A. (1988). Influence of diurnal and inertial boundary layer oscillations on long-range dispersion. *Atmos. Environ.*, *22*, 2445–2462.
McRae, G.J., Goodin, W.R., and Seinfeld, J.H. (1982). Development of a second generation mathematic model for urban air pollution. I. Model formulation. *Atmos. Environ.*, *16*, 679–696.
Mellor, G.L., and Yamada, T. (1974). A hierarchy of turbulence closure models for planetary boundary layers. *J. Atmos. Sci.*, *31*, 1791–1806.
Mellor, G.L., and Yamada, T. (1982). Development of a turbulence closure model for geophysical fluid problems. *Rev. Geophys.*, *20*, 851–875.
Meroney, R.N. (1990). Fluid dynamics of flow over hills/mountains—insights obtained through physical modeling. In *Atmospheric Processes over Complex Terrain* (W. Blumen, ed.), *Meteorological Monographs*, *23*, 145–171. American Meteorological Society, Boston.
Merzkirch, W. (1974). *Flow Visualization*. Academic Press, New York.
Merzkirch, W. (1987). *Flow Visualization*, 2d ed. Academic Press, New York.
Miller, A., and Thompson, J.C. (1970). *Elements of Meteorology*. Merrill Publishing Co., Columbus, OH.
Miller, A., Thompson, J.C., Peterson, R.E., and Haragan, D.R. (1983). *Elements of Meteorology*, 4th ed., Merrill Publishing Co., Columbus, OH.
Moeng, C.-H. (1984). A large-eddy-simulation model for the study of planetary boundary-layer turbulence. *J. Atmos. Sci.*, *41*, 2052–2062.
Moeng, C.H., and Sullivan, P.J. (1994). A comparison of shear- and buoyancy-driven planetary boundary layer flows. *J. Atmos. Sci.*, *51*, 999–1022.
Moeng, C.-H., and Wyngaard, J.C. (1988). Spectral analysis of large-eddy simulations of the convective boundary layer. *J. Atmos. Sci.*, *45*, 3575–3587.
Monin, A.S. (1959a). Smoke propagation in the surface layer of the atmosphere. In *Atmospheric Diffusion and Air Pollution* (F.N. Frenkiel and P.A. Sheppard, eds.), *Advances in Geophysics*, *6*, 331–343. Academic Press, New York.
Monin, A.S. (1959b). Turbulent diffusion in the surface layer under stable stratification. In *Atmospheric Diffusion and Air Pollution* (F.N. Frenkiel and P.A. Sheppard, eds.), *Advances in Geophysics*, *6*, 429–434. Academic Press, New York.
Monin, A.S., and Obukhov, A.M. (1954). Basic turbulent mixing laws in the atmospheric surface layer. *Tr. Geofiz. Inst. Akad. Nauk. SSSR*, *24* (151), 163–187.
Monin, A.S., and Yaglom, A.M. (1971). *Statistical Fluid Mechanics: Mechanics of Turbulence, Vol. 1*. MIT Press, Cambridge, MA.
Monin, A.S., and Yaglom, A.M. (1975). *Statistical Fluid Mechanics: Mechanics of Turbulence, Vol. 2*. MIT Press, Cambridge, MA.
Monj, N. (1972). Budgets of turbulent energy and temperature variance in the transition zone from forced to free convection. Ph.D. thesis, Department of Atmospheric Sciences, University of Washington, Seattle, Washington.
Monteith, J.L., and Unsworth, M.H. (1990). *Principles of Environmental Physics*, 2d ed. Edward Arnold, London.
Moran, M.D., and Pielke, R.A. (1996a). Evaluation of a mesoscale atmospheric dispersion modeling system with observations from the 1980 Great Plains Mesoscale Tracer Field Experiment. Part I: Datasets and meteorological simulations. *J. Appl. Meteorol.*, *35*, 281–307.
Moran, M.D., and Pielke, R.A. (1996b). Evaluation of a mesoscale atmospheric dispersion modeling system with observations from the 1980 Great Plains Mesoscale Tracer Field Experiment. Part II: Dispersion simulation. *J. Appl. Meteorol.*, *35*, 308–329.
Morris, R.E., and Myers, T.C. (1990). *User's Guide to the Urban Airshed Model*, Vol. I-V. U.S. Environmental Protection Agency, Research Triangle Park, NC.
Morton, B.R., and Middleton, J. (1973). Scale diagrams for forced plumes. *J. Fluid Mech.*, *58*, 165–176.
Morton, B.R., Taylor, G.I., and Turner, J.S. (1956). Turbulent gravitational convection from maintained and instantaneous sources. *Proc. Roy. Soc. London*, *A234*, 1–23.
Nappo, C.J. (1981). Atmospheric turbulence and diffusion estimats derived from observations of a smoke plume. *Atmos. Environ.*, *15*, 541–547.
Nappo, C.J., and Rao, K.S. (1987). A model study of pure katabatic flows. *Tellus*, *39A*, 61–71.
National Center for Atmospheric Research (1983). *Regional Acid Deposition: Models and Physical Processes*. NCAR Tech. Note 214+STR, pp. 279–290. National Center for Atmospheric Research, Boulder, CO.
Navarra, J.G. (1979). *Atmosphere, Weather and Climate: An Introduction to Meteorology*. W.B. Saunders Co., Philadelphia.
Neiburger, M., Edinger, J.G., and Bonner, W.D. (1982). *Understanding Our Atmospheric Environment*, 2d ed. W.H. Freeman and Co., San Francisco.
Nieuwstadt, F.T.M. (1980a). An analytic solution of the time-dependent one-dimensional diffusion equation in the atmospheric boundary layer. *Atmos. Environ.*, *14*, 1361–1364.
Nieuwstadt, F.T.M. (1980b). Application of mixed-layer similarity to the observed dispersion from a ground-level source. *J. Appl. Meteorol.*, *19*, 157–162.
Nieuwstadt, F.T.M. (1984a). The turbulence structure of the stable, nocturnal boundary layer. *J. Atmos. Sci.*, *41*, 2202–2216.
Nieuwstadt, F.T.M. (1984b). Some aspects of the turbulent stable boundary layer. *Boundary-Layer Meteorol.*, *30*, 31–55.
Nieuwstadt, F.T.M. (1992a). A large-eddy simulation of a line source in a convective atmospheric boundary layer. I. Dispersion characteristics. *Atmos. Environ.*, *26A*, 485–495.
Nieuwstadt, F.T.M. (1992b). A large-eddy simulation of a line source in a convective atmospheric boundary layer. II. Dynamics of a buoyant line source. *Atmos. Environ.*, *26A*, 497–503.
Nieuwstadt, F.T.M., and de Valk, J.P.J.M. (1987). A large eddy simulation of buoyant and non-buoyant plume dispersion in the atmospheric boundary layer. *Atmos. Environ.*, *21*, 2573–2587.
Nieuwstadt, F.T.M., Mason, P.J., Moeng, C.-H., and Schumann, U. (1992). Large-eddy simulation of the convective boundary layer: A comparison of four com-

puter codes. In *Turbulent Shear Flows* (F. Durst et al., eds.), *8*, pp. 343–367. Springer-Verlag, Berlin.

Nieuwstadt, F.T.M., and van Dop, H. (1982). *Atmospheric Turbulence and Air Pollution Modelling*. D. Reidel Pub. Co., Dordrecht, Holland.

Nieuwstadt F.T.M., and van Ulden, A.P. (1978). A numerical study on the vertical dispersion of passive contaminants from a continuous source in the atmospheric surface layer. *Atmos. Environ.*, *12*, 2119–2124.

Nowacki, P., Samson, P.J., and Silman, S. (1996). Sensitivity of Urban Airshed Model (UAM-IV) calculated air-pollutant concentrations to the vertical diffusion parameterization during convective meteorological situations. *J. Appl. Meteorol.*, *35*, 1790–1803.

Obukhov, A.M. (1941). Energy distribution in the spectrum of a turbulent flow. *Izv. Akad. Nauk. SSSR, Geogr. Geofiz. Ser.*, No. 4–5, 453–466.

Ogawa, Y., Diosey, P.G., Uhera, K., and Ueda, H. (1985). Wind tunnel observation of flow and diffusion under stable stratification. *Atmos. Environ.*, *19*, 65–75.

Ogura, Y. (1952). Theory of turbulent diffusion in the atmosphere. *J. Meteorol. Soc. Japan*, *30*, 53–58.

Ogura, Y. (1959). Diffusion from a continuous source in relation to finite observation interval. In *Atmospheric Diffusion and Air Pollution* (F.N. Frenkiel and P.A. Sheppard, eds.), *Advances in Geophysics*, *6*, 149–159. Academic Press, New York.

Oke, T.R. (1987). *Boundary Layer Climates*, 2d ed. Halsted, New York.

Orgill, M.M., and Schreck, R.I. (1985). An overview of the ASCOT multi-laboratory field experiments in relation to drainage winds and ambient flow. *Bull. Amer. Meteorol. Soc.*, *66*, 1263–1277.

Overcamp, T.J. (1976). A general Gaussian diffusion-deposition model for elevated point sources. *J. Appl. Meteorol.*, *15*, 1167–1171.

Overcamp, T.J., and Ku, T. (1986). Effect of a virtual origin correction on entrainment coefficients as determined from observations of plume rise. *Atmos. Environ.*, *20*, 293–300.

Owen, C.R., and Thompson, W.R. (1963). Heat transfer across rough surfaces. *J. Fluid Mech.*, *15*, 321–334.

Palmen, E., and Newton, C.W. (1969). A*tmospheric Circulation Systems*. Academic Press, New York.

Panofsky, H.A., and Dutton, J.A. (1984). *Atmospheric Turbulence*. Wiley-Interscience, New York.

Panofsky, H.A., Tennekes, H., Lenschow, D.H., and Wyngaard, J.C. (1977). The characteristics of turbulent velocity components in the surface layer under convective conditions. *Boundary-Layer Meteorol.*, *11*, 355–361.

Pasquill, F. (1961). Estimation of the dispersion of windborne material. *Meteorol. Mag.*, *90*, 33–49.

Pasquill, F. (1962). *Atmospheric Diffusion*. D. van Nostrand Co., London.

Pasquill, F. (1966). Lagrangian similarity and vertical diffusion from a source at ground level. *Quart. J. Roy. Meteorol. Soc.*, *92*, 185–195.

Pasquill, F. (1974). *Atmospheric Diffusio*n, 2d ed. Ellis Horwood Ltd., Chichester, England.

Pasquill, F. (1976). Atmospheric dispersion parameters in Gaussian plume modeling. Part II. Possible requirements for change in the Turner Workbook values. Report No. EPA-600/4-76-030b, U.S. Environmental Protection Agency, Research Triangle Park, NC.

Pasquill, F. (1978). Dispersion from individual sources. In *Air Quality Meteorology and Atmospheric Ozone* (A.L. Morris and R.C. Barras, eds.), pp. 235–261. American Society of Testing Materials, Publication 653.

Pasquill, F., and Smith, F.B. (1983). *Atmospheric Diffusion*, 3rd ed. Ellis Horwood Ltd., Chichester, England.

Pedgley, D. (1982)). *Windborne Pests and Diseases: Meteorology of Airborne Organisms*. Halsted Press, New York.

Pedlosky, J. (1979). *Geophysical Fluid Dynamics*. Springer-Verlag, Berlin.

Pendergast, M.M. (1984). Meteorological fundamentals. In *Atmospheric Science and Power Production* (D. Randerson, ed.), pp. 33–79. U.S. Department of Energy, Technical Information Center, Oak Ridge, TN.

Pendergrass, W., and Arya, S.P. (1984). Dispersion in neutral bounday layer over a step change in surface roughness. II. Concentration profiles and dispersion parameters. *Atmos. Environ.*, *18*, 1281–1296.

Physick, W.L., and Abbs, D.J. (1991). Modeling of summertime flow and dispersion in the coastal terrain of southeastern Australia. *Mon. Weath. Rev.*, *119*, 1014–1030.

Physick, W.L., Noonan, J.A., McGregor, J.L., Hurley, P.J., Abbs, D.J., and Manins, P.C. (1993). LADM—A Lagrangian atmospheric dispersion model. Technical Report No. 24, CSIRO Division of Atmospheric Research, Australia.

Pielke, R.A. (1979). Air pollution—a national concern. *Bull. Amer. Meteorol. Soc.*, *60*, 1461.

Pielke, R.A. (1984). *Mesoscale Meteorological Modeling*. Academic Press, Orlando, FL.

Pielke, R.A., Cotton, W.R., Walko, R.L., Tremback, C.J., Nicholls, M.E., Moran, M.D., Wesley, D.A., Lee, T.J., and Copeland, J.H. (1992). A comprehensive meteorological modeling system—RAMS. *Meteorol. Atmos. Phys.*, *49*, 69–91.

Pierce, T.E., Lamb, B.K., and van Meter, A.R. (1990). Development of a biogenic emissions inventory system for regional scale air pollution models. *Proceedings of the 83rd Air and Waste Management Association Annual Meeting*, Pittsburgh.

Powell, D.C., and Elderkin, CE. (1974). An investigation of the application of Taylor's hypothesis to atmospheric boundary layer turbulence. *J. Atmos. Sci.*, *31*, 990–1002.

Prandtl, L. (1925). Bericht uber Untersuchungen zur ausgebildeten Turbulenz. *Zs. angew. Math. Mech.*, *5*, 136–139.

Priestley, M.B. (1981). *Spectral Analysis and Time Series*. Academic Press, San Diego.

Ragland, K.W., and Dennis, R.L. (1975). Point source atmospheric diffusion model with variable wind and diffusivity profiles. *Atmos. Environ.*, *9*, 175–189.

Raupach, M.R., and Legg, B.J. (1983). Turbulent dispersion from an elevated line source: Measurements of wind-concentration moments and budgets. *J. Fluid Mech.*, *136*, 111–137.

Reid, J.D. (1979). Markov chain simulations of vertical dispersion in the neutral surface layer for surface and elevated releases. *Boundary-Layer Meteorol.*, *16*, 3–22.

Reiter, E.R. (1967). *Jet Streams*. Anchor Books, Doubleday & Co., New York.

Reynolds, O. (1894). On the dynamical theory of incompressible viscous fluids and the determinatoin of the criterion. *Phil. Trans. Roy. Soc. London*, *186*, 123–161.

Richardson, L.F. (1920). The supply of energy from and to atmospheric eddies. *Proc. Roy. Soc. London*, *A97*, 354–373.

Richardson, L.F. (1922). *Weather Prediction by Numerical*

Process. Cambridge University Press, Cambridge, England.
Richardson, L.F. (1926). Atmospheric diffusion shown on a distance-neighbour graph. *Proc. Roy. Soc. London, A110*, 709–737.
Roberts, O.F.T. (1923). The theoretical scattering of smoke in a turbulent atmosphere. *Proc. Roy. Soc. London, A104*, 640–654.
Robinson, E. (1977). Effects on physical properties of the atmosphere. In *Air Pollution*, 3rd ed., *Vol. II* (A.C. Stern, ed.). Academic Press, New York.
Rodhe, H.P., Crutzen, P., and Vanderpol, A. (1981). Formation of sulfuric and nitric acid in the atmosphere during long-range transport. *Tellus, 33*, 132–141.
Roselle, S.J. (1991). Sensitivity of modeled ozone concentrations to uncertainties in biogenic emissions estimates. M.S. Thesis, Department of Marine, Earth and Atmospheric Sciences, North Carolina State University, Raleigh.
Roselle, S.J., Pierce, T.E., and Schere, K.L. (1991). The sensitivity of regional ozone modeling to biogenic hydrocarbons. *J. Geophys. Res., 96*, 7371–7394.
Rosenberg, N.J., Blad, B.L., and Verma, S.B. (1983). *Microclimate: The Biological Environment*, 2d ed. Wiley-Interscience, New York.
Roth, M. (1991). Turbulent transfer characteristics over a suburban surface. Ph.D. Thesis, Department of Geography, University of British Columbia, Vancouver.
Rotta, J.C. (1951). Statistische Theorie nichthomogener Turbulenz. *Z. Phys., 129*, 547–572.
Rounds, W. (1955). Solutions of the two-dimensional diffusion equation. *Trans. Amer. Geophys. Union, 36*, 395–405.
Saeger, M., Langstaff, J., Walters, R., Modica, L., Zimmerman, D., Fratt, D., Bulleba, D., Ryan, R., Demmy, J., Tax, W., Sprague, D., Mudgett, D., and Werner, A. (1989). The NAPAP emission inventory (version 2): Development of the annual data and modeler's tapes. Report No. EPA-600/7-89-012a, U.S. Environmental Protection Agency, Research Triangle Park, NC.
Salmon, J.R., Teunissen, H.W., Mickle, R.E., and Taylor, P.A. (1988). The Kettles Hill project: Field observations, wind-tunnel simulations and numerical model predictions for flow over a low hill. *Boundary-Layer Meteorol., 43*, 309–343.
Sawford, B.L. (1993). Recent developments in the Lagrangian stochastic theory of turbulent dispersion. *Boundary-Layer Meteorol., 62*, 197–215.
Sawford, B.L., and Guest, F.M. (1987). Lagrangian stochastic analysis of flux-gradient relationships in the convective boundary layer. *J. Atmos. Sci., 44*, 1152–1165.
Schatzmann, M., Snyder, W.H., and Lawson, Jr., R.E. (1993). Experiments with heavy gas jets in laminar and turbulent cross-flows. *Atmos. Environ., 27A*, 1105–1116.
Scheffe, R.D. (1990). Urban Airshed Model study of five cities. Report no. EPA 450/2-90-006a, U.S. Environmental Protection Agency, Research Triangle Park, NC.
Schere, K.L., and Demerjian, K.L. (1978). A photochemical box model for urban air quality simulation. *Proceedings of the 4th Conference on Sensing of Environmental Pollutants*, pp. 427–433. American Chemical Society, Washington, DC.
Schmidt, H., and Schumann, U. (1989). Coherent structure of the convective boundary layer derived from large-eddy simulation. *J. Fluid Mech., 200*, 511–562.
Schowalter, D.G., DeCroix, D.S., Lin, Y.-L., Arya, S.P., and Kaplan, M.L. (1996). The sensitivity of large-eddy simulation to local and nonlocal drag coefficients at the lower boundary. NASA Contractor Report 198310, NASA Langley Research Center, Hampton, VA.
Schumann, U. (1975). Subgrid-scale model for finite difference simulations of turbulent flows in plane channels and annuli. *J. Comput. Phys., 18*, 376–404.
Seaman, N.L., Ludwig, F.L., Donall, E.G., Warner, T.T., and Bhumralkar, C.M. (1989). Numerical studies of urban planetary boundary-layer structure under realistic synoptic conditions. *J. Appl. Meteorol., 28*, 760–781.
Segal, M., Pielke, R.A., Arritt, R.W., Moran, D., Yu, C.-H., and Henderson, D. (1988). Application of a mesoscale atmospheric dispersion modeling system to the estimation of SO_2 concentrations from major elevated sources in southern Florida. *Atmos. Environ., 22*, 1319–1334.
Sehmel, G.A. (1980). Particle and gas dry deposition: A review. *Atmos. Environ., 14*, 983–1011.
Sehmel, G.A. (1984). Deposition and resuspension. In *Atmospheric Science and Power Production* (D. Randerson, ed.), pp. 533–583. U.S. Department of Energy, Technical Information Center, Oak Ridge, TN.
Seigneur, C., and Saxena, P. (1984). A study of atmospheric acid formation in different environments. *Atmos. Environ., 18*, 2109–2124.
Seinfeld, J.H. (1986). *Atmospheric Chemistry and Physics of Air Pollution*. Wiley-Interscience, New York.
Sherman, C.A. (1978). A mass-consistent model for wind fields over complex terrain. *J. Appl. Meteorol., 17*, 312–319.
Shiotani, M., and Iwantani, Y. (1976). Horizontal space correlations of velocity fluctuations during strong winds. *J. Meteorol. Soc. Japan, 54*, 59–67.
Sillman, S., Al-Wali, K., Marsik, F., Nowacki, P., Samson, P., Rodgers, M., Garland, L., Martinez, J., Stoneking, C., Imhoff, R., Lee, J., Newman, L., Weinstein-Lloyd, J., and Aneja, V. (1995). Photochemistry of ozone formation in Atlanta, GA—Models and measurements. *Atmos. Environ., 29*, 3055–3066.
Singer, I.A., and Smith, M.E. (1966). Atmospheric dispersion at Brookhaven National Laboratory. *Int. J. Air Water Pollut., 10*, 125–135.
Singh, H.B. (1995). *Composition, Chemistry, and Climate of the Atmosphere*. Van Nostrand-Reinhold, New York.
Sklarew, R.C., Fabrick, A.J., and Prager, J.E. (1972). Mathematical modeling of photochemical smog using the PICK method. *J. Air Poll. Cont. Assoc., 22*, 865–869.
Slade, D.H. (1968). *Meteorology and Atomic Energy 1968*. U.S. Department of Energy, Technical Information Center, Oak Ridge, TN.
Slin, W.G.N. (1984). Precipitation scavenging. In *Atmospheric Science and Power Production* (D. Randerson, ed.), pp. 466–532. U.S. Department of Energy, Technical Information Center, Oak Ridge, TN.
Smagorinski, J. (1963). General circulation experiments with the primitive equations. Part I: The basic experiment. *Mon. Weath. Rev., 91*, 99–152.
Smith, F.B., (1957). The diffusion of smoke from a continuous elevated point source into a turbulent atmosphere. *J. Fluid Mech., 2*, 49–76.
Smith, F.B. (1968). Conditioned particle motion in a homogeneous turbulent field. *Atmos. Environ., 2*, 491–508.
Smith, F.B., and Hay, J.S. (1961). Expansion of clusters of particles in the atmosphere. *Quart. J. Roy. Meteorol. Soc., 87*, 82–101.
Smith, M.E. (1951). The forecasting of micrometeorological variables. *Meteorol. Monogr., 4*, 50–55.
Smith, M.E. (1984). Review of the attributes and performance of 10 rural diffusion models. *Bull. Amer. Meteorol. Soc., 65*, 554–558.

Smith, W.H. (1981). *Air Pollution and Forests.* Springer-Verlag, New York.

Snyder, W.H., Khurshudyan, L.H., Nekrasov, I.V., Lawson, R.E., and Thompson, R.S. (1991). Flow and dispersion of pollutants within two-dimensional valleys. *Atmos. Environ.*, *25A*, 1347–1375.

Snyder, W.H., and Lawson, R.E. (1976). Determination of a necessary height for a stack close to a building—a wind tunnel study. *Atmos. Environ.*, *10*, 683–691.

Snyder, W. H., and Lumley, J.L. (1971). Some measurements of particle velocity autocorrelation functions in turbulent flow. *J. Fluid Mech.*, *48*, 41–71.

Snyder, W.H., Thompson, R.S., Lawson, R.E., Castro, I.P., Lee, J.T., Hunt, J.C.R., and Ogawa, Y. (1985). The structure of strongly stratified flow over hills: Dividing-streamline concept. *J. Fluid Mech.*, *152*, 249–288.

Sommeria, G. (1976). Three-dimensional simulations of turbulent processes in an undisturbed trade wind boundary layer. *J. Atmos. Sci.*, *33*, 216–241.

Sorbjan, Z. (1989). *Structure of Atmospheric Boundary Layer.* Prentice Hall, Englewood Cliffs, NJ.

Spangler, T.C., and Dirks, R.A. (1974). Mesoscale variations of the urban mixing height. *Boundary-Layer Meteorol.*, *6*, 423–441.

Stern, A.C. (1976–86). *Air Pollution*, 3rd ed. Academic Press, New York.

Stern, A.C. (1986). *Air Pollution*, Fourth Edition. Academic Press, San Diego, CA.

Stern, A.C., Boubel, R.W., Turner, D.B., and Fox, D.L. (1984). *Fundamentals of Air Pollution*, 2d ed., Academic Press, San Diego.

Stockwell, W.R. (1986). A homogeneous gas phase mechanism for use in a regional acid deposition model. *Atmos. Environ.*, *20*, 1615–1632.

Stockwell, W.R., Middleton, P., Chang, J.S., and Tang, X. (1990). The second generation regional acid deposition model chemical mechanism for regional air quality modeling. *J. Geophys. Res.*, *95*, 16343–16368.

Stout, J.E., and Arya, S.P. (1994). Mean motion and trajectories of heavy particles falling through a boundary layer. *Preprints, Eighth Joint Conference on Applications of Air Pollution Meteorology*, pp. 127–134. American Meteorological Society, Boston.

Stout, J.E., Arya, S.P., and Genikhovich, E.L. (1995). The effect of nonlinear drag on the motion and settling velocity of heavy particles. *J. Atmos. Sci.*, *52*, 3836–3848.

Stull, R.B. (1988). *An Introduction to Boundary Layer Meteorology.* Kluwer Academic Pub., Boston.

Stull, R.B. (1991). Static stability—an update. *Bull. Amer. Meteor. Soc.*, *72*, 1521–1529.

Sullivan, P.J. (1971). Longitudinal dispersion within a two-dimensional turbulent shear flow. *J. Fluid Mech.*, *49*, 551–576.

Sullivan, P.P., McWilliams, J.C., and Moeng, C.-H. (1994). A subgrid-scale model for large-eddy simulation of planetary boundary-layer flows. *Boundary-Layer Meteorol.*, *71*, 247–276.

Sun, W.-Y. (1986). Air pollution in a convective boundary layer. *Atmos. Environ.*, *20*, 1877–1886.

Sun, W.-Y. (1989). Numerical study of dispersion in the convective boundary layer. *Atmos. Environ.*, *23*, 1205–1217.

Sun, W.-Y., and Chang, C.-Z. (1986a). Diffusion model for a convective boundary layer. Part I: Numerical simulation of convective boundary layer. *J. Climate Appl. Meteorol.*, *25*, 1445–1453.

Sun, W.-Y., and Chang, C.-Z. (1986b). Diffusion model for a convective boundary layer. Part II: Plume released from a continuous point source. *J. Climate Appl. Meteorol.*, *25*, 1454–1463.

Sutton O.G. (1932). A theory of eddy diffusion in the atmosphere. *Proc. Roy. Soc. London*, *A135*, 143–165.

Sutton, O.G. (1953). *Micrometeorology.* McGraw-Hill, New York.

Sykes, R.I., and Henn, D.S. (1992). Large-eddy simulation of concentration fluctuations in a dispersing plume. *Atmos. Environ.*, *26A*, 3127–3144.

Taylor, G.I. (1921). Diffusion by continuous movements. *Proc. London Math. Soc.*, *20*, 196–211.

Taylor, G.I. (1935). Statistical theory of turbulence, I-III. *Proc. Roy. Soc. London*, *A151*, 421–464.

Taylor, G.I. (1938). The spectrum of turbulence. *Proc. Roy. Soc. London*, 476–490.

Taylor, G.I. (1959). The present position in the theory of turbulent diffusion. In *Atmospheric Diffusion and Air Pollution* (F.N. Frenkiel and P.A. Sheppard, eds.), *Advances in Geophysics*, *6*, 101–115. Academic Press, New York.

Taylor, J.H. (1965). Project Sand Storm, an experimental program in atmospheric diffusion. *Environ. Res. Papers*, *No. 134*, Report No. AFCRL-65-649, U.S. Air Force Cambridge Research Laboratories.

Taylor, P.A., and Teunissen, H.W. (1987). The Askervein Hill Project: Overview and background data. *Boundary-Layer Meteorol.*, *39*, 15–39.

Tennekes, H., and Lumley, J.L. (1972). *A First Course in Turbulence.* MIT Press, Cambridge, MA.

Therry, G., and Lacarrere, P. (1983). Improving the eddy kinetic energy model for planetary boundary layer description. *Boundary-Layer Meteorol.*, *25*, 63–88.

Thompson, R.S. (1993). Building amplification factors for sources near buildings: A wind tunnel study. *Atmos. Environ.*, *27A*, 2313–2325.

Thomson, D.J. (1971). Numeric calculation of turbulent diffusion. *Quart. J. Roy. Meteorol. Soc.*, *97*, 93–98.

Thomson, D.J. (1987). Criteria for the selection of the stochastic models of particle trajectories in turbulent flows. *J. Fluid Mech.*, *180*, 529–556.

Tirabassi, T., Tagliazucca, M., and Zannetti, P. (1986). KAPPA-G, a non-Gaussian plume dispersion model: Description and evaluation against tracer measurements. *J. Air Pollut. Cont. Assoc.*, *36*, 592–596.

Turner, D. B. (1961). Relationship between 24–hr. mean air quality measurement and meteorological factors in Nashville, TN. *J. Air Pollut. Cont. Assoc.*, *11*, 483–489.

Turner, D.B. (1964). A diffusion model for an urban area. *J. Appl. Meteorol.*, *3*, 83–91.

Turner, D.B. (1970). *Workbook of Atmospheric Dispersion Estimates.* Office of Air Programs Pub. No. AP-26, U.S. Environmental Protection Agency, Research Triangle Park, NC.

Turner, D.B. (1994). *Workbook of Atmospheric Dispersion Estimates*, 2d ed. Lewis Pub., CRC Press, Boca Raton, FL.

Turner, J.S. (1973). *Buoyancy Effects in Fluids.* Cambridge University Press, Cambridge, England.

Uliasz, M. (1994). Lagrangian particle dispersion modeling in mesoscale applications. In *Environmental Modelling II* (P. Zannetti, ed.), Computational Mechanics Publications, pp. 71–102.

University Corporation for Atmospheric Research (1990). Atmospheric trace gases that are radiatively active and of significance to global change. *Earth Quest*, *Vol. 4*, 10–11, Office of Interdisciplinary Studies, UCAR, Boulder, CO.

University Corporation for Atmospheric Research (1992). Our ozone shield. *Reports to the Nation on Our Changing Planet*, No. 2, UCAR, Boulder, CO.

Urone, P. (1976). The primary air pollutants—gaseous; their occurrence, sources, and effects. In *Air Pollution*, 3rd ed., *Vol. I* (A.C. Stern, ed.), pp. 23–75. Academic Press, New York.

Urone, P. (1986). The pollutants. In *Air Pollution*, 3rd ed., *Vol. VI* (A.C. Stern, ed.), pp. 1–59. Academic Press, New York.

U.S. Environmental Protection Agency (1973–1983). Compilation of air pollution emission factors. AP-42 and Supplements, U.S. Environmental Protection Agency, Research Triangle Park, NC.

U.S. Environmental Protection Agency (1979). *Protecting Visibility*. Report No. EPA 450/5-79-008, U.S. Environmental Protection Agency, Research Triangle Park, NC.

U.S. Environmental Protection Agency (1982). Air quality criteria for particulate matter and sulfur oxides. Report No. EPA-600/8-82-029, U.S. Environmental Protection Agency.

U.S. Environmental Protection Agency (1986). Guideline on air quality models (revised). Report No. EPA 450/2-78-027R, U.S. Environmental Protection Agency, Office of Air Quality Planning and Standards, Research Triangle Park, NC.

U.S. Environmental Protection Agency (1996). Appendix W to Part 51—Guideline on Air Quality Models. *Federal Register*, 61, 41840–41894. U.S. Government Printing Office.

U.S. Nuclear Regulatory Commission (1979). Methods for estimating atmospheric transport and dispersion of gaseous effluents in routine releases from light-water-cooled reactors. Regulatory Guide 1.111, Washington, DC.

van Aalst, R.M., and Diederen, H.S.M.A. (1985). Removal and transformation processes in the atmosphere with respect to SO_2 and NO_x. In *Interregional Air Pollution Modelling* (S. Zwerver and J. van Ham, eds.), pp. 83–147. Plenum Press, New York.

van den Hout, K.D., and van Dop, H. (1985). Interregional modelling. In *Interregional Air Pollution Modelling* (S. Zwerver and J. van Ham, eds.), pp. 11–82. Plenum Press, New York.

Van der Hoven, I. (1968). Deposition of particles and gases. In *Meteorology and Atomic Energy—1968* (D. Slade, ed.), pp. 202–207. U.S. Department of Energy, Technical Information Center, Oak Ridge, TN.

van Haren, L., and Nieuwstadt, F.T.M. (1989). The behavior of passive and buoyant plumes in a convective boundary layer, as simulated with a large-eddy model. *J. Appl. Meteor.*, 28, 818–332.

van Ulden, A.P. (1978). Simple estimates for vertical diffusion from sources near the ground. *Atmos. Environ.*, 12, 2125–2129.

van Ulden, A.P., and Holtslag, A.A.M. (1985). Estimation of atmospheric boundary layer parameters for diffusion applications. *J. Climate Appl. Meteorol.*, 24, 1196–1207.

van Ulden, A.P., and Nieuwstadt, F.T.M. (1980). Authors' reply. *Atmos. Environ.*, 14, 269–270.

Venkatram, A. (1992). Vertical dispersion of ground-level releases in the surface boundary layer. *Atmos. Environ.*, 26A, 947–949.

Venkatram, A., Scire, J., and Pleim, J. (1984). ADOM model development program. Vol. 1: Approach to the evaluation of the ADOM model. Report No. ERT P-B 866–100, ENSR Corp., Acton, MA.

Verma, S.B., Rosenberg, N.J., and Blad, B.L. (1978). Turbulent exchange coefficients for sensible heat and water vapor under advective conditions. *J. Appl. Meteorol.*, 17, 330–338.

Vogel, B., Fiedler, F., and Vogel, H. (1995). Influence of topography and biogenic volatile organic compounds emission in the state of Baden-Wurttemberg on ozone concentrations during episodes of high air temperatures. *J. Geophys. Res.*, 100, 22,907–22,928.

Vogel, B., Vogel, H., and Fiedler, F. (1992). Numerical simulation of the interaction of transport, diffusion and chemical reactions in an urban plume. In *Ozone in the Troposphere and Stratosphere*, Vol. 1 (R.D. Hudson, ed.), pp. 97–100. NASA Conf. Publ. 3266.

Weil, J.C. (1988a). Plume rise. In *Lectures on Air Pollution Modeling* (A. Venkatram and J.C. Wyngaard, eds.), pp. 119–166. American Meteorological Society, Boston.

Weil, J.C. (1988b). Dispersion in the convective boundary layer. In *Lectures on Air Pollution Modeling* (A. Venkatram and J.C. Wyngaard, eds.), pp. 167–227. American Meteorological Society, Boston.

Weil, J.C., and Brower, R.P. (1984). An updated Gaussian plume model for tall stacks. *J. Air Pollut. Cont. Assoc.*, 34, 818–827.

Weil, J.C., Sykes, R.I., and Venkatram, A. (1992). Evaluating air-quality models: Review and outlook. *J. Appl. Meteorol.*, 31, 1121–1145.

Wesely, M.L., Hart, R.L., and Spear, R.E. (1985). Measurements and parameterization of particulate sulfur dry deposition over grass. *J. Geophys. Res.*, 90, 2131–2143.

Wesely, M.L., and Hicks, B.B. (1977). Some factors that affect the deposition rates of sulfur dioxide and similar gases on vegetation. *J. Air Pollut. Cont. Assoc.*, 27, 1110–1116.

White, F.M. (1974). *Viscous Fluid Flow*. McGraw-Hill, New York.

Whitten, G.Z., Hogo, H., and Killus, J.P. (1980). The carbon-bond mechanism: A condensed kinetic mechanism for photochemical smog. *Environ. Sci. Technol.*, 14, 690.

Whitten, G.Z., Killus, J.P., and Johnson, R.G. (1985). Modeling of auto exhaust for smog chamber data for EKMA development. U.S. EPA Contract Report 68–02–3735, Systems Applications, Inc., San Rafael, CA.

Williamson, S.J. (1973). *Fundamentals of Air Pollution*. Addison-Wesley, Reading, MA.

Willis, G.E., and Deardorff, J.W. (1974). A laboratory model of the unstable planetary boundary layer. *J. Atmos. Sci.*, 31, 1297–1307.

Willis, G.E., and Deardorff, J.W. (1976). A laboratory model of diffusion into the convective planetary boundary layer. *Quart. J. Roy. Meteorol. Soc.*, 102, 427–445.

Willis, G.E., and Deardorff, J.W. (1978). A laboratory study of dispersion from an elevated source in a modified convective planetary boundary layer. *Atmos. Environ.*, 12, 1305–1311.

Willis, G.E., and Deardorff, J.W. (1981). A laboratory study of dispersion from a source in the middle of the convective mixed layer. *Atmos. Environ.*, 15, 109–117.

Willis, G.E., and Deardorff, J.W. (1983). On plume rise within the convective boundary layer. *Atmos. Environ.*, 17, 2435–2447.

Wilson, J.D., and Sawford, B.L. (1996). Review of Lagrangian stochastic models for trajectories in the turbulent atmosphere. *Boundary-Layer Meteorol.*, 78, 191–210.

Wilson, J.D., Thurtell, G.W., and Kidd, G.E. (1981a). Numerical simulation of particle trajectories in inhomogeneous turbulence. I: Systems with constant turbulent velocity scale. *Boundary-Layer Meteorol.*, *21*, 295–313.

Wilson, J.D., Thurtell, G.W., and Kidd, G.E. (1981b). Numerical simulation of particle trajectories in inhomogeneous turbulence. II: Systems with variable turbulent velocity scale. *Boundary-Layer Meteorol.*, *21*, 423–441.

Wilson, J.D., Thurtell, G.W., and Kidd, G.E. (1981c). Numerical simulation of particle trajectories in inhomogeneous turbulence. III: Comparison of predictions with experimental data for the atmospheric surface layer. *Boundary-Layer Meteorol.*, *21*, 443–463.

Wyngaard, J.C. (1975). Modeling the planetary boundary layer—extension to the stable case. *Boundary-Layer Meteorol.*, *9*, 441–460.

Wyngaard, J.C. (1982). Boundary-layer modeling. In *Atmospheric Turbulence and Air Pollution Modelling* (F.T.M. Nieuwstadt and H. van Dop, eds.), pp. 69–106. D. Reidel Pub. Co., Dordrecht, Holland.

Wyngaard, J.C., and Brost, R.A. (1984). Top-down and bottom-up diffusion of a scalar in the convective boundary layer. *J. Atmos. Sci.*, *41*, 102–112.

Wyngaard, J.C., and Clifford, S.F. (1977). Taylor's hypothesis and high-frequency turbulence spectra. *J. Atmos. Sci.*, *34*, 922–929.

Wyngaard, J.C., Cote, O.R., and Rao, K.S. (1974). Modeling the atmospheric boundary layer. *Advan. Geophys.*, *18A*, 193–211. Academic Press, New York.

Yaglom, A.M. (1972). Turbulent diffusion in the surface layer of the atmosphere. *Izv. Atmos. Ocean. Phys.*, *8*, 333–340.

Yamada, T. (1977). A numerical experiment on pollutant dispersion in a horizontally-homogeneous atmospheric boundary layer. *Atmos. Environ.*, *11*, 1015–1024.

Yamada, T., and Bunker, S. (1988). Development of a nested grid second moment turbulence closure model and application to the 1982 ASCOT Brush Creek data simulation. *J. Appl. Meteorol.*, *27*, 562–578.

Yamada, T., and Mellor, G.L. (1975). A simulation of the Wangara atmospheric boundary layer data. *J. Atmos. Sci.*, *32*, 2309–2329.

Yamamoto, G., Shimanuki, A., and Nishinomiya, S. (1970). Diffusion of falling particles in diabatic atmospheres. *J. Meteorol. Soc. Japan*, *48*, 417–424.

Yeh, G.T., and Huang, C.H. (1975). Three-dimensional air pollutant modeling in the lower atmosphere. *Boundary-Layer Meteorol.*, *9*, 381–390.

Yocom, J.E., and Upham, J.B. (1977). Effects on economic materials and structures. In *Air Pollution*, 3rd ed., *Vol. II* (A.C. Stern, ed.). Academic Press, New York.

Young, J., Aissa, M., Boehm, T., Coats, C., Eichinger, J., Grimes, D., Hallyburton, S., Olerud, D., Roselle, S., van Meter, A., Wayland, R., and Pierce, T. (1989). Development of the Regional Oxidant Model version 2.1. Report No. EPA-600/3-89-044, U.S. Environmental Protection Agency, Research Triangle Park, NC.

Zannetti, P. (1990). *Air Pollution Modeling*. Van Nostrand Reinhold, New York.

Zeman, O. (1981). Progress in the modeling of planetary boundary layers. *Ann. Rev. Fluid Mech.*, *13*, 253–273.

Zeman, O., and Lumley, J.L. (1976). Modeling buoyancy driven mixed layers. *J. Atmos. Sci.*, *33*, 1974–1988.

Zhang, D., and Anthes, R.A. (1982). A high resolution model of the planetary boundary layer sensitivity tests and comparisons with SESAME-79 data. *J. Appl. Meteorol.*, *21*, 1594–1609.

Zhang, Y.Q., Arya, S.P., and Snyder, W.H. (1996). A comparison of numerical and physical modeling of stable atmospheric flow and dispersion around a cubical building. *Atmos. Environ.*, *30*, 1327–1345.

Zwerver, S., and van Ham, J. (1985). *Interregional Air Pollution Modelling*. Plenum Press, New York.

Symbols

A	Total mass advection (Chapter 6)	c	Mass or volumetric concentration
A	Coefficient in the expression for concentration (chapters 6, 12)	c	An empirical constant (Chapter 8)
		c'	Concentration fluctuation
A	Aerodynamic surface area (Chapter 10)	c'	A dimensionless constant
A	Constant (Chapter 10)	c_a	Mole concentration of air (Chapter 1)
A_α	Dimensionless coefficient (Chapter 8)	c_a	Concentration above the PBL (Chapter 12)
a	An empirical constant (Chapter 8)	c_b	Concentration in background flow (Chapter 12)
a	Combination of reference variables (Chapter 6)	c_h	A constant (Chapter 4)
a	Dimensionless coefficient (Chapter 9)	c_i	Mole concentration of species i
a_b	Acceleration due to buoyancy	c_m	A constant (Chapter 4)
a_1	Empirical constant (Chapter 8)	c_{max}	Maximum concentration
B	Bowen ratio (Chapter 4)	c_{mi}	Mass concentration of species i (Chapter 1)
B	Empirical constant (Chapter 5)		
B	Dimensional coefficient (Chapter 6)	c_o	Ground-level concentration
B_α	Dimensionless constant (Chapter 8)	$c_{o max}$	Maximum g.l.c.
b	Combination of reference variables (Chapter 6)	c_p	Specific heat at constant pressure
		CT	Continental tropical air mass
b	Empirical constant in similarity relations (Chapter 8)	c_v	Specific heat at constant volume
		c_{vi}	Volumetric concentration of species i
b	Dimensionless coefficient (Chapter 9)	c_w	Concentration in precipitated water
C	An empirical constant (Chapters 5, 11)	c_y	Cross-wind integrated concentration
C_B	An empirical constant	D	Layer depth or thickness (Chapter 4)
C_D	Surface drag coefficient	D	Molecular diffusivity (chapters 2, 6)
C_D	Drag coefficient of the falling particle (Chapter 10)	D_i	Molecular diffusivity of species i
		D_m	Total mass diffusion
C_{DN}	Drag coefficient in neutral stability	D_w	Molecular diffusivity of water vapor
C_f	An empirical constant	$\dfrac{D}{DT}$	Total derivative
C_H	Heat transfer coefficient		
C_{HN}	Heat transfer coefficient in neutral stability		
C_k	An empirical constant	d	An empirical constant (Chapter 8)
C_s	An empirical constant	d_1	An empirical constant (Chapter 4)
C_w	An empirical constant	d_2	An empirical constant
C_ε	An empirical constant	de_s	Change in the saturation vapor pressure
C_θ	An empirical constant	dH	Heat added to a parcel
C_1	An empirical constant	d_o	Zero-plane displacement
C_2	An empirical constant	dp	Change in the air pressure
c	Specific heat (Chapter 2)	dT	Change in the air temperature

SYMBOLS

dU	Change in the internal energy	h_o	Average height of roughness elements or plant canopy
dW	Change in the work performed		
E	Rate of evaporation or condensation (Chapter 4)	h_s	Height or depth of the surface layer
		i	Intensity of turbulence
E'	Fluctuating or instantaneous turbulence kinetic energy (Chapter 11)	i_u	Turbulence intensity in x direction
		i_v	Turbulence intensity in y direction
E_o	Rate of evaporation or condensation at the surface (Chapter 4)	i_w	Turbulence intensity in z direction
		K	Eddy diffusivity
$E(k)$	Turbulent energy spectrum	K	Kurtosis (Chapter 5)
e	Water vapor pressure (Chapter 2)	K_E	Eddy diffusivity for energy transfer
e	Fluctuating turbulence kinetic energy (Chapters 4, 5)	K_h	Eddy diffusivity of heat
		K_m	Eddy viscosity or diffusivity of momentum
e_o	Water vapor pressure at the surface or at $z = z_o$	K_r	Eddy viscosity at the reference height
		K_w	Eddy diffusivity of water vapor
e_s	Water vapor pressure at saturation	K_x	Eddy diffusivity in x direction
F	Froude number (Chapter 3)	K_y	Eddy diffusivity in y direction
F	Momentum flux (Chapter 4)	K_z	Eddy diffusivity in z direction
F	A function of certain variables (Chapter 10)	K_ε	Eddy diffusivity for dissipation
		k	Thermal conductivity (Chapter 3)
$\underline{F}$	Diffusive mass flux vector	k	The von Karman constant
FB	Fractional bias	L	Obukhov (buoyancy) length
F_b	Effluent buoyancy flux parameter	L	Length of an urban area in the mean flow direction (Chapter 12)
F_c	Deposition flux		
F_D	Drag force on the particle	L_1	Length scale of spatial averaging in x direction (Chapter 4)
F_g	Gravitational force on the particle		
F_m	Effluent momentum flux parameter	L_1	Integral length scale in the direction of mean flow (Chapter 5)
F_w	Wet deposition flux		
F_x	Diffusive flux in x direction (Chapter 6)	L_b	Effluent buoyancy length scale
F_y	Lateral dispersion factor (Chapter 9)	L_e	Latent heat of evaporation/condensation
F_z	Vertical dispersion factor (Chapter 9)	L_m	Effluent momentum length scale
F_z	Diffusive flux in z direction (Chapter 6)	L_y	Large-eddy length scale in y direction
F_*	Dimensionless buoyancy flux parameter	l	Large-eddy length scale
f	Coriolis parameter (chapters 2, 4)	ℓ	Fluctuating mixing length (Chapter 4)
f	A function of variables	ℓ	Large-eddy length scale (Chapter 11)
f	A fluctuating variable (chapters 2, 5)	ℓ_b	Buoyancy length scale (Chapter 4)
f	Normalized frequency	ℓ_f	Filter length scale
f_m	Normalized frequency corresponding to the peak in spectrum	ℓ_h	Mean mixing length for heat transfer
		ℓ_m	Mean mixing length for momentum transfer
f_y	Lateral dispersion function (Chapter 9)		
f_z	Vertical dispersion function	ℓ_o	Mixing length in the outer layer of the PBL
G	Magnitude of geostrophic wind		
$\underline{G}$	Geostrophic wind vector	ℓ_x	Characteristic length scale of large eddies in x direction
G_o	Magnitude of surface geostrophic wind		
g	Acceleration due to gravity	ℓ_y	Characteristic length scale of large eddies in y direction
g	A fluctuating variable (chapters 2, 5)		
H	Sensible heat flux (chapters 3, 4)	ℓ_z	Characteristic length scale of large eddies in z direction
H	Building or hill height (Chapter 3)		
H	Effective source height or release height	ℓ_ε	Dissipation length scale
H^*	Dimensionless release height	M	Mass flux (Chapter 4)
H_G	Ground heat flux to or from the subsurface medium	$\underline{M}$	Mass flux vector (Chapter 6)
		m	Mean molecular mass of air (Chapter 2)
H_L	Latent heat flux	m	Exponent in power-law wind profile equation (chapters 4, 6)
H_o	Sensible heat flux at the surface		
H_s	Stack height (Chapter 10)	m	A coefficient (Chapter 10)
H_s	Height of the dividing streamline (Chapter 3)	m_d	Mean molecular mass of dry air
		m_i	Mean molecular mass of species i
h	Boundary-layer thickness or height	m_w	Mean molecular mass of water vapor
h_i	Internal boundary-layer thickness (Chapter 3)	N	Brunt–Vaisala frequency (chapters 3, 4)
		N	Normalized frequency (Chapter 5)
h_i	Inversion layer thickness (Chapter 4)	N_E	Normalized frequency in Eulerian spectrum

N_L	Normalized frequency in Lagrangian spectrum	S_i	Source or sink of species i
		S_L	Lagrangian spectrum
n	Exponent in the power-law eddy viscosity profile (chapters 4, 6)	S_w	Source or sink of water vapor
		s	Static stability (chapters 2, 4, 10)
n	Wave frequency (chapters 5,7)	s	Any space or time variable (Chapter 2)
n	Number of moles (Chapter 1)	s	Sampling interval (chapters 5, 7)
P	Period of the wave (Chapter 5)	T	Air temperature
P	Probability distribution function (Chapter 5)	T	Length of record or sampling duration (chapters 5, 7)
Pr	Prandtl number	$\underline{T}$	Flow transport vector (Chapter 6)
p	Air pressure (chapters 1, 2, 4, 11)	T_i	Integral time scale
p	Probability density function (chapters 5, 7, 8, 11)	T_{iE}	Eulerian integral time scale
		T_{iL}	Lagrangian integral time scale
p	Exponent (Chapter 9)	T_o	Air temperature at the reference state
p_o	Pressure of the reference state	T_u	Urban air temperature
p_r	Rate of precipitation	T_v	Virtual temperature of moist air
Q	Source strength of a continuous point source	T_{vo}	Virtual temperature at the reference state
		T_{vp}	Virtual temperature of the parcel
Q_a	Area source strength	T_*	Convective temperature scale
Q_ℓ	Line source strength	t	Time
Q_o	Kinematic heat flux at the surface	U	Lagrangian (particle) velocity in x direction
q	Specific humidity (chapters 2, 4)		
q	Exponent (Chapter 9)	U'	Lagrangian velocity fluctuation in x direction
q'	Specific humidity fluctuation		
q_*	Specific humidity scale for the surface layer	U_o	Mean velocity in the approach flow
		u	Instantaneous velocity component in x direction
q_s	Specific humidity at saturation		
R	Chemical reaction term (chapters 2, 6, 12)	u	The characteristic velocity scale of turbulence (Chapter 5)
R	Mean radius of the earth (Chapter 3)		
R	Specific gas constant (Chapter 2)	u'	Fluctuating velocity component in x direction
R	Normalized autocorrelation function (chapters 5, 7)		
		u_*	Friction velocity
R'	Autocorrelation function (Chapter 5)	u_e	Effective advection velocity (Chapter 5)
R_E	Eulerian autocorrelation function	u_f	Local free-convective velocity scale
Re	Reynolds number	u_g	Geostrophic velocity component in x direction
Rf	Flux Richardson number		
R_i	Chemical reaction term for species i (chapters 2, 6)	u_p	Effective plume transport velocity
		u_r	Velocity at the reference height z_r
Ri	Richardson number	V	Volume of gas (Chapter 1)
Ri$_B$	Mixed Richardson number (Chapter 3)	V	Wind speed (Chapter 4)
Ri$_b$	Bulk Richardson number (Chapter 4)	V	Lagrangian (particle) velocity in Y direction
Ri$_c$	Critical Richardson number		
Ri$_h$	Bulk Richardson number for the PBL	V'	Lagrangian velocity fluctuation in Y direction
R_L	Lagrangian autocorrelation function		
R_N	Net radiation	$\underline{V}$	Wind vector
R^*	Absolute gas constant	V_h	Wind speed at the top of the PBL
r	Radial distance	V_m	Layer-averaged wind speed (Chapter 4)
r_a	Aerodynamic resistance to mass transfer	V_m	Mean molecular speed (Chapter 6)
$\underline{r}_o$	Initial separation between two particles	V_r	Wind speed at the reference height
r_s	Radius of stack at exit	V'_r	Relative Lagrangian velocity component between two particles
r_s	Sublayer resistance to mass transfer (Chapter 10)		
		V_s	Wind speed at the top of the surface layer
r_t	Transfer resistance to mass transfer	v	Instantaneous velocity component in y direction
S	Skewness (Chapter 5)		
S	Source or sink term (chapters 2, 6, 12)	v'	Fluctuating velocity component in y direction
S	Normalized spectrum function (chapters 5, 7)		
		v_d	Deposition velocity
S'	Spectrum function	v_g	Geostrophic velocity component in y direction (Chapter 2)
S_E	Eulerian spectrum		
S_H	Source or sink of heat	v_g	Gravitational settling velocity (Chapter 10)

SYMBOLS

v_t	Transfer velocity
v_w	Wet deposition velocity
W	Weighting function (Chapter 5)
W	Lagrangian (particle) velocity in z direction
W'	Lagrangian velocity fluctuation in z direction
W_*	Convective velocity scale
w	Instantaneous velocity component in z (vertical) direction
w'	Fluctuating velocity component in z (vertical) direction
w_e	Entrainment velocity
w_h	Vertical velocity at the top of the PBL
w_r	Washout ratio
$\overline{w}_s$	Mean effluent velocity at stack exit
X	Particle displacement in X direction
$\underline{X}$	Particle displacement vector
X_*	Dimensionless distance from the source in x direction
X_o	Initial particle position in X direction
X_r	Relative displacement of two particles in X direction
x	Coordinate position along x axis
x	Distance from the source in the mean flow direction
x_f	Distance from the source where final plume rise occurs
x_h	Distance from the source where plume reaches the top of the PBL
x_{max}	Distance to the maximum ground-level concentration
x_{o*}	Dimensionless virtual source location upwind of the actual source (Chapter 8)
Y	Particle displacement in Y direction
Y_o	Initial particle position in Y direction
Y_r	Relative displacement of two particles in Y direction
y	Coordinate position along y axis
y_m	Lateral displacement of the meandering plume axis
y_o	Initial separation between two particles
y_p	Half width of a puff or plume in y direction
Z	Particle displacement in Z direction
Z_o	Initial particle position in Z direction
Z_r	Relative displacement of two particles in Z direction
Z_*	Dimensionless particle displacement in the vertical direction
z	Coordinate position along z axis
z	Height above the surface or an appropriate reference plane
$\bar{z}$	Mean plume height
z'	Height above the surface
z'_c	Height of the plume centerline above stack top
z_i	Height of the inversion base
z_m	Geometric mean of the two heights z_1 and z_2 (Chapter 4)
z_m	Vertical displacement of the meandering plume axis (Chapter 7)
z_o	Roughness length parameter
z_p	Half plume or puff thickness in z-direction
z_r	Reference height
α	Specific volume (Chapter 2)
α	Cross-isobar angle (Chapter 4)
α	A constant (Chapter 4)
α	Kolmogorov's constant (Chapter 5)
α	Concentration profile exponent (chapters 6, 8)
α_o	Cross-isobar angle at the surface
α_1	Kolmogorov's constant for longitudinal turbulence (Chapter 5)
α_2	Kolmogorov's constant for lateral turbulence (Chapter 5)
β	Ratio of Lagrangian and Eulerian integral time scales (chapters 5, 7)
β	Exponent in concentration distribution in x direction (chapters 6, 8)
β_b	Empirical entrainment coefficient in buoyant plume rise
β_m	Empirical entrainment coefficient in momentum rise
Γ	Adiabatic lapse rate
Γ_s	Saturated adiabatic lapse rate
γ	Empirical similarity constant (Chapter 8)
γ_c	An empirical constant (Chapter 4)
Δ	Finite difference (chapters 4, 10, 11)
Δ	Grid length scale (Chapter 11)
$\underline{\Delta}$	Gradient operator (Chapter 2, 6)
ΔA	Elemental area
Δh	Thickness of the layer between two pressure surfaces
ΔH	Final plume rise
ΔH_s	Rate of energy storage
ΔT_{u-r}	Difference between urban and rural air temperatures
Δt	Time difference or step
Δu	Difference in mean velocities at two heights in the surface layer
$\Delta \theta$	Difference between potential temperatures at two heights (chapters 2, 4)
$\Delta \theta$	Change in wind direction across the plume depth (chapters 8, 10)
Δx	Small increment in x, or grid size in x direction
Δy	Small increment in y, or grid size in y direction
Δz	Small increment in z, or grid size in z direction
Δz	$z_2 - z_1$ (Chapter 4)
∇^2	Laplacian operator
ε	Rate of energy dissipation
ε_u	Rate of dissipation of u-component energy
ε_v	Rate of dissipation of v-component energy
ε_w	Rate of dissipation of w component energy
ζ	Monin–Obukhov stability parameter
η	Kolmogorov's microscale (length)
θ	Horizontal wind direction

θ	Potential temperature of air
θ_f	Local free-convective temperature scale
θ_m	Mixed-layer–averaged potential temperature
θ_o	Potential temperature at $z = z_o$
θ_r	Air potential temperature at reference height
θ_v	Virtual potential temperature
θ'	Fluctuating potential temperature (chapters 2, 4, 11)
θ'	Horizontal wind direction fluctuation (chapters 4, 6, 9)
θ_*	Characteristic temperature scale
κ	Thermal diffusivity (chapters 2, 4)
κ	Wave number (Chapter 5)
κ_1	Wave number in the direction of mean flow
Λ_L	Length scale
Λ_m	Mean free-path length
Λ_w	Washout coefficient
λ	Wavelength (Chapters 3, 5)
λ	Large-eddy length scale (Chapter 11)
λ_m	Wavelength corresponding to spectral maximum
μ	Dynamic viscosity
ν	Kinematic viscosity
Π	Dimensionless group
π	Ratio of circumference to diameter of a circle
ρ	Mass density of air or other fluid
ρ_o	Density of the reference state
ρ_p	Mass density of settling particle (Chapter 10)
ρ_p	Mass density of the air parcel (Chapter 2)
ρ_s	Density of effluent at stack exit
σ	Dispersion parameter or standard deviation of concentration distribution
σ	Diffusivity ratio (Chapter 11)
σ_q	Standard deviation of specific humidity fluctuations
σ_u	Standard deviation of velocity fluctuations in x direction
σ_v	Standard deviation of velocity fluctuations in y direction
σ_w	Standard deviation of velocity fluctuations in z direction
σ_x	Standard deviation of concentration distribution in x direction
σ_x	Particle dispersion parameter in x direction (chapters 7, 8)
σ_y	Standard deviation of concentration distribution in y direction
σ_y	Particle dispersion parameter in y direction (chapters 7, 8)
σ_z	Standard deviation of concentration distribution in z direction
σ_z	Particle dispersion parameter in z direction (chapters 7, 8)
$\sigma_{z'}$	Standard deviation of particle displacement from its mean height
σ_θ	Standard deviation of temperature fluctuations (Chapter 4)
σ_θ	Standard deviation of horizontal wind direction fluctuations (chapters 4, 7, 9)
σ_ϕ	Standard deviation of vertical wind direction fluctuations
τ	Time lag (Chapter 5)
τ	Shearing stress
$\underline{\tau}$	Shear stress vector
τ_D	Dissipation time scale
τ_o	Surface shear stress or drag
τ_s	Taylor's microscale (time)
ϕ	Latitude (Chapter 2)
ϕ	Vertical wind direction (Chapter 4)
ϕ'	Vertical wind direction fluctuation
ϕ_h	Dimensionless potential-temperature gradient
ϕ_m	Dimensionless wind shear
ϕ_w	Dimensionless specific-humidity gradient
ϕ_ε	Dimensionless rate of energy dissipation
ψ_h	The M–O similarity function for normalized potential temperature
ψ_m	The M–O similarity function for normalized velocity
ψ_w	The M–O similarity function for normalized specific humidity
ψ_ε	Dimensionless rate of energy dissipation
Ω	Rotational speed of the earth
$\underline{\Omega}$	Earth's rotational velocity vector
ω	Wave frequency in radians
ξ	Time lag in Lagrangian particle motion
¯	(Overbar) Average quantity or variable
'	Fluctuating variable

Index

Absorption spectrum, 26–27
Acceleration of air parcel, 32–33
Acidic deposition, 18–19
Acids, 7–8, 19
Adiabatic
  lapse rate, 30–32
  process, 30
Advection
  cold air, 51–54
  heat, 35
  moisture, 51–54
  momentum, 35
  warm air, 51–54
Aerodynamic
  resistance, 231–34
  roughness, 91–92
Aerosol, 11–13
Agricultural/rural sources, 3–4
Air masses, 51–54
  Arctic, 52
  continental, 52–54
  marine, 52–54
  tropical, 52–54
Air parcel, 30–33
Air pollutants, 4–14, 23–24
  acid rain, 23–24
  gaseous, 4–11, 23
  hazardous, 13–14, 23
  particulates, 11–14, 23
  primary, 4–11
  secondary, 4–10
  toxics, 13–14
Air pollution, 1–25
  control, 3, 21–24
  effects, 14–21
  global, 1–2
  indoor, 2
  local, 1–3, 5–14
  problems, 1–3
  regional, 1–2, 5–14
  urban, 1–3, 5–14
Air quality/dispersion models, 127–54,
    185–87, 197–219, 245–86
  box, 278–80

Gaussian, 197–219, 276–77
  gradient transport, 127–54, 185–87
  large-eddy simulation, 249–60
  random-walk, 260–67
  regional, 269–76, 281–86
  second-order closure, 245–49
  urban, 269–81
Air toxics, 13–14
Albedo, 43
Anthropogenic sources, 3–4
Applications, 40, 51, 59, 68–69, 73–75,
    96–103, 124–25, 152, 175, 194–95,
    216–17, 237, 256–67, 285–86
  atmospheric dynamics, 40
  boundary layer meteorology, 73–75,
    96–103
  diffusion theories, 152, 175, 194
  dispersion models, 175, 194, 216–17,
    237, 256–67, 285–6
  macrometeorology, 51
  mesoscale meteorology, 68–69
  micrometeorology, 73–75, 96–103
  synoptic meteorology, 59
  turbulence, 124–25
Area sources, 130, 133–34, 136, 143–44,
    147–48, 276–86
  continuous, 136, 276–86
  instantaneous, 133–34, 143–44,
    147–48
Atmospheric boundary layer. See
    Boundary layer
Atmospheric composition, 4–5, 26
Atmospheric dynamics, 33–40, 241–46,
    249–56, 260–67
  continuity equation, 34
  energy equation, 35
  geostrophic winds, 36
  momentum equations, 35–40, 241
  thermal wind equations, 36
  turbulence modeling, 38–40, 241–46,
    249–56, 260–67
Atmospheric effects of air pollution,
    17–21
Atmospheric layers/strata, 27–29, 77–78

Atmospheric moisture, 29, 31–32
  specific humidity, 29, 31–32
Atmospheric motions/systems, 35–76
  large-scale, 35–37
  macroscale, 42–51
  mesoscale, 59–69
  microscale, 69–75
  small-scale, 37–40
  synoptic scale, 51–59
Atmospheric pressure, 29–30, 35–36, 39
  gradient, 35–36, 39
  hydrostatic, 30
  reference, 30
Autocorrelation functions, 108–12,
    124–25, 157–58, 160–62, 165,
    260–61
  Eulerian, 108–12, 160, 165
  Lagrangian, 124–25, 157–58, 160,
    162, 260–61
  spatial, 109–10, 165
  temporal, 108–9, 124–25, 157–58,
    160, 162, 260–61
Average, 38–39, 85–86, 105, 120–22
  ensemble, 38–39, 105
  spatial, 85–86
  temporal, 85–86, 120–22
Averaging, 38–39, 85–86, 120–22
  conditions, 39
  effects, 120–22
  rules, 39
  time, 85–86

Balloon trajectories, 170–72
Baroclinic
  atmosphere, 36
  boundary layer, 36, 83–84
Barotropic
  atmosphere, 36
  boundary layer, 69, 77–78, 83–84,
    100
Bias, 215
Blackbody, 43
Bluff body, 227–28

305

Boundary layer, 69–75, 77–104
  atmospheric, 69–71
  convective, 81–82
  height, 69–71, 77–78, 81–82
  internal, 71–72
  neutral, 91–92
  planetary, 69–71
  stable, 81–82
  temperatures, 80–82, 84
  turbulence, 37–40, 84–96
  urban, 62–63
  winds, 69–70, 83–84
Boundary-layer parameterization, 96–104, 124–25, 241, 243–48, 264–67
  mixing height, 98
  stability, 100–103
  surface stress and heat flux, 97–98
  turbulence, 101–103, 124–25, 241–48, 264–67
  wind profile, 98–100
Box model, 278–80
Boussinesq approximation, 35, 39
Bowen ratio, 79, 98
Brunt-Vaisala frequency, 65–67, 254
Building
  aspect ratio, 73
  dispersion, 73
  flow around, 72–73
  wake, 72–73
Bulk transfer approach, 97
Buoyancy
  acceleration, 32–33
  effects, 80–85, 90–91, 92–103
  force, 32
  length, 91, 93
  parameter, 32, 84
  production, 243
Buoyancy effects of release, 220–26
Buoyancy flux, 223–26

Canopy layer, 78, 91–92
Capping inversion, 81–82, 142–43, 147–48
Carbon compounds, 7–8
  inorganic, 7
  organic, 8
Chemical
  mechanisms, 273–74
  reactions, 5, 9
  transformation, 5–9, 273–74
Chemistry
  air pollution, 4–14, 273–74
  atmospheric, 4–14, 273–74
Clean Air Act, 22–24
  acid rain, 23–24
  enforcement, 24
  hazardous air pollutants, 23
  mobile sources, 22
  National Ambient Air Quality Standards, 22–23
  stratospheric ozone, 24
Climate and climate change, 20–21, 44–51
Concentrations, 5–6, 85–86, 127–48, 182, 185–91, 198–99
  fluctuation, 85–86
  instantaneous, 85–86, 128–31
  mean, 85–86, 137–48, 182, 185–91, 198–99
  units, 5–6

Conduction of heat, 34
Coning, 74–75
Conjugate power law, 101, 145
Conservation laws, 33–36, 128–31, 137–40, 240–41, 243–44, 272
  energy, 33–34, 241
  mass, 34, 128–31, 137–40, 240, 272
  momentum, 35–36, 241
  turbulence kinetic energy, 243
Control of air pollution, 21–24
Convection, 93–95
  forced, 93–94
  free, 94–95
Convective
  circulation, 190–91, 248, 257–60
  conditions, 81, 84, 94–96
  velocity scale, 94–96, 187–88
Convective boundary layer, 69–70, 80–84, 93–96, 98–103, 117–20, 124–25, 257–60
  height, 81–82, 98
  mean temperatures, 80–82
  mean winds, 69–70, 83–84, 93, 98–100
  turbulence, 93–96, 101–03, 117–20, 124–25, 257–60
Convergence/divergence, 80
  radiative flux, 80
  sensible heat flux, 80
Coriolis
  acceleration, 35, 83
  force, 35, 83
  parameter, 35–36, 83
Correlation
  coefficients, 88
  functions, 108–12, 124–25, 157–58, 160–62, 165, 260–61
Covariance, 39, 88–90
Counter-gradient flux, 90, 248
Critical Richardson number, 85, 93
Cross-wind integrated concentration, 145, 148
Cross-wind spread, 140–43, 145–47, 150–52, 156–63, 171–74
Cyclones and anticyclones, 56–59

Dense gas plumes, 221
Density, 29–32
  atmospheric, 29–32
  parcel, 32
  stratification, 32–33
Deposition, 228–37, 240, 274–76
  flux, 230–31
  gaseous pollutant, 228–34
  models, 234–37, 240
  particle, 228–31, 240
  velocity, 224–34, 240, 274–76
Diffusion
  absolute, 155–63, 170–77
  equation, 129, 137, 240–41, 272
  Fickian, 139–43
  molecular, 128–36
  parameters, 140–41, 202–10
  parameterization, 202–10
  random-walk, 260–67
  relative, 163–72
  turbulent, 137–38
Diffusion experiments, 200–10
Diffusion theories/models, 127–219, 239–86
  Fickian, 139–43

Gaussian models, 139–43, 197–219
  gradient transport, 127–54, 185–87, 239–43
  large-eddy simulation, 249–260
  numerical models, 239–86
  random-walk models, 260–67
  regional dispersion, 269–76, 281–86
  similarity theories, 178–96
  statistical theories, 155–77
Diffusivity, 89–91, 128, 137–38, 143–44, 147, 163, 240–44, 272
  eddy, 89–91, 137–38, 143–44, 147, 163, 240–44
  molecular, 89, 128
Dimensional analysis, 91–96, 180–88, 223–24
Dimensionless parameters, 91–96, 180–88, 223–25
Dispersion-deposition models, 234–37
  source depletion, 235–36
  surface depletion, 236–37
Dispersion parameterization schemes, 202–10
  Briggs' rural/urban, 205–7, 210
  Brookhaven National Laboratory, 203–4
  Pasquill-Gifford, 202–3
  Statistical Theory, 208–10
  Tennessee Valley Authority, 204–5
  Urban, 205–7
Dissipation
  length-scale, 243–44
  turbulence kinetic energy, 113–15, 243–46
Diurnal variations, 77–84
Divergence flux, 80
Downdraft, 190–91, 248, 257–60
Downslope wind, 64–66
Downwash, 221–22
Drag
  coefficient, 146
  force, 227–28, 231
Drainage flow, 64–65
Dry adiabatic lapse rate, 30
Dry deposition, 228–34, 274–75
  flux, 230–31
  velocity, 230–34, 274–75

Earth-atmosphere exchanges, 78–80
  energy, 78–79
  mass, 79–80
  momentum, 80
Earth's rotational speed, 35
Eddy
  correlation, 88–90
  large-scale, 37–39, 90, 108–110
  small-scale, 37–39, 108–9
  structure, 108–112
  viscosity, 89–91
Eddy diffusivity, 89–91, 137–38, 143–44, 147, 163
  heat, 89–91
  mass, 89–90, 137–38, 143–44, 147, 163
  momentum, 89–91
Effects of air pollution, 14–21
  animals, 16–17
  atmospheric, 17–21
  climate, 20–21
  health, 14–15
  materials, 17
  vegetation, 15–16
Effective release height, 212

INDEX

Effective source height, 212
Effluent
  buoyancy, 212, 220–26
  momentum, 212, 220–26
  velocity, 223
Elevated sources, 140–43, 145–48, 178–79, 180–81, 188–89, 191–94, 198–99, 210–11
  maximum concentration, 140–42, 145–46, 210–11
  wind shear effect, 178–79, 180–81
Emissions inventory, 269–71
Energy budget, 43–44, 78–79
  earth-atmosphere-sun system, 43
  earth's surface, 43, 78–79
Energy balance method, 79, 98
Energy dissipation, 113–15, 243–46
Energy fluxes, 78–79
  latent heat, 79
  net radiation, 79
  sensible heat, 79
Entrainment velocity, 98
Episode, air pollution, 14
Euler's equations of motion, 35–36
Eulerian approach, 127–28
  diffusion, 127–28
  reference system, 127
Evaporation, 80, 82
  latent heat, 80
  rate, 80, 82
Exchange
  heat/energy, 78–79
  mass, 79–80
  molecular, 89
  momentum, 80
  turbulent, 78–80, 88–91
Exchange coefficient. *See* Eddy diffusivity

Fanning, 73–74
Fickian diffusion, 131–42
Fick's law, 128
Filtering, 121–22, 159
Flow separation, 72
Fluctuations, 37–40, 85–87
  concentration, 85–86
  temperature, 85–87
  velocity, 85–87
Fluctuating plume model, 169–70
Fluid flows, 37–40
  inviscid, 35–37
  laminar, 37
  turbulent, 37–40
  viscous, 37–40
Flux Richardson number, 114, 241
Fourier's law of heat conduction, 89
Frequency spectrum, 110–11, 117–19
Friction, 80, 83
Friction velocity, 91
Fronts, 54–56
Froude number, 65–66
Frozen turbulence hypothesis, 112–13
Fumigation, 73–74

Gaussian
  diffusion/dispersion models, 132–36, 140–43, 197–219
  distribution, 106–107, 132–36, 140–43, 150–51, 161–63, 198–99
  plume dispersion, 135–36, 141–43, 198–99
  puff diffusion, 131–34, 140–41

General circulation, 44–51
  models, 44–46
  observed features, 46–51
Geostrophic
  balance, 36–37
  flow, 36
  shear, 36
  winds, 36
Gradient transport, 89–91, 127–52
  approach, 89–90, 137–38
  hypothesis, 89–90
  theory, 89–90, 127–52
Gravitational settling, 226–28
Gravity waves, 67
Ground level sources, 143–46, 172–74, 180–89, 200
  concentration, 143–46, 172–74, 185–88
  wind shear, 180–89
Greenhouse
  gases, 20–21
  warming, 20–21

Halogen compounds, 10
Hay-Pasquill approach, 160–61
Hazardous air pollutants, 13–14
Health effects of air pollution, 14–15
Heat capacity, 79
Heat flux, 43–44, 78–80, 89–90, 97–98
  latent, 43–44, 78–79
  sensible, 43–44, 78–80, 89–90, 97–98
Heat transfer, 34–35, 43–44, 78–80, 89–90
  molecular, 34, 89
  turbulent, 34–35, 43–44, 78–80, 89–90
Homogeneous turbulence, 113–4
Horizontal homogeneity, 69–70, 80–84
Humidity, 29, 31–32, 82–83
  diurnal variation
  specific, 29, 31–32, 82–83
Hydraulic jump, 66
Hydrocarbons, 8

Image source, 139–40
Industrial sources, 3
Inertial subrange, 116–17
Inhomogeneity
  flow, 71–73
  surface roughness, 71–72
  surface topography, 72–73
Instability, 32–33, 85
  dynamic, 85
  inviscid, 85
  static, 32–33, 85
Instantaneous equations, 33–35
Integral equations of mass conservation, 130–31
Integral scales, 108–10, 112, 157–58, 163, 167
  length, 110, 112
  time, 108–9, 112, 157–58, 163, 167
Intensity of turbulence, 87–88, 102, 161
  lateral, 87–88, 102, 161
  longitudinal, 87
  vertical, 87–88, 102
Internal boundary layer, 71–72
Inversion of temperature, 73–75, 81–82, 84, 142–43
  elevated, 74–75, 81, 142–43
  surface, 73–75, 81–82, 84

Inviscid flows, 35–36
Isotropic turbulence, 113–17

Jet streams, 47–51

Kinetic theory, 128
Kolmogorov's similarity hypotheses, 115–17
Kurtosis, 107

Lagrangian
  approach/description, 122–25, 155
  dispersion models/theories, 155–77, 180–92, 260–7
  particle motion, 122–25
  reference frame, 155
  spectrum, 122–25, 159–62
  time scale, 122–25, 157–58
Lagrangian-Eulerian relations, 160–63
Laminar flow, 37, 89
Land-sea breezes, 61–62
Langevin equation, 261–62
Lapse rate, 30–32
  dry adiabatic, 30
  moist adiabatic, 32
Latent heat flux, 79–80
Latent heat of evaporation, 78–80
Large-eddy simulation, 38, 249–60
  dispersion, 249–60
  turbulence, 38
Lateral
  component of velocity, 85–88
  spread, 140–43, 156–163, 172–74
  turbulence intensity, 87–88, 102, 161
Lee waves, 66–67
Leaf area index, 230
Length-scale of turbulence, 90–91, 110, 112, 116–20, 161, 241, 243
  large-eddy, 90–91, 110, 112, 116–20, 161, 241, 243
  microscale, 112, 116
Limitations of
  Gaussian models, 216
  gradient transport theories, 148–150, 242–43
  second-order closure models, 248–49
  similarity theories, 195
  statistical theories, 175–76
  stochastic models, 266–67
Line sources, 130, 132–33, 136, 142, 144–45, 182, 185–87, 198–99, 217
  continuous, 130, 136, 142, 144–45, 182, 185–87, 198–99, 217
  instantaneous, 130, 132–33
Local and urban air pollution, 1–3, 5–16
Local free convection similarity theory, 94–95
Local isotropy/similarity, 115–17
Lofting, 74–75
Logarithmic wind profile, 91–92
Log-linear profile, 93
Long-range dispersion, 51, 59
Long-range transport, 51, 59
Longitudinal
  component of velocity, 85–87, 96, 101
  correlation, 108–10, 117
  spectrum, 110–13, 118–20
  turbulence intensity, 87
Longwave radiation, 43–44
Looping plume, 74
Low-level jet, 69

Macroscale systems, 42–51
 dispersion, 51
 energy budget, 43–44
 general circulation, 44–51
Marine tropical air mass, 52–54
Markov equation, 260–3
Mass exchange, 79–80
 carbon dioxide, 79–80
 particulates, 79–80
 water vapor, 79–80
Maximum ground-level concentration, 140–42, 145–46, 210–11
 continuous line source, 142, 145
 continuous point source, 141, 146, 210–11
 instantaneous sources, 140–41
Mean diffusion equation, 39, 137, 240–1
Mean equations of motion, 39, 241
Mesoscale dispersion models, 266–7
Mesoscale systems, 59–69
 effects on dispersion, 68–69
 katabatic winds, 64–65
 lee waves, 66–67
 monsoon circulation, 62
 mountain-valley winds, 63–64
 mountain wakes, 67–68
 sea and land breezes, 61–62
 thermal circulations, 59–65
 urban heat island, 62–63
Micrometeorology, 69–75, 77–104
 atmospheric boundary layer, 69–71, 77–104
 boundary layer height, 69–71, 77–78, 81–82
 internal boundary layer, 71–72
 maximum mixing height, 71
 plume shapes, 73–75
 surface layer, 77–78, 91–95
Microscale systems, 69–75
 atmospheric boundary layer, 69–71
 effects on dispersion, 73–75
 internal boundary layers, 71–72
 obstacle wakes, 72–73
Mixed layer, 81–83
Mixed-layer similarity theory, 95–96, 188–94
 basic hypothesis, 95, 188
 convective time scale, 188
 convective velocity scale, 95, 188
 mixing height, 95, 98, 188
 similarity relations for dispersion, 188–94
 similarity relations for turbulence, 95–96
Mixing height, 70–71, 98
Mixing layer, 81–83
Mixing length, 89–91
Model performance/evaluation, 213–16
Model uncertainties, 213–16
Molecular diffusion, 128–36
Molecular diffusivity, 128
Momentum conservation, 35–36
Momentum effect of release, 220–26
Momentum exchange/flux, 80, 223–26
Monin-Obukhov similarity theory, 92–95, 183–87
 basic hypothesis, 92–93
 buoyancy length scale, 93
 Lagrangian, 183–85
 similarity relations, 93–95
 similarity parameter, 93, 184–87

Natural sources, 4
Navier-Stokes equations, 35
Neutral boundary layer, 91–92, 98–103, 124–25
 height, 98
 mean winds, 91–92, 98–100
 turbulence, 92, 101–03, 124–25
Nitrogen compounds, 6–7
Nocturnal boundary layer. See Stable boundary layer
Nonhomogeneous turbulence, 114–115, 117–20
Numerical dispersion models, 239–86
 gradient transport, 127–54, 185–87, 239–43
 large-eddy simulation, 249–60
 random-walk, 260–67
 regional scale, 269–76, 281–86
 second-order closure, 245–49
 turbulence kinetic energy, 243–45
Numerical modeling of turbulence, 38–40, 240–67
 direct numerical simulation, 38
 gradient transport, 240–42
 Lagrangian stochastic, 260–67
 large-eddy simulation, 38, 249–56
 second-order closure, 245–49
 turbulence kinetic energy, 243–45

Obukhov length, 93
Orographic flows, 63–68
Oxidation reactions/products, 4–5, 8–9
Oxygenated hydrocarbons, 8–9
Ozone, 8–10, 19–20, 24

Particles, 11–14
 coarse, 11–12
 fine, 11–12
 viable, 13
Particle deposition velocity, 228–31, 240
Particle dispersion, 156–72, 180–85, 187–89, 228–29
 mean trajectory/height, 180–85, 187–89, 228–29
 parameters, 156–72, 182–83
 probability distribution, 162, 181–82
Pasquill's stability classes, 101–103, 202–3, 205–7
Peak to average concentration, 169–70
Photochemical
 air pollutants, 5–6, 8–10
 process, 5–6
 reactions, 5, 8–10
 smog, 11
Planetary boundary layer, 69–75, 77–104
 convective, 81–82
 height, 69–71, 77–78, 81–82
 neutral, 91–92
 stable, 81–82
 temperature profiles, 80–82, 84
 turbulence, 37–40, 84–96, 101–3
 wind profiles, 69–70, 83–84
Plume dispersion, 130, 134–36, 139–48, 161–63, 169–76, 182–87, 198–99
 area source, 130, 136
 line source, 130, 136, 144–45,147, 161–63, 182, 185–87, 198
 point source, 130, 134–36, 139–43, 145–48, 161–63, 169–76, 182, 185–87, 198–99

Plume rise, 220–26
 buoyancy effect, 220–26
 formulas, 223–26
 momentum effect, 220–26
 turbulence effect, 225–26
 wind shear effect, 225–26
Plume shapes, 73–75
Point sources, 130–32, 134–36, 139, 141–43, 145–48, ,161–63, 167–70, 181–82, 189–94, 198–201, 210–12, 235–37, 240–42, 246–48, 257–60
 continuous, 134–36, 139, 141–43, 145–48, 161–63, 169–70, 181–82, 189–94, 198–201, 210–12, 235–37, 240–42, 246–48
 instantaneous, 130–32, 167–69, 199, 201
Pollutants. See Air pollutants
Pollution. See Air pollution
Potential temperature, 30–31
Power-law profile, 99–101, 144
Pressure. See Atmospheric pressure
Probability functions, 105–8
 density, 105–7
 distribution, 106
Problems and exercises, 24–25, 40–41, 75–76, 103–4, 125–26, 152–54, 176–77, 195–96, 218–19, 237–38, 267–68, 286
Profile method, 97
Puff diffusion, 130–34, 140–49, 167–72, 175, 199, 207–8
 area source, 133–34, 143–44, 147
 line source, 132–33, 207–8
 point source, 131–32, 140–41, 167–72, 199, 207–8

Radiation, 34–35, 43–44, 78–79
 longwave, 43–44, 78
 net, 43, 79
 shortwave, 43–44, 78
 solar, 43, 78
Radiative cooling/warming, 78–82
Radiative effects, 18–21, 26–27
Radiative properties, 35, 43
 albedo, 35, 43
 emissivity, 35
Radioactive substances, 10–11
Random-walk modeling, 260–67
Receptor, 2–3
Regional air quality models, 269–76, 281–86
Relative diffusion, 163–72
 concept, 163–64
 observations, 170–72
 theory, 163–70
Resistance, 231–34
 aerodynamic, 231–34
 surface, 231–34
 transfer, 231–34
Reynolds averaging, 39, 105
Reynolds decomposition, 39
Reynolds number, 115–16, 227–28
Reynolds tress, 39, 88–89
Richardson number, 84–85, 93–94, 97, 100, 102, 114, 241
 bulk, 97, 100
 critical, 85, 114
 flux, 114, 241
 gradient, 84–85, 93–94, 97, 100, 102, 114

Roughness characteristics, 91–92
  density, 92
  height, 91
  parameter (length), 91–92
  zero-plane displacement, 92

Sampling, 120–22, 161–62
  duration/time, 121–22, 161–62
  frequency/rate, 120–22
Scales, characteristic, 90–96, 108–10, 118–20, 123–25
  length, 90–95, 110–12, 118–20
  temperature, 93–95
  time, 108–12, 118–20, 123–25
  velocity, 91–95
Scales of turbulence, 90–91, 108–10, 118–20, 123–25, 260–62
  integral, 108–10, 112, 123–25, 260–62
  large-eddy, 90–91, 118–20
  microscale, 108–9, 112
Sea breeze circulation, 61–62
Secondary pollutants, 4–10
Second-order closure models, 245–49
Sensible heat flux, 43–44, 78–80, 89–90, 97–98
Separation of flow, 72
Shearing stress, 83, 88
  turbulent, 88
  viscous, 89
Short-range dispersion, 73–75
Shortwave radiation. See Radiation
Similarity scaling, 91–101, 180–88
  local free convection, 94–95, 187–88
  mixed layer, 95–96, 101, 188
  neutral surface layer, 91–92
  stratified surface layer, 92–94, 97–98, 180–88
Similarity functions for
  dispersion, 189–94
  mean flow, 93–94, 97–98
  turbulence, 93–96, 101
Similarity theories, 91–96, 165–67, 180–92
  Lagrangian, 165–67, 180–92
  local free convection, 94–95, 187–88
  mixed layer, 95–96, 188–92
  stable boundary layer, 91–94
  surface layer, 91–95, 180–88
Skewness, 107
Slender-plume approximation, 135, 141, 199
Smog, 11
Smoke, 11
Smoothing, 120–22
Solar constant, 43
Solar radiation, 43
Source depletion model, 235–6
Sources, idealized, 130–36, 161–63
  area, 133–34, 136
  line, 132–33, 136
  point, 131–32, 134–36, 161–63
  volume, 130
Sources of air pollution, 3–4
  agricultural and rural, 3–4
  natural, 4
  urban and industrial, 3
Specific heat, 30
Specific humidity, 31, 82–83
Spectrum function, 110–19, 123–25
  Eulerian, 110–19

Lagrangian, 123–25
  observed, 117–20
Stability, 32–33, 84–85, 93–94, 100–103
  classes, 101–103
  dynamic, 85
  local, 32, 84–85
  nonlocal, 32–33
  parameters, 32, 84–85, 93–94, 100–101
  static, 32–33
Stability effects on
  dispersion, 73–75, 179–80, 183–88, 202–8
  mean flow, 93–94, 98–100
  plume rise, 223–26
  turbulence, 93–96, 101–3
Stable boundary layer, 69–70, 80–85, 93–94, 98–103, 117–20, 124–25
  height, 81–82, 98
  mean temperatures, 80–82
  mean winds, 69–70, 83–84, 93, 98–100
  turbulence, 85, 93–94, 101–3, 117–20, 124–25
Standard deviations of fluctuations
  concentration, 87–88
  temperature, 87–88
  velocity components, 87–88, 94–96, 101
  wind directions, 87–88
Standard deviation of
  concentration distribution, 132
  particle displacements, 157
Stationarity, 86
Statistical description of turbulence, 37–40, 84–96, 105–25, 157–58
  correlation functions, 108–12, 157–58
  covariances, 39, 88–90
  probability distribution, 105–8
  spectrum functions, 110–19, 122–25, 159–62
  standard deviations, 87–88, 94–96, 101
  variances, 87, 94–96
Statistical theory of diffusion, 155–57, 182–83, 208–10
  absolute/one-particle, 155–63, 170–77, 208–10
  relative/two-particle, 163–72
  wind shear effect, 182–83, 208–10
Stochastic models, 252–54, 260–67
  dispersion, 260–67
  subgrid scale, 252–54
  turbulence, 260–67
Stokes law, 227
Stratosphere, 28
Sulfur compounds, 6–7
  acidic, 7
  oxides, 6–7
  sulphates, 7
Subgrid-scale models, 250–55
  dynamic, 252
  first-order, 250–52
  higher order, 254–55
  Smagorinski, 250–52
  stochastic, 252–54
Superposition principle, 131–35
Surface
  absorption, 138
  drag, 77
  energy budget, 43–44, 78–79

fluxes, 78–79, 97–98
  reflection, 138
  stress, 91
Surface depletion model, 236–37
Surface layer, 78
  energy fluxes, 78–80
  scaling parameters, 91–94
  similarity theory, 91–94
Synoptic weather systems, 51–59
  air masses, 51–54
  cyclones and anticyclones, 56–59
  fronts, 54–56
  influence on dispersion, 59

Taylor's hypothesis, 112–13
Taylor's theory of diffusion, 155–58
Temperature, air, 29–32, 34–35, 80–82, 84
  energy equation, 34–35
  inversion, 31
  lapse rate, 31–32
  profile, 81, 84
Terminal velocity of particles, 227–8
  gravitational settling, 226–8
Thermally induced circulations, 59–65
  basic circulation, 60–61
  katabatic winds, 64–65
  mountain-valley winds, 63–65
  monsoons, 62–63
  sea and land breezes, 61–62
  urban heat island, 62–63
Thermal properties, 30–35, 79, 89
  conductivity, 34
  heat capacity, 79
  specific heat, 30
  thermal diffusivity, 35, 89
Thermal wind, 36–37
  baroclinicity, 36
  equations, 36
Thermodynamic relations, 29–33
  Clausius-Clapeyron equation, 31
  equation of state, 6, 30
  first law, 30
  hydrostatic equation, 30
  Poisson equation, 30
  potential temperature, 30
  static stability, 32–33
  virtual temperature, 31
Tilted plume model, 228–9
Time scales, 108–9, 188, 278
Topographic effects, 63–68
Topographically forced circulations, 65–69
  downslope winds, 65–66
  lee waves, 66–67
  mountain wakes, 67–68
  upslope winds, 65–66
Touch-down model, 226
Trajectories, particle, 180–85, 187–89, 228–29
Transfer resistance, 232
Transfer velocity, 230–31
Transition
  laminar-turbulent flows, 84–85
  periods, 79
Transport of pollutants, 42–75
  macroscale, 42–51
  mesoscale, 59–69
  microscale, 69–75
  synoptic scale, 51–59
Trapping, 75

Turbulence, 37–40, 84–96, 101–25
  basic characteristics, 37–38
  closure problem, 39
  correlation functions, 108–10
  covariances, 39, 88–90
  intensities, 87–88
  kinetic energy, 87, 113–19
  probability functions, 105–8
  spectra, 110–19, 123–25
  standard deviations, 87–88, 94–96, 101
  theories/models, 38–40, 91–96, 113–17
  variances, 87, 94–96
Turbulence kinetic energy, 87, 113–19, 243–45, 251–55
  buoyancy generation, 114–15, 243
  dissipation, 113–15, 243–45
  models, 243–45
  shear production, 114–15, 243
  subgrid scale, 251, 254–55
  transport, 114–15, 243
Turbulent diffusion/dispersion theories, 127–96
  Fickian, 139–43
  gradient transport, 127–54
  similarity, 165–67, 180–96
  statistical, 155–77
  variable K, 143–48
Turbulent fluxes, 39, 88–90
  heat, 39, 88–90
  mass, 39, 88–90
  momentum, 39, 88–90
  pollutants, 39, 88–90

Uncertainties in dispersion models, 213–16

Unstable boundary layer. *See* Convective boundary layer
Urban air quality models, 197–219, 269–77
  chemical transformations, 273–74
  diffusion parameterization, 205–7, 271–73
  emissions inventory, 269–71
  Gaussian diffusion model, 276–77
  photochemical models, 278–81
  removal processes, 274–76
Urban diffusion/air quality models, 205–7, 269–80
  box, 278–80
  EKMA, 279–80
  Gaussian, 205–7, 276–77
  three-dimensional grid, 280–81
  UAM, 280–81
Urban flow, 62–63
  boundary layer, 63
  heat island, 62–63

Validation/verification of
  Gaussian models, 214–16
  gradient transport theories, 150–52
  large-eddy simulations, 256–58
  second-order closure models, 246–48
  similarity theories, 192–94
  statistical theories, 170–75
  stochastic models, 263–266
Vapor pressure. *See* Water vapor pressure
Variances, 87–88
Velocity profile, 69–70, 83–84, 91–92, 98–100
  logarithmic law, 91–92
  power law, 99–100

Vertical diffusion/spread, 140–52, 174–75, 208–10
Virtual temperature, 31–33
Viscous flow, 37
Volatile organic compounds, 8
von Karman constant, 91

Wakes of hills/buildings, 67–68, 72–73
  cavity, 72
  far wake, 73
  near wake, 72
Washout coefficient, 275
Water vapor pressure, 31
Wavelength, 111–12, 118–19
Wavenumber, 111–12
Wet deposition, 275–6
Wind direction, 69–70, 100, 179
  profile, 100
  shear, 179
  veering, 69–70
Wind distribution. *See* Velocity profile
Wind hodograph, 69–70
Wind shear, 69–70, 98–100, 178–79, 241–42
  direction shear, 69–70, 100, 178–79, 241–42
  effect on dispersion, 178–79, 241–42
  speed shear, 69–70, 98–100, 178–79
Wind speed, 69–70, 83–84, 98–100
  distribution, 69–70, 83–84, 98–100
  effect on maximum concentration, 212
  effect on plume rise, 223–6
Wind tunnel modeling, 150–51
Wind veering, 69–70

Zero-plane displacement, 92